Hydrosocial Territories and Water Equity

Bringing together a multidisciplinary set of scholars and diverse case studies from across the globe, this book explores the management, governance and understandings around water, a key element in the assemblage of hydrosocial territories. Hydrosocial territories are spatial configurations of people, institutions, water flows, hydraulic technology and the biophysical environment that revolve around the control of water. Territorial politics finds expression in encounters of diverse actors with divergent spatial and political–geographical interests; as a result, water (in)justice and (in)equity are embedded in these socio-ecological contexts. The territory-building projections and strategies compete, superimpose and align to strengthen specific water-control claims of various interests. As a result, actors continuously recompose the territory's hydraulic grid, cultural reference frames and political–economic relationships. Using a political ecology focus, the different contributions to this book explore territorial struggles, demonstrating that these contestations are not merely skirmishes over natural resources, but battles over meaning, norms, knowledge, identity, authority and discourses.

The articles in this book were originally published in the journal *Water International*.

Rutgerd Boelens is a Professor of Political Ecology of Water at CEDLA, University of Amsterdam, the Netherlands, and in the Department of Environmental Sciences, Wageningen University, the Netherlands.

Ben Crow is a Professor of Sociology in the Sociology Department, University of California, Santa Cruz, CA, USA.

Jaime Hoogesteger is an Assistant Professor in the Department of Environmental Sciences, Wageningen University, the Netherlands.

Flora Lu is an Associate Professor in the Department of Environmental Studies and Provost of College Nine and College Ten, University of California, Santa Cruz, CA, USA.

Erik Swyngedouw is a Professor of Geography in the School of Environment, Education and Development, University of Manchester, UK.

Jeroen Vos is an Assistant Professor of Water Governance in the Department of Environmental Sciences, Wageningen University, the Netherlands.

Routledge Special Issues on Water Policy and Governance

https://www.routledge.com/series/WATER

Edited by
Cecilia Tortajada *(IJWRD), Third World Centre for Water Management, Mexico*
James Nickum *(WI), International Water Resources Association, France*

Most of the world's water problems, and their solutions, are directly related to policies and governance, both specific to water and in general. Two of the world's leading journals in this area, the *International Journal of Water Resources Development* and *Water International* (the official journal of the International Water Resources Association), contribute to this special issues series, aimed at disseminating new knowledge on the policy and governance of water resources to a very broad and diverse readership all over the world. The series should be of direct interest to all policy makers, professionals and lay readers concerned with obtaining the latest perspectives on addressing the world's many water issues.

The Private Sector and Water Pricing in Efficient Urban Water Management
Edited by Cecilia Tortajada, Francisco González-Gómez, Asit K. Biswas and Miguel A. García-Rubio

Water for Food Security
Challenges for Pakistan
Edited by Claudia Ringler and Arif Anwar

Water Management and Climate Change
Dealing with Uncertainties
Edited by Cecilia Tortajada, Asit K. Biswas and Avinash Tyagi

Integrated Water Resources Management
From Concept to Implementation
Edited by Cecilia Tortajada

Water Infrastructure
Edited by Cecilia Tortajada and Asit K. Biswas

Frontiers of Land and Water Governance in Urban Regions
Edited by Thomas Hartmann and Tejo Spit

The Water-Energy-Food Nexus in the Middle East and North Africa
Edited by Martin Keulertz and Eckart Woertz

Sustainability in the Water-Energy-Food Nexus
Edited by Anik Bhaduri, Claudia Ringler, Ines Dombrowsky, Rabi Mohtar and Waltina Scheumann

Transboundary Water Cooperation
Principles, Practice and Prospects for China and Its Neighbours
Edited by Patricia Wouters, Huiping Chen and James E. Nickum

Water Reuse Policies for Potable Use
Edited by Cecilia Tortajada and Choon Nam Ong

Hydrosocial Territories and Water Equity
Theory, Governance, and Sites of Struggles
Edited by Rutgerd Boelens, Ben Crow, Jaime Hoogesteger, Flora Lu, Erik Swyngedouw and Jeroen Vos

Hydrosocial Territories and Water Equity

Theory, Governance, and Sites of Struggle

Edited by
Rutgerd Boelens, Ben Crow, Jaime Hoogesteger, Flora Lu, Erik Swyngedouw and Jeroen Vos

LONDON AND NEW YORK

First published 2017
by Routledge
2 Park Square, Milton Park, Abingdon, Oxon, OX14 4RN, UK

and by Routledge
52 Vanderbilt Avenue, New York, NY 10017

First issued in paperback 2019

Routledge is an imprint of the Taylor & Francis Group, an informa business

British Library Cataloguing in Publication Data
A catalogue record for this book is available from the British Library

ISBN 13: 978-0-367-20752-6 (pbk)
ISBN 13: 978-1-138-28884-3 (hbk)

Typeset in TimesNewRomanPS
by diacriTech, Chennai

Publisher's Note
The publisher accepts responsibility for any inconsistencies that may have arisen
during the conversion of this book from journal articles to book chapters, namely
the possible inclusion of journal terminology.

Disclaimer
Every effort has been made to contact copyright holders for their permission to
reprint material in this book. The publishers would be grateful to hear from any
copyright holder who is not here acknowledged and will undertake to rectify any
errors or omissions in future editions of this book.

Contents

PART 3
Hydrosocial struggles

CONTENTS

Citation Information

The following chapters were originally published in *Water International*. When citing this material, please use the original volume number, issue number, date of publication, and page numbering for each article, as follows:

Chapter 2
Defining, researching and struggling for water justice: some conceptual building blocks for research and action
Margreet Z. Zwarteveen and Rutgerd Boelens
Water International, volume 39, issue 2 (2014) pp. 143–158

Chapter 3
Hydrosocial territories: a political ecology perspective
Rutgerd Boelens, Jaime Hoogesteger, Erik Swyngedouw, Jeroen Vos and Philippus Wester
Water International, volume 41, issue 1 (2016) pp. 1–14

Chapter 4
What kind of governance for what kind of equity? Towards a theorization of justice in water governance
Tom Perreault
Water International, volume 39, issue 2 (2014) pp. 233–245

Chapter 5
What is water equity? The unfortunate consequences of a global focus on 'drinking water'
Matthew Goff and Ben Crow
Water International, volume 39, issue 2 (2014) pp. 159–171

Chapter 6
PES hydrosocial territories: de-territorialization and re-patterning of water control arenas in the Andean highlands
Jean Carlo Rodríguez-de-Francisco and Rutgerd Boelens
Water International, volume 41, issue 1 (2016) pp. 140–156

Chapter 7

Losing the watershed focus: a look at complex community-managed irrigation systems in Bolivia
Cecilia Saldías, Rutgerd Boelens, Kai Wegerich and Stijn Speelman
Water International, volume 37, issue 7 (2012) pp. 744–759

Chapter 8

Examining the emerging role of groundwater in water inequity in India
Veena Srinivasan and Seema Kulkarni
Water International, volume 39, issue 2 (2014) pp. 172–186

Chapter 9

The colonial roots of inequality: access to water in urban East Africa
Brian Dill and Ben Crow
Water International, volume 39, issue 2 (2014) pp. 187–200

Chapter 10

Popular participation, equity, and co-production of water and sanitation services in Caracas, Venezuela
Rebecca McMillan, Susan Spronk and Calais Caswell
Water International, volume 39, issue 2 (2014) pp. 201–215

Chapter 11

Creating equitable water institutions on disputed land: a Honduran case study
Catherine M. Tucker
Water International, volume 39, issue 2 (2014) pp. 216–232

Chapter 12

Democratizing discourses: conceptions of ownership, autonomy and 'the state' in Nicaragua's rural water governance
Sarah T. Romano
Water International, volume 41, issue 1 (2016) pp. 74–90

Chapter 13

Adjudicating hydrosocial territory in New Mexico
Eric P. Perramond
Water International, volume 41, issue 1 (2016) pp. 173–188

Chapter 14

Downspout politics, upstream conflict: formalizing rainwater harvesting in the United States
Katie M. Meehan and Anna W. Moore
Water International, volume 39, issue 4 (2014) pp. 417–430

Chapter 15

Disputes over territorial boundaries and diverging valuation languages: the Santurban hydrosocial highlands territory in Colombia
Bibiana Duarte-Abadía and Rutgerd Boelens
Water International, volume 41, issue 1 (2016) pp. 15–36

Chapter 16

Diverging realities: how framing, values and water management are interwoven in the Albufera de Valencia wetland in Spain
Mieke Hulshof and Jeroen Vos
Water International, volume 41, issue 1 (2016) pp. 107–124

Chapter 17

Disputes over land and water rights in gold mining: the case of Cerro de San Pedro, Mexico
Didi Stoltenborg and Rutgerd Boelens
Water International, volume 41, issue 3 (2016) pp. 447–467

Chapter 18

Territorial pluralism: water users' multi-scalar struggles against state ordering in Ecuador's highlands
Jaime Hoogesteger, Rutgerd Boelens and Michiel Baud
Water International, volume 41, issue 1 (2016) pp. 91–106

Chapter 20

Water scarcity and the exclusionary city: the struggle for water justice in Lima, Peru
Antonio A. R. Ioris
Water International, volume 41, issue 1 (2016) pp. 125–139

Chapter 21

Inclusive recognition politics and the struggle over hydrosocial territories in two Bolivian highland communities
Miriam Seemann
Water International, volume 41, issue 1 (2016) pp. 157–172

Chapter 22

Virtual water trade and the contestation of hydrosocial territories
Jeroen Vos and Leonith Hinojosa
Water International, volume 41, issue 1 (2016) pp. 37–53

Chapter 23

From Spain's hydro-deadlock to the desalination fix
Erik Swyngedouw and Joe Williams
Water International, volume 41, issue 1 (2016) pp. 54–73

Chapter 24

Santa Cruz Declaration on the Global Water Crisis
Rutgerd Boelens, Ben Crow, Brian Dill, Flora Lu, Constanza Ocampo-Raeder and Margreet Z. Zwarteveen
Water International, volume 39, issue 2 (2014) pp. 246–261

For any permission-related enquiries please visit:
http://www.tandfonline.com/page/help/permissions

Notes on Contributors

Michiel Baud is the Director of CEDLA (Centre for Latin American Research and Documentation) and a Professor in the Department of Geography, Planning and International Development Studies, University of Amsterdam, the Netherlands.

Rutgerd Boelens is a Professor of Political Ecology of Water at CEDLA, University of Amsterdam, the Netherlands, and in the Department of Environmental Sciences, Wageningen University, the Netherlands.

Calais Caswell is a graduate student in the School of International Development and Global Studies, University of Ottawa, Ottawa, Canada.

Ben Crow is a Professor of Sociology in the Sociology Department, University of California, Santa Cruz, CA, USA.

Brian Dill is an Associate Professor in the Department of Sociology, University of Illinois, Urbana-Champaign, IL, USA.

Bibiana Duarte-Abadía is a PhD student at CEDLA, University of Amsterdam, the Netherlands.

Matthew Goff is a Research Associate in the Sociology Department, University of California, Santa Cruz, CA, USA.

Leonith Hinojosa is a researcher at the Earth and Life Institute, Universite Catholique de Louvai, Belgium.

Jaime Hoogesteger is an Assistant Professor in the Department of Environmental Sciences, Wageningen University, the Netherlands.

Mieke Hulshof is a Hydrologist at Acacia Water, the Netherlands.

Antonio A. R. Ioris is a human geography lecturer at the School of Geosciences, The University of Edinburgh, UK.

Seema Kulkarni is a Senior Fellow at SOPPECOM (Society for Promoting Participative Ecosystem Management), Pune, India.

Flora Lu is an Associate Professor in the Department of Environmental Studies and Provost of College Nine and College Ten, University of California, Santa Cruz, CA, USA.

Rebecca McMillan is a graduate student in the School of International Development and Global Studies, University of Ottawa, Ottawa, Canada.

Katie M. Meehan is an Associate Professor in the Department of Geography, University of Oregon, Eugene, OR, USA.

Anna W. Moore is a graduate student in the Department of Geography, University of Oregon, Eugene, OR, USA.

Constanza Ocampo-Raeder is an Assistant Professor of Anthropology in the Department of Sociology and Anthropology, Carleton College, Northfield, MN, USA.

Eric P. Perramond is an Associate Professor and the Director of Southwest Studies & Environmental Programs, Colorado College, Colorado Springs, CO, USA.

Tom Perreault is a Professor in the Department of Geography, Syracuse University, Syracuse, NY, USA.

Sarah T. Romano is an Associate Professor of Political Science and International Affairs, University of Northern Colorado, Greeley, CO, USA.

Jean Carlo Rodríguez-de-Francisco is a Researcher in the Environmental Policy and Natural Resources Management Department, German Development Institute/Deutsches Institut für Entwicklungspolitik (DIE), Bonn, Germany.

Cecilia Saldías is a Professor in the Department of Agricultural Economics, Ghent University, Belgium.

Miriam Seemann is consultant at GFA Consulting Group GmbH, Hamburg, Germany.

Néstor L. Silva is a doctoral candidate in the Department of Anthropology at Stanford University, CA, USA.

Stijn Speelman is a Professor at the Department of Agricultural Economics, Ghent University, Belgium.

Veena Srinivasan is the programme leader in the Water, Land and Livelihoods Programme, Ashoka Trust for Research in Ecology and the Environment, Bangalore, India.

Susan Spronk is an Associate Professor in the School of International Development and Global Studies, University of Ottawa, Ottawa, Canada.

Didi Stoltenborg is a Lecturer in the Department of Environmental Sciences, Wageningen University, the Netherlands.

Erik Swyngedouw is a Professor of Geography in the School of Environment, Education and Development, University of Manchester, UK.

Catherine M. Tucker is an Associate Professor in the Department of Anthropology, Indiana University, Bloomington, IN, USA.

Jeroen Vos is an Assistant Professor of Water Governance in the Department of Environmental Sciences, Wageningen University, the Netherlands.

Kai Wegerich is a staff member at Institut für Geowissenschaften und Geographie, Martin-Luther-Universität Halle-Wittenberg, Germany.

Philippus Wester is a Faculty Member in the Department of Environmental Sciences, Wageningen University, the Netherlands, and the chief scientist of Water Resources Management at the International Centre for Integrated Mountain Development, Kathmandu, Nepal.

Joe Williams holds a PhD in Geography from Manchester University, UK, and is a Teaching Fellow in Human Geography in the Department of Geography, Durham University, UK.

Margreet Z. Zwarteveen is a Professor in the Department of Geography, Planning and International Development Studies, University of Amsterdam, the Netherlands.

Introduction: interweaving water struggles, the making of territory and social justice

Jaime Hoogesteger[a], Jeroen Vos[a], Rutgerd Boelens[b], Ben Crow[c], Flora Lu[d] and Erik Swyngedouw[e]

[a]Department of Environmental Sciences, Wageningen University, the Netherlands; [b]CEDLA, University of Amsterdam, the Netherlands; [c]Sociology Department, University of California, Santa Cruz, CA, USA; [d]Department of Environmental Studies and Provost of College Nine and College Ten, University of California, Santa Cruz, CA, USA; [e]School of Environment, Education and Development, University of Manchester, UK

This edited book seeks to contribute to the understanding and theorization of the hydrosocial relations and processes that shape water governance and its outcomes in terms of water justice, equity and sustainability. Its 25 chapters bring together new and earlier published articles, in particular those that have been published in two special issues of the journal *Water International*: 'Towards Equitable Water Governance' (Vol. 39 (2), 2014) and 'Hydrosocial Territories: A Political Ecology Perspective' (Vol. 41 (1), 2016). All these chapters closely relate to the abovementioned themes through review and original research articles focusing on specific case studies, their analysis and theorization. These chapters offer a rich array of interdisciplinary approaches that interpret the divergent realities and outcomes of water governance. Many of the contributions also point out the challenges and opportunities for advancing more inclusive, democratic and equitable forms of water governance (Boelens et al., 2014; Goff and Crow, 2014). An underlying premise of the chapters is that water governance entails deeply contested political process (Lu, 2012; Perreault, 2014; Zwarteveen and Boelens, 2014; Swyngedouw and Williams, 2016). Therefore, we examine water governance as the processes that shape 'how organization, decisions, order and rule are achieved in heterogeneous and highly differentiated societies' (Bridge and Perreault, 2009, p. 476). In doing so, we acknowledge that water governance reflects and projects social, economic, administrative and political power through decisions about the design, manipulation and control of water, reflected in different forms of resource allocation, use, development and management.

The chapters of the special issue 'Towards Equitable Water Governance' focus on understanding these relations and finding ways forward in advancing democracy, equity and justice in the water sector. In water policies, stakeholder participation is often advocated as 'the' means to advance democracy and justice. From this perspective, participation is often conceptualized as a tool that can help solve different policy problems through reducing conflict, increasing the knowledge base and improving the legitimacy and efficacy of policies, among others. Participation is seen as a means to empower and give voice to groups that have often been excluded from policy processes. Nonetheless, as many contributions in this book show, many mechanisms that have been designed to increase participation of stakeholders – such as the acknowledgement of legal pluralism and politics of recognition in water governance – have played mere lip service to the principles of democratic decision-making, and moreover, have become mechanisms to streamline decision-making for more 'effective' (often top-down) policy implementation (Sze et al., 2009). Taken a step further, participation can become an instrument to advance and legitimize the de-politicization of water governance through the establishment of a so-called even playing field that neutralizes dissent by means of the co-optation of stakeholders and the development of foreclosed consensual modes of policy making that operate in a setting of generally agreed objectives (such as sustainability, consensus building and good governance) that do not question vested positions and interests (Swyngedouw, 2011).

Although under liberal premises, civil society is often portrayed as a homogenously constituted democratic force in which (all) individuals can fend for their interests, the cases show clearly that different groups in this 'civil society' have different power to influence water governance (Hoogesteger and Wester, 2015; Mehta et al., 2012). Peasant and other marginalized water users are mostly excluded from decision-making processes, because they are either not considered legitimate participants (Boelens, 2015) or because their lack of required knowledge, skills and resources curtails their capacity to influence the outcomes of the decision-making processes (Turnhout et al., 2010). Therefore, top-down policies to include stakeholders rarely mark the transformation of existing inequities, deep-seated resentment, socio-environmental conflicts and sometimes violent encounter. In such contexts, the strategies of social protest, alliance building and networking on the part of the excluded stakeholders form the basis for the advancement of emergent forms of inclusion, equity and justice in water governance (Hoogesteger and Verzijl, 2015). The fact that through these struggles powerful, vested interests such as state institutions, large corporations and landlords are contested and challenged forms the basis for a re-negotiation of water distribution, its allocation and management. These struggles also challenge the legitimacy of formal authorities and dominant discourses and practices. Resistance and struggle are dynamic processes that bring about changes that can materialize in the form of amended legal terms, new institutional practices, new policies, different decision-makers and/or the transformation of water flows. As many of the chapters show, grassroots struggles for water stand at the core of the advancement of more just and equitable forms of water governance.

Water governance is about how nature, technology and the social are entwined and intrinsically interrelated (Linton and Budds, 2014; Meehan 2013). Through their actions, people are strongly involved in the everyday production and reproduction of the society and environment they live in – although not necessarily in the ways they foresee, plan or desire (Winner, 1986). Here, power relationships shape water flows, and vice versa. This is clearly manifested in how water is distributed and used in river basins, where and to whom water flows are directed, how and by whom water use systems are designed and managed and how whole hydrological cycles are transformed by human interventions. Therefore, a fundamental question is how to conceptualize boundaries between 'nature' and 'society', how the contents of – and interlinkages among – particular natural, social and technological elements are established, by which actors and with what interests and consequences.

It is precisely this question that underlies the basic inquiry of the special issue 'Hydrosocial Territories: A Political Ecology Perspective'. The fundamental point of departure for this special issue is the recognition that water is not only a fluid that flows through a territory, but rather that water *is* quintessentially territorial. It not only flows through geographical spaces and places; through its passage it shapes them. Where and how water flows – either naturally or through human-made technologies such as pipes and canals – determines where people live, how food production takes place, where industrial production is done, how landscapes are organized, what kind of livelihoods strategies develop, and what kind of power structures emerge for its control. To capture and better understand these interrelationships between water, nature, space, technology and society, the notion of hydrosocial territories is defined by Boelens et al. (2016, p. 2) as:

> the contested imaginary and socio-environmental materialization of a spatially bound multi-scalar network in which humans, water flows, ecological relations, hydraulic infrastructure, financial means, legal-administrative arrangements and cultural institutions and practices are interactively defined, aligned and mobilized through epistemological belief systems, political hierarchies and naturalizing discourses

Territories, although often considered natural, are actively produced through the interactions between society, technology and nature. The cultural and historical origins that shape hydrosocial territories, as well as the social institutions and technologies that (re) produce or challenge and transform them, thus need to be understood. As many of the chapters show, the state and its associated institutions have played a central role in defining and ordering a specific notion and coupled materiality of hydrosocial territories.

The notion that water is the property of a nation state, imbued with the power to order its flows as it deems necessary for its development, is intrinsically territorial. This assertion, commonly imbricated with state policies that promote the firm involvement of (trans) national capitalist enterprises and the workings of market forces, has laid the basis for a very specific form of water resource development in most countries around the world (cf. Bakker, 2010; Boelens et al., 2015; Boelens et al., 2014; Perreault, 2014). A consequence is that those poised as the legitimate and central heirs and decision-makers over water allocations – governing rules and rights in water management – are an alliance of neoliberal market-based and state legal, administrative and institutional structures and actors. This alliance has laid the basis for the construction of large-scale infrastructure for the utilization of water 'for the greater good of the nation' and its transnational positioning. This intrinsically political notion is nonetheless claimed as objective and neutral by its proponents and beneficiaries while it continuously violates alternative (often local) notions of hydrosocial territorial ordering. These violations of alternative notions of water governance often go hand in hand with the dislocation of local powers, authorities, institutions and established water allocation, distribution and right systems; dislocations that entail severe implications in terms of socio-environmental justice and equity.

Yet, local populations are not passive in the face of new hydrosocial territorial ordering. Through struggles, protests and the creation of alliances and networks, existing and alternative hydrosocial territories are defended and recreated in the battlefield of territorial pluralism, one in which water technologies, resources and the flow of water itself are constantly negotiated and transformed at multiple scales and in different arenas of contestation (Swyngedouw and Williams, 2016; Vos and Hinojosa, 2016). In this context, in which power differences are often large and persistent, water justice is often hard to find; yet it is in the day-to-day water battles in which vested powers are challenged that the seeds for change are sown, providing the hope that at different scales and through the commitment and perseverance of those who struggle for justice, equity and inclusive forms of water governance; struggles that we as academics bring to light through documentation, analysis and dissemination.

Structure and content of the book

The book is organized in to three parts that elaborate on the themes addressed earlier: 1) Theories of the Hydrosocial and Water Equity, 2) Water Governance and 3) Hydrosocial Struggles.

Part 1: Theories of the Hydrosocial and Water Equity brings together four theoretical chapters that lay the conceptual foundations that have inspired and informed many of the remaining chapters. In Chapter 2, Margreet Zwarteveen and Rutgerd Boelens provide a framework for understanding water problems as problems of justice. They make a call for explicitly accepting water problems as implicitly contested and recognizing that water justice is embedded and specific to historical and socio-cultural contexts relating it not only to questions of distribution, but also to cultural recognition, political participation and the integrity of ecosystems. In Chapter 3, Rutgerd Boelens, Jaime Hoogesteger, Erik Swyngedouw, Jeroen Vos and Philippus Wester define and explore *hydrosocial territories* as spatial configurations of people, institutions, water flows, hydraulic technology and the biophysical environment that revolve around the control of water. Using a political ecology approach, they argue that territorial struggles go beyond struggles over natural resources as they involve disputes over meaning, norms, knowledge, imaginaries, identity, authority and discourses. In Chapter 4, Tom Perreault critically reviews literatures related to the concepts of water and hydrosocial relations; water governance and spatial scale; and equity, justice and rights. He argues that ecological governance and environmental justice can be addressed only by viewing water and society as simultaneously social and natural; issues of democratization, human welfare and ecological conditions must inform the institutional arrangements for governing water. In Chapter 5, Matthew Goff and Ben Crow examine the ideas of equity in household water and argue that the dominant focus on improving the potability of water has muted attention to the wider consideration of domestic water and its impact on livelihoods and poverty.

Part 2: Water Governance bundles eight case study contributions that deal with issues of water governance, its implementation and the often contested outcomes on the ground. In Chapter 6, Jean Carlo Rodríguez-de-Francisco and Rutgerd Boelens explore how payment for environmental services (PES) approaches envision, design and constitute new hydrosocial territories by reconfiguring local water control. Two cases from the Ecuadorian highlands are used to clarify how PES implementation weakens local hydrosocial territories in favor of dominant interests. In Chapter 7, Cecilia Saldías, Rutgerd Boelens, Kai Wegerich and Stijn Speelman analyze water allocation practices in peasant communities of the Bolivian inter-Andean valleys. They show how historical claims, organizational capacity, resource availability and geographical position and infrastructure influence current water allocation. They argue that examining the historical background and context-based conceptualizations of space, place and water system development are crucial to understanding local management practices and informing water policies. In Chapter 8, Veena Srinivasan and Seema Kulkarni use two groundwater case studies from India – one agricultural (Kukdi) and one urban (Chennai) – to demonstrate how gaps in planning, design and policy exacerbate inequity. They suggest that better monitoring, inter-agency coordination and rethinking water entitlements and norms are needed. In Chapter 9, Brian Dill and Ben Crow show that for residents in Dar es Salaam and Nairobi, hauling water is particularly difficult in the informal settlements that cover significant portions of both cities. They show that this inequality between rich and poor is rooted in the segregation of colonial rule and is sustained by the continuing injustice of land policies and the multiple complications involved with upgrading urban settlements. In

Chapter 10, Rebecca McMillan, Susan Spronk and Calais Caswell argue that the technical water committees in Venezuela are an example of co-production of public service delivery between state and citizen. They show how this experiment in urban planning promotes participation as empowerment, because the committees are part of a wider political agenda, engage citizens in a broader process of social change, promote rethinking of the concept of citizenship and have thus far avoided elite capture. In Chapter 11, Catherine Tucker explores the decade-long process by which village-level water committees established a reserve in 2002 to protect communal mountain springs in the Montaña Camapara region of Honduras. In so doing, she considers the conditions under which shared dependence on water resources may motivate cooperation and foster equitable access to water in the face of difficult challenges posed by conflicts over land and water right claims and degradation of the resource. In Chapter 12, Sarah Romano argues that the effectiveness of discourses of ownership, autonomy and state roles and responsibilities in the water sector are essential in supporting water committees' goals of political inclusion and legal recognition in Nicaragua. The case demonstrates how discourses 'from below' can have a democratizing effect on water governance by helping to carve out space for marginalized actors' policy interventions. In Chapter 13, Eric Perramond explains how New Mexico has redefined and territorialized water rights as private property through the adjudication process and administrative governance rules. He argues that state adjudication of water rights disrupts horizontal social relations, rescales water governance and gives the state new ways to govern users vertically through water-crisis measures.

Part 3: Hydrosocial Struggles bundles ten case study–based chapters that deal with issues of social struggles for access to water, the protection of territory, voice in decision-making, autonomy in management and the establishment of alternative discourses and knowledges. These chapters advance the notion that water and its management are hydrosocial processes that are closely related to questions of territory, authority and struggles over imaginaries, discourses and rule-making. In Chapter 14, Katie Meehan and Anna Moore examine the formalization of rainwater harvesting and the implications of new policy trends for water governance in the United States. Their analysis indicates three trends: 1) the 'codification' of water through administrative rather than public law, 2) the institutionalization of rainwater harvesting through market-based tools and 3) the rise of policies at different spatial scales, resulting in greater institutional complexity, new bureaucratic actors and potential points of friction. Drawing on the cases of Colorado and Texas, they argue that states with diverse legal traditions of water enable more successful regulatory environments for downspout alternatives. In Chapter 15, Bibiana Duarte-Abadía and Rutgerd Boelens examine the divergent modes of conceptualizing, valuing and representing the *páramo* highlands of Santurban, Colombia. Using game theory, they argue that efforts that wish to reconcile diverging interests using a universalistic territorial representation generate a hydrosocial imaginary that makes actors' power differentials invisible and enables subtle reallocations of water rights. In Chapter 16, Mieke Hulshof and Jeroen Vos explore two divergent framings of the coastal wetlands of the Albufera of Valencia, Spain. They show how different stakeholders deploy highly diverging realities to pursue their interests in a highly unequal political playing field and pose that recognition and empowerment are the first steps toward more sustainable water management. In Chapter 17, Didi Stoltenborg and Rutgerd Boelens analyze different visions and positions in a conflict between the developer of the open-pit mine Cerro de San Pedro, Mexico, and project opponents. They use the echelons of rights analysis (ERA) framework that distinguishes four layers of dispute: contested resources, contested contents of rules and regulations, disputed decision-making power and conflicting discourses. They show how communities' land and water rights are circumvented by governmental bodies and ambivalent regulations, and argue that multi-actor, multi-scale alliances may offer opportunities to foster environmental and social justice solutions. In Chapter 18, Jaime Hoogesteger, Rutgerd Boelens and Michiel Baud analyze how Ecuadorian state policies and institutional reforms have territorialized water since the 1960s and how peasant and indigenous communities have challenged this ordering since the 1990s by creating multi-scalar federations and networks. They argue that water governance is formed in contexts of territorial pluralism that revolves around the interplay of divergent interests in defining, constructing and representing hydrosocial territory. In Chapter 19, Flora Lu and Néstor Silva show how in Ecuadorian Amazonia, the Waorani, a group known for their critical attitude toward outsiders, have been subject to territorial circumscription and practices of governance with the goal of pacification, sedentism and geographically concentrating once seminomadic populations. In this context, the authors present two contrasting potable water projects in the Waorani community of Gareno: one by a state corporation tasked with using a percentage of oil profits to modernize neglected extraction-site communities, the other by a non-governmental organization (NGO) focused on the installation of household-level rainwater catchment and filtration systems. The authors analyze the new forms of hydrosociality that each project offers to the Waorani. In Chapter 20, Antonio Ioris, based on a case study of Lima, argues that conditions and discourses of water scarcity entwine with inadequate and unfair allocation, use and conservation of water. He shows how the association between investment priorities, political agendas and corruption scandals, which lead to selective abundances, discriminatory practices, uneven development, environmental injustices and persistent scarcities, are perpetuated and deepened in specific hydroterritorial configurations. In Chapter 21, Miriam Seemann examines the dominant human–nature interactions that underlie recent formalization policies and the (re)configuration of hydrosocial territories in the Tiraque Valley, Bolivia. She examines how hydrosocial territories are (re)configured by Bolivia's representative and inclusive discourses and forms of water 'governmentalities' that can lead to unequal distribution of resources, water rights and decision-making power. Her results challenge 'pro-indigenous' and inclusive discourses that promote formal recognition of customary 'water territories'. In Chapter 22, Jeroen Vos and Leonith Hinojosa analyze how national and international water regulations reshape communities' hydrosocial territories by changing water governance structures to favor export commodity sectors. They show how transnational companies formulate and enforce global water governance arrangements oriented toward strengthening export production chains, often through asymmetrical relationships with local groups in water-export regions. They argue that these arrangements compromise political representation and water security for local communities and companies. In Chapter 23, Erik Swyngedouw and Joe Williams show how desalination became the subject of a delicate consensus that strategically aligned disparate actors in Spain's water politics. This strategy, they argue, represents an attempt to remove political dissent from the sphere of water governance, and to build regional and national consensus around a re-imagined productionist logic for Spain's hydraulic development. The authors outline six contradictions of desalination that could form a potential terrain for a repoliticization of the Spanish waterscape. Chapter 24 is the Santa Cruz Declaration (Boelens et al., 2014). It is a broadly disseminated and debated chapter that argues that the global water crisis is not primarily driven by water scarcity, but fundamentally one that is steered by social and power relationships that express and are based on injustice and inequality. The declaration expands and elaborates on this understanding pointing out directions for addressing the water issue. The book closes with a short concluding Chapter 25 that reflects on the major issues addressed in the book and lessons that can be learned from the rich and diverse case studies and analyses that are presented.

References

Bakker, K. (2010). *Privatizing water. Governance failure and the world's urban water crisis*. Ithaca, NY: Cornell University Press.

Boelens, R. (2015). *Water, power and identity. The cultural politics of water in the Andes*. London, UK: Routledge.

Boelens, R., Crow, B., Dill, B., Lu, F., Ocampo-Raeder, C., & Zwarteveen, M. Santa Cruz declaration on the global water crisis. *Water International* 39 (2), 246–261.

Boelens, R., Hoogesteger, J., & Baud, M. (2015). Water reform governmentality in Ecuador: Neoliberalism, centralization, and the restraining of polycentric authority and community rule-making. *Geoforum* 64, 281–291.

Boelens, R., Hoogesteger, J., Swyngedouw, E., Vos, J., & Wester, P. (2016). Hydrosocial territories: A political ecology perspective. *Water International* 41 (1), 1–14.

Bridge, G., & Perreault, T. (2009). Environmental governance. In N. Castree, et al (Eds.), *Companion to environmental geography* (pp. 475–397). Oxford, UK: Blackwell.

Goff, M., & Crow, B. (2014). What is water equity? The unfortunate consequences of a global focus on 'drinking water'. *Water International* 39 (2), 159–171.

Hoogesteger, J., & Verzijl, A. (2015). Grassroots scalar politics: Insights from peasant water struggles in the Ecuadorian and Peruvian Andes. *Geoforum* 62, 13–23.

Hoogesteger, J., & Wester, P. (2015). Intensive groundwater use and (in)equity: Processes and governance challenges. *Environmental Science and Policy* 51, 117–124.

Linton, J., & Budds, J. (2014). The hydrosocial cycle: Defining and mobilizing a relational-dialectical approach to water. *Geoforum* 57, 170–180.

Lu, F. (2012). Petroleum extraction, Indigenous people and environmental injustice in the Ecuadorian Amazon. In F. Gordon & G. Freeland (Eds.), *International environmental justice: Competing claims and perspectives* (pp. 71–95). Hertfordshire, UK: ILM Publishers.

Meehan, K. (2013). Disciplining de facto development: Water theft and hydrosocial order in Tijuana. *Environment and Planning D* 31, 319–336.

Mehta, L., Veldwisch, G.J., & Franco, J. (2012). Introduction to the special issue: Water grabbing? Focus on the (re)appropriation of finite water resources. *Water Alternatives* 5 (2), 193–207.

Perreault, T. (2014). What kind of governance for what kind of equity? Towards a theorization of justice in water governance. *Water International* 39 (2), 233–245.

Swyngedouw, E. (2011). Interrogating post-democratization: Reclaiming egalitarian political spaces. *Political Geography* 30 (7), 370–380.

Swyngedouw, E., & Williams, J. (2016). From Spain's hydro-deadlock to the desalination fix. *Water International* 41 (1), 54–73.

Sze, J., London, J., Shilling, F., Gambirazzio, G., Filan, T., & Cadenasso, M. (2009). Defining and contesting environmental justice: Socio-natures and the politics of scale in the Delta. *Antipode* 41 (4): 807–843.

Turnhout, E., Van Bommel, S., & Aarts, N. (2010). How participation creates citizens: Participatory governance as performative practice. *Ecology and Society* 15 (4), 26. Retrieved from http://www.ecologyandsociety.org/vol15/iss4/art26/.

Vos, J., & Hinojosa, L. (2016). Virtual water trade and the contestation of hydrosocial territories. *Water International* 41 (1), 37–53.

Winner, L. (1986). *The whale and the reactor: A search for limits in an age of high technology*. Chicago, IL: Chicago University Press.

Zwarteveen, M.Z., & Boelens, R. (2014). Defining, researching and struggling for water justice: Some conceptual building blocks for research and action. *Water International* 39 (2), 143–158.

Defining, researching and struggling for water justice: some conceptual building blocks for research and action

Margreet Z. Zwarteveen[a,b]* and Rutgerd Boelens[b,c,d]**

[a]Department of Geography, Planning and International Development Studies, University of Amsterdam, the Netherlands; [b]Department of Environmental Sciences, Wageningen University, the Netherlands; [c]Centre for Latin American Research and Documentation, University of Amsterdam, the Netherlands; [d]Department of Social Sciences, Catholic University of Peru, Lima

This article provides a framework for understanding water problems as problems of justice. Drawing on wider (environmental) justice approaches, informed by interdisciplinary ontologies that define water as simultaneously natural (material) and social, and based on an explicit acceptance of water problems as always contested, the article posits that water justice is embedded and specific to historical and socio-cultural contexts. Water justice includes but transcends questions of distribution to include those of cultural recognition and political participation, and is intimately linked to the integrity of ecosystems. Justice requires the creative building of bridges and alliances across differences.

Introduction

The distribution of rights to access water and participate in decision making on water management and governance is extremely skewed in many countries of the world. This has always been so, but risks worsening because of growing competition caused by increasing water demand and decreasing water availability (because of ecosystem degradation and climate change). It is ironic that contemporary water policies and legislative measures to address problems of water scarcity risk further widening the gap between the water 'haves' and 'have nots'. In particular, the water rights and water-based livelihoods of smallholder irrigator communities in many countries in the global South are under constant threat by bureaucratic administrations, market-driven policies, desk-invented legislation and top-down project intervention practices, which tend to steer water flows in the direction of supposedly more productive uses and users (Isch, Peña, & Boelens, 2012; Molle, Mollinga, & Wester, 2009; Swyngedouw, 2005). Indeed, in arid and semi-arid areas, wealth differences between farmers increasingly are as much or more a function of people's differential access to water as they are of differential access to land. Likewise, the dynamics of market-led land reforms are importantly governed by access to water (see Liebrand, Zwarteveen, Wester, & van Koppen, 2012). The question of how to fairly distribute material water access rights and political water decision-making rights therefore deserves attention. In this article, we suggest some concepts and theoretical tools in support of an agenda for research and action on water justice. The article is informed by our interactions and discussions within the Justicia Hídrica/Water Justice alliance, and by the many studies done in the context of the alliance.[1]

The question of water justice combines, in complex and sometimes paradoxical ways, demands for more just socio-economic distribution and for more or better cultural-political recognition (cf. Fraser, 2000; Schlosberg, 2004). Understanding water justice requires creative analyses that link geo-hydrological and climatological insights into water availability patterns with understandings of the socio-technical and legal-cultural determinants of how available water flows are accessed and allocated. The evolving research and action field constituting the political ecology of water (e.g. Ahlers & Zwarteveen, 2009; Bakker, 2004; Boelens, 2009; Budds, 2004; Loftus, 2009; Martínez-Alier, 2012, 2013; Perreault, Wraight, & Perreault, 2011; Swyngedouw, 2005) is inspiring here, as it explicitly begins from an understanding of nature, technology and society as mutually constitutive, forming 'hydrosocial networks' that establish how water is (to be) distributed.

In the rest of this article, we propose and discuss some further theoretical concepts and ideas that are useful for identifying, exposing and challenging water-based injustices. One important criterion in our choice of theoretical tools is that they need to be suitable for recognizing the power and politics of water use, management and governance. This starts with the recognition that power and politics are everywhere, and not confined to the formally designated realms of decision making and official political arenas. Nor is power expressed only in explicit laws, rules and hierarchies, but it also importantly operates through less visible norms that often present themselves as natural or inevitable. These are often implicit in perceptions about what is 'normal' and in cultural codes of conduct and behaviour (Foucault, 1975, 2008).

After this introduction, we first provide a brief sketch of the current water policy discourse, arguing that this discourse requires critical scrutiny in terms of how it actively produces water inequities. We continue with a section in which we discuss and propose a relational and grounded definition of water justice, after which we suggest several theoretical concepts and ideas useful for researching and analyzing water injustices. We then briefly explore how demands for water justice can be formulated and advanced, and end with some concluding remarks.

Challenging the mainstream

On the waves of a heightened political consciousness of the scarcity of water, the world is witnessing a rapidly growing body of scientific and policy literature that presents models, guidelines, tool boxes and rule systems to govern water affairs. A quick review of these shows that there is a rather strong consensus about what the problems are as well as about what to do about them. Confining it to water for food and agriculture, this consensus has it that there is too little water for feeding the world's population. Although this water scarcity is attributed to a multitude of causes, one that stands out and receives relatively much attention is the wastefulness of farmers, who use far too many 'drops per crop' – something that is linked to the fact that 60–70% of the world's freshwater resources (estimates vary) are used in agriculture. Suggested ways to make farmers use water more efficiently include water-saving technologies (such as drip irrigation) and the pricing of water (or making its allocation subject to market or quasi-market principles), which is expected to induce farmers to use it more cautiously as well as leading to the redirection of flows to where marginal returns are highest.

More generally, solutions that are currently favoured to combat the water crisis combine three sets of beliefs: a belief in markets, a belief in participatory processes of deliberation and a belief in engineering (Sneddon & Fox, 2007). Together, these three beliefs culminate in 'integrated water resources management' (IWRM) tool boxes and recipes that neatly prescribe how water should most efficiently and effectively be used, managed and governed. These, and the larger IWRM discourse that they form part of, are actively disseminated through research and knowledge centres, international banks and funding agencies, government bureaucracies and development agencies (Goldman, 2001). When implemented, they produce new forms of 'water governmentality' that entail the repatterning of water spaces and territories; the reshaping of rules and authorities and of labour and production relations; and the rearranging of water user groups and families in new water power hierarchies (Boelens, 2013; Perramond, 2013).

IWRM is seen and presented as a break with the technocratic, supply-driven and construction-oriented paradigm of the past, to signal a new era in which economic, social, environmental and social concerns are addressed simultaneously and in their mutual interactions. As a growing body of political ecology and water justice studies have shown, however, IWRM, although flagging sustainability and democracy, is often used to hide or sanction processes of dispossession and accumulation of water, processes that are far from democratic or participatory (Allan, 2006; Molle et al., 2009). Often, for example, water scarcity is presented as a global and natural phenomenon that threatens humanity as a whole. Yet, not everyone is equally threatened by water scarcity (Bakker, 2004; Ioris, 2012), and accumulation by some often goes hand in hand with deepening scarcity as experienced by others (Arroyo & Boelens, 2013).

IWRM discourses and the allocations they sanction create rankings of water uses and users on the basis of specific calculations of efficiency, with the most efficient uses and users being awarded the premium of modernity and water citizenship. 'Modern' users – such as large-scale commercial enterprises, agribusiness firms, private drinking-water companies, and mining and hydropower conglomerates – thus become the example to be followed, representing the ideals of water use efficiency and water market rationality that science preaches (Boelens & Vos, 2012). In contrast, people who use traditional irrigation systems for growing their own food crops come to be seen as 'backward'. For water scarcity problems to be overcome, they either need to disappear or they need to correct their water misbehaviour to join 'progress' and 'development' (Castro, 2007; Vera & Zwarteveen, 2008).

The critical examination of these discourses, and of the concepts of efficiency and modernity they employ, forms an essential part of the effort to understand and fight water injustices. It hinges on attempts to unravel the politics and political implications of proposed reforms in water governance and regulation. This entails scrutiny of prevailing modes of water distribution and water authority, as well as of the discourses, institutions and technologies through which these become articulated. It involves for instance the examination of how supposedly neutral efficiency terms rely for their implementation on reallocations of water (Boelens & Vos, 2012; van der Kooij, Zwarteveen, Boesveld, & Kuper, 2013; Van Halsema & Vincent, 2012). It also includes an assessment of which and whose histories, world-views, knowledge systems, norms and practices prevail and why; an analysis, in other words, of the politics of disciplining – the modalities and strategies of power that are (consciously or unconsciously) used to generate a set of values, beliefs and behaviours (e.g. Meehan, 2013; Rodríguez de Francisco, Budds, & Boelens, 2013). On the other hand, it entails understanding and (re)valuing the strategies of deviance and resistance by those water user groups and communities who are targeted, incorporated or excluded by dominant water policy and governance cultures (Boelens, 2009; Perreault et al., 2011; Vera & Zwarteveen, 2008).

Defining water (in)justice: (re)distribution, (mis)recognition and voice

Many political-philosophical theories have aimed to conceptualize justice as a universal and transcendent notion, focusing 'on what justice *should be*' (Lauderdale, 1998, p. 5). Definitions of justice in dominant libertarian or entitlement theories (e.g. Nozick, 1974) for instance stress the connection between individual freedom (vis-à-vis state control) and private property rights, and posit these as key universal principles of humanity and human society. Neoliberal interpretations of justice (for instance as articulated or implied by Hayek, 1944 and Friedman, 1962) build on and extend these definitions, stressing both that individuals must have the 'freedom' to pursue the maximization of their own interests and that all individuals are 'equal' through their inclusion as participants and players in the market game. In these philosophical-theoretical perspectives, large economic and distribution inequalities are compatible with 'justice' because these are the outcomes of people's own aspirations and strivings.

Also part of the positivist tradition, and in line with the postulates of the political-philosophical founder of utilitarianism, Jeremy Bentham, liberal utilitarian principles of justice consist of those societal orders that bring the greatest happiness to the greatest number of citizens. Hence, the rights and happiness of some individuals may be sacrificed if this would enhance the well-being of most others. These ideas were expressed in a new, uniform language to "establish a system that aims to construct happiness societally by means of reason and law" (Bentham, 1988(1781), pp. 1–2). The calculation of happiness was in the hands of moral and justice experts, since common people were not considered rational enough to oversee the interests of all. Later utilitarian elaborations were more 'participatory' in subtly including the people in (and excluding the 'irrational deviants' from) this empire of liberal justice.

Alternative liberal theories emphasize not equal distribution but 'fair procedures', to guarantee that justice can take place according to autonomous decisions based on ethical principles. Rawls's influential *A Theory of Justice* (1971), for example, uses the metaphor of a "veil of ignorance" behind which people are supposed to make decisions on justice (and in particular, universal efficiency) without knowing the impact these decisions will have on themselves. Although these definitions and ideas presuppose the equality of all, they work to justify distributive planning and decision making in arenas where people are not at all equal but divided along lines of class, gender, education and ethnicity.

Most *legal* justice constructions display variations of these liberal ideas and ideals of justice. They proclaim uniform values of justice and a uniform property framework, based on the proclaimed equality of all citizens before the law. Water laws are no exception: they are commonly presented as objective, rational systems for designing societal life, rather than as deeply cultural phenomena and political products (Roth, Boelens, & Zwarteveen, 2005). For this, the legal systems sustaining water policies emphasize unity and uniformity (the same water rules and regulations apply to all), with the state enjoying a monopoly on water rule making and dispute resolving, subjugating all other tribunals or rights frameworks (Boelens, 2009). In practice, the 'equality of all' that such uniform frameworks presuppose works to deny or ignore existing social hierarchies and differences (such as those based on class, ethnicity or gender), with the reference for the proclaimed equality being (implicitly) based on the class, gender and cultural characteristics (and normative standards and interests) of a small but powerful minority (Vos, Boelens, & Bustamante, 2006).

Precisely because equality cannot be assumed or simply proclaimed, homogeneous concepts of justice based on abstract, universal criteria tend to poorly correspond (and respond) to the experiences of and claims made by the 'non-equals': marginalized indigenous and peasant societies, for instance, or women. In addition, and as argued by Young (1990), Fraser (2000) and Schlosberg (2004), theories that focus only on (universal) distributive models and procedures are poorly equipped to "examine the social, cultural, symbolic and institutional conditions *underlying* poor distributions in the first place" (Schlosberg, 2004, p. 518). We suggest therefore that definitions and understandings of justice cannot be based only on abstract notions of 'what should be', but also need to be anchored in how injustices are *experienced*. They need to be related both to the diverse 'local' perceptions of equity and to the discourses, constructs and procedures of formal justice. As Lauderdale suggests, this requires a relational, grounded, comparative and historical approach. "The study of justice includes an analysis of the fair distribution of benefits and burdens, including rights, obligations, desserts and needs. The approach includes analyses of public plans and policies set up to implement ideas of justice" (p. 9). Harvey similarly proposes conceptualizing justice as "a socially constituted set of beliefs, discourses, and institutionalizations expressive of social relations and contested figurations of power that have everything to do with regulating and ordering material social practices within places for a time" (1996, p. 330). These definitions emphasize the historical and place-based specificity of justice, using it as a way to examine how specific modes of ordering are rooted in specific societies and the effects this has on the distribution of property, wealth and authority (cf. Zwarteveen, 2006).

Drawing on work which looks at how processes of environmental change work to reallocate incomes, resources and power, Schlosberg (2004) follows the suggestions of Fraser (2000) in a lucid attempt to conceptualize environmental justice. He proposes a "trivalent conception of justice" (Schlosberg, 2004, p. 521) which includes, along with distribution, the dimensions of recognition (e.g. of specific cultural identities, rights and practices) and participation (in decision making).[2] According to Schlosberg, justice "requires not just an understanding of unjust distribution and a lack of recognition, but, importantly, the way the two are tied together in political and social processes" (p. 528). Like Schlosberg, we think it is important to add dimensions of (cultural) recognition and procedural democracy to those of (re)distribution. Next, given the life-securing and life-threatening nature of the resource and its embedded-ness in delicate and dynamically shaped socio-natural environments, and the need to sustain livelihood security for current as well as future generations, a fourth sphere of water justice struggle may be referred to as 'socio-ecological justice' (socio-natural or socio-ecological integrity).

Understanding (in)justice, then, encompasses the examination of both *formally accre-dited justice* (formal schemes of interpretation and legitimization, and legal-positivist constructs of 'rightness') and *socially perceived justice or equity* (location-, time-, and group-specific constructs of 'fairness' – see Boelens, 2009) that are used by different societal groups.[3] In addition, an analysis is needed of why certain views on justice or equity gain prominence while others are ignored, and how this works to reproduce or challenge prevailing social hierarchies and relations of power.

An overview of key concepts

The following section presents a number of important concepts and terms which we think are crucial to identify, understand, analyze and react to water-based forms of injustice. We acknowledge that not everyone working in the diverse 'water worlds' is familiar with all these terms, but we think that they open opportunities to deepen the understanding of the particular and entwined political, socio-economic, technical-biophysical and cultural dynamics that contribute to overt and covert injustices.

Situated knowledges. Determining what is unfair, inaccurate, or incomplete cannot be done from a transcendent outside position but always implies engagements and identifications with those whose lives and worlds are the objects of inquiry (see Baviskar, 2007). This knowledge position starkly contrasts with the one implied by much water knowledge, which continues to be based on the belief in the possibility of 'objective' truth that can be obtained through the unclouded gaze of a detached observer. Indeed, much water knowledge speaks 'as if' from nowhere, from a value-free and god-like position, by someone without interests or background, representing the universal good. Sceptical of the possibility of producing such universally valid statements about reality, the powers of reason, and the subject–object split (Baviskar, 2007; Butler, 1995; cf. Donahue & Johnston, 1998; Foucault, 2008; Haraway, 1991), we instead see meanings, discourses and (the production of) truths as internal to inequitable water orders, rather than external: they come about through situated perspectives that need to be made as explicit as possible. Awareness of the specificity and positionality (or situatedness) of all knowledge also prompts a heightened vigilance about the political effects of certain discursive representations, in particular when they travel from one site to another.

Thus, truths, concepts and language are never 'neutral' denominators of objective realities that are out there waiting to be discovered, but co-constitute – or are an intrinsic part of – such realities. They emerge through social processes in which agreement, persuasion, belief, culture and world-view play a role. Research and analysis mediate between different, yet mutually conditioned, views – those of the researcher(s) and those of the people and environments who are being studied – each forming part of their own socio-natural environments. Rather than the latter being simply the 'objects' of research, they 'talk back', interacting with researchers and co-developing meanings, truths and interpretations. In recognition of the relational dialectics between researchers and researched, a self-conscious research attitude is needed.

Self-consciousness, accountability and reflexivity are important for all research, but are of special importance for researching and understanding questions of justice, because understandings of justice more obviously combine 'facts' (about water availabilities, for instance) with opinions and values (about what is fair or just). Facts and values to name and judge specific socio-natural orders often come together in, and are expressed through, particular discourses. For Foucault (1975), discourses comprise groups of related statements which govern the variety of ways in which it is possible to talk about something and which thus make it difficult, if not impossible, to think and act outside them. Discursive practices are characterized by "a delimitation of a field of objects, the definition of a legitimate perspective for the agent of knowledge, and the fixing of norms for the elaboration of concepts and theories" (Foucault, 1977, p. 199). Because discursive practices are mixed up with power, certain representations of reality serve certain interests and interest groups better than others.

Explicit attentiveness to the ways in which realities, problems and solutions are discursively framed is therefore important. The current preferred language for thinking about water is clearly neoliberal in flavour. Although seldom made explicit, it reflects a specific political (but objectified) ideology with very particular ideas about the nature of human beings and the preferred direction of development (see Achterhuis, Boelens, & Zwarteveen, 2010; Boelens & Zwarteveen, 2005). Efforts to identify and expose injustices in water need to critically question such established water discourses in order to arrive at a *repoliticization* and a *contextualization*: at visualizing the workings of power in and through discourse, at showing how particular ways of phrasing and techniques of governance serve to hide contentious distributional and representational questions, and at exposing the specificity of time, place and positionality of the knower(s).

Socio-natures. A second set of theoretical starting points for naming and understanding water (in)justices are those that question the boundaries between nature, technology and society (or humans) by positing that such boundaries are themselves the products of human minds and social conventions. This is an important insight, because the act of relegating phenomena to the realms of nature – naturalization – is a well-known and much-used strategy to depoliticize water problems, placing contentious questions of distribution outside of the domain of public debate. Water scarcity is for instance often referred to as a *natural* problem caused by climate change and changing weather conditions, rather than as a problem of distribution or of power relations (e.g. Bakker, 2004; Ioris, 2012). Frequent calls for using river basins as 'the natural unit for water management' can for instance be seen as a way to depoliticize water management by recourse to the "naturalizing metaphor" (Bakker, 1999) of the river basin (see Saldías, Boelens, Wegerich, & Speelman, 2012; Wester, 2008). As pointed out by Blomquist and Schlager, "the definition of a watershed and the selection of boundaries are matters of *choice*. As soon as the matter of choice is present, there is a role for politics" (2005, pp. 104–10). Notions of water scarcity obviously may have 'absolute' subsistence and survival properties, but they are always deeply mediated by humans and determined by power relationships that construct 'scarcity' far beyond just the wickedness of nature.

The ambition to conceive nature and society as co-constituted can draw upon science and technology studies, where vocabularies of hybridity are used by actor-network theory scholars (see Latour, 1993; Law & Hassard, 1999) as well as by feminist science studies (Haraway, 1991). Words like "waterscapes" (Baviskar, 2007; Swyngedouw, 2003), "naturecultures" (Haraway, 1991) and "hydrosocial networks" (Boelens, 2013; Swyngedouw, 2003; Wester, 2008) all convey the idea that infrastructural and institutional water developments develop 'part natural part social', as material dynamic reflections of historic and never-ending socio-political-geographical struggles (Ahlers & Zwarteveen, 2009; Swyngedouw, 2003; Zwarteveen, 2006).

Contestation. As the previous sections have indicated, existing and emerging ways of using, accessing and distributing water tend to be contentious. The coproduction of waterscapes or hydrosocial networks is constituted by, and simultaneously constitutes, the political economy of access and control over resources (Budds, 2004; Harvey, 1996; Swyngedouw, 2003). Water reforms thus unquestionably also imply changes in access to and control of this resource; as water is a finite resource, those who receive more generally do so at the expense of others who receive less. Water, in other words, is an intrinsically contested resource. We distinguish four main echelons of water contestation:

- First, the very distribution of the *resource* is contested: Who has access to water, to hydraulic infrastructure, to the material and financial means to use and manage water resources?
- Second, conflicts and disagreements also and importantly occur over the *contents of rules, norms and laws* that determine water distribution and allocation.
- A third way in which water rights are contested relates to struggles over *authority*. Who decides about questions of water distribution? Who is entitled to participate in water law and policy making? Whose opinions and norms are listened to and accommodated? Whose definitions, priorities and interests prevail?
- A fourth and last important area of contestation lies in the *discourses* used to articulate water problems and solutions. What are the accepted languages and practices for framing and shaping water laws, and what are the preferred ways of conceptualizing water problems? How do different regimes of representation characterize the relations among actors, the social and technical environment, and water access and control; and how do they devise or promote institutions, techniques, strategic artefacts and practices to realize their views and objectives?

These echelons of water struggles directly relate to each other and are shaped in mutual interaction. For example, a particular discourse will also entail a particular way of organizing decision making, and work to legitimize some forms of authority over others; it will also favour some rules and priorities for resource management and allocation; this then fosters the ways in which available resources are being distributed and used – privileging some groups over others.

Complexity. Water allocation and management involve often contradictory and complex (or 'wicked') problems: that is, clusters of interrelated problems, characterized by high levels of uncertainty and a diversity of competing values and decision stakes. Typically, 'knowing' and representing wicked problems, let alone proposing solutions, is a highly controversial matter, in which many different accounts of reality compete with each other (Wester, de Vos, & Woodhill, 2004; Whatmore, 2009).

One possible way to deal with complexity is to develop ever more sophisticated expert devices that allow mapping environmental phenomena – such as water pollution, floods and droughts – into knowledge and incorporating them into 'evidence-based' management strategies. Increasingly, such devices are based on and make use of remote sensing and GIS, and include predictive models, risk indicators, monitoring instrumentation and ways to calculate environmental services. Without disputing the wonders that can be done with new observation techniques and models, their use does involve a risk: it may work to strengthen the faith in the possibility of objectively 'knowing' and rationally managing water problems. It allows, in other words, proceeding as if water problems were largely about ordered events and as if it were possible to produce one singular best account of their causes, effects and solutions.

We instead suggest that there is merit in acknowledging that most water problems belong to the domain of the 'unordered', where decisions are based on power, perception, and situated perspectives and understanding (cf. Kurtz & Snowden, 2003). Therefore, we argue for the need to remain vigilant about the temptation to unequivocally use 'science' and the objectification it entails in dealing with water's complexity. Knowledge about water will always and necessarily be uncertain and provisional. Relaxing the search for the one most accurate and reliable account of water problems and realities usefully opens the door to accepting diverse and plural knowledges about processes of water-related change – including those based on the experiences and knowledge of people who live in changing environments.

Water rights. Water injustices are importantly about structural water scarcities caused by resource capture, and the resulting patterns of unequal access to water and decision-making spaces (Ahlers, 2010; Wester, 2008; Zwarteveen, 2006). Understanding how injustices are produced or how to support greater fairness or democracy, therefore, hinges on insights in the dynamics of water allocation: on how water rights are defined or understood. We propose an ontological definition of water rights that departs from more mainstream conceptions, which tend to see 'clear and enforceable' water rights as a tool and condition to make water managers and users mutually accountable or market-based trading of water possible. In line with our definition of water justice, we base our understanding of water rights instead on the explicit acknowledgment of their historical specificity and embeddedness in particular ecological and cultural settings. In this under-standing, locally existing norms and water control practices, and the power relations that inform and surround them, are deeply *constitutive* of water rights (see Boelens & Zwarteveen, 2005).

Because of the variable availability and fluid characteristics of the resource of water, and because of the difficulties in rigorously monitoring and controlling water flows, there is a lot of scope for users at different levels to act in ways that diverge from distributional agreements as stipulated in state laws, regulations, infrastructural layouts and technolo-gies. To capture the difference between 'rights on paper' and actual water control and distribution, we have suggested the following distinction of categories of rights: reference rights, activated rights and materialized rights. As we have explained in more detail elsewhere, these categories can be seen as different manifestations of rights (Boelens & Zwarteveen, 2005). In most water-use contexts, water rights exist in conditions of legal pluralism. This implies that multiple rules, norms and principles of different origins and sources of legitimization coexist. Therefore, even when there appears to be legal and administrative uniformity, water rights' complexity in practice can be huge. Understanding justice likewise requires insights into how water rights and rule systems are being shaped in everyday water-use practices; the complex and often divergent ways in which they interact with various socio-legal frameworks and power structures (at different scales); and the potential and actual conflicts among different rights systems (over water use, rules, authority, and discourses or ideologies).

Scale and scalar politics. Naming, defining and understanding water (in)justice is intrin-sically scale-sensitive, with judgements of whether a situation is just or not changing with the units of time and place used. Appreciating the fairness of water distribution within an irrigation system, for instance, critically depends on how the boundaries of the command area of the system are drawn. Intra-system fairness may be achieved over time by gradually reducing the area irrigated, to the detriment of those whose lands are situated in the parts that are no longer included in the system. Temporal and geographical scales are always socially constructed, and hence contingent and dynamic (see McCarthy, 2005), with the choice and definition of scales and scalar configurations sometimes themselves being contested in struggles over what is fair or equitable (Brown & Purcell, 2005). Indeed, 'jumping' scales can be an effective strategy to make injustices disappear. That mining companies dispossess peasant communities of their water rights, for instance, is unjust from the perspectives of these communities, but such injustices tend to be seen as minor by governments and the general public when measured against the financial contributions of mining companies to national development.

As Swyngedouw and Heynen assert:

> The priority, both theoretically and politically, ... never resides in a particular social or ecological or geographical scale; instead it resides in the socio-ecological process through which particular social and environmental scales become constituted and subsequently re-constituted. In other words, socioecological processes give rise to scalar forms of organisation – such as states, local governments, interstate arrangements and the like – and to a nested set of related and interacting socioecological spatial scales.... The continuous reorganisation of spatial scales is an integral part of social strategies to combat and defend control over limited sources and/or a struggle for empowerment. (2003, pp. 912–13)

What this means for understanding questions of water justice is that these need to include an explicit understanding of how scales are used, constructed and entwined in hydrosocial dynamics and networks, among others, through political struggle (cf. Brown & Purcell, 2005). In this respect, it is particularly important to critically examine terms such as 'local' and 'global', because so-called 'local' phenomena often consist of specific manifestations of supra-local processes and powers. Harvey (2003), for example, shows how processes of dispossession, appropriation or theft form an integral part of the reorganization of capital on a global scale.

Connecting struggles for redistribution, recognition, participation and socio-natural integrity

Where authors such as Harvey (1996) insist on environmental-justice movements to transcend particularity and pluralism and work towards the singular and universal in order to be able to confront globalizing capitalist injustice, Schlosberg argues that an environmental-justice movement can be unified but it cannot be uniform. "If Foucault taught us anything, it is that power is multiple, and arises everywhere in everyday situations and must be constantly resisted where it is experienced. It is no different with (in)justice" (2004, p. 534; see also Martínez-Alier, 2012). In environmental and water justice movements, there is (and only can be) the possibility of unity when there is no uniformity regarding how the notion is or should be defined. In line with Schlosberg, we suggest therefore that while accepting and acknowledging diversity, difference and plurality it is important to examine how bridges can be established among the diverse ways of viewing and struggling for water justice: engagement and alliances across contexts, continents, scales and differences (cf. Mouffe, 2007; Schlosberg, 2004).

Injustices in water may sometimes produce very open conflicts, with people who feel treated unjustly actively protesting, for instance in cases of resistance against the introduction of water privatization policies or when water grabbing or large-scale pollution are the evident outcomes of such new policies (Bebbington, Humphreys Bebbington, & Bury, 2010). Although such injustices attract much attention, most injustices occur in less spectacular ways and involve more subtle and long-winding processes of struggle, in which officially endorsed and unofficial water cultures confront each other to create clashes between social, political and economic water values, meanings and decision-making powers. Some injustices never produce open disputes or struggles but instead consist of the silent sufferings. These are often provoked by the water reallocations and dispossessions that accompany the erosion of existing water cultures and forms of knowledge (see Ahlers, 2010; Vos et al., 2006). Indeed, the strong policy push to make water rights transferable through the uniformization of rights systems, which some scholars interpret as a new round of 'enclosures' of the commons or processes of 'accumulation by dispossession', is more likely to create such silent take-overs instead of provoking spectacular water wars.

Proposals to improve the water security of marginal water user communities often dangerously go along with such calls for uniformization, by demanding the formal recognition of plural legal and normative systems. However, whether such recognition indeed means improvement is a question, the answer to which is not straightforward. And is recognition enough? Recognition, the cultural dimension of justice, refers to acknowledging and respecting various forms of dealing with, organizing around, and talking about water. This has to do with diversity, identity and culture, and relates primarily to forms of injustice that deny or discriminate against particular socially and culturally embedded rules and practices of water management and control (e.g, Crow & Odaba, 2010; Zwarteveen, 2010). Granting autonomy to groups of people or water user communities to devise and apply their own water rules addresses a form of cultural recognition, as does the acceptance and recognition of women, indigenous and peasant users and leaders as legitimate water actors. This also relates to representational justice – the issue of political participation in control and decision making, of sharing in water authority – both at local management levels and at broader scales of water governance.

Water justice questions may also extend to public water investments and often include the socialization of technology and reform of land tenure. Cultural recognition, participation and redistribution are related and influence each other in complex ways. They are intimately tied up with questions of power and hegemony, and reinforce each other dialectically. Cultural norms that are unfairly biased against some groups (on the basis of class, ethnicity, gender, caste or a combination of those) tend to become institutionalized in the state and the economy, and serve to justify their lesser access to water. Meanwhile, their economic disadvantage impedes equal participation in the making of water-allocation rules and laws, and in actual water-distribution decisions. The result is a downward spiral of economic and cultural subordination.

Yet, despite the entanglement between socio-economic, cultural-political and representational justice, there is merit in distinguishing them, because the remedies to address these kinds of injustices are different, and sometimes even conflicting (see Fraser, 2000; Zwarteveen, Roth, & Boelens, 2005). For example, claims for recognition often take the form of calling attention to the supposed specificity of some group and then affirming its value. In contrast, redistribution claims often call for abolishing economic arrangements that underpin group specificity. Instead of calling for the right to be different, these call for the right to be equal. Distinguishing the different kinds of justice allows asking questions about the relation between claims for recognition, claims for participation and claims for redistribution, and about the interferences that arise when these claims are made simultaneously. It also draws attention to the politics involved in claiming rights for specific groups, or in calls for redressing historical injustices and inequities.

In sum, demands for greater water justice require a critical view that acknowledges its cultural, political and material dimensions – which, moreover, are all embedded in socio-natural environments asking for stewardship. Good scholarship and policy making, as well as balanced activist and grass-roots action, require conceptualizing cultural recognition, political participation, socio-economic equality and care for socio-ecology in forms that support rather than undermine one another. And they require clarifying the political dilemmas that arise when trying to combat these injustices simultaneously. Therefore, we realize that strategies of empowerment, resistance and users' appropriation of water control that challenge existing water distributions, laws, authorities and expertise are only successful when headed by those who demand more control: groups of water users (e.g., Bebbington et al., 2010; Hoogesteger, 2012). Through struggle and well-organized representation at negotiating platforms, they can define and negotiate their water rules and defend and enforce their water rights, and influence the formulation of the rules of play.

Conclusions

In some countries in the global North, and in many countries in the global South, growing demand and decreasing availability of water of sufficient quality are leading to intensifying competition and conflict among different water uses and users. Globalization and a neoliberal policy climate tend to help certain powerful actors – local, national and often transnational – accumulate water resources and rights at the expense of the economically less powerful. New competitors, including megacities and mining, forestry and agribusiness companies, claim ever larger shares of available surface-water and groundwater resources. Unequal water distribution and related pressures on land resources and territories, both legally condoned and through large-scale extralegal appropriation practices, generate misery and poverty among smallholder families and in rural communities, while posing profound threats to environmental sustainability and national food security.

This constitutes the backdrop of this article, in which we have outlined some basic theoretical notions and reflections which, when taken together, provide a loose conceptual framework for starting to name, define and understand water (in)justice and the mechanisms and processes producing it. The embeddedness and situatedness of (in)justice, and of the possible ways to remedy it, is central to the approach we propose. It brings with it an acknowledgment of diversity and plurality – in views, knowledges, rights systems, ideas and norms about fairness, etc. This does not mean that we embrace a cultural relativist stance or that we deny the larger similarities across specific instances of injustice, or the parallels in the processes and mechanisms that produce them. One obvious similarity between very diverse cases of water injustice is that they often entail transfers of water from supposedly less to more productive uses, and more specifically from lower-value food crops to high-value export crops or industries. Such transfers are actively promoted and legitimized through the uniformization of rights systems that are part of neoliberal (policy) discourses, which have become quasi-hegemonic in thinking about and acting on water.

We propose a conceptualization of 'water justice' that explicitly thematizes its relational character and contextuality, and that recognizes both its material and economic dimensions ('redistribution') and its cultural and political dimensions ('recognition' and 'participation', respectively; cf. Fraser, 2000; Schlosberg, 2004), while taking place in the arena of struggles for socio-ecological justice ('socio-natural integrity'). For this, we argue that it is important to recognize that politics and power as much pervade the allocation of water as ways of thinking and talking about it. Changing water allocations – whether through reform policies, new technologies, or markets – implies complex processes of political contestation, negotiation and struggle. These happen around the water resources themselves, but are also about the rules, norms and laws that form the basis of distribution processes and about who has (or should have) the political authority and legitimacy to decide these questions. Contestations and struggles also occur over the discourses and knowledge used to frame or legitimize water policies or ways of distributing water. In addition, we argue for a post-positivist and constructivist epistemology and a reflexive research attitude; a conception of nature and society as mutually constitutive; an understanding of water control as multi-layered, complex and 'wicked'; an ontological definition of 'water rights' as reflecting and co-constituting locally and historically specific constellations of property relations, expressing and embedded in social relations of power; and an understanding of the scalar dimensions of resistance and civil-society action.

We are aware that water justice will not happen as a result of accurate theories and well-intentioned philosophies, and that it cannot be 'legally engineered' or 'donated' by policy makers. It instead calls for the transdisciplinary co-creation of knowledge, involving mutuality and reciprocity among water users, policy makers, activists and scientific communities. It starts with taking seriously, and developing awareness of, the many manifestations of injustice, from brutal water grabs to much more subtle politics of disciplining and normalization. It involves the critical questioning of 'official water truths' and their claims to rationality, efficiency, democracy and equity. In the end, though, change will only happen through critical engagement and solidarity with those who experience injustice. It importantly consists of attempts to creatively link demands for redistribution with those for cultural recognition; of efforts to improve the political participation of those who are excluded or whose voices are silenced; and of actively interweaving diverse struggles for water justice across context, differences and scales.

Notes

1. The Justicia Hídrica/Water Justice alliance (www.justiciahidrica.org) is a research and action network that sets out to support water policies that contribute to an equitable distribution of water and democratic allocation procedures. As a broad alliance of researchers, policy makers, professionals and grass-roots organizations, its activities combine: interdisciplinary research on the dynamics and mechanisms of processes of water accumulation and conflicts; training and awareness-raising of a critical mass of water professionals, leaders and policy makers; and support for civil-society strategies that engage with the questions, needs and opportunities of marginalized groups.
2. A crucial difference between liberal theorists like Rawls and Miller on the one hand and Young and Fraser on the other is that the former, in their liberal search for perfect justice, assume and subsume recognition "within the distributive or procedural spheres of justice" (Schlosberg, 2004, p. 520).
3. These fairness or equity perceptions differ enormously; therefore, they cannot be reified or romanticized and constitute a power relation in themselves.

References

Achterhuis, H., Boelens, R., & Zwarteveen, M. (2010). Water property relations and modern policy regimes: Neoliberal utopia and the disempowerment of collective action. In: R. Boelens, A. Guevara & D. Getches (Eds.), *Out of the mainstream.* London & Washington, DC: Earthscan.

Ahlers, R., & Zwarteveen, M. (2009). The water question in feminism: Water control and gender inequities in a neo-liberal era. *Gender, Place and Culture, 16*(4), 409–426. doi:10.1080/09663690903003926

Ahlers, R. (2010). Fixing and nixing: The politics of water privatization. *Review of Radical Political Economics, 42*(2), 213–230. doi:10.1177/0486613410368497

Allan, T. (2006). IWRM: The new sanctioned discourse?. In: P. Mollinga, A. Dixit & K. Athukorala (Eds.), *Integrated water resources management* (pp. 38–63). New Delhi: Sage.

Arroyo, A. & R. Boelens (Eds.) (2013). *Aguas Robadas. Despojo hídrico y movilización social.* Abyayala, Quito: Alianza Justicia Hídrica.

Bakker, K. (1999). The politics of hydropower: Developing the Mekong. *Political Geography, 18* (2), 209–232. doi:10.1016/S0962-6298(98)00085-7

Bakker, K. (2004). *An uncooperative commodity: Privatizing water in England and wales.* Oxford: Oxford University Press.

Baviskar, A. (2007). *Waterscapes: The cultural politics of a natural resource.* Delhi: Permanent Black.

Bebbington, A., Humphreys Bebbington, D., & Bury, J. (2010). Federating and defending: Water, territory and extraction in the Andes. In: R. Boelens, D. Getches & A. Guevara, *Out of the mainstream. Water rights, politics and identity* (pp. 307–327). London & Washington, DC: Earthscan.

Bentham, J. (1988 (1781)). *The principles of morals and legislation.* Amherst, NY: Prometheus Books.

Boelens, R., & Vos, J. (2012). The danger of naturalizing water policy concepts: Water productivity and efficiency discourses from field irrigation to virtual water trade. *Agricultural Water Management, 108*, 16–26. doi:10.1016/j.agwat.2011.06.013

Boelens, R., & Zwarteveen, M. (2005). Prices and politics in Andean water reforms. *Development and Change, 36*(4), 735–758. doi:10.1111/j.0012-155X.2005.00432.x

Boelens, R. (2009). The politics of disciplining water rights. *Development and Change, 40*(2), 307–331. doi:10.1111/j.1467-7660.2009.01516.x

Boelens, R. (2013). Cultural politics and the hydrosocial cycle: Water, power and identity in the Andean highlands. *Geoforum,* doi:10.1016/j.geoforum.2013.02.008

Brown, J. C., & Purcell, M. (2005). There's nothing inherent about scale: Political ecology, the local trap and the politics of development in the Brazilian Amazon. *Geoforum, 36*, 607–624. doi:10.1016/j.geoforum.2004.09.001

Budds, J. (2004). Power, nature and neoliberalism: The political ecology of water in Chile. *Singapore Journal of Tropical Geography, 25*(3), 322–342. doi:10.1111/j.0129-7619.2004.00189.x

Butler, J. (1995) Contingent foundations. In: S. Benhabib, J. Butler, D. Cornell & N. Fraser, *Feminist contentions. A philosophical exchange* (pp. 35–58). New York, NY: Routledge.

Castro, J. E. (2007). Poverty and citizenship: Sociological perspectives on water services and public–private participation. *Geoforum, 38*(5), 756–771. doi:10.1016/j.geoforum.2005.12.006

Crow, B., & Odaba, E. (2010). Access to water in a Nairobi slum: Women's work and institutional learning. *Water International, 35*(6), 733–747. doi:10.1080/02508060.2010.533344

Donahue, J. M., & Johnston, B. R. (1998). *Water, culture and power, local struggles in a global context.* Washington, DC: Island Press.

Foucault, M. (1975). *Discipline and punish. The birth of the prison.* New York, NY: Vintage Books.

Foucault, M. (1977). Language, counter-memory, practice. In: D. F. Bouchard (Ed.), *Language, counter-memory, practice: Selected essays and interviews by Michel Foucault.* Ithaca, NY: Cornell University Press.

Foucault, M. (2008). *The birth of biopolitics.* New York, NY: Palgrave Macmillan.

Fraser, N. (2000). Rethinking recognition. *New Left Review,* May/June, 107–120.

Friedman, M. (1962). *Capitalism and freedom.* Chicago: University of Chicago Press.

Goldman, M. (2001). The birth of a discipline: Producing authoritative Green Knowledge, World Bank-Style. *Ethnography, 2*(2), 191–217. doi:10.1177/14661380122230894

Haraway, D. (1991). *Simians, cyborgs and women: The reinvention of nature.* New York, NY: Routledge.

Harvey, D. (1996). *Justice, nature & the geography of difference.* Oxford: Blackwell Publishers.

Harvey, D. (2003). *The new imperialism.* Oxford: Oxford Press.

Hayek, F. A. (1944). *The road to serfdom.* London: George Routledge.

Hoogesteger, J. (2012). Democratizing water governance from the grassroots: The development of Interjuntas-Chimborazo in the Ecuadorian Andes. *Human Organization, 71*(1), 76–86.

Ioris, A. (2012). The geography of multiple scarcities: Urban development and water problems in Lima, Peru. *Geoforum, 43*(3), 612–622. doi:10.1016/j.geoforum.2011.12.005

Isch, E.F. Peña & R. Boelens (Eds.). (2012). *Agua, Injusticia y Conflictos.* Cusco: CBC.

Kurtz, C. F., & Snowden, D. J. (2003). The new dynamics of strategy: Sense-making in a complex and complicated world. *IBM Systems Journal, 42*(3), 462–483. doi:10.1147/sj.423.0462

Latour, B. (1993). *We have never been modern.* Cambridge, MA: Harvard University Press.

Law, J., & J. Hassard (Eds.) (1999). *Actor network theory and after.* Oxford: Basil Blackwell.

Lauderdale, P. (1998). Justice and equity: A critical perspective. In: R. Boelens and G. Dávila (Eds.), *Searching for equity* (pp. 5–10). Assen: Van Gorcum.

Liebrand, J., Zwarteveen, M., Wester, P., & van Koppen, B. (2012). The deep waters of land reform: Land, water and conservation area claims in Limpopo Province, Olifants Basin, South Africa. *Water International, 37*(7), 773–787. doi:10.1080/02508060.2012.740613

Loftus, A. (2009). Rethinking political ecologies of water. *Third World Quarterly, 30*(5), 953–968. doi:10.1080/01436590902959198

Martínez-Alier, J. M. (2012). Environmental justice and economic degrowth: An alliance between two movements. *Capitalism Nature Socialism, 23*(1), 51–73. doi:10.1080/10455752.2011.648839

Martínez-Alier, J. M. (2013) Injusticias hídricas: el agua corre hacia el poder. La Jornada, 16/01/2013. Retrieved from http://www.jornada.unam.mx/2013/01/16/opinion/024a1pol

McCarthy, J. (2005). Scale, sovereignty and strategy in environmental governance. *Antipode,* 731–753. doi:10.1111/j.0066-4812.2005.00523.x

Meehan, K. (2013). Disciplining de facto development: Water theft and hydrosocial order in Tijuana. *Environment and Planning D: Society and Space, 31,* 319–336. doi:10.1068/d20610

Molle, F., Mollinga, P., & Wester, F. (2009). Hydraulic bureaucracies and the hydraulic mission: Flows of water, flows of power. *Water Alternatives, 3*(2), 328–349.

Mouffe, C. (2007). Artistic activism and agonistic spaces. *Art & Research, 1*(2), 1–5.

Nozick, R. (1974). *Anarchy, state, and Utopia.* New York: Basic Books.

Perramond, E. P. (2013). Water governance in New Mexico: Adjudication, law, and geography. *Geoforum, 45,* 83–93. doi:10.1016/j.geoforum.2012.10.004

Perreault, T., Wraight, S., & Perreault, M. (2011) The Social Life of Water. Histories and geographies of environmental injustice in the Onondaga Lake watershed, New York. Justicia Hídrica Research Document. www.justiciahidrica.org.

Rawls, J. (1971). *A theory of justice.* Cambridge and London: The Bellknap Press of Harvard University Press.

Roth, D.R. Boelens & M. Zwarteveen (Eds.) (2005). *Liquid relations. Contested water rights and legal complexity.* New Brunswick, NJ: Rutgers University Press.

Rodríguez de Francisco, J. C., Budds, J., & Boelens, R. (2013). Payment for environmental services and unequal resource control in pimampiro, Ecuador. *Society and Natural Resources, 26,* 1217–1233. doi:10.1080/08941920.2013.825037

Saldías, C., Boelens, R., Wegerich, K., & Speelman, S. (2012). Losing the watershed focus: A look at complex community-managed irrigation systems in Bolivia. *Water International, 37*(7), 744–759. doi:10.1080/02508060.2012.733675

Schlosberg, D. (2004). Reconceiving environmental justice: Global movements and political theories. *Environmental Politics, 13*(3), 517–540. doi:10.1080/0964401042000229025

Sneddon, C., & Fox, C. (2007). River basin politics and the rise of 'ecological' democracy in Southeast Asia and Southern Africa. Paper presented at the seminar. *Water, Politics and Development,* Stockholm World Water Week, 12 August 2007.

Swyngedouw, E., & Heynen, N. (2003). Urban political ecology, justice and the politics of scale. *Antipode,* 898–918. doi:10.1111/j.1467-8330.2003.00364.x

Swyngedouw, E. (2003). Modernity and the production of the Spanish waterscape 1890–1930. In: K. Zimmerer & T. J. Bassett (Eds.), *Political ecology* (94–114). New York: Guildford Press.

Swyngedouw, E. (2005). Dispossessing H2O: The contested terrain of water privatization. *Capitalism Nature Socialism, 16*(1), 81–98. doi:10.1080/1045575052000335384

van der Kooij, S., Zwarteveen, M., Boesveld, H., & Kuper, M. (2013). The efficiency of drip irrigation unpacked. *Agricultural Water Management, 123,* 103–110. doi:10.1016/j.agwat.2013.03.014

Van Halsema, G. E., & Vincent, L. (2012). Efficiency and productivity terms for water management: A matter of contextual relativism versus general absolutism. *Agricultural Water Management, 108,* 9–15. doi:10.1016/j.agwat.2011.05.016

Vera, J., & Zwarteveen, M. (2008). Modernity, exclusion and resistance: Water and indigenous struggles in Peru. *Development, 51,* 114–120. doi:10.1057/palgrave.development.1100467

Vos, H. D., Boelens, R., & Bustamante, R. (2006). Formal law and local water control in the andean region: A fiercely contested field. *International Journal of Water Resources Development, 22* (1), 37–48. doi:10.1080/07900620500405049

Wester, P. (2008) Shedding the Waters. Institutional Change and water Control in the Lerma-Chapala Basin, México. PhD Thesis, Wageningen University.

Wester, P., de Vos, H., & Woodhill, J. (2004) The enabling environment. Discussion Paper. FAO Conference on Water for Food and Ecosystems.

Whatmore, S. (2009). Mapping knowledge controversies: Environmental science, democracy and the redistribution of expertise. *Progress in Human Geography, 33*(5), 587–599.

Young, I. M. (1990). *Justice and the politics of difference*. Princeton, NJ: Princeton University Press.

Zwarteveen, M. (2006) Wedlock or deadlock. Feminists' attempts to engage irrigation engineers. PhD Thesis, Wageningen University.

Zwarteveen, M. (2010). A masculine water world: The politics of gender and identity in irrigation expert thinking. In: R. Boelens, D. Getches & A. Guevara (Eds.), *Out of the mainstream* (75–98). London & Washington, DC: Earthscan.

Zwarteveen, M., Roth, D., & Boelens, R. (2005). Water rights and legal pluralism: Beyond analysis and recognition. In: D. Roth, Boelens R. and M. Zwarteveen (Eds.), *Liquid relations* (pp. 254–268). New Brunswick, NJ: Rutgers University Press.

Hydrosocial territories: a political ecology perspective

Rutgerd Boelens[a,b], Jaime Hoogesteger[b], Erik Swyngedouw[c], Jeroen Vos[b] and Philippus Wester[b,d]

[a]Centre for Latin American Research and Documentation, and Department of Geography, Planning and International Development Studies, University of Amsterdam, the Netherlands; [b]Water Resources Management Group, Department of Environmental Sciences, Wageningen University, the Netherlands; [c]School of Environment, Education and Development, University of Manchester, UK; [d]International Centre for Integrated Mountain Development, Kathmandu, Nepal

ABSTRACT

We define and explore *hydrosocial territories* as spatial configurations of people, institutions, water flows, hydraulic technology and the biophysical environment that revolve around the control of water. Territorial politics finds expression in encounters of diverse actors with divergent spatial and political-geographical interests. Their territory-building projections and strategies compete, superimpose and align to strengthen specific water-control claims. Thereby, actors continuously recompose the territory's hydraulic grid, cultural reference frames, and political-economic relationships. Using a political ecology focus, we argue that territorial struggles go beyond battles over natural resources as they involve struggles over meaning, norms, knowledge, identity, authority and discourses.

In this introduction we present a conceptual framework for exploring *hydrosocial territories*: socially, naturally and politically constituted spaces that are (re)created through the interactions amongst human practices, water flows, hydraulic technologies, biophysical elements, socio-economic structures and cultural-political institutions. In this special issue, we aim to explore how processes of territorialization around water are intrinsically linked to different and often divergent water governance systems and their contestation. Our aim is to develop a better understanding of how the relations between society, nature, territory and governance play out specifically in the water domain.

This issue bundles articles that were presented and discussed at the International Irrigation Society Landscape Conference in Valencia, Spain, between 25 and 27 September 2014. This conference brought together an interdisciplinary group of scholars from around the world to discuss how the notion of hydrosocial territories can help advance a better understanding of interrelated local, regional, national and international processes of water governance and the issues of equity and justice in water control. This resulted in a rich collection of articles, approaches and insights with regard to how the lens of hydrosocial territories can help unravel different water-centred processes.

Drawing on the broader literature, the contributions to this special issue (with most authors applying a political ecology approach), and the insights generated during the above-mentioned conference, this introduction explores how and challenges why actors commonly portray water territories as mere biophysical 'nature'. This makes water problems and their solutions appear as politically neutral, technical and/or managerial issues which can be 'objectively' solved according to technical knowledge, 'rational water use' and 'good governance'. Contrasting such a conception, which is often used as a veil to legitimize deeply political choices that protect and stabilize specific political orders, we call for a repoliticization, that is the recognition of the political nature, of hydrosocial territories through the study of everyday water use praxis.

To examine this theoretical field and its implication for the interpretation of the empirical, this special issue deals with the contradictions, conflicts and societal responses generated by the configuration of hydrosocial territories. It examines how socionatural arrangements and water politics either enhance or challenge the unequal distribution of resources and decision-making power in water governance – the mechanisms, structures, knowledge systems and discourses underpinning their operation. In addition, the range of articles in this issue seek to understand and identify alternatives that contribute to the creation of proposals that respond to questions of socio-economic fairness, political democracy and ecological integrity.

This introductory article is structured as follows. We begin by defining hydrosocial territories and their constituting elements. Then, we outline four conceptual themes that are intrinsically related to the constitution of hydrosocial territories: first, hydrosocial networks and territorialization; second, the politics of scalar territorial reconfiguration; third, the governmentalization of territory; and fourth, territorial pluralism. Finally, in the concluding section, we offer an overview of the presented issues.

Defining hydrosocial territories

Territories, although often considered natural, are actively constructed and historically produced through the interfaces amongst society, technology and nature. They are the outcomes of interactions in which the contents, presumed boundaries and connections between nature and society are produced by human imagination, social practices and related knowledge systems. This is clearly manifested in how river basin management, water flows, water use systems and hydrological cycles are mediated by governance structures and human interventions that entwine the biophysical, the technological, the social and the political. We therefore conceptualize a 'hydrosocial territory' as

> the contested imaginary and socio-environmental materialization of a spatially bound multi-scalar network in which humans, water flows, ecological relations, hydraulic infrastructure, financial means, legal-administrative arrangements and cultural institutions and practices are interactively defined, aligned and mobilized through epistemological belief systems, political hierarchies and naturalizing discourses.[1]

Hydrosocial territories (imagined, planned or materialized) have contested functions, values and meanings, as they define processes of inclusion and exclusion, development and marginalization, and the distribution of benefits and burdens that affect different groups of people in distinct ways. For instance, prevailing water governance and intervention projects commonly respond to growing urban water needs, globalizing commercial export agriculture and industrial growth sectors (see the contributions in this issue; Duarte-Abadía, Boelens, & Roa-Avendaño, 2016; Swyngedouw, 2015). This leads to processes of resources accumulation and the simultaneous dispossession of vulnerable groups of their livelihoods (Crow et al., 2014; Martínez-Alier, 2002; Vos & Boelens, 2014), creating social and environmental inequities (Bridge & Perreault, 2009; Harris & Roa-García, 2013; Roa-García, 2014). Therefore, the question of how, by which actors, through which strategies and with what interests and consequences the 'natural' and the 'social' boundaries of hydrosocial territories are conceptualized and materialized through interlinked natural, social and technological elements, is fundamental (Baviskar, 2007; Damonte-Valencia, 2015; cf. Bakker, 2010; Latour, 1993).

Socionature, hydrosocial networks and territorialization

The notion that society and nature are intrinsically linked and interdependent is common amongst geographers (e.g. Castree, 2008; Perreault, 2014; Swyngedouw, 2007). People are strongly involved in the everyday production and re-production of the environment they live in – although not necessarily in the ways they foresee, plan or desire (cf. Agnew, 1994; Baletti, 2012; Winner, 1986). As Duarte-Abadía and Boelens (2016), Hulshof and Vos (2016), and Seemann (2016) show in this issue, people inscribe their life worlds, in particular biophysical environments, by using, inhabiting and/or managing these according to their ideologies, knowledge and socio-economic and political power. In doing so, people generate environments, environmental knowledge systems, and territory.

Creating hydrosocial territories involves humanizing nature and building humanized waters based on social, political and cultural visions of the world-that-is and the world-that-should-be (Boelens, 2015; Swyngedouw, 2015). Therefore the (re)creation of hydrosocial territories (and water) needs to be analyzed in the context of their historical, cultural and political settings (see also Bury et al., 2013; Lansing, 1991; Orlove & Caton, 2010). Consequently, thinking of hydrosocial territories and the processes of their constitution and (re)configuration requires going beyond dichotomizing presentations that separate (or 'purify' – Latour, 1993) nature from society. Rather, these should be seen as hybrids that simultaneously embody the natural and the social; the biophysical and the cultural; the hydrological and the hydraulic; the material and the political. As Haraway (1991), Latour (1993), Smith (1984) and Swyngedouw (1999, 2007), amongst others, have elaborated, this also goes beyond a perspective of profound interrelatedness between the realms of nature and society. In fact, "the dialectic between nature and society becomes an internal one" (Swyngedouw, 1999, p. 446), rendering nature as an undivided part of the process of societal or rather "socionatural" production (Haraway, 1991; Latour, 1993; Lefebvre, 1991).

Water and water technologies entwine ecology and society. Water flows through landscapes, technologies and cities, connecting places, spaces and people. The natural and/or human-induced variations in its flow create, transform or destroy social linkages, lived spaces and boundaries as they produce new social, land and water configurations (cf. Hoogesteger, 2013; Mosse, 2008). These in turn create and transform social/political hierarchies, conflicts, and forms of collaboration. Therefore, water, technologies, society and nature are intrinsically interrelated and mutually determining elements that together organize as specific socionatural networks. The networks of relations constituting hydrosocial territories can be termed "hydrosocial networks" (Wester, 2008, p. 21). These networks are intentionally and recursively shaped around water and its use; they are precarious and reversible outcomes of modes of ordering (Law, 1994). Bolding (2004) defines two critical characteristics of hydrosocial networks: span and durability. Span refers to the spatial, social, material and institutional reach or extent of a hydrosocial network and can run from a single small canal to the interlinking of several river basins. This depends on the scale of analysis and the associations that are being traced. Durability refers to the strength of a hydrosocial network, to how strong and stabilized the associations are amongst the heterogeneous elements forming the network. It also refers to the time dimension of the network, to how long the network sticks together before it falls apart. Without water the network literally falls dry.

In the terms of Latour (1993), water as well as water technologies are *actants* in an actor-network. And it is common to find worldviews and epistemological positions that express how water possesses many properties and faces: powerful, productive, destructive; engineered, natural and supernatural (see e.g. Boelens, 2014; Illich, 1986; Linton & Budds, 2014). Water is thus simultaneously a physical and a social actant in cultural and political processes and can for instance "be and become a border, a resource for regeneration, a foundation for empire, a means of nation building, and a material linkage between past and present" (Barnes & Alatout, 2012, p. 485). Therefore, the examination of water flows, water distribution, hydraulic infrastructure, water-based production, water security, and the historical, geographical and technical-political processes that created and transformed them gives profound insights into who – and based on what imaginaries and knowledge systems – designs(ed), controls(ed) and has(d) the power to (re)produce specific hydrosocial networks and territories (Boelens & Post Uiterweer, 2013; Kaika, 2005; Meehan, 2013; Wester, Merrey, & De Lange, 2003).

To argue that hydrosocial territories are 'humanized nature' or 'socionatures' is to insist that they are not fixed, bounded, and spatially coherent territorial entities. Rather, it poses that territory and the processes of territorialization are – and should be examined as – spatially bound, subject-built, socionatural networks that are produced by actors who collaborate and compete around the definition, composition and ordering of this networked space (Rodriguez-de-Francisco & Boelens, 2016; Swyngedouw & Williams, 2016; see also Agnew, 1994; Elden, 2010; Escobar, 2008). Therefore, "territory is not external to the society that formed it, but rather is its substance, it also embodies the contradictions, conflicts and struggles of that society" (Baletti, 2012, p. 578).

The notions and strategies of *how* to make territory profoundly diverge amongst actors, just like the 'territorialities' that are produced. For this reason, as Hoogesteger, Baud, and Boelens (2016), Perramond (2016), Romano (2016) and Seemann (2016) show in this issue, the challenge for grass-roots collectives that strive to build and defend their water-based territories, such as local watersheds and irrigation and drinking water systems, is often complex. Apart from the threats posed by powerful outsiders (i.e. state agencies, agro-export chains, mining companies), they face the need to solve water conflicts inside their collectives. In building and defending their hydrosocial territory, a water users collective, although internally differentiated, requires a collective identity connected to its water sources and socio-technical infrastructure system – a shared normative system and a physical, natural and human-bounded territorial water-control space (Boelens, 2015; Hoogesteger, 2013). Grass-roots territorialization is therefore a struggled process that builds on and re-creates mutual dependency through cooperation and the mobilization of its parts towards a common resource control objective (cf. Hoogesteger & Verzijl, 2015).

The politics of scalar territorial reconfiguration

A focus on hydrosocial networks highlights the social relations that connect local human actors and nonhuman actants to broader political, economic, cultural and ecological scales. These scales are neither natural nor fixed but are produced through frictions between social practice, environmental processes and structural forces (Bridge & Perreault, 2009; Heynen & Swyngedouw, 2003). Spatial scales – that is, the geographically constituted 'levels' of social interactions and interconnectedness (e.g. household, community, watershed, region, nation, globe) – are produced, contested and reconfigured through myriad state, market, civil-society and individual actions and everyday practices (Neumann, 2009; Swyngedouw, 1999; Warner, Wester, & Hoogesteger, 2014).

Hydrosocial territories at a specific scale exist and are deeply enmeshed in other territories that exist and operate at broader, overlapping, counterpoised and/or hierarchically embedded administrative, cultural, jurisdictional, hydrological and organizational scales. In the (trans)formation of hydrosocial territories, scales and the ways they connect require continual re-production and are therefore subject to negotiation and struggle (e.g. Ferguson & Gupta, 2002; Molle, 2009; Saldías, Boelens, Wegerich, & Speelman, 2012). Groups with divergent territorial interests struggle to define, influence and command particular scales of resource governance, and to determine the ways in which these mutually connect in a given sociospatial conjuncture. As Swyngedouw (2004, p. 33) observes, "Spatial scales are never fixed, but are perpetually redefined, contested and restructured in terms of their extent, content, relative importance and interrelations." Whether the repatterning of the scale of territories actually takes place in accordance with the desires and interests of a particular group of actors depends not only on the quality of the territorial proposals, but also crucially on the support and power of an interlocked multi-scalar coalition that provides the technical, scientific and discursive support to this reconfiguration (Swyngedouw, 2007, 2015).

Different scalar plans and projections about how to organize hydrosocial territory envision very different ways of patterning local livelihoods, production and regional economic and socionatural development. These projections of how these territories, their water and their people are and ought to be organized may commonly lead to the empowerment of certain groups of actors while disempowering others, and offer arenas for claim-making and contestation. Hoogesteger et al. (2016) show how organizational scales of administrators and water users in the Ecuadorian Highlands determine how the water users claim participation in decision making about how and by whom water is managed in different hydrosocial territories. Vos and Hinojosa (2016) show how, in contexts of growing importance of export production chains and international virtual water trade, new forms of water regulation at local and national scales reshape the communities' hydrosocial territories. The resulting hydrosocial configurations compromise the political representation, water security, and property structures of local communities and private companies in strongly divergent ways.

These cases illustrate that although the impacts of deterritorialization and repatterning hydrosocial territories may be felt mostly by individuals, households, or water-use collectives and organizations at the local level, the processes are deeply and dynamically interconnected at various scales. Therefore, as also Romano, Hulshof and Vos, Perramond, and Seemann show in their contributions, hydrosocial territories at different interrelated scales are sites of political contestation whereby the production of new (and the defence of existing) socionatural relationships is crucial; the transformation of existing technological, legal, institutional and symbolic arrangements is at stake. In other words, these hydroterritorial struggles and conflicts respond to site-specific processes "through which symbolic formations are forged, social groups enrolled, and natural processes and 'things' entangled and maintained" (Swyngedouw, 2007, p. 10).

Governmentalization of territory: from humanized nature to 'naturalized natures'

As shown for the Colombian highland territories (*páramos*) by Duarte and Boelens (2016), the Nicaraguan rural water-use communities by Romano, the Spanish desalination plans and infrastructure by Swyngedouw and Williams, the Ecuadorian Highlands by Rodriguez-de-Francisco and Boelens, and the *acequias* in New Mexico by Perramond, dominant hydrosocial territories blend society and nature in ways that correspond with particular water truths and knowledge claims. In other words, powerful hydrosocial territories envision to position and align humans, nature and thought within a network that aims to transform the diverse socionatural water worlds into a dominant governance system (cf. Baviskar, 2007; Escobar, 2008; Kaika, 2005; Lansing, 1991; Mosse, 2008; Zwarteveen & Boelens, 2014), with 'dominance' often characterized by divisions along ethnic, gender, class or caste lines, frequently sustained by modernist water-scientific conventions. In the words of Foucault (1991/1978), such hydro-territorial projects and imaginaries aim to "conduct the conduct" of specific subject populations (what he framed as "governmentality" – government mentality and/or rationality; see also Scott, 1998).

The processes that 'governmentalize' territory, and so produce space with new or reinforced hierarchical relationships between water governors and subject water actors and actants, has profound socio-environmental and political consequences. The new territorial configurations commonly entwine technological, industrial, state-administrative and scientific knowledge networks that enhance local–global commodity transfers, resource extraction and development/conservation responding to non-local economic and political interests (Büscher & Fletcher, 2014; Yacoub, Duarte-Abadía, & Boelens, 2015). To do so, they commonly curtail local sovereignty and create a political order that makes these local spaces comprehensible, exploitable and controllable (Bebbington & Bury, 2013; Rodriguez de Francisco & Boelens, 2015).

Territorial governmentalization projects seek to fundamentally alter local water users' identification with community, neighbourhood, kinship or federative solidarity organization in order to change water users' ways of belonging and behaving, according to new identity categories and hierarchies. Making such 'new subjects' requires these water users to frame their worldviews, needs, strategies and relationships differently, building and believing in new models of agency, causality, identity and responsibility. Simultaneously, such frames exclude other options and thus "delimit the universe of further scientific inquiry, political discourse, and possible policy options" (Jasanoff & Wynne, 1998, p. 5). As shown by Duarte-Abadía et al. (2015), Hulshof and Vos (2016), and Swyngedouw and Williams (2016), to governmentalize territories through 'new' discourses and ideologies creates specific forms of consciousness that are called upon (presumably in a self-evident manner) in order to defend particular water policies, authorities, hierarchies, and management practices.

Subtle imposition (or less subtle indoctrination) of particular perspectives on hydro-social territories can be seen as constituting a politics of truth which legitimates certain water knowledges, practices and governance forms and discredits others. They separate 'legitimate' forms of water knowledge, rights and organization from 'illegitimate' forms (Forsyth, 2003; Foucault, Sellenart, & Burchell, 2007). As a result, the production of water knowledge and truths – and the ways these inform the shaping of particular water artefacts, rules, rights and organizational structures – concentrates on the issue of how to align local users and livelihoods to the imagined multi-scalar water-power hierarchies (Boelens, 2015). Discourses about 'hydrosocial territory' join power and knowledge (Foucault, 1980) to ensure a specific political order as if it were a naturalized system, by making fixed linkages and logical relations amongst a specified set of actors, objects, categories and concepts that define the nature of problems as well as the solutions to overcome them.

Hydrosocial territoriality, as a battle of divergent (dominant and non-dominant) discourses or narratives, has consolidating a particular order of things as its central stake. Though thoroughly mediated in everyday praxis, ruling groups strategically deploy discourses that define and position the social and the material in a human-material-natural network to leave the political order unchallenged and stabilize their ways of "conducting subject populations' conduct" (Foucault, 1980, 1991/1978).

As various contributions in this issue demonstrate, territorial governmentality projects do not necessarily aim to obliterate alternative territorialities. Most often, modern tactics of territorialization aim to 'recognize', incorporate and discipline local territorialities, integrating local norms, practices and discourses into its mainstream government rationality and its spatial/political organization. This subtle strategy to incorporate and marginalize locally existing territorialities in mainstream territorial projects makes use of 'managed' or 'neoliberal multiculturalism': through 'participatory' strategies it recognizes the 'convenient' and sidelines 'problematic' water cultures and identities.

Territorial pluralism, contested 'territories-in-territory' and alternative ways of ordering

New hydrosocial territories result from the intersection and confrontation of divergent territorial projects and the realization of contested political-economic, socio-environmental imaginaries. Such imaginaries, here, can be understood as the socio-environmental world views and aspirations held by particular social groups, as the wished-for patterning of the material and ecological territorial worlds with and through the corresponding values, symbols, norms, institutions and social relationships. As Steger and Paul (2013, p. 23) suggest, "imaginaries are patterned convocations of the social whole. These deep-seated modes of understanding provide largely pre-reflexive parameters within which people imagine their social existence." They are the historical constructs through which particular actors define and aim to shape their desired territorial whole, often in confrontation with the contrasting images adhered to by competing subject groups. As a consequence, everyday politics over territorial order finds expression in the encounters of diverse political and geographical projects, such as forms of state organization, spatial control over water, and the power relations amongst national and global political and economic alliances. All of these compete, superimpose, and foster their territorial interests to strengthen their water control. Thereby, they continuously transform the territory's hydraulic grid, cultural reference frames, economic base structures, and political relationships. These overlapping hydropolitical projects tend to generate 'territorial pluralism' and make diverse 'territories-in-territory' – that is, overlapping, often contested, and interacting hydroterritorial configurations in one and the same space, but with differing material, social and symbolic contents and different interlinkages and boundaries.

As Hoogesteger et al. (2016) show, the complex interplay amongst, for example, state-defined territories and the hydrosocial territories of local user collectives may express outright confrontation, docile alignment, or obedience, but also dynamic mutual recognition. With respect to the latter, in many places around the world, official and customary water management strategies are deeply intertwined in a "shotgun marriage" (Boelens, 2009, p. 315). State and customary modes of territorial ordering depend on each other in complicated (and often confrontational) ways. Unable to provide water for all sectors of society, the state relies on informal/illegal norms, infrastructure and organizations that have the capacity to provide water to the citizenry, as shown for instance by Ioris (2016), Meehan (2013), Romano (2016), and Vos, Boelens, and Bustamante (2006). 'Recognition' of customary hydrosocial infrastructure and its context-specific solutions guarantees the state's legitimacy and stability. Therefore, in everyday water-governance politics, outright repression of local, vernacular and illegal hydrosocial territoriality coexists with strategic allowance and recognition. Some local rules, rights and illegal infrastructure are institutionalized, at the expense of most others and at the cost of intensifying the repression of more contentious, defiant and disloyal norms and hydro-territorial institutions.

In hydro-territorialization policies, it is common to find that this simultaneous legalization and delegitimation of local rights removes important protections for local collectives and occasionally massive resource transfers to newly intervening actors (see e.g. Boelens & Seemann, 2014; Perramond, 2016; Seemann, 2016). Territorial struggles, therefore, entwine battles over natural resources with struggles over meaning, norms, knowledge, decision-making authority, and discourses. For this reason, the struggles of local territorial collectives are about water and economic resources to sustain their livelihoods as much as they are about the discourses that support their claims to self-define their own water rules, nature values, territorial meanings and user identities.

With intensifying universal state formalization policies that aim to 'recognize' and reorganize local rights systems, and increased market-based efforts to expand into new territories, local 'customary' hydrosocial territories increasingly become sites of political mobilization and resistance to external domination. Given that state agents, agro-commercial enterprises, mining companies, hydropower conglomerates and other dominant players expand their activities into 'new' areas that are often intensely used by their local inhabitants, 'local' communities and associations also look for responses that extend beyond their home domains. Increasingly, they organize and pursue their objectives at a variety of scales.

The politics of dominant players (who try to align user communities to their frames, rules and scalar hierarchies of power) as well as the resistance strategies of local groups (who aim to localize resource access and decision-making power) are fundamentally related to their power to compose or manipulate patterns of multiple scales (Swyngedouw, 2004, 2009; see also Bebbington, Humphreys-Bebbington, & Bury, 2010; Hoogesteger & Verzijl, 2015). Marginalized water user collectives therefore often challenge the 'manageable scales' to which they are confined, "attempting to liberate themselves from these imposed scale constraints by harnessing power and instrumentalities at other scales. In the process, scale is actively produced" (Jonas, 1994, p. 258, quoted by Swyngedouw, 2004, p. 34). For example, Hoogesteger et al. (2016) and Boelens et al. (2014) show how community and regionally based peasant and indigenous organizations in the Ecuadorian Highlands have been able to advance their claims to water because of their connections to multi-scalar networks of development, environmental, and human rights organizations. Their hydrosocial networks, in part, become "counter-geographies" (Brenner, 1998, p. 479; see also Bridge & Perreault, 2009; Hoogesteger, 2012; Romano, 2016). As a consequence, the permanent reorganization of territories, their configurations and spatial scales "are integral to social strategies and serve as the arena where struggles for control and empowerment are fought" (Swyngedouw, 2004, p. 33).

In looking at struggles, most attention is given to blatant water conflicts the encroachment of resources and decision-making powers. But everyday social action may be far more influential (cf. Scott, 1998). Many user collectives extend informal networks as largely invisible undercurrents that actively challenge domination. These 'undertows' enable action on broader political scales, constituting flexible trans-local networks. "They evade patrolling by dominant, formal powers, while materially practicing and extending their own water rights and discursively constructing their counter-narratives ... to defend local rights and contest encroachment, surveillance and repression" (Boelens, 2015, p. 250). This creation of locally embedded hydrosocial territory is at the heart of collective action in many water-control places and spaces, subtly giving 'water', 'territory', 'rights' and 'identity' (new) local meanings. As 'root-stocks', such forms of hydroterritoriality connect underground and produce shoots above and roots below – alternatingly operating in the open and under the surface – making them difficult to understand, contain and grab for officialdom and other dominant powers (Boelens, 2015; see also Bebbington, 2012; Meehan, 2013). The outcomes of these hydroterritorial intersections, conflicts and reorderings are not predetermined and, as Swyngedouw (2007, p. 24) explains, "celebrate the visions of the elite networks, reveal the scars suffered by the disempowered and nurture the possibilities and dreams for alternative visions".

Concluding remarks

As we have explored here, and as further illustrated and scrutinized by the diverse contributions to this special issue, understanding water governance and territorial planning systems as based on socionatural politics provides opportunities to critically examine the power-laden contents of prevailing hydrosocial regimes and networks. It also offers insights into alternative ways of conceptualizing and building nature–society–power relations, enabling more equitable governance forms that, amongst others, build on transdisciplinary knowledge and more bottom-up modes of decision making.

This article and most of the contributing authors to the special issue use a political ecology focus. This enhances the understanding of how the formulation and implementation of new modes of water governance and the reconstruction of hydrosocial territories may often result in unequal costs and benefits for different actors. It also supports comprehending how the dominant ways of conceptualizing these socionatural configurations and of 'knowing environmental problems and solutions' actively depoliticize forms of socio-economic inequality, misrecognition, and political exclusion. The contributions therefore seek to show the political nature of the mechanisms of water access and distribution that are built into hydroterritorial planning, the relations that shape rights and rules regarding water decision making, and the discourses that underpin water policies and hydrosocial territorial reform.

As various articles show, most resource and territorial struggles in water-control systems are rooted in how new water governance proposals undermine, transform, incorporate and/or reorder existing local forms of collective self-governance and territorial autonomy. Classic 'exclusion-oriented' and modern 'inclusion-oriented' policies – and hybrids – aim to involve local water user communities and territories in ruling groups' hydroterritorial projections and rationalities and so shape or reinforce the dominant hydroterritorial order. Alignment with these supposedly more rational and efficient schemes generally legitimizes the authority and cultural supremacy of external political-economic power groups and deepens unequal water distribution as well as unsustainable extraction of surpluses and resources from local communities.

However, the articles also show that many 'local' (vernacular) or marginalized resource users and management collectives actively challenge and respond to the norms, knowledge, distribution patterns, governance forms, and identities that are imposed on them. Often, a fundamental component of these struggles is the effort to 'redesign' and reshape the hydraulic grids, units and artefacts that underlie the structure and logic of dominant hydrosocial territories – the latter frequently being based on gender, ethnic, class, caste or other inequalities and contradictions. This entails transforming the world of technology-embedded cultural and distributional norms and political relations, including the corresponding definitions of proper functioning, social aptness and technical efficiency. Next, such struggles for alternative territoriality often involve building and engaging in new multi-scalar networks, which link local communities to trans-local actors and alliances. Through scalar politics, grass-roots collectives employ material and discursive practices to contest dominant reterritorialization politics and stake claims for economic redistribution, cultural recognition, political legitimacy, and democracy. As such, vernacular, non-dominant hydrosocial territories often are physical, cultural, socio-legal and political spaces that enable water users to manoeuvre in local water worlds as well as in broader political webs that determine water control. Whether, to what extent, and in what ways the dominant or opposing agents are successful in producing, reinforcing, or reordering the hydrosocial territories they envision depends on their capacity to mobilize and exercise power, enforcing negotiation and change through strategic alliances.

Note

1. *Epistemological belief systems* express the nature and scope of knowledge; they conceptualize what knowledge is and how it can be acquired. *Naturalizing discourses* entwine knowledge claims and social and material practices with power and legitimacy in order to shape particular 'truths' (or 'truth regimes') and so strategically 'represent reality'; they aim to convincingly explain (as if it were 'natural') how socionatural reality needs to be understood and experienced, thus obliterating alternative modes of representing reality.

Acknowledgements

The research, debates and reflections on which this article and special issue are based form part of the activities organized by the international Justicia Hídrica/Water Justice alliance (www.justiciahidrica.org). The authors wish to thank the alliance's academic, activist and grass-roots members for sharing their experiences, insights and reflections, which have importantly contributed to this publication.

The views and interpretations in this publication are those of the authors and are not necessarily attributable to their organizations.

References

Agnew, J. (1994). The territorial trap: The geographical assumptions of international relations theory. *Review of International Political Economy, 1*(1), 53–80. doi:10.1080/09692299408434268

Bakker, K. (2010). *Privatizing water. Governance failure and the world's urban water crisis.* Ithaca, NY: Cornell University Press.

Baletti, B. (2012). Ordenamento Territorial: Neo-developmentalism and the struggle for territory in the lower Brazilian Amazon. *Journal of Peasant Studies, 39*(2), 573–598. doi:10.1080/03066150.2012.664139

Barnes, J., & Alatout, S. (2012). Water worlds: Introduction to the special issue of Social Studies of Science. *Social Studies of Science, 42*(4), 483–488. doi:10.1177/0306312712448524

Baviskar, A. (2007). *Waterscapes. The cultural politics of a natural resource.* Delhi: Permanent Black.

Bebbington, A. (2012). Underground political ecologies: The second annual lecture of the cultural and political ecology specialty group of the Association of American Geographers. *Geoforum, 43*(6), 1152–1162. doi:10.1016/j.geoforum.2012.05.011

Bebbington, A., Humphreys-Bebbington, D., & Bury, J. (2010). Federating and defending: Water, territory and extraction in the Andes. In R. Boelens, D. H. Getches, & J. A. Guevara-Gil (Eds.), *Out of the mainstream: Water rights, politics and identity* (pp. 307–328). London: Earthscan.

Bebbington, A., & Bury, J. (eds.). (2013). *Subterranean struggles: New dynamics of mining, oil, and gas in Latin America.* Austin: University of Texas Press.

Boelens, R. (2009). The politics of disciplining water rights. *Development and Change, 40*(2), 307–331. doi:10.1111/j.1467-7660.2009.01516.x

Boelens, R. (2014). Cultural politics and the Hydrosocial cycle: Water, power and identity in the Andean highlands. *Geoforum, 57,* 234–247. doi:10.1016/j.geoforum.2013.02.008

Boelens, R. (2015). *Water, power and identity. The cultural politics of water in the Andes.* London: Earthscan, Routledge.

Boelens, R., Hoogesteger, J., & Rodriguez-de-Francisco, J. C. (2014). Commoditizing water territories: The clash between Andean water rights cultures and payment for environmental services policies. *Capitalism Nature Socialism, 25*(3), 84–102. doi:10.1080/10455752.2013.876867

Boelens, R., & Post Uiterweer, N. C. (2013). Hydraulic heroes: The ironies of utopian hydraulism and its politics of autonomy in the Guadalhorce Valley, Spain. *Journal of Historical Geography, 41*, 44–58. doi:10.1016/j.jhg.2012.12.005

Boelens, R., & Seemann, M. (2014). Forced engagements: Water security and local rights formalization in Yanque, Colca Valley, Peru. *Human Organization, 73*(1), 1–12. doi:10.17730/humo.73.1.d44776822845k515

Boelens, R., & Vos, J. (2014). Legal pluralism, hydraulic property creation and sustainability: The materialized nature of water rights in user-managed systems. *COSUST, 11*, 55–62. doi:10.1016/j.cosust.2014.10.001

Bolding, A. (2004). *In hot water* (PhD dissertation). Wageningen University, The Netherlands.

Brenner, N. (1998). Between fixity and motion: Accumulation, territorial organization and the historical geography of spatial scales. *Environment and Planning D: Society and Space, 16*, 459–481. doi:10.1068/d160459

Bridge, G., & Perreault, T. (2009). Environmental governance. In N. Castree, et al. (Eds.), *Companion to environmental geography* (pp. 475–397). Oxford, UK: Blackwell.

Bury, J., Mark, B. G., Carey, M., Young, K. R., McKenzie, J. M., Baraer, M. French, A., & Polk, M. H. (2013). New geographies of water and climate change in Peru: Coupled natural and social transformations in the Santa river watershed. *Annals of the Association of American Geographers, 103*(2), 363–374. doi:10.1080/00045608.2013.754665

Büscher, B., & Fletcher, R. (2014). Accumulation by conservation. *New Political Economy, 20*(2), 273–298. doi:10.1080/13563467.2014.923824

Castree, N. (2008). Neoliberalising nature: The logics of deregulation and reregulation. *Environment and Planning A, 40*(1), 131–152. doi:10.1068/a3999

Crow, B., Lu, F., Ocampo-Raeder, C., Boelens, R., Dill, B., & Zwarteveen, M. (2014). Santa cruz declaration on the global water crisis. *Water International, 39*(2), 246–261. doi:10.1080/02508060.2014.886936

Damonte-Valencia, G. (2015). Redefiniendo territorios hidrosociales: control hídrico en el valle de Ica, Perú (1993–2013), *Cuadernos de Desarrollo Rural. 12*(76), 109–133. doi:10.11144/Javeriana.cdr12-76.rthc

Duarte-Abadía, B., & Boelens, R. (2016). Disputes over territorial boundaries and diverging valuation languages: The Santurban hydrosocial highlands territory in Colombia. *Water International, 41*(1), 15–36. doi:10.1080/02508060.2016.1117271

Duarte-Abadía, B., Boelens, R., & Roa-Avendaño, T. (2016). Hydropower, encroachment and the re-patterning of hydrosocial territory: The case of Hidrosogamoso in Colombia. *Human Organization, 74*(3), 243–254. doi:10.17730/0018-7259-74.3.243

Elden, S. (2010). Land, terrain, territory. *Progress in Human Geography, 34*(6), 799–817. doi:10.1177/0309132510362603

Escobar, A. (2008). *Territories of difference: Place, movements, life, redes.* Durham NC: Duke University Press.

Ferguson, J., & Gupta, A. (2002). Spatializing states: Toward an ethnography of neoliberal governmentality. *American Ethnologist, 29*(4), 981–1002. doi:10.1525/ae.2002.29.4.981

Forsyth, T. (2003). *Critical political ecology. The politics of environmental sciences.* London: Routledge.

Foucault, M. (1980). Power/knowledge. In C. Gordon, (Ed.). *Foucault. Power/knowledge: Selected interviews and other writings 1972–1978.* New York, NY: Pantheon Books.

Foucault, M. (1991[1978]). Governmentality. In G. Burchell, C. Gordon, & P. Miller (Eds.), *The Foucault effect: Studies in governmentality* (pp. 87–104). Chicago: University of Chicago Press.

Foucault, M., Sellenart, M., & Burchell, G. (2007). *Security, territory, population.* New York, NY: Palgrave Macmillan.

Haraway, D. (1991). *Simians, cyborgs, and women: The reinvention of nature.* New York, NY: Routledge.

Harris, L., & Roa-García, M. C. (2013). Recent waves of water governance: Constitutional reform and resistance to neoliberalization in Latin America (1990–2012). *Geoforum, 50*, 20–30. doi:10.1016/j.geoforum.2013.07.009

Heynen, N., & Swyngedouw, E. (2003). Urban political ecology, justice and the politics of scale. *Antipode, 34*(4), 898–918.

Hoogesteger, J. (2012). Democratizing water governance from the Grassroots: The development of Interjuntas-Chimborazo in the Ecuadorian Andes. *Human Organization, 71*(1), 76–86. doi:10.17730/humo.71.1.b8v77j0321u28863

Hoogesteger, J. (2013). Trans-forming social capital around water: Water user organizations, water rights, and nongovernmental organizations in Cangahua, the Ecuadorian Andes. *Society & Natural Resources, 26*(1), 60–74. doi:10.1080/08941920.2012.689933

Hoogesteger, J., Baud, M., & Boelens, R. (2016). Territorial pluralism: Water users' multi-scalar struggles against state ordering in Ecuador's highlands. *Water International, 41*(1), 91–106. doi:10.1080/02508060.2016.1130910

Hoogesteger, J., & Verzijl, A. (2015). Grassroots scalar politics: Insights from peasant water struggles in the Ecuadorian and Peruvian Andes. *Geoforum, 62*, 13–23. doi:10.1016/j.geoforum.2015.03.013

Hulshof, M., & Vos, J. (2016). Diverging realities: How frames, values and water management are interwoven in the Albufera de Valencia wetland, Spain. *Water International, 41*(1), 107–124. doi:10.1080/02508060.2016.1136454

Illich, I. (1986). *H20 and the waters of forgetfulness*. London: Marion Boyars.

Ioris, A. (2016). Water scarcity and the exclusionary city: The struggle for water justice in Lima, Peru. *Water International, 41*(1), 125–139. doi:10.1080/02508060.2016.1124515

Jasanoff, S., & Wynne, B. (1998). Science and decision making. In S. Rayner & E. L. Malone (Eds.), *Human choice and climate change* (pp. 1–87). Columbus, OH: Battelle Press.

Jonas, A. (1994). The scale politics of spatiality. *Environment and Planning D: Society and Space, 12*(3), 257–64.

Kaika, M. (2005). *Cities of flows. Modernity, nature and the city*. London: Routledge.

Lansing, S. (1991). *Priests and programmers: Technologies of power in the engineered landscape of Bali*. Princeton: Princeton University Press.

Latour, B. (1993). *We have never been modern*. Cambridge, MA: Harvard University Press.

Law, J. (1994). *Organizing modernity*. Oxford: Blackwell Publishers.

Lefebvre, H. (1991). *The production of space*. Oxford: Blackwell.

Linton, J., & Budds, J. (2014). The hydro-social cycle: Defining and mobilizing a relational-dialectical approach to water. *Geoforum, 57*, 170–180. doi:10.1016/j.geoforum.2013.10.008

Martínez-Alier, J. (2002). *The environmentalism of the poor*. Cheltenham: Edward Elgar.

Meehan, K. (2013). Disciplining de facto development: Water theft and hydrosocial order in Tijuana. *Environment and Planning D: Society and Space, 31*, 319–336. doi:10.1068/d20610

Molle, F. (2009). River-basin planning and management: The social life of a concept. *Geoforum, 40*, 484–494. doi:10.1016/j.geoforum.2009.03.004

Mosse, D. (2008). Epilogue: The cultural politics of water A Comparative Perspective. *Journal of Southern African Studies, 34*(4), 939–948. doi:10.1080/03057070802456847

Neumann, R. P. (2009). Political ecology: Theorizing scale. *Progress in Human Geography, 33*(3), 398–406. doi:10.1177/0309132508096353

Orlove, B., & Caton, S. C. (2010). Water sustainability: Anthropological approaches and prospects. *Annual Review of Anthropology, 39*, 401–415. doi:10.1146/annurev.anthro.012809.105045

Perramond, E. (2016). Adjudicating hydrosocial territory in New Mexico. *Water International, 41*(1), 173–188. doi:10.1080/02508060.2016.1108442

Perreault, T. (2014). What kind of governance for what kind of equity? Towards a theorization of justice in water governance. *Water International, 39*(2), 233–245. doi:10.1080/02508060.2014.886843

Roa-García, M. C. (2014). Equity, efficiency and sustainability in water allocation in the Andes: Trade-offs in a full world. *Water Alternatives, 7*(2), 298–319.

Rodríguez de Francisco, J. C., & Boelens, R. (2015). Payment for environmental services: Mobilising an epistemic community to construct dominant policy. *Environmental Politics, 24*(3), 481–500. doi:10.1080/09644016.2015.1014658

Rodriguez-de-Francisco, J. C., & Boelens, R. (2016). PES hydrosocial territories: De-territorialization and re-patterning of water control arenas in the Andean highlands. *Water International, 41*(1), 140–156. doi:10.1080/02508060.2016.1129686

Romano, S. (2016). Democratizing discourses: Conceptions of ownership, autonomy, and 'the state' in Nicaragua's rural water governance. *Water International, 41*(1), 74–90. doi:10.1080/02508060.2016.1107706

Saldías, C., Boelens, R., Wegerich, K., & Speelman, S. (2012). Losing the watershed focus: A look at complex community-managed irrigation systems in Bolivia. *Water International, 37*(7), 744–759. doi:10.1080/02508060.2012.733675

Scott, J. C. (1998). *Seeing like a state*. New Haven: Yale University Press.

Seemann, M. (2016). Inclusive recognition politics and the struggle over hydrosocial territories in two Bolivian highland communities. *Water International, 41*(1), 157–172. doi:10.1080/02508060.2016.1108384

Smith, N. (1984). *Uneven development*. Blackwell: Oxford.

Steger, M. B., & Paul, J. (2013). Levels of subjective globalization: Ideologies, imaginaries, ontologies. *Perspectives on Global Development and Technology, 12*(1–2), 17–40. doi:10.1163/15691497-12341240

Swyngedouw, E. (1999). Modernity and hybridity: Nature, regeneracionismo, and the production of the Spanish waterscape, 1890-1930. *Annals of the Association of American Geographers, 89*(3), 443–465. doi:10.1111/0004-5608.00157

Swyngedouw, E. (2004). Globalisation or 'glocalisation'? Networks, territories and rescaling. *Cambridge Review of International Affairs, 17*(1), 25–48. doi:10.1080/0955757042000203632

Swyngedouw, E. (2007). Technonatural revolutions: The scalar politics of Franco's hydro-social dream for Spain, 1939?1975. *Transactions of the Institute of British Geographers, 32*(1), 9–28. doi:10.1111/tran.2007.32.issue-1

Swyngedouw, E. (2009). The political economy and political ecology of the hydrosocial cycle. *Journal of Contemporary Water Research & Education, 142*, 56–60. doi:10.1111/jcwr.2009.142.issue-1

Swyngedouw, E. (2015). *Liquid power: contested hydro-modernities in 20th century spain*. Cambridge, MA: MIT Press.

Swyngedouw, E., & Williams, J. (2016). From Spain's hydro-deadlock to the desalination fix. *Water International, 41*(1), 54–73. doi:10.1080/02508060.2016.1107705

Vos, H. D., Boelens, R., & Bustamante, R. (2006). Formal law and local water control in the Andean region: A fiercely contested field. *International Journal of Water Resources Development, 22*(1), 37–48. doi:10.1080/07900620500405049

Vos, J., & Boelens, R. (2014). Sustainability standards and the water question. *Development and Change, 45*(2), 205–230. doi:10.1111/dech.12083

Vos, J., & Hinojosa, L. (2016). Virtual water trade and the contestation of hydrosocial territories. *Water International, 41*(1), 37–53. doi:10.1080/02508060.2016.1107682

Warner, J., Wester, P., & Hoogesteger, J. (2014). Struggling with scales: Revisiting the boundaries of river basin management. *Wiley Interdisciplinary Reviews: Water, 1*(5), 469–481. doi:10.1002/wat2.2014.1.issue-5

Wester, P. (2008). *Shedding the waters: institutional change and water control in the Lerma-Chapala Basin, Mexico* (PhD-thesis), Wageningen University.

Wester, P., Merrey, D. J., & De Lange, M. (2003). Boundaries of consent: Stakeholder representation in river basin management in Mexico and South Africa. *World Development, 31*(5), 797–812. doi:10.1016/S0305-750X(03)00017-2

Winner, L. (1986). *The whale and the reactor: A search for limits in an age of high technology*. Chicago: Chicago Univ. Press.

Yacoub, C., Duarte-Abadía, B., & Boelens, R. (Eds.). (2015). *Agua y Ecología Política*. Quito: Abya-Yala.

Zwarteveen, M., & Boelens, R. (2014). Defining, researching and struggling for water justice: Some conceptual building blocks for research and action. *Water International, 39*(2), 143–158. doi:10.1080/02508060.2014.891168

REVIEW ARTICLE

What kind of governance for what kind of equity? Towards a theorization of justice in water governance

Tom Perreault

Department of Geography, Syracuse University, Syracuse, New York, USA

This article critically reviews literatures related to the core concepts of this special issue: water and hydrosocial relations; water governance and spatial scale; and equity, justice and rights. It argues that only by viewing water and society as simultaneously social and natural can we address both ecological governance and environmental justice. It argues that the institutional arrangements we employ for governing water must address issues of democratization, human welfare and ecological conditions. The article illustrates these arguments with reference to the social and environmental effects of mining activity and associated water contamination on the Bolivian Altiplano.

Introduction

Residents of the Huanuni Valley, on the Bolivian Altiplano (high plain), live with acute water scarcity. This is not a result of 'natural' or even socially produced water shortage but because the plentiful water that flows in the Huanuni River and collects in the nearby Uru Uru and Poopó Lakes are contaminated in the extreme by mining operations located in the upper reaches of the watershed. Ironically, it is those communities located furthest downriver, and most distant from the mines, that suffer the greatest contamination. It is here, where the river's floodplain forms the *pampa* (plain) of Alantañita, that sediments containing cadmium, lead, mercury and other heavy metals have accumulated, at depths of over a metre, and cover agricultural fields with a layer of toxic effluvium. The Huanuni River, which once nourished people, animals and crops alike, now flows gun-metal grey and is little more than a conduit for the toxic discharge of the mines upstream. The floodplain at Alantañita, like the ones in the communities of Paco Pampa and Sora Sora, is littered with thousands of plastic bottles, washed down from the mining centre of Huanuni, which has no other form of solid waste disposal, let alone sewage treatment. People in Alantañita gather water where they can, and most end up carrying it from the town of Machacamarca, over an hour away by foot. Residents of Paco Pampa have somewhat better access to fresh water, and most households have viable wells, or share a well with neighbours. But their water access would be more secure if the community's most reliable well was not concessioned to the mill operated by the Huanuni mining company. Water from this well flows in an open and leaky sluice, past abandoned fields, to the minerals processing plant some 5 km downriver. Not surprisingly, water scarcity has resulted in dangerously low levels of water consumption. On average, residents of this area consume less than half of what the World Health Organization recommends for *minimum* daily consumption, a factor that contributes to the valley's low agricultural production and high levels of poverty, human and animal illness, and out-migration.

Mining and agriculture are equally dependent on water as a factor of production – without it, neither could exist. But the spatial scales of water governance in agriculture and mining are distinctly different. In the Huanuni Valley, as in so many other places in the Andes, the reconfiguration of the scales of governance towards the national and global have direct bearing on questions of equity, rights and social justice. The privileged position afforded to mining in Bolivia means that competing water users – in this case semi-subsistence indigenous peasant farmers – suffer from inadequate access to clean water. Such relations highlight the fundamental relationship between water governance, spatial scale and social justice. This article is an attempt to theorize these concepts, which together form the core of this special issue. My starting-point is both epistemological and political. I contend that how we define and deploy such concepts as water, governance, scale, equity, justice and rights *matters* for how we enact and experience the social relations of water governance. The article does not present original empirical material, nor is it meant to build new theory. Rather, it critically reviews the literatures on water and the hydrosocial; environmental governance and spatial scale; and equity, justice and rights – placing these concepts in the same analytical frame and probing the relationships between them. The arguments presented here will be familiar terrain for many readers. I encourage them to move on to the other articles in this special issue. The article is primarily intended for the majority of water resources professionals, scholars and students who are *not* critical human geographers or political ecologists, and for whom the concepts of (and relationships between) environmental governance, hydrosocial relations and social justice represent new and exciting ideas. The next section of the article critically reviews the political ecology literature on water and hydrosocial relations. This is followed by a consideration of water governance and spatial scale. I then consider the concepts of equity, justice and rights, and their relationship to water governance. The article's conclusion highlights the connections between these concepts, and the lessons we can learn from them.

Water and the hydrosocial

It has become something of a truism to speak of water in terms of the 'hydrosocial' (see, *inter alia*, Bakker, 2002, 2003; Budds & Hinojosa, 2012; Loftus, 2007, 2009; Swyngedouw, 1999, 2004). Water is neither purely 'natural' nor purely 'social' but simultaneously and inseparably both: a hybrid 'socio-nature'. Bakker's (2002) distinction between H_2O and 'water' is useful here. "Whereas H_2O circulates through the hydro-logical cycle, water *as a resource* circulates through the hydrosocial cycle – a complex network of pipes, water law, meters, quality standards, garden hoses, consumers, [and] leaking taps" – in addition to the processes of rainfall, evaporation and runoff associated with the hydrological cycle (Bakker, 2002, p. 774). Water, in this sense, exists apart from human influence (as rainfall, in aquifers and oceans, as soil moisture and evaporation, etc.), and simultaneously is produced and enacted through human labour and social action within a given mode of production (as irrigation systems, fountains, water law, sewer systems, thirst, customary rights, etc.). Water is given meaning through cultural beliefs, historical memory, and social practice, and exists as much in discourse and symbolism as it does as a physical, material thing. It is, as Swyngedouw (2004) notes, a product of historically sedimented social actions, institutions, struggles and discourses, which in turn help shape the social relations through which it is produced and enacted. Water and society are, as Swyngedouw, Loftus (2009) and many others have pointed out, mutually constitutive. Water lubricates social functions and life itself; it is both a factor of production and a product of social labour, and – unlike virtually all other natural resources – is as universally necessary for individual bodies as it is for civilizations (Bakker, 2003). But as socially produced nature, water is not politically neutral. Rather, it both reflects and reproduces relations of social power.

Similar arguments have been in place for some time. In his 1974 critique of the neo-Malthusianism of Paul Ehrlich, Garret Hardin and others, Harvey (1974, p. 265) presents a historical materialist view of natural resources:

> A 'thing' cannot be understood or even talked about independently of the relations it has with other things. For example, 'resources' can be defined only in relationship to the mode of production which seeks to make use of them and which simultaneously 'produces' them through both the physical and mental activity of the users. There is, therefore, no such thing as a resource in the abstract or a resource which exists as a 'thing in itself'.

In this view, water as a natural resource can only be understood relative to the social relations of production and consumption that give it meaning and that shape its characteristics: flow, quality, quantity, etc. As Swyngedouw (2004: 28) puts it, "Water is a hybrid thing that captures and embodies processes that are simultaneously material, discursive, and symbolic." In this view, water is not an inert object of nature, but rather an active participant, or 'actant' (Latour, 1987), whose materiality and geo-ecological properties shape social relations, even as those social relations act on and transform water's materiality. For example, the processes by which people abstract water from a river, lake or aquifer and channel it through hydraulic infrastructure both transform the water and its ecology (through processes of diversion, filtration, storage, delivery, etc.) *and* transform society (through socially differentiated processes of water provision, sanitation, class formation, luxury consumption, etc.). In this sense, water is at once naturally occurring and socially produced, both embodiment of and precondition for social power (Loftus, 2009; Perreault et al. 2012); or as Budds and Hinojosa (2012, p. 120) put it, "Water's materiality and social relations constitute and express each other."

While the notion of the hydrosocial cycle has proven both productive and provocative, I would caution that not everything associated with water circulates in the same way. Water itself – H_2O – circulates, while other elements of hydrosocial systems tend to accumulate, including water rights, forms of water-related knowledge, biotoxins, and the toxic sediments that have covered the *pampa* of Alantañita (Perreault, 2013). Water's materiality plays a crucial but under-theorized role in what Harvey (2003) refers to as accumulation by dispossession, including the accumulation of water and water rights through such processes as water grabbing, service privatization, and pollution. It bears asking, then, what difference water makes to processes of capital accumulation, and moreover, how the materiality of water, sediment, minerals and biotoxins shape processes of dispossession. As Sneddon (2007, p. 186) has argued, "Processes of accumulation always necessarily involve transformations of nature. These proceed in diverse ways, and understanding the specific biophysical relationships that are sustained or disrupted via such transformations is prerequisite for understanding the conflicts that so often follow on the heels of environmental change." Examining processes of contamination, sedimentation and bioaccumulation, as well as water and land grabbing, illuminates the myriad ways in which water's materiality intersects with, embodies and reproduces forms of social power.

What does this mean in terms of our analysis here? It is worth highlighting some of the central insights of this literature and reflecting on what they might contribute to interdisciplinary and trans-scalar debates regarding water and its governance. Perhaps most importantly, this work insists that we cannot consider water apart from the social relations that produce it and give it meaning, and that those social relations are always historically constituted and exist within a context of uneven power. Such a perspective both accounts for material, historically rooted processes and is attentive to the symbolic meanings with which social relations and nature are imbued. This view of water – as social and natural and thoroughly *political* – matters for the forms of governance we establish (or struggle against). If water is understood as *only* 'natural', if its social history and political character are ignored (as is all too often the case in technical reports and policy statements regarding water management), then water governance is more easily "rendered technical" (Li, 2007); that is, the practices and institutions of decision making may serve to veil the power relations inherent in water's production. The institutional arrangements and social relations involved in water governance are the subject of the next section.

Water governance and scale

What are we talking about when we talk about governance? I am concerned that, like 'sustainable development' and 'social capital,' the term 'environmental governance' has gained broad acceptance, and is often deployed, without the benefit of rigorous critique (see Bridge & Perreault, 2009). Indeed, the vagueness and malleability of the term may serve to obscure political interest and ideological positions, as in the World Bank's formulaic calls for 'good governance', a position that is surely hard to argue with. After all, who wants *bad* governance? But such calls help conceal the political and economic interests that lie behind the institutional arrangements, social relations, material practices and scalar configurations involved in so-called 'good governance'. If we are to employ this concept, then it is imperative we do so critically, carefully elucidating the political nature inherent in the institutional arrangements and socio-environmental relationships to which it refers.

The concept of 'governance' has emerged in recent decades to address questions of economic and political coordination, and refers to the ways that institutional stability – rules, social order, rights, norms, etc. – is achieved in society (Bridge & Perreault, 2009). Bakker (2010, p. 44) defines governance somewhat broadly as "a process of decision making that is structured by institutions (laws, rules, norms, and customs) and shaped by ideological preferences". *Environmental* governance, then, has been deployed in a variety of theoretical perspectives and academic disciplines to examine the institutional diversification of environmental and resource management as an aspect of political-economic restructuring under neoliberal capitalism – a process commonly referred to as the shift from 'govern*ment* to govern*ance*', or toward 'glocalization' (Swyngedouw, 1997). Whichever shorthand one chooses, the concept of governance serves as a broad conceptual framework for analyzing the interplay of institutional arrangements, spatial scales, organizational structures and social actors involved in making decisions regarding nature and natural resources, particularly under conditions of neoliberal capitalism (Himley, 2008). Clearly, governance refers to the functions of government, but also, and importantly, the relations among government, quasi-governmental and non-governmental actors and agencies. Environmental governance is particularly concerned with the act of governing resources and environments, and the ensemble of organizations, institutional frameworks, norms and practices, operating across multiple spatial scales, through which such governing occurs (McCarthy & Prudham, 2004).

A central feature of this approach is that, while it recognizes the importance of spatial scale as a component of water governance, there is no prescribed, privileged scale at which governance should take place. The concept of water governance may be applied to a diversity of scalar arrangements: watershed-based forms of management; canal-based irrigators' associations; municipal service providers; international water forums; etc. But if analysts of water governance do not specify a preferred scale of action, they are insistent that hydrosocial scale is central to policy implementation and political action. Scale is central to water governance to the extent that governance schemes unfailingly target or privilege given spatial scales of hydrology or water management: canal-based irrigation systems, municipal drinking water systems, and transnational compacts are just a few examples (Bridge & Perreault, 2009). Little wonder, then, that the watershed appears as a seemingly natural scale for managing water. Watersheds appear natural and tangible, and therefore as the ideal scale for managing complex hydrosocial relations. But as Molle (2009) and others have pointed out, watershed management is shot through with problems, not the least of which is the spatial malleability of the concept itself. Which of the multiple nested watersheds forms the proper scale for governance? And the concept seems to lose analytical purchase altogether when we consider the effects of trans-basin diversions, the drilling of deep wells, or desalination (Cohen & Davidson, 2011).

Scale cannot be understood apart from a theory of space; and like space, scale is socially produced. That is, particular spatial scales emerge out of the historically sedimented frictions of social relations, and as such are inherently political. In other words, like space itself, scale is socially produced, as an outcome of social practices, perceptions and relationships over time. Moreover, scale is a relational concept, which only has meaning relative to other scales. As McCarthy (2005, p. 738, emphasis in original) points out in parsing the politics of scale, it is

> impossible to separate out the delineation *of* any single scale from relationships *among* scales. More precisely, the establishment of scales as spatially organised and differentiated units of socio-spatial organization ... unavoidably occurs in relationship to other scales: the delineation or elimination of any particular scale *as* an arena, locale, place, or so on is always done *relative to* other scales and the relationships among them, and necessarily introduces changes into their ordering and hierarchies.

In other words, the politics involved in the production and differentiation of scale (e.g. efforts to establish water users' associations within a city, or for rights to irrigation within a river basin) are inseparable from the relationships between scales established through processes of scale ordering (e.g. the delineation of use rights on the part of particular groups relative to the use rights of others).

As debates concerning watershed management have demonstrated, the choice of scale for water governance is not politically neutral (Cohen & Davidson, 2011; Molle, 2009). To the extent that particular scales for water governance are seen as 'natural' and immutable (e.g. watershed management, canal-scale irrigation), they run the risk of obscuring the politics that lie behind the production of such scales. This, then, is a scalar and spatial expression of Tania Murray Li's notion of "rendering technical": scale choice in this sense becomes a technique of government, a conceptual machine for manufacturing consent while treating political struggles and power relations as mere technical problems to be resolved through the right mix of administrative policy and hydraulic infrastructure.

It is here that the notion of 'waterscape' serves as an analytical corrective to the simplistic scalar assumptions inherent in much water governance policy. Budds and Hinojosa (2012) examine the waterscape not as an alternative scale to the 'watershed' but rather as a co-produced socio-natural entity, "in which social power is embedded in, and shaped by, both water's material flow and its symbolic meanings, and which becomes embodied in, and manifested through, a wide array of physical objects and forms of representation" (p. 124). Budds and Hinojosa define 'waterscape' as "the ways in which flows of water, power and capital [and here we might productively add 'labour'] converge to produce uneven socioecological arrangements over space and time, the particular characteristics of which reflect the power relations that shaped their production" (p. 124). The concept of waterscape permits analysis of the metabolic relationship between water and society within a given socio-spatial milieu. Importantly, a waterscape does not exist at a fixed, pre-given spatial scale. Rather, as Budds and Hinojosa (2012) assert, a waterscape is a "sociospatial configuration" constituted by the interrelationships between social and geo-ecological processes that incorporate but in most cases extend beyond any given watershed. As such, waterscapes may entail social or natural processes, social relations, institutions or artefacts not physically proximate to the watershed in question. Examples might include investment capital for the construction of dams and canals, legislation granting or prohibiting rights to access, social arrangements such as regional water management boards or irrigators' associations, or built infrastructure such as wells, canals, water meters, dams, or sewage treatment facilities. In contrast to the fixed, administratively predetermined spatial scale of the watershed, the scale of a waterscape is analytically flexible, and thus attentive to uneven relations of power. This networked view of hydrosocial relations highlights the place-based material effects of processes, relations and phenomena that may be spatially or temporally distant (Budds & Hinojosa, 2012; Loftus, 2006; Swyngedouw, 2004). Crucially, a waterscape perspective highlights the power relations that flow through, are reflected in, and are reproduced by these complex assemblages (Perreault et al. 2012). This discussion leads us to the next, which addresses equity, justice, and rights, in relation to water governance.

Equity, justice and rights

This special issue revolves around the concept of equity in water governance. The *American Heritage Dictionary* (1992) defines 'equity' in part as "the state, quality, or ideal of being just, impartial, and fair; Justice applied in circumstances covered by law yet influenced by principles of ethics and fairness". Similarly, as Boelens (1998, p. 16) notes, "Equity is about fairness, about 'social justice', about the 'acceptability' of something.... Equity is directly related to rules and rule-making processes *and* to the exchange and distribution of material or immaterial resources in specific social settings." Though brief and incomplete for our purposes here, these definitions go some way toward opening a line of inquiry for understanding equity. Equity is fairness. Equity is impartiality. Equity is something defined in law, and yet informed by deeper ethical principles. Equity is justice. Indeed I would argue that the notion of equity in water governance cannot be understood *apart* from a theory of justice. A call for equity alone does not get us very far in understanding the complexities of water governance. As a stand-alone concept, equity speaks most immediately to the present, and in the context of water governance can refer primarily to distributional issues: fairness in access to drinking water and sewerage, to water for irrigation, in exposure to pollution. But equity falls short when considering historical processes of exclusion or social struggle, or in the context of political claims and culturally rooted understandings of water. How do we determine equity in the face of incommensurable claims to water – for instance, between the abstraction of water for large-scale mining operations versus small-scale agriculture (such as in the Huanuni Valley)? How do we weigh subsistence in poor households against electricity generation in regional hydroelectric schemes? When these uses are in direct contradiction, the language of equity is only partially useful.

I would argue, then, that we must link our discussions of equity to a theory of justice and a critical understanding of rights. Social and environmental justice can be thought of in many ways, and a full review of the concept of justice is far beyond the scope of the present article. Perhaps the two most commonly mobilized conceptualizations of justice are distributional justice and procedural justice. These are, I think, the forms of justice that most resemble the concept of 'equity': fairness achieved through pre-established rules of distribution and procedure. Justice, in this sense, can be achieved by ordering society from behind Rawls' (1971) 'veil of ignorance', or by assuring access to due process and the rule of (mutually recognized and universally applied) law. Surely these are necessary conditions, but far from sufficient for achieving social justice. Here, other forms of justice emerge as necessary. Justice as 'recognition' – akin to the 'right to have rights' – takes into account the need for the socially excluded to be acknowledged as legitimate claimants, to be recognized as having valid political, social and cultural standing, as a precondition for other forms of justice (cf. Boelens, 2009). These are the so-called 'third-generation rights' to collective recognition for indigenous peoples, ethnic and sexual minorities, etc.[1]

Amartya Sen's notion of capabilities is particularly helpful in this regard. For Sen (2001), 'capability' refers to an individual's capacity to achieve certain basic needs in society, as mediated by institutional frameworks such as law, rights and societal norms. A capabilities approach emphasizes 'positive freedoms' – the right, ability, or capacity to do or achieve something – as opposed to 'negative freedoms', or freedom from an external constraint. Sen argues that governments should be measured according to their citizens' actual capabilities – their freedoms to achieve desired ends – as opposed to the idealized formal rights they are legally accorded. In this view, justice is the maximization of everyone's human potential, achieved by both material provision of basic needs (water, food, shelter) and the social institutions necessary for everyone to attain them. As such, the provision of water (and food and shelter and other basic necessities) is a means to an end – not the end itself (Bakker, 2010). This, for Sen, constitutes the very meaning of development – the freedom to fulfil one's capabilities infers other bundles of rights, and cannot be viewed in isolation from the institutional arrangements through which rights are allocated, nor the physical infrastructure through which resources such as water are delivered. This, then, is a relational view of rights.

Similarly, social justice can be viewed as a kind of relationship between people – a social relation. We call for justice in relation to someone and something, within the context of societal norms, practices and institutional arrangements. As Young (1990, p. 25) puts it, "Rights refer to doing more than having, to social relations that enable or constrain action." As such, it makes little sense to speak of justice apart from the social relations of production and reproduction that shape our individual and collective lives. Further, our discussion of justice, equity and rights must be rooted in an understanding of political economy. From this perspective, equity in water governance must be examined from the standpoint of critical engagement with the institutional arrangements of the market, the state and civil society (to take an overly simplistic model of society) through which water is allocated and accessed. As is painfully evident in the Huanuni Valley, a legal right to water (formally accorded to all Bolivian citizens under the 2009 constitution) means little to those who lack adequate infrastructure, money, or political influence. As the Bolivian case illustrates, the specificities of regional political economies need to be taken into account in any analysis of the right to water.

This recognition brings us to the question of rights, and the relationship between rights, justice and equity. As Mirosa and Harris (2012) have pointed out, there are many ways of discussing rights to water. The 'right to water' refers to formal, legal recognition of an individual or group's right to water; this differs somewhat from the 'human right to water', which recognizes the right of all people to water sufficient to satisfy basic needs and human dignity, regardless of their citizenship. And of course these must not be confused with 'water rights', which are more limited and individualized rights to water, linked directly to the property relation. As Mirosa and Harris point out, specific water rights for some may in effect deny basic human rights to water for others – particularly marginalized and vulnerable populations. There is little need here to review the arguments for the human right to water, or the critiques levelled against it, as they are well rehearsed in the literature (see, *inter alia*, Bakker, 2007, 2010; Mirosa & Harris, 2012). It is worth noting, however, the problems of 'rights talk' more generally, and the tendency of rights, as commonly conceptualized, to be individualized, atomizing, universalistic, state-centric, and anthropocentric. In the liberal tradition, rights are generally subsumed by, and subordinated to, liberal and neoliberal logics and the exigencies of capital (Harvey, 2008). As Bakker (2007, 2010) and others have pointed out, there is nothing in the human right to water that precludes water's commodification, commercialization or privatization. The individualized nature of the human right to water also presents a (potential) contradiction to visions of collective rights to water (and other resources) held by some indigenous peoples. Indeed, the World Bank has used the human right to water as a justification for market-based reforms in the water sector, arguing that the market is the most efficient means by which the poor can gain sufficient access to water. For both proponents and opponents of water service privatization, the point remains that designating water as a human right does little to define the institutional arrangements through which water is allocated and accessed, and thus leaves open the thorny issues of governance. But, as Bakker (2010), Mirosa and Harris (2012) and others have argued, the claim of water as a human right does have symbolic and tactical importance.

In her critique of the 'human right to water' debates, Bakker (2007) suggests that viewing water as a collective good and a commons resource, rather than as an economic good, shifts both the terms of debate and the possibilities for conceptualizing equity in water governance. For advocates of water service privatization, equity means economic efficiency, such that consumers should pay the total cost of the burden they impose on the system, otherwise known as full cost recovery. In contrast, those who view water as a collective good define pricing in social terms, according to the ability to pay. This implies a need for cross-subsidy and a recognition of the right to social reproduction (Bakker, 2010). In Sen's terms, cross-subsidizing water service for the poor would help secure their capabilities, an end whose attainment would surely be aided by the tactical designation of water access as a human right (Bakker, 2010; Mirosa & Harris, 2012). Privatization implies the creation of some sort of property; and property – as a social relation – is above all the ability to exclude. By contrast, a human right denotes universal need, access and entitlement, a concept fundamentally opposed to the exclusion implied in the property relation. Notwithstanding the critiques of 'rights talk' from postmodernists and others, we would do well to acknowledge the political power of the notion of rights, and the discursive and material potency of rights claims. As Mitchell (2003, p. 25, emphasis in original) has written, "Rights establish an important *ideal* against which the behavior of the state, capital, and other powerful actors must be measured – and held accountable." The ideal of rights – however partial, whatever its limitations – is a *political* ideal: "at once a means of organizing power, a means of contesting power, and a means of adjudicating power, and these three roles are frequently in conflict" (p. 22). It is within the spaces produced through this conflict that water is governed equitably or not, that needs are met or not, that justice is attained or not.

In considering water rights and the right to water, we may benefit from an engagement with the literature on the 'right to the city' (see, *inter alia*, Attoh, 2011; Harvey, 2008; Lefebvre 1996[1968]); Mitchell 2003). Claiming a right to the city demands that we unpack the concept of rights, and the sorts of rights under consideration. As Attoh (2011, p. 669, emphasis in original) asks, "What *kind* of right is the right to the city?" This is no simple question. For Harvey (2008), whose primary concern is the radical democratization of urban space, the right to the city is more than just a right to participate in decisions regarding the allocation and administration of goods and services; it extends to control over the city and the process of urbanization as the motor of capitalist development. For Harvey, then, the right to the city is ultimately the right to control the means of production and the distribution of surplus. For Mitchell (2003), whose primary concern is access to and control of urban public space, the right to the city has as much to do with protection *from* democracy – that is, to secure the rights of dissident minorities from what is often revanchist majority rule (the contested right of homeless persons to benches, parks and public urination is illustrative here). For Lefebvre, the right to the city is multiform, encompassing the right to produce urban life free of the constraints and alienation imposed by capitalist relations of production. The right to the city, of necessity, is a complex bundle of inter-related rights: "a superior form of rights: rights to freedom, to individualization in socialization, to habitat and to inhabit. The right to the *ouvre*, to participation and *appropriation* (clearly distinct from the right to property), are implied in the right to the city" (1996[1968], p. 174, emphasis in original). Thus, for Lefebvre, the right to the city is the right to *inhabit* – to social and individual life in all its complexity. Like Sen's notion of capabilities, the right to the city is a means to a larger end, that of meaningful social life and social reproduction. Harvey (2008, p. 23) echoes this senti-ment: "The question of what kind of city we want cannot be divorced from that of what kind of social ties, relationship to nature, lifestyles, technologies and aesthetic values we desire. The right to the city is far more than the individual liberty to access urban resources: it is a right to change ourselves by changing the city." For Harvey, then, the right to the city is inherently a collective (rather than individual) right. And as with any rights, the right to the city is a site of struggle.

The diverse and at times contradictory nature of the ways that rights to the city have been conceptualized is something of a double-edged sword. For Attoh (2011, p. 670), the "radical openness" of the concept has the potential to bind together multiple struggles for justice, even as it exposes fundamental contradictions between majoritarian impulses and the need to protect minorities from democratic tyranny. How the right to the city is defined shapes the forms of justice to which it leads, as well as the kinds of cities it produces. Similarly, how we think about the right to water, about water equity and water justice, will shape the forms of governance we enact and the spatial and temporal scales at which we do so.

I do not mean to push this analogy too far, or draw parallels where none exist. Obviously, water and cities are not equivalent things: water is a resource, while cities are places, spaces and complex social relations, of which the flow of water is but one component. There are, nevertheless, vital points of convergence between the right to the city and the right to water. Perhaps most obviously, the right to the city presupposes the right to water, to human dignity, to healthy environments. Moreover, there are important lessons to be learned in the way we think about the constellation of social relations that inhere in cities and water: governance, water, justice and equity are all relational concepts, and, as Iris Marion Young notes, are more about doing than about having. Like the right to the city, the 'right to water' is a deceptively complex, multiform and in some respects contradictory set of rights. As social relations, rights are simultaneously institutional arrangements for organizing power and sites of social struggle. How these rights are defined shapes the forms and scales of governance enacted and the flows of water to which they lead. The right to water, no less than the right to the city, *matters* inasmuch as struggles over rights are key moments in broader processes of the production of space (Mitchell, 2003).

Within this recognition, however, lies another: the inherent tension between, on the one hand, rights to democratic participation in the appropriation of the means of production and social reproduction, and, on the other hand, the right of minorities to protection from democratic tyranny. This, in essence, is the problem of 'legal pluralism': the contradiction between collective rights to commons resources and a liberal, individualized conceptualization of resource rights. In such contexts, how rights to water are defined and how equity is conceptualized and enacted will in large part determine the forms of justice and injustice in our waterscapes (Boelens, 1998, 2009). Water governance, as sets of institutions, laws and rights that structure social action and social relations, produces both space and scale; and, in dialectical fashion, these spaces and scales shape the hydrosocial relations and forms of social life we experience.

Conclusion

In what ways does this discussion help us understand the hydrosocial relations in Bolivia's Huanuni Valley, where this article began? Water governance in Bolivia, as in so many other places, is comprised of a patchwork of institutional arrangements, norms, traditional uses, legal grey areas and sector-specific practices. As often as not, formal law and water rights are clear and fairly progressive. Bolivia has remarkably strong environmental legislation. But day-to-day reality is at best an imperfect reflection of legal ideals, and in practice, water flows to the powerful and the privileged, while all too often the urban and rural poor do their best to cope with water scarcity, water pollution, or frequently both. The water problems experienced in the rural Huanuni Valley are of a different character than those of the urban areas most commonly discussed in the water governance literature. Much of this literature addresses governance problems related to urban drinking-water systems: privatization and commercialization; inadequate service extension to peri-urban neighbourhoods; prepaid water meters; etc. In these studies, water is treated as an environmental good – an essential factor of production and social reproduction – the core problem of which is access. Seldom do these studies explore the issues of water quality and pollution – water as an environmental *liability*, where exposure, rather than access, is the primary concern (but see Sultana, 2011). This is an unfortunate lacuna. Water contamination is an enormous concern for the urban as well as the rural poor. Urban water is subject to a witch's brew of industrial chemicals, hydrocarbons, sewage, solid waste and other pollutants, while rural water sources are often contaminated by agricultural runoff, human and animal waste, and, as in cases such as the Huanuni Valley, acid drainage and toxic effluent from mining operations.

The residents of the Huanuni River valley are steadily being separated from their means of production and social reproduction by the inexorable accumulation of toxic sediments on their agricultural fields, in their animals, and in their own bodies – what I have elsewhere referred to as a process of "dispossession by accumulation" (Perreault, 2013). The acute water pollution and resulting scarcity in the Huanuni Valley and elsewhere highlight the need for greater attention to what Bakker (2010, ch. 7) has referred to as "ecological governance", which is simply a recognition that social and ecological processes are intimately interrelated and that environmental governance should incorporate an ethic of ecological care. The power imbalance between the state mining firm and the indigenous *campesino* communities downstream, and the socio-ecological catastrophe to which this has led, merit a rethinking of our key terms. A focus on 'governance' or 'governance failure' or even 'equity' will surely fall short. Here, I would suggest that what is needed is greater attention to ecological processes and environmental justice. In contrast to the literature on water governance, the problem of socially differentiated exposure to pollution has been a cornerstone of environmental justice, as both an academic field and a sphere of political activism. I do not see this approach as a substitute for careful attention to water governance but rather as an important and too often neglected component of such analysis. In other words, we must attend to what Bond (2012, p. 197) has referred to as the "hydro-socio-ecological connections" that underpin claims to water rights as well as social and spatial organization more generally. These connections are at once ecological and political, and need to be understood in terms of social and environmental justice. Just as the right to the city, as a complex bundle of rights, is the right to *inhabit* the city – to the resources and services necessary for dignified life, and to appropriation of the means of production and social reproduction – so too the institutional arrangements and spatial scales we employ for governing water must address questions of individual and collective fulfilment, democratization, material relations of production, and ecological sensitivity. Our forms of water governance must address the human right to water not as an end in itself but as a means to attaining the broader objective of a just society. Only by viewing water and society as simultaneously social *and* natural, and thoroughly political – as a densely woven hydro-socio-ecological fabric – can we address both ecological governance and environmental justice. These ideas are taken up in various forms by the other articles in this special issue. Through empirically rich analyses, the authors examine the relationships among water rights, governance, equity and justice. My hope is that this article helps provide a conceptual frame for these analyses.

Acknowledgements

I am grateful for the support of colleagues with the Centro de Ecología y Pueblos Andinos and the Coordinadora para la Defensa del Río Desaguadero y Lagos Uru Uru y Poopó in Oruro, Boliva. This article was inspired in part by my interaction with the Justicia Hídrica (Water Justice) network, based in Wageningen, the Netherlands. Special thanks to Ben Crow, Flora Lu, Constanza Ocampo-Raeder and Sarah Romano for their encouragement, and for inviting me to participate in the workshop on equitable water governance which formed the basis for this special issue.

Note

1. 'First-generation rights' refers to individual liberties to speech, assembly, religion, etc. established in Enlightenment-era documents like the US Constitution and Bill of Rights and Thomas Paine's *Rights of Man*. 'Second-generation rights' refers to the socio-economic rights to a fair wage, collective bargaining and the like, established in the context of industrial capitalism and

the labour struggles it engenders. These were widely established and codified in law during the early twentieth century, particularly in the wake of the Bolshevik Revolution and the Great Depression. 'Third-generation rights' to collective identity and recognition have been established through the UN and its agencies such as the International Labour Organization (e.g. ILO (1989) Convention No. 169 on the Rights of Indigenous Peoples), and more recently by state governments. See Attoh (2011) for a fuller discussion.

References

American Heritage Dictionary (3rd edition). (1992). Boston: Houghton Mifflin.

Attoh, K. (2011). What *kind* of a right is the right to the city? *Progress in Human Geography, 35*, 669–685.

Bakker, K. (2002). From state to market? Water *mercantilización* in Spain. *Environment and Planning A, 34*, 767–790.

Bakker, K. (2003). *An uncooperative commodity: Privatizing water in England and Wales*. Oxford, NY: Oxford University Press.

Bakker, K. (2007). The 'commons' versus the 'commodity': Alter-globalization, anti-privatization and the human right to water in the global South. *Antipode, 39*, 430–455.

Bakker, K. (2010). *Privatizing water: Governance failure and the world's urban water crisis*. Ithaca: Cornell University Press.

Boelens, R. (1998). Equity and rule-making. In R. Boelens and G. Dávila, (Eds.), *Searching for equity: Conceptions of justice and equity in peasant irrigation* (pp. 16–34). Assen, The Netherlands: Van Gorcum.

Boelens, R. (2009). The politics of disciplining water rights. *Development and Change, 40*, 307–331.

Bond, P. (2012). The right to the city and the eco-social communing of water: Discursive and political lessons from South Africa. In F. Sultana and A. Loftus (Eds.), *The right to water: Politics, governance, and social struggles* (pp. 190–205). New York: Earthscan.

Bridge, G., & Perreault, T. (2009). Environmental governance. In N. Castree, Demeritt, D.Liverman, D. and Rhoads, B. (Eds.), *Companion to environmental geography* (pp. 475–497). Oxford, NY: Blackwell.

Budds, J., & Hinojosa, L. (2012). Restructuring and rescaling water governance in mining contexts: The co-production of waterscapes in Peru. *Water Alternatives, 5*, 119–137.

Cohen, A., & Davidson, S. (2011). The watershed approach: Challenges, antecedents, and the transition from technical tool to governance unit. *Water Alternatives, 4*, 1–14.

Harvey, D. (1974). Population, resources, and the ideology of science. *Economic Geography, 50*, 256–277.

Harvey, D. (2003). *The new imperialism*. London: Oxford University Press.

Harvey, D. (2008). The right to the city. *New Left Review, 53*, 23–40.

Himley, M. (2008). Geographies of environmental governance: The nexus of nature and neoliberalism. *Geography Compass, 2*, 433–451.

International Labour Organization (ILO). (1989). *Indigenous and tribal peoples convention*, C169, 27 June 1989. Retrieved from http://www.ilo.org/indigenous/Conventions/no169/lang–en/index.htm.

Latour, B. (1987). *Science in action: How to follow scientists and engineers through society*. Milton Keynes: Open University Press.

Li, T. M. (2007). *The will to improve: Governmentality, development and the practice of politics*. Durham, North Carolina: Duke University Press.

Loftus, A. (2006). Reification and the dictatorship of the water meter. *Antipode, 38*, 1023–1045.

Loftus, A. (2007). Working the socio-natural relations of the urban waterscape. *International Journal of Urban and Regional Research, 31*, 41–59.

Loftus, A. (2009). Rethinking political ecologies of water. *Third World Quarterly, 30*, 953–968.

McCarthy, J. (2005). Scale, sovereignty, and strategy in environmental governance. *Antipode, 37*, 731–753.

McCarthy, J., & Prudham, S. (2004). Neoliberal nature and the nature of neoliberalism. *Geoforum, 35*, 275–283.

Mirosa, O., & Harris, L. M. (2012). Human right to water: Contemporary challenges and contours of a global debate. *Antipode, 44*, 932–949.

Mitchell, D. (2003). *The right to the city: Social justice and the fight for public space*. New York, NY: Guilford.

Molle, F. (2009). River-basin planning and management: The social life of a concept. *Geoforum, 40*, 484–494.

Perreault, T. (2013). Dispossession by accumulation? Mining, water, and the nature of enclosure on the Bolivian Altiplano. *Antipode, 45*(5), 1050–1069.

Perreault, T., Wraight, S., & Perreault, M. (2012). Environmental justice in the Onondaga lake waterscape, New York. *Water Alternatives, 5*, 485–506.

Rawls, J. (1971). *A theory of justice*. Cambridge, MA: Harvard University Press.

Sen, A. (2001). *Development as freedom*. New York, NY: Alfred Knopf.

Sneddon, C. (2007). Nature's materiality and the circuitous paths of accumulation: Dispossession of freshwater fisheries in Cambodia. *Antipode, 39*, 167–193.

Sultana, F. (2011). Suffering for water, suffering from water: Emotional geographies of resource access, control and conflict. *Geoforum, 42*, 163–172.

Swyngedouw, E. (1997). Neither global nor local: 'Glocalization' and the politics of scale. In K. R. Cox (Ed.), *Spaces of globalization: Reasserting the power of the local* (pp. 137–166). New York, NY: The Guilford Press.

Swyngedouw, E. (1999). Modernity and hybridity: Nature, *regeneracionismo*, and the production of the Spanish waterscape, 1890–1930. *Annals of the Association of American Geographers, 89*, 443–465.

Swyngedouw, E. (2004). *Social power and the urbanization of water: Flows of power*. Oxford, NY: Oxford University Press.

Young, I. M. (1990). *Justice and the politics of difference*. Princeton, NJ: Princeton University Press.

Lefebvre, H. (1996[1968]). E. Kofman and Lebas, E. (Ed.). *Writings on cities*. Oxford, NY: Blackwell.

What is water equity? The unfortunate consequences of a global focus on 'drinking water'

Matthew Goff[a] and Ben Crow[b]

[a]Research Associate, University of California, Santa Cruz, USA;
[b]Sociology Department, University of California, Santa Cruz, USA

In recent years, 'equity' has become a goal of water governance. Yet, the indices and policy guidelines for household water, published by the WHO and UNICEF and adopted globally, focus on either 'drinking water' or a limited interpretation of the 'human right to water'. We examine ideas of equity in household water and argue that the dominant focus on improving the potability of water has muted attention to the wider consideration of domestic water and its impact on livelihoods and poverty. A focus on the many capabilities enabled by domestic water illuminates some of these issues.

Framing the problem

Water is the original solution. In one sense of the word, it is the liquid into which microbes, chemicals, particulate matter and much else is dissolved. But in water we also find the potential for social power, viable livelihoods, healthy bodies, transformed landscapes, energy generation, wealth, hope, and the existence of life itself. The multi-faceted uses of water directly impact daily life and human survival. Different levels and circumstances of water access have the potential to elevate people out of poverty or condemn them to it. Control over water can and often does afford individuals and groups significant social and economic power (Swyngedouw, 1997). Water is a prerequisite for life and for many types of economic and domestic activity. For all these reasons, it is appropriate to question the equity of current access to and distribution of water.

Water has similarities to money. It circulates; it is fungible, storable, investible and desirable. It flows across space and time, sometimes changing hands (literally or figuratively), all the while imbuing, defining, redefining, and regenerating social power (Swyngedouw, 2004). Rights to irrigation water enable water to be turned into money (through both the value of crops and the increased value of irrigated or water-rich land). Easy, reliable access to water is a prerequisite for livelihoods and small businesses, much as the availability of investible funds is an entry requirement. So, like money, access to water is a prerequisite for many people to get a better life. As a consequence, water equity can be a matter of concern for justice, just as equity in incomes is.

The research for this paper was undertaken for the NSF-funded workshop, Multi-Scalar and Cross-Disciplinary Approaches Towards Equitable Water Governance, held at the University of California, Santa Cruz, in February 2013. The paper was initially Goff's award-winning undergraduate thesis, and has been substantially revised by the two authors.

In recent years, scholars (e.g. Phansalkar, 2007) have adopted 'equity' as a term that seems useful. In their text *Water, Place, and Equity*, Whiteley, Ingram, and Perry (2008) compile research that makes a case for the inclusion of equity as a necessary consideration in all water resource management. They argue that the long-term success of water resource management may depend on how equity concerns are integrated with the contemporary utilitarian perspective. Budds (2008, p. 68) claims that theories of the hydrosocial cycle (to be discussed later in this paper) are only useful insofar as they "promote social equity and environmental sustainability". Perhaps more importantly, international agencies like the WHO/UNICEF Joint Monitoring Program (JMP) (WHO, 2011b) have adopted equity as a major goal.

We seek to address a number of questions. First, how do scholars and international agencies conceive of 'equity' in the context of domestic or drinking water? We focus on domestic, as opposed to agricultural and industrial, uses of water. Separating domestic from agricultural and industrial uses of water can be challenging, since domestic water can come from irrigation as well as municipal systems (Bakker, Margaretha, Barker, Meinzen-Dick, & Konradsen, 1999). For the goals of this paper, however, this distinction is less problematic. We explain below the wider implications of understanding household water as 'drinking water'. Is equity connected to 'improved' water source access, or to the quantity of potable water households can collect? If these ideas are not sufficient, then what else must be considered for water governance to move toward equity for all? In short, what is water equity?

The paper is organized as follows. In the next section, we examine some ways of thinking about equity that have emerged from the JMP. Then, in the third section, we discuss the human right to water, which may be a driver of water development. We suggest, in the fourth section, how Sen's concepts of freedoms and capabilities could guide a more holistic model of water development, and we discuss the historical production of inequity in this context. Our conclusion argues that water equity is justiciable because domestic water plays such a key role in household work, reproduction, livelihoods and prospects.

How does the WHO conceive of equity?

The main way in which the JMP measures water equity is through indicators showing the proportion of a population having access to 'improved water sources'. In this section, we discuss this measure and some new goals that have been suggested by the JMP.

Indices of access to 'improved water sources' – drinking, not domestic

The indices of improved water and sanitation access generated by the JMP are the most widely used measures of basic urban and rural water development. Since the establishment of the JMP in 1990, the measure of the proportion of national populations having access to 'improved sources' of water has been included, for example, in the World Bank's annual World Development Reports, as well as the reports of many other international agencies. Most books and articles on the topic of water access make reference to these global figures, and often that reference is at the beginning, providing justification for writing about water (e.g. Fishman, 2011). This access index has become the consensus measure accepted worldwide. There are nearly 7 million references on the Internet to "access to improved water and sanitation" and 142,000 citations in articles and books on

Google Scholar. This measure has been generally accepted as a meaningful indicator of global injustice.

As further indication of the acceptance, and practical consequences, of the JMP measure of access to safe water, it was adopted as part of Goal 7 in the Millennium Development Goals (UN, 2014). Furthermore, in 2012, the UN News Centre reported that the goal had been met:

> The goal of reducing by half the number of people without access to safe drinking water has been achieved, well ahead of the 2015 deadline for reaching the globally agreed development targets aimed at ridding the world of extreme poverty, hunger and preventable diseases, the United Nations said today. Between 1990 and 2010, over two billion people gained access to improved drinking water sources, such as piped supplies and protected wells, according to a joint report by the UN Children's Fund (UNICEF) and the UN World Health Organization (WHO).

While the UN was celebrating the achievement of this Millennium Development Goal, some caveats were expressed, even from within the UN. The Office of the High Commissioner for Human Rights reported the special rapporteur for safe drinking water and sanitation, Catarina de Albuquerque, as saying that despite this achievement, "significant disparities still exist and the most vulnerable people in the world have not benefited from progress" (UN OHCHR, 2012). In Zambia, the *Post Newspapers* focused on the case of a resident in a settlement outside Lusaka: "Kaula says people in her area have continued to maintain unprotected ground wells that they dug decades ago because they have to rely on a communal water point, that is closed at certain times of the day by those managing it" (Post Newspapers Zambia, 2012). In Canada, the Calgary Centre for Affordable Water and Sanitation Technology (CAWST, 2012) commented: "These results do not draw attention to regional disparities nor do they address the measurement of water quality – access to an improved water source does not equate to safe drinking water in the home."

There are at least two problems, in addition to these well-founded caveats, with this declaration of victory (see also Albuquerque with Roaf, 2012). First, the JMP index of access to water misses a set of access characteristics. Second, recent research (Onda, LoBuglio, & Bartram, 2012) suggests that the JMP's measure of improved water seriously underestimates the number of people with unsafe water and overestimates the efficacy of 'improved water sources'. We describe both of these problems below.

Missing characteristics

The JMP has from the start been concerned with 'drinking water'. It is reasonable for a body created by the World Health Organization and the United Nations Children's Fund to be concerned with health aspects of water. The focus on safe drinking water, however, has overshadowed the importance of the other aspects of domestic water that may be crucial for households to get out of poverty. 'Domestic' and 'drinking' refer to the same household water, but the focus on drinking water has fostered a single-minded concern for the health aspects of water, and the acceptance of an access index that fails to measure other important characteristics for domestic water. These include water collection time, reliability, social arrangements, and cost. The failure of the index to cover these characteristics means that an opportunity to reduce poverty may be overlooked. A narrow focus on health has encouraged international agencies to believe that the goal of access has been achieved. But in low-income urban areas and many rural areas, this water access does not provide what is needed to enable paths out of poverty (Noel et al, 2010). Easy access to domestic water is key to releasing the creative energy of women and children, enabling small businesses run out of the household, and providing for the domestic work that is necessary for the dignity and social networking of household members (Koolwal & van de Walle, 2013). None of these factors are included in the JMP indices. In other words, the success of the JMP focus on access to drinking water has overshadowed, or crowded out, other important measures of access to domestic water and the key role of domestic water in poverty reduction (Crow, Swallow, & Asamba, 2012; Noel et al, 2010).

The JMP index focuses on a set of 'improved' water sources: piped water into dwelling; piped water into yard; public tap or standpipe; protected dug well or spring; and rainwater. In principle, all of these sources can provide access to safe water, though in practice they may fall short of that. But only the piped water into a dwelling and sometimes piped water into the yard give the level of access likely to enable routes out of poverty. In practice, the 'public tap or standpipe', the common form of 'improved' water provision in low-income urban areas, has been the foundation for the sale of high-cost and unreliable water. This source is often controlled by large-scale vendors or even organized crime (Crow & Odaba, 2010). More crucially, these taps are too few, and the water supply too uncertain, for domestic water provision.[1] While the JMP has made it clear in certain reports (WHO, 2011b) that water piped onto premises is preferable to other improved sources, the adoption in the MDGs of the 'improved sources' measure has supplanted other useful goals. The fact that standpipes controlled by powerful local figures can fall into the same category of improved sources as a direct pipe into the home is unsatisfactory.

Underestimate of those using unsafe water

Research by Onda et al. (2012) suggests that the JMP indicator underestimates the proportion of the global population using unsafe water. Instead of the JMP estimate of 780 million people using unsafe water, they suggest the number is closer to 1.8 billion. This more than doubling of the estimate arises because between 750 million and 1.6 billion of the people using 'improved water sources' actually get water that is not safe to drink, mostly because of faecal contamination. Onda et al.'s (2012) research suggests that even on the limited ground of drinking water access, the JMP index is deeply unsatisfactory.

We suggest in a later section that a more satisfactory measure of access to water could focus on the capabilities afforded to households and individuals. One of these could be the ability to lead a healthy life, the intended focus for which the indices of improved water and sanitation are a proxy.

The Joint Monitoring Program's ideas for future equity targets

The goals set by the JMP for post-MDG water development may become significant determinants of UN agency, aid agency, national government and city government policy. We do not have access to the internal documents of the JMP. Publicly available documents (WHO, 2013), however, summarize the results of a large-scale consultation program (4 groups of 40–60 academics) rethinking the targets and indices for water, sanitation and hygiene. And, when we asked a senior UNICEF specialist, with involvement in the JMP, "whether recent developments in the Joint Monitoring Program have begun to take account of ... 'domestic water' issues for equity in household access" we received a clear response. The answer in short is that the JMP has not taken into account 'domestic water' and it is unlikely to do so in the very near future.

> The focus of WHO and UNICEF on the JMP has indeed been on meeting basic drinking water needs – defined as water for drinking, food preparation and basic personal hygiene. Under the discussions that took place over the past 2.5 years on WASH under the post-2015 Development agenda it was briefly considered to expand beyond basic drinking water needs – but there was a quick consensus that guided by the principles of the Human Right to Water and Sanitation, the JMP would focus on qualitative aspects of basic access (availability within a 30 minute water collection round trip) and on aiming for – what was dubbed – intermediate access (water quality, supply on-premises and a measure of reliability) rather than for expanding into access to water for other domestic purposes. (Rolf Luyendijk, Senior Statistics and Monitoring Specialist, UNICEF – New York, personal communication 29 October 2013)

Luyendijk explained further:

> Our discussions were focused on what the WASH sector could argue for under the post-2015 development agenda. And with well over 700 million people still without access to an improved drinking water source and probably close to 2 billion without access to a micro-biologically safe drinking water source – it was broadly felt that we should keep the focus on access to "drinking water" as defined above rather than include access to domestic water.

This clear and thoughtful email outlines the pragmatic decisions and emerging priorities from the JMP consultation: a focus on basic 'drinking water' needs. The suggested post-MDG targets for 2040 and 2030 reflect this focus on basic access. Nevertheless, the 2040 targets seek halving of households using 'intermediate water service' and that will likely reduce the labour, uncertainties and time demands of water access in ways that could enable the range of capabilities arising from domestic water provision.

The main target proposed by the JMP for 2040 is sensible: "the proportion of the population not using an intermediate drinking water service at home has been reduced by half and inequalities in access have been progressively reduced". This is saying that households should have water directly on their premises, without intervening water traders and other middlemen, and noting that access inequalities should be reduced. The indicator they propose to measure this is also sensible: "percentage of population using an improved drinking water source on premises with discontinuity less than 2 days in the last 2 weeks; with less than 10 CFU *E. coli*/100 mL year round at source; accessible to all members of the household at the times they need it".

The proposed goal for 2030 is more problematic. This is that "everyone uses a basic drinking water supply and inequalities in access have been progressively elimi-nated". As the document recognizes, there is need for some specification of inequal-ities in access. What they propose as an indicator to support the 2030 goal is very problematic: "percentage of population using an improved drinking water source with a total collection round trip time of 30 minutes or less, including queuing". A 30-minute round trip is burdensome. Almost all households need to make multiple trips to get sufficient water for their laundry. A household of 4–7 people may need 6–10 containers of water on the days when they do laundry. Six trips of 30 minutes each suggests a water collection time of not less than 3 hours. Can this be called significant progress in reducing access inequalities? Both the 2040 and 2030 measures of equity are still focused primarily on the provision of safe 'drinking' water, though they are beginning to move toward wider concerns.

The human right to water and the WHO

Along with the JMP plans, United Nations statements are important in laying foundational conceptions of water equity. The human right to water (HRW) was ratified by the UN in 2011. Studying the formulation of the HRW is helpful because it is an important driver of development goals for domestic water (Mirosa & Harris, 2012).

In the 1992 Dublin Statement on Water and Sustainable Development, the UN stated their view that all humans have the right to access clean water and sanitation at an affordable price (Ingram, Whiteley, & Perry, 2008). In 2010, the UN strengthened the statement, saying that safe, clean water and sanitation is a "human right essential to the full enjoyment of life and all other human rights" (WHO, 2011a, p. 17). A few months later, the UN passed Resolution A/HRC/RES/15/9, affirming that the human right to water is legally binding upon states (UN, 2011), as well as stating:

> The right to water and sanitation is derived from the right to an adequate standard of living and inextricably related to the right to the highest standard of physical and mental health, and the right to life and human dignity. Human rights principles define various characteristics against which the enjoyment of the right can be assessed, namely: availability, safety (with reference to the WHO drinking water quality guidelines), acceptability, accessibility, affordability, participation, non-discrimination and accountability. (WHO, 2011a, 17)

The inclusion of acceptability, participation, non-discrimination and accountability takes the HRW beyond a minimum quantity and collection time and distance. These categories imply that the human right to water is a socially mediated commitment, established by specific societal standards. Yet, the JMP seems to have a simplified view of the human right to water. In their summary of the HRW, they suggest that 50–100 litres per capita per day is sufficient to fulfil basic needs; they also set a chemical composition guideline for the minimum required quality (WHO, 2011a). They add that "the water source has to be within 1000 metres of the home and collection time should not exceed 30 minutes" (WHO, 2011a). Their suggestions (WHO, 2011a) can be summarized in a formula that looks something like this :

> HRW = 50–100 L per person per day from an improved water source at ≤5% of family income + ≤1000 m distance + ≤30 minutes collection time + culturally acceptable sanitation facilities

Aside from the inclusion of "culturally acceptable sanitation facilities" (and there is no direct explanation of how acceptability would be ascertained), the WHO's calculus of water rights is materially and technically based and includes little room for social mediation. The UN's suggested characteristics of participation, non-discrimination and accountability are notably absent. National and local government standards often vary from these generally suggested (and not legally binding) minimums. For example, South African law requires only 25 L per person per day, but closer, within 200 meters of the home (Pienaar & van der Schyff, 2007). The WHO formula lacks the crucial adaptability to the varying contexts in which water is used.

Statements regarding the human right to water in UN publications rarely specify a quantity requirement. One UN statement suggests "adequate amounts of clean water, for such basic needs as drinking, sanitation and hygiene" (1998, p. 3). Similar language, requiring 'adequate' or 'sufficient' water, can be found in state policy (e.g. the South

African Constitution, the State of California's Assembly Bill 685). This seemingly unclear rhetoric is an ambiguity that may be problematic but does provide space to include wider concerns about access. There are rhetorical possibilities to reinterpret this law with more flexible and context-specific implications than stricter concepts of the HRW allow. 'Sufficient' implies a malleability that may be necessary for water governance to account for unique circumstances and needs.

So, despite the UN's more complex and adaptable formulation of the HRW, the JMP has maintained a technical, formulaic view that leaves little room for participation. The JMP conception of the HRW, at its simplest form, is a right to a basic amount of affordable, accessible water. This simple, clear formulation is certainly valuable for policy and legal arguments (Mirosa & Harris, 2012). But in the long run, it may not be conducive to achieving greater water equity.

The strengths of the HRW include its adaptability, and its recognition that the achievement of access will be iterative. But one of its weaknesses is its maintenance of the WHO and JMP tradition that the main and sometimes exclusive focus should be on the health aspects of 'drinking' water. As we have argued earlier, this focus overlooks critical aspects of a household's access to domestic water, notably the opportunity costs of collection and uncertainty, the range of domestic uses of water beyond drinking, and the role of water as an entry requirement to most business. Water equity, in other words, needs to go beyond both the simple indices of the JMP and the formulation of the HRW. We suggest in the next section that equity may be best conceived as not a quantity of water, nor access to improved water sources, but a set of capabilities and freedoms made possible by easy access to domestic water.

Equity in water access understood through freedoms and capabilities

Sen's proposed freedoms and capabilities approach (1999) suggests ways to think about water equity. The goal of social change, according to this approach, is to expand people's abilities to be and do things that they value (Alkire, 2002, p. 2). The freedoms and capabilities approach is particularly helpful for understanding equity in access to and use of water because water has a wide range of uses, social roles and symbolic meanings. This approach is focused on the social relationships that enable individual abilities or capabilities, and it provides analytical purchase on a wide range of human potential, including those most basic capacities required for survival and flourishing (see also Linton, 2012).

So, for example, many women in low-income urban areas make lunch and snack items to sell to people on the go. We can ask: does a woman wanting to start such a business have the social relations giving access to water? These social relationships include proximate connections (options: ability to buy from a nearby vendor's tap; owning a tap; having a contract with the water utility; living in a community where infrastructure has been built) and resources (money to pay the vendor; a household member with time to make trips to the water source; containers for collecting and storing the water; purification facilities if the water is not clean). The capabilities approach suggests that the social relations of proximity and of resources are key requirements for this woman to have the capability to make lunch and snack items for sale.

Sen (1999, p. 10) argues that freedoms are both the ends and the means of intentional social change, and he illustrates the idea of freedoms as enabling social change with this range of five instrumental freedoms (Table 1). In the table, ways each of these freedoms might influence access to water are suggested. Sen's itemization of the portfolio of

Table 1. Sen's instrumental freedoms and their application to water.

Instrumental freedom	General example	Illustration for water
Political freedoms help to promote economic security	Free speech and elections	Representation of the poor and marginalized, along with democratic pressure, may be required for water delivery infrastructure to be built in low-income areas
Economic facilities help to generate personal abundance and public resources	Opportunities for participation in trade and production	Access to water is a prerequisite for many informal-sector businesses
Social opportunities facilitate economic participation	Education and health facilities	Substantial and uncertain water collection times constrain the social opportunities of women and children, including their ability to get education and health care
Transparency guarantees	To prevent corruption, illicit deals and financial irresponsibility	Open books and open responsibilities in water utilities can help reduce corruption and favouritism in water supply for the rich and powerful (in irrigation as in domestic water)
Protective security	To provide a social safety net against famine and other causes of extreme distress	Provision of adequate access to safe and sufficient water is key for livelihoods which may provide security in times of deprivation

instrumental freedoms can be envisaged as a checklist of the wide range of social conditions that impinge upon a household's access to water. These include representation, or voice, in order to get appropriate infrastructure; transparency and accountability, to curtail corruption; and the protective security provided by access to water.

Sen proposes equity as access to a range of freedoms and capabilities established in comparison with those routinely enjoyed by others in a community or globally. In the Santa Cruz Declaration on the Global Water Crisis (in this issue), it is suggested that water equity can be judged as equity within water uses (agriculture, industry, domestic/residential) and deliberated fairness between uses. Equity within uses (or hydrosocial cycles, as we explain below) can be interpreted most simply as comparable quantities of water available to users within a set of farmers, residents or small-business owners. A more sophisticated interpretation could focus on comparison of *capabilities* rather than *quantities*.

Capabilities could include: starting a business; maintaining a household; cooking; raising a crop on the land with secure tenure; being able to fish; being able to trade; having good health; and not having to waste much time getting water. Then, equity in capabilities, between rich and poor or between ethnic groups, can be formulated for a particular use or hydrosocial cycle. For example, the consideration of equity could ask: What conditions of water access (e.g. time and uncertainty of collection; storage ability) are required to maintain a business?

Comparing fairness between uses (from a capabilities perspective) is somewhat trickier. Most simply, we could compare water allocation quantities. This is simple enough. But a more complete comparison would include health, job creation, well-being and other social goals (perhaps using the Human Development Index) across different sectors. As demands rise from urban areas, for example, agriculture in many parts of the world is being required to reduce the amount of water it uses for irrigation. The fairness of this negotiation centres on the quantities of water used, but it is the wider consequences for society, the capabilities that are promoted by water use, that have to be assessed and balanced. Will diversion of water from agriculture to urban areas, for example, generate more jobs, better health and well-being, or greater social productivity?

Sen's approach to capabilities and freedoms does not address collective capabilities and social power effectively (Evans, 2002; Selwyn, 2011). Selwyn concludes a sympathetic reading of Sen's *Development as Freedom* (1999) with a summary of some limits to Sen's approach:

> By wedding his vision to a conception of capitalist markets as spheres of freedom, Sen undermines his own attempts at seeking out alternative routes to human fulfillment. His framework (methodological individualism, a Weberian conception of class and a limited Smithian understanding of the market) disables him from getting at the roots of causes of mass deprivation. (2011, p. 75)

This reminds us that the search for equitable water access should not be confined to a focus on individual capabilities. The control of water may reflect and reinforce the power of individuals and social groups. This can be seen in, amongst other places, the historical production of water inequalities.

Hydrosocial cycles and the historical production of inequality

What we have called 'uses' of water, such as agriculture, domestic activities or energy production, can also be imagined as different hydrosocial cycles. In contrast to the hydrological cycle postulated by hydrology, the hydrosocial cycle (Bakker, 2002; Linton, 2010) recognizes that different social uses constitute particular relations to water and specific cycles of human use, return to nature and reuse. Then, the formulation of equity within uses and fairness between uses (Lele et al., 2013) can be seen as equity within one hydrosocial cycle and deliberated fairness between cycles.

Phansalkar (2007) writes that *equality* in water access signifies an exact comparison of circumstances between parties – e.g. everyone gets 50 L piped into their house, regardless of circumstance. But he argued that *equity* is a more socially mediated concept, one that is relational and value laden. Equity in water access would mean that each party attains the water access that is socially acceptable to them and their society. The idea of equity within uses and fairness between uses parallels this distinction. From our perspective, this means that people would have the water access that enables freedoms and capabilities comparable to others within their hydrosocial cycle.

The idea of a cycle that combines the particulars of natural and human processes in one location or region has utility. Water scarcity, for example, has frequently been seen as the result of natural processes affected by rainfall, landforms, rivers and climate, amongst others. Several social scientists have found cases where scarcity has been generated or propagated by social processes (Mehta 2001; Kaika, 2006) with inequitable consequences. The notion of a hydrosocial cycle can help to illuminate the coproduction of scarcity through natural and social means. Somewhat similarly, water uses and water equity can be understood to occur within coproduced (natural and social) water cycles (see Figure 1 in the Introduction to this issue).

Many water inequities, however, do not have their origins in the water sector. Water inequities can emerge due to a combination of historical events, biophysical circumstances, and power dynamics that surpass the realm of water provision. Swyngedouw's (1997, 2004) path-breaking study of conditions in Guayaquil, Ecuador, illustrated how the pre-existing dynamics of social power may underpin conditions of exclusion and inequity within water governance.

Similarly, Dill and Crow (this issue) argue that the historical production of water inequality in Nairobi and Dar es Salaam arises substantially from the continuation of the racial segregation of space in those cities, dating back to colonial times, and the sustained zoning that separates the city into formal and informal areas. In these cities, spatially distinct populations, marked by some combination of race, ethnicity, standing and income, may be separated by city boundaries and the extension of city services into areas with good and less good water access and those with wholly inadequate water access.

Some of the factors producing water inequities are illustrated in Table 2. It is difficult to visualize or illustrate graphically the interaction of inequality generated on historical time scales, the contemporary spatial distribution of unequal access to water, and the scarcities and deprivations experienced by households on daily, weekly and seasonal time scales. Even so, it is clear that water inequities do not always originate in the hydrological water cycle. And thus, water inequities may sometimes not be remedied through only the provision of more water. That would frequently help. But to address the complex roots of inequity, a more sophisticated cross-disciplinary strategy will be needed, one that accounts for the production of inequity with a greater awareness of historical and political factors.

Table 2. Historical construction of some contemporary water inequities.

Era	Practices leading to water inequalities	Consequence
Colonial	Urban and rural residential segregation by race Provision of services to settler areas in the countryside and to European areas in cities Construction of water infrastructure (irrigation, city supply networks) serving particular favoured areas	Uneven spatial development and water and sanitation provision
Post-colonial	Officials of post-independence governments nevertheless see informal and squatter settlements as illegal Continuation of zoning decisions excluding some rural areas and city sectors	Exclusion of some areas, periodic attempts to demolish settlements New institutions (including organized crime) emerge to distribute water and other services
Contemporary	Water allocations and infrastructure tend to exclude low-income areas WHO/UNICEF focus on 'drinking water'	Sets of capabilities, including livelihoods, education and home maintenance are constrained for low-income women and children

Conclusions

The ideas of water equity that have come to dominate popular discussions of deprivation focus primarily on the quantity and potability of water. Even the newly accepted idea of a human right to water concentrates on the health aspects of water supply. The new goals for water improvement under discussion at the JMP still focus on drinking water and thus fail to capture key aspects of water required for people's livelihoods and the reproduction of families and households. We argue that this is an inadequate conception of water equity because it leaves out many other aspects of domestic water access. Domestic water is not just drinking water; it is a prerequisite for many other aspects of life, including livelihoods and maintaining the home.

Indicators of access frame global understanding and global goals for water equity. The current indicators ignore important aspects of the provision of domestic water, notably some that will be useful in combating poverty and improving livelihoods. These indicators tend to support a hydrologically founded understanding of water inequity that we believe is insufficient. Water inequity is not born solely out of poor water provision. There are many other factors involved, and so if equity is a goal of contemporary water governance, then we need a more holistic set of indicators and a more cross-disciplinary perspective.

What is water equity? It is a claim that water access and water allocation deserve to be justiciable. By justiciable we mean subject to assertions of injustice deserving examination against global and local norms of equality. We suggest that access to water is justiciable for at least three reasons. First, water shares with income and wealth, themselves widely accepted as matters for claims of injustice, a role as a prerequisite for business and livelihood opportunities. Second, water plays a key role in domestic work. Without easy access to water, the tasks of maintaining a home and bringing up children are constrained. Third, the time devoted to water collection deprives a particular section of the population, women and children in low-income communities, of opportunities to live lives that they value.

Water access is equitable if residents are able to live healthy lives that they value. Water equity cannot be judged purely by the material circumstances of access, such as quantity of potable water available. Equity must be understood more holistically. It is the capabilities enabled by water access that really matter. We suggest that the freedom and capabilities approach provides particular insight into water inequities because it illuminates the way different conditions may enable or limit key capabilities required for lives, livelihoods and freedoms.

Note

1. A JMP task force began to discuss some of these questions in 2011, and post-2015 goals have emerged from large-scale consultation (WHO, 2013). We discuss this below.

References

Albuquerque, C. de, with Roaf, V. (2012). *On the right track: Good practices in realizing the rights to water and sanitation*. Water and Sanitation Program.

Alkire, S. (2002). *Valuing freedoms: Sen's capability approach and poverty reduction*. Oxford, UK: Oxford University Press.

Bakker, K. (2002). From state to market? Water *Mercantilización* in Spain. *Environment and Planning A, 34*, 767–790.

Bakker, Margaretha, Barker, R., Meinzen-Dick, R., & Konradsen, F. (1999). "Multiple uses of water in irrigated areas: A case study from Sri Lanka." IFPRI-IWMI System-Wide Initiative on Water Management.

Budds, J. (2008). Whose scarcity? The hydrosocial cycle and the changing waterscape of La Ligua River Basin, Chile. In M. K. Goodman, M. T. Boykoff, & K. T. Evered, (Ed.), *Contentious geographies: Environmental knowledge, meaning, scale* (pp. 59–68). Ashgate Studies in Environmental Policy and Practice. Farnham, UK: Ashgate.

Calgary Center for Affordable Water and Sanitation Technology. (2012). World Meets Target for Affordable Water But". March 6th. http://www.cawst.org/en/about-us/news/417-world-meets-target-for-access-to-improved-water-sources-but-neglects-focus-on-safe-drinking-water

California Legislature. (2012). State Assembly Bill 685 Chapter 524. Approved by Governor September 25, 2012.

Crow, B., & Odaba, E. (2010). Access to water in a Nairobi slum: Women's work and institutional learning. *Water International, 35*(6), 733–747.

Crow, B., Swallow, B., & Asamba, I. (2012). Community organized household water increases rural incomes, but also men's work. *World Development, 40*(3), 528–541. March 2012.

Evans, P. (2002). Collective capabilities, culture and amartya sen's *development as freedom. Studies in Comparative Development, 37*(2), 54–60.

Fishman, C. (2011). *The big thirst: The secret life and turbulent future of water.* New York, NY: Free Press.

Ingram, H., Whiteley, J. M., & Perry, R. (2008). The importance of equity and the limits of efficiency in water resources. In J. M. Whiteley, H. Ingram, & R. W. Perry (Eds.), *Water, place, & equity.* Cambridge, MA: MIT Press.

Kaika, M. (2006). The political ecology of water scarcity: The 1989–1991 Athenian drought. In N. Heynen, et al. (Eds.), *In the nature of cities: Urban political ecology and the politics of urban metabolism.* New York, NY: Routledge.

Koolwal, G., & van de Walle, D. (2013). Access to water, women's work, and child outcomes. *Economic Development and Cultural Change, 61*(2), 369–405. (January 2013).

Lele, S., Srinivasan, V. Jamwal, P., Thomas, B. K., Eswar, M., & Zuhail, T. Md. (2013). Water management in Arkavathy basin: A situation analysis. *Environment and Development Discussion Paper No.1.* Bengaluru: Ashoka Trust for Research in Ecology and the Environment.

Linton, J. (2010). *What is water? The history of a modern abstraction.* Vancouver: UBC Press.

Linton, J. (2012). The human right to what? Water, rights, humans and the relation of things. In F. Sultana & A. Loftus, *The right to water: Politics, governance and social struggles.* New York, NY: Earthscan.

Mehta, L. (2001). The manufacture of popular perceptions of scarcity: Dams and water-related narratives in Gujarat, India. *World Development 29*(12), 2025–2041.

Mirosa, O., & Harris, L. M. (2012). Human right to water: Contemporary challenges and contours of a global debate. *Antipode, 44*(3), 66–4812.

Noel, S. et al. (2010). The impact of domestic water on household enterprises: Evidence from Vietnam. *Water Policy, 12*, 237–247.

Onda, K., LoBuglio, J., & Bartram, J. (2012). Global access to safe water: Accounting for water quality and the resulting impact on MDG progress. *International Journal of Environmental Research and Public Health, 9*(3), 880–894.

Phansalkar, S. J. (2007). Water, equity and development. *International Journal of Rural Management, 3*(1), 1–25.

Pienaar, G. J., & van der Schyff, E. (2007). The reform of water rights in South Africa. *Law, Environment and Development Journal, 3/2*, p. 179. Available at http://www.lead-journal.org/content/07179.pdf

Post Newspapers Zambia. (2012). "MDG on water: Fides Kaula's story". 11 March 2012, http://www.postzambia.com/post-print_article.php?articleId=25709

Selwyn, B. (2011). Liberty limited? A sympathetic re-engagement with amartya sen's *development as freedom. Economic and Political Weekly, XLVi* (37), September 10 2011.

Sen, A. (1999). *Development as freedom.* Oxford, UK: Oxford University Press.

Swyngedouw, E. (1997). Power, nature, and the city. The conquest of water and the political ecology of urbanization in Guayaquil, Ecuador: 1880–1990. *Environment and Planning A, 29*, 311–332.

Swyngedouw, E. (2004). *Social power and the urbanization of water: Flows of power.* Oxford, UK: Oxford University Press.

UN (2014). Millenium Development Goals. Retrieved from http://www.un.org/millenniumgoals/

UN News Center. (2012). World meets goal of boosting access to clean water but lags on better sanitation – UN 6 March http://www.un.org/apps/news/story.asp?NewsID=41465.

United Nation Office of the High Commissioner for Human Rights. (2012). UN Millennium development goals: Some targets met, but much remains to be done. Geneva 10 July 2012. http://www.ohchr.org/en/NewsEvents/Pages/DisplayNews.aspx?NewsID=12339&LangID=E

United Nations. (2011). "The human right to water and sanitation: Milestones." UN-Water Decade Programme on Advocacy and Communication (UNW-DPAC).

Whiteley, J. M.H. Ingram, & R. W. Perry (Eds). (2008). *Water, place, and equity.* Cambridge: MIT Press.

World Health Organization (WHO). (2011a). "The human right to water and sanitation: Media brief." June 1, 2011.

World Health Organization (WHO). (2011b). *Drinking water equity, safety, and sustainability: Thematic report on drinking water 2011.* New York, NY: WHO & UNICEF.

World Health Organization (WHO). (2013). *Progress on sanitation and drinking water 2013 update.* Geneva: World Health Organization & Unicef.

PES hydrosocial territories: de-territorialization and re-patterning of water control arenas in the Andean highlands

Jean Carlo Rodríguez-de-Francisco[a,b] and Rutgerd Boelens[b,c]

[a]Environmental Policy and Natural Resources Management Department, German Development Institute/ Deutsches Institut für Entwicklungspolitik (DIE), Bonn, Germany; [b]Water Resources Management Group, Department of Environmental Sciences, Wageningen University, the Netherlands; [c]Department of Geography, Planning and International Development, and CEDLA Centre for Latin American Research and Documentation, University of Amsterdam, the Netherlands

ABSTRACT

This article explores how payment for environmental services (PES) approaches envision, design and actively constitute new hydro-social territories by reconfiguring local water control arenas. PES aims to conserve watershed ecosystems by repatterning and commoditizing the link between 'water service providers' upstream and 'water consuming' populations downstream. Two case illustrations from the Ecuadorian highlands are used to clarify how PES implementation – though presented as if it were apolitical and neutral – weakens locally crafted hydrosocial territories in favour of dominant interests. If consolidated, this depoliticized PES implementation fosters the consolidation of new (market-environmentalist) territories, subjects and interactions, further marginalizing the less powerful upstream communities' livelihood strategies.

Introduction

The peasants were driven further and further up …, and not only the servants but also the free communities. Good irrigated land, fertile valleys and gentle mountain slopes – there where the Incas built terrace gardens: the good land was occupied by the masters. Communities got dry, barren land, and as the Indians tamed this wild land, irrigated it or cultivated it, timing their crops to the variable rainfall, the hacienda owners pushed them even higher, and spread their own plantation boundaries, just as they pleased. (Arguedas, 1980, pp. 32–33, own translation)

Water access and control rights have always been contested matters for peasant and indigenous communities in the Andes. Disputes over water control, with and among a growing universe of users and water-use sectors, do not relate only to the liquid water itself. They also involve negotiations and conflicts over designing and controlling its infrastructure, over defining water access and control institutions, and over the legitimacy of the authorities defining these institutions. Next, they relate to conflicts among water governance discourses, constituting fierce battlefields in which many groups vie for particular water policies and governance frameworks (Boelens, Hoogesteger, & Baud, 2015; Gelles, 2000; Lynch, 2012).

The Andean highlands figure prominently in many of these struggles and debates. Highlands that in the recent past, in the eyes of powerful actors, were considered merely wilderness, barren or unproductive no man's land, are now at the centre of many development initiatives and water-resource interventions. Such initiatives promise societal development, for instance by extracting minerals, producing hydro-electric energy, or conserving environmental services in local watershed territories in order to secure drinking-water provision and agribusinesses production (Duarte-Abadia, Boelens, & Roa-Avendaño, 2015; Rodríguez-de-Francisco, Budds, & Boelens, 2013). The latter watershed conservation programmes, which are increasingly popular, commonly adopt a payment for environmental services (PES) approach, wherein upstream land managers are to change or maintain certain land uses so that downstream water users have more reliable water supply and improved water quality. To this end, PES are said to be voluntary transactions where downstream water users pay upstream land managers to conserve ecosystems considered to yield more and better-quality water downstream.

This article argues that, under PES market logic, water rights and nature's socio-ecological properties are redefined and grounded in economic incentive structures, deeply transforming the locally existing political and cultural definitions of water control and collective property (Rodríguez-de-Francisco & Boelens, 2014; see also Büscher, Sullivan, Neves, Igoe, & Brockington, 2012; Castree, 2008; Fairhead, Leach, & Scoones, 2012; Kosoy & Corbera, 2010; Mosse, 2008).

Therefore, this article aims to explain the tensions that arise between peasant/indigenous hydrosocial territories in the Andean highlands and the production of hydrosocial territory under PES rationality schemes. As the article shows, such depoliticized conservation programmes seek to secure water resources as a commodity for capital accumulation downstream, while imposing heavy-handed, ill-compensated land use limitations on marginalized highland communities. The article argues instead for interactive intervention approaches that recognize the political dimensions and impacts of conservation efforts, which explicitly need to address environmental change as the result of local-national-global interactions among interest groups, including everyday power struggles over natural resource control. Depoliticized PES conservation commonly tends to neglect the further marginalization of upstream communities and the ways in which upstream conservation may constitute a water control strategy for the better-off.

This article is structured as follows. After this introduction, a theoretical framework is outlined regarding the major hydrosocial territorial transformations that take place under a PES regime. Thereafter, as a brief illustration, background information is presented on the history of PES introduction in Ecuador's Pimampiro territories. Afterwards, the concept of a hydrosocial territory is scrutinized from the perspective of highland peasant communities, comparing this to the PES-envisioned hydrosocial territory. The final section puts forward conclusions regarding the contradictions and tensions resulting when an externally empowered market-environmentalist territorial discourse and a monetized conservation rationality introduce a new conception of hydro-territoriality, which overrides the pre-existing notions of natural resource control and the management institutions and cultural practices embedded in a locally controlled territory.

Control-localizing hydrosocial networks and the challenges of market-environmentalism in the Andean highlands

More than a physical space, territory is a complex of socio-historic and geopolitical constructs that encompass the material and symbolic space where the economic, environmental and cultural activities of a society/collective evolve. Many have challenged the concept of 'territory' as a single, fixed, geographically or politically bounded unit, and notions of territorial construction as a process-based, subject-built assemblage of socio-material practices are common nowadays. Such territorial shaping is based on the diverse, divergent interests of a variety of human actors in interaction with nonhuman actants (Agnew, 1994; Baletti, 2012; Elden, 2010; Rincón-Gamba, 2013). Actors have different views of 'territory' and unequal power to materialize these conceptualizations; this makes territory, its contents, scalar configurations, values and meaning contingent upon political struggle and power asymmetries (Delaney & Leitner, 1997; Houdret, Dombrowsky, & Horlemann, 2014; Lebel, Garden, & Imamura, 2005).

The same goes for the contested properties and definitions of what constitutes 'hydrosocial territoriality'. By nature, the definition of such a socio-natural or hydrosocial spatial network is contested, differing according to discourse and discipline (Boelens, Hoogesteger, Swyngedouw, Vos, & Wester, 2015). For example, as we have elaborated on elsewhere (Boelens, Hoogesteger, & Rodríguez-de-Francisco, 2014, p. 1), "Governments tend to define such a territory as a geopolitical space clearly delimited by their water bureaucracy's unit boundaries. Positivist science disciplines tend to conflate the concept with the units comprised by watershed or catchment boundaries. Others (e.g., indigenous or anthropological currents) would emphasize the fact that it constitutes a socio-cultural construct, based on people's historical, context-based appropriation of a space and place constituted by socio-hydrological relations." As a result, rather than speaking of what a hydrosocial territory 'is' or 'should be', we argue for the need to understand and examine them as divergent material and discursive representations. In all their empirical and conceptual diversity, they give insight into the contested field of water control, water power structures and the socio-cultural and political legitimacy of particular angles of water knowledge. Territorialization, both materially and discursively, is a cultural and political-economic practice and phenomenon (see also Agnew, 1994; Baletti, 2012; Perramond, 2013; Saldías, Boelens, Wegerich, & Speelman, 2012).

For many common resource management collectives, the ideological and material construction of 'territory' has developed on the basis of the ecosystems and natural resources available within their territories, in dynamic interaction with supralocal forces and networks (Paasi, 2009; Sack, 1986). Fishermen, pastoralists and riparian societies, desert tribes and forest people are some illustrations. For water control collectives in the Andes, Boelens (2008) explains how they conceive of the intrinsic connection that binds local water control systems to particular localities and communities, in terms of agroecological properties, political and cultural embeddedness and historical processes of rights generation, transformation and defence. These intimately entwine water rights with processes of hydraulic and territorial construction and identity formation (Gelles, 2000, 2010; Hoogesteger, 2013; Lynch, 2012; Zwarteveen & Boelens, 2014).

Water control in the highlands, because of water's complex physical-ecological and adverse political-economic operating settings, forces users to cooperate intensively. It builds on mutual dependence. Key tasks in water organization, such as system operation and maintenance, resource mobilization, decision making, communication and conflict resolution, are enlaced with bonds of rights and obligations. They express and depend on both effective and affective relationships. As a result, within wider, multi-scale political economy settings and local-national-global processes of identification, water user families commonly identify deeply with their water sources, user organization and hydrosocial territory.

These human–water–territory linkages, though essential bonds, are not essentialistic or predefined; neither can they be romanticized. They tend to be the result of processes of internal struggle, negotiation and consensus. As 'territory-embedded' networks of socio-natural relationships, including strategic links with human/nonhuman elements of the broader political economy, they are also basic to the collective defence of water vis-à-vis third parties (e.g., landlords, mining corporations, neighbouring communities, and state agencies).

Historically and nowadays, such constructs of (predominantly 'control-localized') hydrosocial territory have been fiercely challenged by national and supranational elites and power groups, seeking to control or annex them to their own purview, in order to access and control their water and other natural resources. Beyond obtaining dominion over natural, material, financial and human resources, territorial control enables ruling groups to reinforce and reproduce the dominant social order (Foucault, Sellenart, & Burchell, 2007; Meehan, 2013; Sack, 1986), for instance in terms of water governance and water-power hierarchies. This implies, among others, the discursive and material definition and commensuration of hydro-territorial boundaries, of institutions, rights and responsibilities, and of the appropriate norms for acting, behaving and thinking in relation to water.

Here, the redefinition and alternative interlinking of scales is one strategic political instrument, as it re-establishes the territorial boundaries and strategies of governance and intervention, including or excluding actors from decision-making processes with respect to water resources management and control (Lebel et al., 2005; Norman, Bakker, & Cook, 2012; Saldías et al., 2012). Swyngedouw (2004, p. 26), in this respect, stated that "the emergence of new territorial scales of governance and the redefinition of existing scales (such as the nation-state) change the regulation and organisation of social, political and economic power relations". Through scalar politics, hydrosocial configurations and forms of territorial water governance are deterritorialized and reterritorialized. Water governance and conflict mediation used to be concentrated at the level of the national state. "This has been and still is an important scale for the regulation and negotiation of social, economic and cultural life and for the articulation of the processes of de-territorialisation/re-territorialisation. Yet, historical geographic analyses have illustrated how capitalism has always made existing forms of territorial organisation porous, unstable and prone to transgressions and transformations. The production of space, by perpetually reworking the networks of capital circulation and accumulation flows, discards existing spatial configurations and scales of governance, while new ones are produced" (p. 32).

National governments in Ecuador, Latin America and elsewhere, jointly with international developmental organizations, have aggressively worked towards rescaling their water governance structures in the past decades, upwards to transnational governance scales and simultaneously downwards to local governments operating in public–private partnerships. Commonly the aim was to open local hydrosocial territories to extraction and production by 'efficient, competitive entrepreneurs' operating within the free market–based world economy, for which it was central to install private, commoditized and transferable water rights (e.g. Bauer, 2004; Boelens & Seemann, 2014; Duarte-Abadía & Boelens, 2016). As Foucault (1991) argued when examining "government mentality" or the "rationality and strategies of governors over the conduct of subjects", rather than just law itself, rule makers increasingly developed a range of multiform government tactics which were based not just on legal force but particularly on the need to productively/economically *manage and direct* society. This "governmentality" transition, from a legal-bureaucratic regime dominated by structures of sovereignty to a water society governed by multiscale government techniques, did not replace nation-state rule making; rather, the State and its territorial spaces were "governmentalized". In other words, rather than imprinting the State structure on local territories, it entailed a process of what Foucault would call "governmentalization of the State" (p. 103), involving the effort to install new, neoliberal governance practices to manage local territories.

However, in the Andean countries, these new neoliberal governance techniques, installed through blunt neoliberal water reforms and property rights privatization programmes since the 1980s,[1] met with massive popular resistance (see e.g. Gelles, 2000; Mayer, 2002). For this reason, in the last decade, new – far more subtle – governance techniques were invented to continue and deepen the neoliberal governmentality project: these forms fostered the commoditization of water resources by installing different forms of market-environmentalism and so re-organizing local territories. Important examples of this neoliberal conservation governmentality can be found in many PES schemes, currently very popular among policy makers (e.g. Büscher et al., 2012; Castree, 2008; Duarte-Abadía & Boelens, 2016; Fairhead et al., 2012; McAfee & Shapiro, 2010). Mainstream PES governmentality projects seek to delegate the water-controlling tasks and techniques to private and microterritorial power structures, in alliance with international policy and development institutes. The depoliticized PES model fundamentally implicates (individual and collective) water users as individual agents and private holders of commoditized, transferable water rights – similar to commercial water entities (Boelens et al., 2014; Rodríguez-de-Francisco et al., 2013; cf. Bakker, 2005; Farley & Costanza, 2010).

Below, this depatterning of existing water territories is further scrutinized via "government through community" (Rose, 1999), whereby 'community' and its socio-natural institutions and properties are, simultaneously, profoundly transformed, through the workings of PES market-based alignment of water flows. PES as a water control tool – by means of regulating and controlling land and its uses[2] – seeks to guarantee the water flows that sustain capital production downstream. This article will illustrate how PES extends the rationality and the domain of dominant hydrosocial territoriality, to gain control of land and water conservation in the Andes highlands, and how this control is at odds with upstream communal hydrosocial territories.

Methodological notes

This article is based on ethnographical research with the communities involved in the two PES schemes implemented in the municipality of Pimampiro, Ecuador. Research methods included archival and literature research, participatory action-research workshops, focus group discussions, and semi-structured interviews with PES intermediaries (e.g. the municipality of Pimampiro), the service buyers (e.g. Water Utility of Pimampiro) and the communities and private owners in Mariano Acosta participating and not participating in the PES deal (e.g. Nueva América Association). In this article, the two cases have an illustrative role. Further contextual and methodological information can be found in Rodríguez-de-Francisco et al. (2013) and Rodríguez-de-Francisco and Boelens (2014).

An illustration: the repatterning of hydro-territoriality in Pimampiro's highlands

Payments for contested waters

The race for water between Pimampiro's urban centre versus the community of Mariano Acosta (the highlands of Pimampiro) started in the mid-nineteenth century, when the haciendas around Pimampiro lost power and became commercial agriculture areas (Preston, 1990). The principal managers and winners of this water race were the powerful mestizos of Pimampiro, who were connected to the municipality and who started the Irrigation Water Board. Since 1981, this race was made far easier for Pimampiro, after successful scalar politics and subsequent municipal reconfiguration promoted by the wealthy white-mestizos. Pimampiro's urban centre was officially recognized as a municipality and, despite heavy resistance by the people of Mariano Acosta, the latter's territory was annexed to the new municipality. Notwithstanding its vernacular continuation as a peasant community, jurisdictionally Mariano Acosta fell under the political control of Pimampiro (CESA, 1998).

By that time, Pimampiro had already accumulated most of Mariano Acosta's water sources to suit the irrigation and drinking water needs of people in the main urban centre, leaving Mariano Acosta with no irrigation rights at all. For many years Mariano Acosta peasants have been demanding water rights, among others, to access part of the water that originates in and flows from their own territory. But Pimampiro's need for water has not decreased; it has grown because of greater drinking water needs for the growing urban centre, but most of all, because of the commercial agricultural intensification downstream.

For these reasons, as the PES discourse proliferated in Andean countries during the 2000s, Pimampiro was one of the first villages to opt in. Following environmental studies, according to the municipality of Pimampiro (GMP, 2010), the forests and páramos (highland moors) along the Ecuadorian Andes are to be considered natural reserves of water, which could be made available through environmental services provision, favouring society as a whole. However, there is the constant threat of deforestation caused by settlers seeking to make a living from farming, livestock and logging (GMP, 2010).

Starting in 2000–2001, the municipality of Pimampiro organized its first PES scheme to secure downstream water provision, with the support of an NGO and international donors. In 2008, another PES scheme was installed. These two PES schemes are located around the drinking and irrigation water intakes upstream of the the Chamachán and Pisque watersheds (Figure 1). Pimampiro charges water consumers a 20% surcharge and gets additional external funds to give payments to landowners who, in exchange,

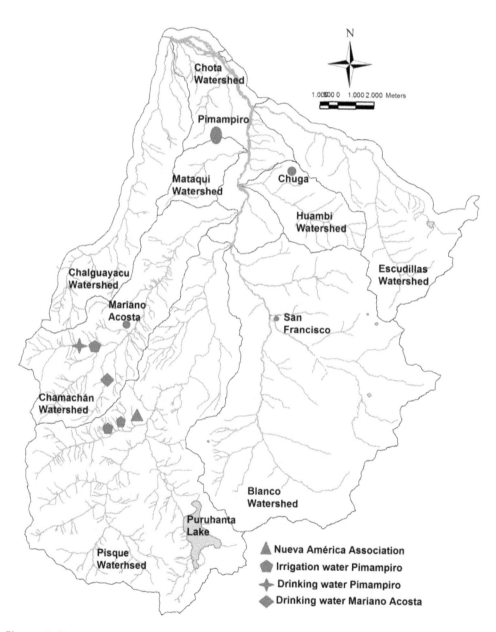

Figure 1. Pimampiro, Ecuador: watersheds and drinking and irrigation water intakes. Source: Rodríguez-de-Francisco et al. (2013).

must stop expanding the agricultural frontier into natural forest, páramos and land undergoing natural regeneration. According to the municipality of Pimampiro (GMP, 2010), low productivity and low agricultural prices are the main causes of deforestation pushing the agricultural frontier further into the highlands without attention to proper cultivation practices. By creating compensation for conservation, the municipality says it will engage and support poor upstream farmers to safeguard water provision and reduce the municipality's water stress (water quality and dry-season quantity).

Payments in both schemes are given according to the following classification: USD 6 per hectare per year for not cultivating "disturbed forest or páramo"; USD 8 per hectare per year for "mature secondary forest"; and USD 12 per hectare per year for "primary forest or páramo" (Guerrero, 2010; Wunder & Albán, 2008; see Rodríguez-de-Francisco et al., 2013, for an explanation of how these payments were negotiated).

The users of these watershed services are, mainly, the water utility of Pimampiro and the Pimampiro Irrigation Board. While drinking-water users are paying to conserve the upstream Nueva América forest in the Pisque watershed,[3] irrigation-water users are not paying any contribution so far. Therefore, the PES scheme in Chamachán is funded, among others, via the EU's Proderena programme, and contributions from the Province of Imbabura and Birdlife Ecuador (Guerrero, 2010).

What is crucial to understand is that, unlike PES theory and discursive rationality, the PES schemes in Chamachán and Pisque (as often happens in the Andean countries) were not at all based on voluntary decisions and transactions among up- and down-streamers. In Rodríguez-de-Francisco et al. (2013) and Rodríguez-de-Francisco and Boelens (2014), we showed that both schemes were profoundly permeated by forceful power plays and manipulation, and the use of both coercive modes of power and Foucauldian disciplinary power (see also Rodríguez-de-Francisco & Boelens, 2015). This made many upstream PES participants feel that they were forced to accept the reconfiguration of their hydrosocial territories and to align with this new socio-nature.

Peasant-indigenous versus PES hydrosocial territoriality: divergent schemes of representation

In the recent past, as mentioned above, power groups in Pimampiro and elsewhere considered the highlands areas unproductive, inhospitable lands, where the only important resource was timber. This was due to harsh climatic and topographic conditions that limited agricultural activity. In this context, communal organizations have always played an important role in building institutions of reciprocal support to guarantee the subsistence of their mutually dependent members, in engaging in collective action, and in mediating the internal and external power asymmetries related to natural resource access and control. Mariano Acosta is a territory that resembles the tenacity of peasant indigenous groups in their struggle for livelihoods and autonomy. In the early nineteenth century they arrived here in search of land that would enable them to escape from their semi-serf status in the neighbouring haciendas. But this territory was not granted to them. They built a 'territory of their own' through a long physical and legal struggle, subsequently defending it from ever-increasing threats. In the most literal sense, their territory was founded upon processes of political-economic and cultural creation and imagination, generating material subsistence and meaning in the context of unequal power structures (Roseberry, 1989, p. 14). Mariano Acosta's community territory entwined site-specific means of livelihood and survival, through territory-rooted worldviews, strategies and properties. In this respect, De la Cadena (1989, pp. 77–78) stated: "Since the development of capitalism in Andean peasant economies has not managed to secularize all facets of reproduction, the community is an institution interweaving diverse aspects: technological rituals, magical and administrative authorities, and commercial ceremonies. Economics, politics and rituals 'braid' together in peasant reproduction and are manifested in their institutions.... Community borders are not only territorial but also social and political, the boundaries within which rights and obligations are exercised under the community organization's sanction."

As we have argued, such peasant-indigenous territorialization through reciprocal relations is far from romantic, while at the same time, noncommoditized production and reproduction serves as a fundamental livelihood strategy for collective subsistence, in particular for the poorest groups without sufficient financial means. Although Mariano Acosta has lost its collective access to irrigation water – a battle that is not over yet, although official powers have joined the wealthier Pimampiro irrigators – access to drinking water is organized collectively, a fiercely defended bastion. And while land in Mariano Acosta's territory is largely owned individually these days, collective labour parties are fundamental, as are numerous mutual-support exchanges among indigenous and peasant community members. For example, in Mariano Acosta it is common to find *mingas* (noncommoditized collective work parties for the benefit of community livelihood), *maquimañachi* ('hand-lending', work-for-work exchanges among community members), 'grass exchanges' (pasture exchanges for cattle feeding), etc. These forms of collaboration interlink cultural and economic livelihood reproduction with the (re)creation of self-identity and territorial belonging.

Such collaborative strategies are also complemented by market interaction. An Andean household optimizes its intervention in the societal process of production by both generating income through the market and obtaining other nonmercantile support (Golte & de la Cadena, 1986). Nevertheless, while these two spheres are mutually interdependent and "subsidize each other", they can also deplete each other (Mayer, 2002). As we have argued and shown for many localities in the Andean highlands, recent scale politics and intensified pursuit of commoditization have rapidly changed and sometimes disrupted existing boundaries or balances between community and market spheres. In Boelens et al. (2014, p. 9) we state that "in spite of capitalist market penetration and ongoing accumulation, dispossession, and theft, non-mercantile exchange and interactions in the Andean communities have resisted – and will resist in the future – displacement by predominantly commoditized relationships. This becomes particularly clear in the field of community water control." Important reasons are that neither peasant families and communities nor their water control systems can sustain their livelihoods amidst exclusively mercantile relationships; and that many exchanges, activities, relations and ways of thinking in Andean livelihood strategies and irrigation systems simply cannot be reduced to exchange values and market-economy issues.

As we argued elsewhere: "Peasant economy is part of, subordinated to, and simultaneously a bastion against commoditized market exploitation. The awareness that 'community' and 'socio-territory' form a central axis for the defence and effective use of a community's productive resources, both collective and individual, is a powerful mainstay of peasant livelihoods. Households perceive that non-commodity relationships ensure long-term reproduction and offer a protective framework against the vicious circles of poverty, debt, and exploitation" (Boelens et al., 2014, p. 9).

PES narrative largely contrasts with this peasant-indigenous territoriality. It seeks to highlight the link between upstream landholders and downstream water users, promoting an apparently inclusive, neutral approach. As in the words of Pimampiro officials, it claims that, since everyone benefits, the challenge of securing water provision and quality is the responsibility of all: downstream water users by paying, and upstream land users by recognizing the value of nature and accepting compensated restrictions on their land use.

In this sense, PES is based on the idea that to save nature it is necessary to sell it (McAfee & Shapiro, 2010). As reflected in statements by the implementing agencies in Pimampiro, a key issue in the PES discourse is the idea of inclusive sustainable development: these areas have historically been neglected by state support, so paying them for conserving hydrosocial territory is a way to get them to stop using their natural resources inappropriately and a way to lift them out of poverty. They argue that the payment culture justifies the land use restrictions that highland farmers face as the result of conservation, and will guarantee that communities can finally become integrated, rightful participants in the capitalist market (Rodríguez-de-Francisco et al., 2013). PES territoriality makes it possible to transform the natural resource (water) into a form of commoditized natural capital. Its foundational logic, significantly, stems from new institutional economics, in which individual and collective organizational practices, working rules, and the alignment of hydrosocial, biophysical-political components and interactions follow from rational choices made by individual agents, each reasoning in line with their (mostly material) self-interest. Fundamentally, the form and quality of collective action in hydrosocial territories is seen as the outcome of multiple economic cost–benefit analyses and is triggered by positive or perverse incentives.

As the Pimampiro cases witness, PES-driven hydro-territorial construction and commodification of nature involve, as explained by Gómez-Baggethun and Ruiz-Pérez (2011), first, economically framing water/nature as a commodity; second, assigning an economic value to water/nature by economic valuation; third, appropriating water/nature by formalizing mostly private property rights to specific 'ecosystem services' and to the land 'producing these services'; and fourth, commercializing these ecosystem services by creating a territorial-institutional setting that defines groups of buyers and sellers (in terms of interlinked rules, assets, authorities, hydrosocial relationships, identities, boundaries, etc.). This commodification process facilitates appropriation of land and water resources for hydro/environmental ends and to fortify the PES territory. Sullivan (2009) also described this as a modern form of "enclosing the commons".

Tensions between PES and Andean territories

"Pimampiro has already stolen our water and now, with laws and miserly payments for conservation, they want to control the land and forest that we have always simply used." This quote from a peasant from Mariano Acosta summarizes the opinion of many inhabitants of this area regarding the way PES has been designed and implemented in the area. The miniscule payment they receive, the stricter land use limitations and stricter enforcement by community forest rangers, and the fact that they "have to just watch water flow down to Pimampiro" without being able to use it for agricultural purposes, explain the disappointment expressed by interviewees.

Communities in Mariano Acosta also criticize the fact that they are now being labelled as nature destroyers: "The people from Pimampiro do not seem to understand that it was our great-grandparents, parents and we ourselves who have conserved the forest that it is now standing." Rather than seeing rural landscapes as humanly coproduced environments, PES implementing and managing organizations from Pimampiro see rural landscapes as humanly degraded lands needing intervention, and their inhabitants as people who need to be taught how to deal with the environment.

Moreover, as shown in Rodríguez-de-Francisco et al. (2013), PES implementation seems to disregard traditional conservation practices. The case of fallow land, located just adjacent to and in between the PES territory, is illustrative. Leaving land fallow has been a traditional strategy to 'renew' land productivity, while re-creating biodiversity habitats and other environmental services. However, land undergoing processes of natural regeneration is considered 'conservation land' (with severe restrictions or zoned for nonusage under the forestry law, which is more strictly enforced since PES introduction), so peasants are paradoxically inclined to work the land more intensively, to prevent this conservation classification.

Families living in PES target areas expressed resentment that the municipality is increasingly dictating and controlling land management, in the interests of conservation and with little regard for their livelihoods, traditional practices and territorial identities (Rodríguez-de-Francisco et al., 2013). Peasants find pride in being farmers and want to continue being so, despite the fact that the PES officials in local government tell them that they need to rearrange their ways of living and become conservation renters or tour guides: "We are peasants and we don't want to live from the rents of conservation" (Nueva América peasant landholder, personal communication, 5 October 2010).

The poorest peasants with land only in PES target areas suddenly lose their livelihood opportunities (since nobody possessing only one hectare can live on just 6–12 USD per year as 'compensation' for nonusage); in fact they experience 'community enclosure'. Meanwhile, PES may provide opportunities for the better-off upstream community members. The latter have other, nonagricultural income sources or have sufficient land outside PES intervention areas. As in other communities, this challenges peasant organizations, creating divisions, challenging their legitimacy or breaking down community institutions.

In addition, the PES designers (in accordance with private utility-maximization rationality) decided to give payment only to those who have individual land titles. This made some community members, as in the case of Chamachán, decide to opt out of communal land by privatizing portions of it (Rodríguez-de-Francisco & Boelens, 2014). What may be even more shocking is the municipal administration's attempt, without even informing or consulting the people affected beforehand, to transform the territory of a peasant association into a municipal natural reserve, since it had "unclear land tenure". A municipality official defended this radical alignment decision by stating, "we don't want to reward illegality" (Rodríguez-de-Francisco & Boelens, 2014).

Discussion: repatterning territories and reconstituting subjects

Conservation policies and strategies imply control over nature–society relations and have transformative effects on socio-natural landscapes (Himley, 2009; Robbins, 2004). As in the case of PES, this deeply involves political choices; conservation authorizes access to and control over resources for some while denying it to others. In the PES schemes we have examined in Pimampiro (which, by the way, are presented as success stories in international scientific forums and journals – Rodríguez-de-Francisco et al., 2013), ill-compensated land use restrictions are imposed on upstream farmers, while the 'water monopoly' of downstream water users is further enhanced. Through PES, the marginalized groups who first transformed the 'forgotten lands' into living territories now come under the spotlight due to outsiders' combined water-extractive and environmental interests, and become annexed to the dominant hydrosocial territories of powerful local/international alliances.

The annexation of Mariano Acosta to the municipality of Pimampiro illustrates the politics of scale and expresses how the apparently neutral character of municipality scale is strategically used by Pimampiro authorities to gain jurisdictional control over Mariano Acosta and its water sources. Likewise, the apparently neutral scale of 'watershed', defined as in need of protection and able to deliver commoditized 'environmental services', serves to support a watershed approach that seeks upstream conservation, disregarding the livelihood necessities of the upstream communities. It imposes land use restrictions that for many poor families in the headwater areas of Mariano Acosta are livelihood-threatening and which for most are poorly compensated, to say the least.

Furthermore, this reconfiguration of the hydrosocial territorial scales and relationships provides insight into how new actor groups are created that resemble a market-exchange setting (buyers and sellers), how nature is priced and depicted as a commodity, how the introduction of monetary payments becomes the driver of social and material interaction, and how land that was previously commonly managed (not a commodity) now becomes privatized. These changes affect the very fabric of Andean reciprocity and existing forms of hydro-territoriality.

PES hydro-territoriality is based on the assumption that political decisions are the product of individual agents' interactions, each rationally pursuing individual material self-interest. Despite its integrated discourse and the fact that its governmentality project is to be analyzed as 'government through community', the intention is to deconstruct and re-align the notion of community and territory. Mayer (2002, p. 5) puts forward a rightful critique of its new institutional narrowness which (besides neglecting the actual existence of unequal power relationships as in Pimampiro) bypasses both the community and the household as crucial actors in Andean waterscapes "in order to focus directly on the individuals within it – individuals who make cost-benefit choices, within the context of the household's means and needs, between rewards and punishments and between investments and payoffs. The household in this model is a miniature marketplace where rational actors trade in everything – food, affection, authority, leisure, pleasure – and compete with each other." The community and the territory are seen as the aggregate outcomes of the choices and behaviours of individuals.

Directly related to this, in the political domain, PES hydro-territoriality pays little or no attention to social differentiation, while political processes are seen as the conflict and sum of individual rational decisions (Rodríguez-de-Francisco & Boelens, 2014). Next, cultural, religious, metaphysical and psychological factors influencing water control tend to be denied, or explained in economic terms. The day-to-day contestation among multiple interest groups with unequal power, in a context-specific process of negotiation, conflict and collaboration, is neglected. This makes it possible to come up with universal guidelines for establishing effective and efficient hydrosocial configurations, presented as neutral and natural, but in fact corresponding to profound political economic interests.

Given the fact that subjects are co-produced in the process of 'making territory', PES actively produces new subjects as new hydrosocial territories are being constituted (Baletti, 2012; Foucault, 1991). These new human subjects, as the cases exemplify, would have market-aligned roles and identities, constituting just rationally interacting individuals and homogeneous social groups with singular functional roles. Instead, communities, households and individuals – rooted in a shared past, present and future – are linked through, embedded in, and moved by webs of power, symbols and meaning, only parts of which are economic. For these reasons, many of them refuse to accept their reconstitution by PES and market-environmentalism.

Conclusion

As shown in this article, depoliticized PES implementation produces new hydrosocial territories and territorial conceptualization, in which both the nonhuman and the human – the natural, the technical and the social – are simultaneously and jointly transformed and aligned in a different way. Concretely, a new, socially disembedded view of water resources, a private appropriation and exchange value–driven property regime, a market-functionalist perspective of 'natural but human-spoiled' territory, and a narrow conception of human agency and rational choice–based decision making are pushed forward through strict market conservation.

Furthermore, we have discussed how labelling PES as 'conservation for society' obscures the fact that upstream landowners are further restrained in favour of water provision for capital accumulation downstream. PES aims to secure environmental service provision and enlarge water quality and quantity flows 'for the benefit of all' but forgets to examine how its socio-natural transformations and newly constituted hydrosocial territories may enlarge and deepen power asymmetries, poverty and inequality. As the PES schemes that are implemented in Pimampiro show, confining the notion of water control to the economic and biophysical domains results in developing hydrosocial territories as anti-politics schemes, that misrecognize and extend the skewed distribution of rights over these 'environmental services' and that disregard the intricate regionally and historically rooted cultural ecologies of water.

PES, as a depoliticized natural resource management intervention for water control, seeks to embed upstream territories under the rationale and institutions of the hydrosocial territory of those who have the economic and political power to pay for water. In Pimampiro's highlands, as in many other places, this means that territories based on reciprocal subsistence strategies are pressured by rules, surveillance, fines and sometimes violence to become functional environmental infrastructure for the capitalist productive infrastructure downstream. This restructuring and repatterning of territory in the name of conservation hides the socio-cultural disciplining and further impoverishment of many communities and especially the poorest households. Dominant PES hydro-territorial representation, with the highlands as extractive sources for the water commodities they produce, clearly conflicts with hydrosocial conceptions that see territory, ideologically and politically, as history-based homes of cultural belonging and socio-productive spaces for recreating livelihoods.

The article also makes manifest how expanding PES-based dominant hydrosocial territory to the highlands goes beyond transforming water control itself: it envisions and strategizes to bring order, that is, localized universalistic market(-like) order, to the highlands and to normalize peasant livelihoods by repatterning them as custodians of nature – the commoditized nature that 'environmental service buyers' want to see in these areas. In other words, PES hydrosocial territoriality forcefully aims to redefine, at once, water rights, water management, systems of locally established prestige and authority, and user–community–state relationships, as well as the rooted patterning of water subsistence rationality. The 'government through community' governmentality project of PES goes far beyond earlier notions of decentralization, privatization and subsidiarity, even beyond 'government through community', since it requires profound deconstruction, resignification and repatterning of the existing, practised and lived notion of 'community', including the deterritorialization/reterritorialization of local hydrosocial networks. This governmentalization of newly constructed space, thereby, also erases and reconstitutes the very conceptualization of 'water' (its values, meaning, substance, flows and relations) and of the subject 'water user/water caretaker'. As growing resistance in the Andean highlands shows, however, the latter do not passively accept this radical reconfiguration of their waters, territories, cultures and beings but struggle to defend their territories and reshape transformation efforts.

Notes

1. Neoliberal water governance techniques that were based on the Friedman/Hayek 'Chicago Boys' ideology, implemented through Pinochet's 1981 Chilean Water Code, and subsequently imposed on Latin American countries by the World Bank as 'the model' throughout the 1990s (see e.g. Bauer, 1998, 2004; Budds, 2010; de Vos, Boelens, & Bustamante, 2006).
2. Land control includes, among others, land use restrictions, changes in land tenure forms (privatization of commons, or vice versa), reinforcement of legal control, force and violence (or the threat of them), and eviction (Peluso & Lund, 2011).
3. See Echavarría, Vogel, Albán, and Meneses (2004), Rodríguez-de-Francisco et al. (2013) and Wunder and Albán (2008) for more information on the Nueva América PES scheme in Pimampiro.

References

Agnew, J. (1994). The territorial trap: The geographical assumptions of international relations theory. *Review of International Political Economy, 1*(1), 53–80. doi:10.1080/09692299408434268

Arguedas, J. M. (1980). *Todas las Sangres. Volumen I y II*. Lima: Editorial Milla Batres.

Bakker, K. (2005). An uncooperative commodity: Privatizing water in England and Wales. *Annals of the Association of American Geographers, 95*(3), 542–565. doi:10.1111/j.1467-8306.2005.00474.x

Baletti, B. (2012). Ordenamento Territorial: Neo-developmentalism and the struggle for territory in the lower Brazilian Amazon. *Journal of Peasant Studies, 39*(2), 573–598. doi:10.1080/03066150.2012.664139

Bauer, C. (1998). Slippery property rights: Multiple water uses and the neoliberal model in Chile, 1981–1995. *Natural Resources Journal, 38*(1), 109–155.

Bauer, C. (2004). *Siren song: Chilean water law as a model for international reform.* Washington, DC: RFF Press.

Boelens, R. (2008). Water rights arenas in the Andes: Upscaling networks to strengthen local water control. *Water Alternatives, 1*(1), 48–65.

Boelens, R., Hoogesteger, J., & Baud, M. (2015). Water reform governmentality in Ecuador: Neoliberalism, centralization, and the restraining of polycentric authority and community rule-making. *Geoforum, 64*, 281–291. doi:10.1016/j.geoforum.2013.07.005

Boelens, R., Hoogesteger, J., & Rodríguez-de-Francisco, J. C. (2014). Commoditizing water territories: The clash between Andean water rights cultures and Payment for Environmental Services policies. *Capitalism Nature Socialism, 25*(3), 84–102. doi:10.1080/10455752.2013.876867

Boelens, R., Hoogesteger, J., Swyngedouw, E., Vos, J., & Wester, F. (2016). Hydrosocial territories: A political ecology perspective. *Water International, 41*(1), 1–14. doi:10.1080/02508060.2016.1134898

Boelens, R., & Seemann, M. (2014). Forced engagements: Water security and local rights formalization in Yanque, Colca Valley, Peru. *Human Organization, 73*(1), 1–12. doi:10.17730/humo.73.1.d44776822845k515

Budds, J. (2010). Water rights, mining and indigenous groups in Chile's Atacama. In R. Boelens, D. Getches, & A. Guevara (Eds.), *Out of the mainstream* (pp. 197–212). London: Earthscan.

Büscher, B., Sullivan, S., Neves, K., Igoe, J., & Brockington, D. (2012). Towards a synthesized critique of neoliberal biodiversity conservation. *Capitalism Nature Socialism, 23*(2), 4–30. doi:10.1080/10455752.2012.674149

Castree, N. (2008). Neoliberalising nature: The logics of deregulation and reregulation. *Environment and Planning A, 40*(1), 131–152. doi:10.1068/a3999

CESA. (1998). *Diagnóstico integral de Pimampiro.* Quito: Central Ecuatoriana de Servicios Agrícolas.

de la Cadena, M. (1989). Cooperación y Conflicto. In E. Mayer & M. de la Cadena (Eds.), *Cooperación y conflicto en la comunidad Andina: Zonas de producción y organización social* (pp. 77–112). Lima: IEP.

de Vos, H., Boelens, R., & Bustamante, R. (2006). Formal law and local water control in the Andean Region: A fiercely contested field. *International Journal of Water Resources Development, 22*(1), 37–48. doi:10.1080/07900620500405049

Delaney, D., & Leitner, H. (1997). The political construction of scale. *Political Geography, 16*(2), 93–97. doi:10.1016/S0962-6298(96)00045-5

Duarte-Abadía, B. & Boelens, R. (2016). Disputes over territorial boundaries and diverging valuation languages: The Santurban hydrosocial highlands territory in Colombia. *Water International, 41*(1), 15–36. doi:10.1080/02508060.2016.1117271

Duarte-Abadia, B., Boelens, R., & Roa-Avendaño, T. (2015). Hydropower, encroachment and the re-patterning of hydrosocial territory: The case of Hidrosogamoso in Colombia. *Human Organization, 74*(3), 243–254. doi:10.17730/0018-7259-74.3.243

Echavarría, M., Vogel, J., Albán, M., & Meneses, F. (2004). *The impacts of payments for watershed services in Ecuador.* London: IIED.

Elden, S. (2010). Land, terrain, territory. *Progress in Human Geography, 34*(6), 799–817. doi:10.1177/0309132510362603

Fairhead, J., Leach, M., & Scoones, I. (2012). Green Grabbing: A new appropriation of nature? *Journal of Peasant Studies, 39*(2), 237–261. doi:10.1080/03066150.2012.671770

Farley, J., & Costanza, R. (2010). Payments for ecosystem services: From local to global. *Ecological Economics, 69*(11), 2060–2068. doi:10.1016/j.ecolecon.2010.06.010

Foucault, M. (1991). Governmentality. In G. Burchell, C. Gordon, & P. Miller (Eds.), *The Foucault effect: Studies in governmentality.* Chicago, IL: University of Chicago Press.

Foucault, M., Sellenart, M., & Burchell, G. (2007). *Security, territory, population: Lectures at the Collège de France 1977–1978*. New York, NY: Palgrave Macmillan.

Gelles, P. (2000). *Water and power in Highland Peru: The cultural politics of irrigation and development*. New Brunswick, NJ: Rutgers University Press.

Gelles, P. (2010). Cultural identity and indigenous water rights in the Andean highlands. In R. Boelens, D. Getches, & A. Guevara (Eds.), *Out of the mainstream. Water rights, politics and identity* (pp. 119–144). London: Earthscan.

GMP. (2010). *Plan de desarrollo y ordenamiento territorial 2011-2031 del cantón San Pedro de Pimampiro*. Pimampiro: Gobierno Municipal de Pimampiro.

Golte, J., & de la Cadena, M. (1986). La codeterminación de la organización social Andina. *Allpanchis*, 22(19), 7–34.

Gómez-Baggethun, E., & Ruiz-Pérez, M. (2011). Economic valuation and the commodification of ecosystem services. *Progress in Physical Geography*, 35(5), 613–628. doi:10.1177/0309133311421708

Guerrero, A. (2010). *La experiencia del proyecto GPI-PRODERENA y los beneficiarios de la microcuenca del Río Chamachán, en la parroquia Mariano Acosta, cantón Pimampiro*. Ibarra: GPI, Proderena and GMP.

Himley, M. (2009). Nature conservation, rural livelihoods, and territorial control in Andean Ecuador. *Geoforum*, 40(5), 832–842. doi:10.1016/j.geoforum.2009.06.001

Hoogesteger, J. (2013). Trans-forming social capital around water: Water user organizations, water rights, and nongovernmental organizations in Cangahua, the Ecuadorian Andes. *Society & Natural Resources: An International Journal*, 26(1), 60–74. doi:10.1080/08941920.2012.689933

Houdret, A., Dombrowsky, I., & Horlemann, L. (2014). The institutionalization of River Basin Management as politics of scale – Insights from Mongolia. *Journal of Hydrology*, 519(C), 2392–2404. doi:10.1016/j.jhydrol.2013.11.037

Kosoy, N., & Corbera, E. (2010). Payments for ecosystem services as commodity fetishism. *Ecological Economics*, 69(6), 1228–1236. doi:10.1016/j.ecolecon.2009.11.002

Lebel, L., Garden, P., & Imamura, M. (2005). The politics of scale, position, and place in the governance of water resources in the Mekong Region. *Ecology and Society*, 10(2), 18.

Lynch, B. D. (2012). Vulnerabilities, competition and rights in a context of climate change toward equitable water governance in Peru's Rio Santa Valley. *Global Environmental Change*, 22(2), 364–373. doi:10.1016/j.gloenvcha.2012.02.002

Mayer, E. (2002). *The articulated peasant: Household economies in the Andes*. Boulder, CO: Westview Press.

McAfee, K., & Shapiro, E. (2010). Payments for ecosystem services in Mexico: Nature, neoliberalism, social movements, and the state. *Annals of the Association of American Geographers*, 100(3), 579–599. doi:10.1080/00045601003794833

Meehan, K. (2013). Disciplining de facto development: Water theft and hydrosocial order in Tijuana. *Environment and Planning D*, 31(2), 319–336. doi:10.1068/d20610

Mosse, D. (2008). Epilogue: The cultural politics of water – A comparative perspective. *Journal of Southern African Studies*, 34(4), 939–948. doi:10.1080/03057070802456847

Norman, E. S., Bakker, K., & Cook, C. (2012). Introduction to the themed section: Water governance and the politics of scale. *Water Alternatives*, 5, 52–61.

Paasi, A. (2009). Territory. In J. Agnew, K. Mitchell, & G. Toal (Eds.), *A companion to political geography* (pp. 109–122). Oxford: Blackwell.

Peluso, N. L., & Lund, C. (2011). New frontiers of land control: Introduction. *Journal of Peasant Studies*, 38(4), 667–681. doi:10.1080/03066150.2011.607692

Perramond, E. P. (2013). Water governance in New Mexico: Adjudication, law, and geography. *Geoforum*, 45, 83–93. doi:10.1016/j.geoforum.2012.10.004

Preston, D. (1990). From hacienda to family farm: Changes in environment and society in Pimampiro, Ecuador. *The Geographical Journal*, 156(1), 31–38. doi:10.2307/635433

Rincón-Gamba, J. (2013). Territorios, culturas y jerarquización socioespacial en la migración contemporánea. *Cuadernos de Geografía*, 22(1), 81–92. doi:10.15446/rcdg

Robbins, P. (2004). *Political ecology: A critical introduction*. Malden, MA: Blackwell publishing.

Rodríguez-de-Francisco, J. C., & Boelens, R. (2014). Payment for environmental services and power in the Chamachán watershed, Ecuador. *Human Organization, 73*(4), 351–362. doi:10.17730/humo.73.4.b680w75u27527061

Rodríguez-de-Francisco, J. C., & Boelens, R. (2015). Payment for Environmental Services: Mobilising an epistemic community to construct dominant policy. *Environmental Politics, 24*(3), 481–500. doi:10.1080/09644016.2015.1014658

Rodríguez-de-Francisco, J. C., Budds, J., & Boelens, R. (2013). Payment for environmental services and unequal resource control in Pimampiro, Ecuador. *Society & Natural Resources, 26*(10), 1217–1233. doi:10.1080/08941920.2013.825037

Rose, N. (1999). *Powers of freedom: Reframing political thought.* Cambridge: Cambridge University Press.

Roseberry, W. (1989). *Anthropologies and histories: Essays in culture, history and political economy.* New Brunswick, NJ: Rutgers University Press.

Sack, R. D. (1986). *Human territoriality: Its theory and history.* Cambridge: Cambridge University Press.

Saldías, C., Boelens, R., Wegerich, K., & Speelman, S. (2012). Losing the watershed focus: A look at complex community-managed irrigation systems in Bolivia. *Water International, 37*(7), 744–759. doi:10.1080/02508060.2012.733675

Sullivan, S. (2009). An ecosystem at your service? *The Land, Winter 2008/2009,* 21–23.

Swyngedouw, E. (2004). Globalisation or 'glocalisation'? Networks, territories and rescaling. *Cambridge Review of International Affairs, 17*(1), 25–48. doi:10.1080/0955757042000203632

Wunder, S., & Albán, M. (2008). Decentralized payments for environmental services: The cases of Pimampiro and PROFAFOR in Ecuador. *Ecological Economics, 65*(4), 685–698. doi:10.1016/j.ecolecon.2007.11.004

Zwarteveen, M., & Boelens, R. (2014). Defining, researching and struggling for water justice: Some conceptual building blocks for research and action. *Water International, 39*(2), 143–158. doi:10.1080/02508060.2014.891168

Losing the watershed focus: a look at complex community-managed irrigation systems in Bolivia

Cecilia Saldías[a], Rutgerd Boelens[b], Kai Wegerich[c] and Stijn Speelman[d]

[a]Department of Agricultural Economics, Ghent University, Belgium; [b]Environmental Sciences Group, Wageningen University, the Netherlands; [c]International Water Management Institute, Uzbekistan; [d]Department of Agricultural Economics, Ghent University, Belgium

Water policies tend to misrecognize the complexity of community-managed irrigation systems. This paper focuses on water allocation practices in peasant communities of the Bolivian interandean valleys. These communities manage complex irrigation systems, and tap water from several surface sources, many of them located outside the watershed boundaries, resulting in complex hydro-social networks. Historical claims, organizational capacity, resources availability, and geographical position and infrastructure are identified as the main factors influencing current water allocation. Examining the historical background and context-based conceptualizations of space, place and water system development are crucial to understanding local management practices and to improving water policies.

Introduction

Irrigation systems in the Andean valleys and highland zones are essentially community managed and often overlapping; members of a community may belong to different systems concurrently (Boelens 2009). Moreover, in Bolivia, communities in the interandean valleys often rely on water transfers from neighbouring watersheds and multiple sources as a strategy to cope with water scarcity (see e.g. Gupta and van der Zaag 2008). These sources each have their own organization for management (Gutierrez 2005).

In this complex Andean context, the relevance of the integrated water resources management (IWRM) framework, which proposes river basins as the key unit for water management, is highly questionable. Yet Bolivia's Plan Nacional de Cuencas (PNC) establishes the watershed as the planning unit for water management (Centro AGUA 2012). By unravelling the water management of peasant communities of Abanico Punata in Bolivia, this paper demonstrates the failure to capture the complexity on the ground, where water management and governance cannot be circumscribed to physical boundaries, using watersheds. It also shows how the new irrigation law seems to fail in capturing the existing complexity.

The analysis is framed in the following section. The next provides background information on the study area. Then, the systems and the current water allocation practices among the communities are described in terms of water *quantity*[1] using data extracted from the Pucara Database.[2] Next, the origin of the systems is examined to understand water access, water rights and the functioning of the user organizations. The factors influencing water access are revealed – i.e. historical claims, organizational capacity, resource availability, geographical position and infrastructure (cf. Perreault 2008). This part is based on qualitative data concerning the process of water allocation gathered through in-depth interviews with former technical staff of the irrigation development project and community leaders,[3] and on secondary data. The focus on historical development enables comprehension of current water management complexity and insight into the problems and frictions that might be caused by the new irrigation law. This understanding is essential to moving towards policies that better fit local realities. This analysis is then embedded in the context of legal and customary rights. Finally, discussion and conclusions are presented.

Challenging the IWRM approach

The IWRM paradigm suggests that to achieve effective water management, the most appropriate planning unit is the watershed (Newson 1997, Grigg 2008). Under this framework, water users need to be organized mainly in accordance with the hydrological boundaries of the water supply infrastructure. While this approach might be useful within engineer-developed irrigation systems, the management of traditional or community-managed irrigation systems is often not that straightforward (Wegerich 2010). Next to the physical unit, there also exists a social unit, formed by groups of interrelated people sharing water sources, infrastructure, agro-productive livelihood systems and water management institutions, forming "hydro-social networks" (see e.g. Wester 2008). Often, hydrological boundaries do not match with administrative boundaries or with community territories (Boelens 2009) – or with the networks of power influencing water management (Molle 2009).

The appropriateness of the IWRM framework for such complex systems is questionable. Moss (2004) refers to problems of "spatial fit" related to water management around natural boundaries. Shah *et al.* (2000) see it as a problem of "contextual fit". In fact, water management at the watershed level seems to be a political choice rather than a natural choice (Warner *et al.* 2008, Molle *et al.* 2010). This paper, rather than applying the IWRM framework, challenges the approach with respect to its adequacy in capturing complex systems that dynamically include the overlapping of water systems, water transfers, multiple sources, several management organizations, etc. Here, shifting the perspective from natural to socio-natural watershed boundaries, compound water analysis and control fields (Swyngedouw 1997, Moss 2004, Wester 2008) might be a more appropriate approach.

Beccar *et al.* (2002, p. 2) state that irrigation systems constitute

a complex set-up to control water, combining and interrelating *physical* elements (water sources and flows, the places where it is applied and the hydraulic infrastructure to catch, conduct and distribute it), *normative* elements (rules, rights, and obligations related with access to water and other necessary resources), *organizational* elements (human organization to govern, operate, and sustain the system) and *agro-productive* elements (soils, crops, technology, capital, labour force and the capacities and knowledge of the art of irrigation).

Moreover, for the Andean region, Boelens (2009, p. 311) argues that for "most communities, the conceptualization of a water use system is entirely different from those in conventional irrigation or water management manuals". He concurs with Hendriks (2006, p. 84), who argues that according to local perceptions a water use system, far from being a mere physical hydraulic system with a functional organization, is considered "a system of rights, of obligations, and of (cultural) management regarding one or more water sources, shared among a given pool of users".

This complexity reflects the socio-technical nature of irrigation and water use systems, where natural, technical and social aspects interact and influence each other (Mollinga 1997, Swyngedouw 1997). In this respect, Uphoff (1986) distinguishes three types of activities related to irrigation in his "matrix of irrigation activities": (1) water use activities; (2) control structure activities; and (3) organizational activities. The matrix provides a tool for analyzing the linkages among activities. A shortcoming of the matrix is that it looks at irrigation activities but does not address the way in which stakeholders perform them – the implicit suggestion is that the activities are performed by one irrigation community or organization responsible for all the tasks. The present case study highlights that one would have to consider multiple matrixes next to each other, representing the different communities that make joint decisions about water delivery infrastructure, but at the same time use different water sources.

Often, the complexity regarding water management in community-managed irrigation systems in Andean countries is misunderstood or ignored by legislators, policy makers and intervention projects (Boelens and Zwarteveen 2005, Boelens 2010). Thus, to establish policies that fit the local context and users' realities, it is fundamental to consider the history and contextual rationale regarding water management.

Study area

Physical context

Pucara watershed belongs to the area known as Valle Alto situated in the interandean valleys of the department of Cochabamba. Its area is estimated at 440 km². It is subdivided into five smaller watersheds (Figure 1). Pucara watershed is characterized by having four ecological zones according to the altitudinal variation from 2800 to 4600 metres above sea level (Cruz 2009). Several natural lakes have been converted into reservoirs and water is transferred to irrigate downstream communities. The region is classified as semi-arid, with greater water availability in the highlands (Gerbrandy and Hoogendam 2002). The main river is the Pucara River (Figure 2). It gives origin to the so-called *Abanico Punata* (the Punata Fan) when it splits into several branches after entering the valley. It provides water for irrigation during the rainy season and is the main source of aquifer recharge. The area is suitable for agriculture thanks to good soils and suitable temperature conditions. The irrigable area is estimated at 4500 ha, distributed over 80 peasant communities (del Callejo and Vasquez 2007).

Socio-political and economic context

Peasant communities in the Valle Alto have a socio-territorial organization. Organized around peasant unions, they follow the structure of labour unions, adding traditional elements of reciprocity and community life. The institutional structure is not age-old; peasant unions originated in the 1952 National Revolution. Subsequently, they supported the Agrarian Reform of 1953 and became the community representatives (Gandarillas *et al.* 1994). These organizations installed the compulsory affiliation of community members, including responsibilities, obligations and benefits.

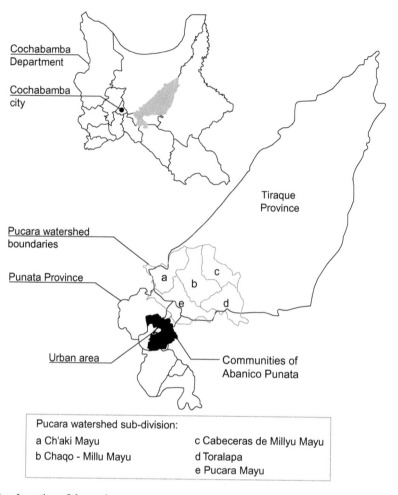

Figure 1.   Location of the study area.

Source: Saldías (2009).

Historically, the Valle Alto – including Punata province – has provided food (mainly grains) to the city of Cochabamba and other urban centres in Bolivia. Nowadays, Punata remains an important region for food production and plays a role in the food security of both the farmers and the urban centres (Rojas and Montenegro 2007). Although land has been fractured to small-holdings (1 ha or less), agriculture remains an important economic activity, representing 51% of the household income, while 31% corresponds to "agricultural transformation"[4] (FAO-CISTEL 1998).

Water availability is an important limiting factor for agricultural development (see e.g. FM Bolivia 2010). According to FAO-CISTEL (1998), of the total area of Valle Alto (around 82,631 ha), only 6.9% has permanent irrigation and 34.4% has occasional irrigation. To promote agricultural development and reduce rural poverty, several new large-scale irrigation projects have been proposed, which include water transfers. However, they have not yet been executed, due among other reasons to conflicts among communities (personal communication, key informant). Nevertheless, the State and international cooperation are promoting small-scale irrigation projects (Prudencio 1998).

Mapping the complexity

To illustrate the complexity of local water management, the identified irrigation systems are described. Next, the water allocation pattern among the communities managing the systems is examined.

Irrigation systems: a description

A total of seven irrigation systems using surface water were identified in the study area. According to the flow regime three are regulated, using reservoir water – Totora Khocha (TK), Laguna Robada (LR) and Lluska Khocha–Muyu Loma (LK-ML) – and four are non-regulated, i.e. water is taken from the river in scheduled distribution modalities – these are known as Mit'a, Rol, Riada and Pilayacu. The different systems vary in terms of interval of supply, discharge, period and type of use (Table 1). Reservoirs are preferred for land preparation because they have low water supply frequency and high discharge. In contrast, wells, which are not considered in this article, have high water supply frequency and low discharge and are thus preferred for irrigation of cash crops, animal watering and washing (del Callejo and Vasquez 2007).

The TK and LK-ML reservoirs are situated within the hydrological boundaries of Pucara watershed; LR lies outside these boundaries. The TK reservoir gets water from adjacent watersheds (Cuenca A, Cuenca B and Cuenca C). Water stored in these reservoirs is shared by the communities of Tiraque and Punata, the latter located outside the watershed boundaries (Figure 2). In terms of infrastructure, a distinction is made between 'intake and conveyance infrastructure' and 'delivery infrastructure'. The LR system has its own conveyance infrastructure while the systems of LK-ML and TK share part of their conveyance infrastructure. The delivery infrastructure – within Abanico Punata – is common to all systems including the river systems; it consists of two main canals (starting at the Paracaya Intake) and secondary canals. Given their own history, the systems developed their own water management.

How is water allocated?

Water quantity evaluation examines the volume of water allocated per community in relation to its area or population and is expressed in mm/y and m^3/y per family, respectively. The volume estimated is based on the system's average discharge and the time allotment, for the systems with available data (LR, LK-ML, TK, Mit'a and Pilayacu). The time allotment is an expression of the water right, i.e. the sum of the time allotments of all community members with water right entitlements provides the total time allotment per community. These calculations give an overview of water allocation among the communities. It is however important to recognize that actual practices might differ from the 'theoretical' water rights, as informal exchange and sale of water turns among irrigators do occur.

The results show that around 11% of the 65 communities held water access rights accumulating to quantities between 640 and 326 mm/y, whereas around 42% of them were allocated between 115 and 11 mm/y. This shows that access to water is uneven. In terms of allocation per family, the majority of the communities receive an amount of water above 900 m^3/y (Figure 3). The communities showing the highest values for both measurements are concentrated upstream. Obviously, there is a relation between the quantity of water and the number of systems a given community has access to. Moreover, there is an overlap of

Table 1. Main features of the irrigation systems.

Water type	Flow	Source	Name/modality	Communities served	Volume (hm³)	Discharge at system level (L/s)	Period available	Use
Surface	Regulated	Dam	Totora Khocha	65	6.5	1600	June–Dec.	Irrigation, land preparation
			Laguna Robada	10	1.6	400	Mar.–Dec.	
			Lluska Khocha–Muyu Loma	11	1.8	180–200	Mar.–Dec.	
	Non-regulated	Spring	Pilayacu	1	0.31	2–20	Dec.–May	Irrigation, animal watering, laundry
		River	Mit'a	33	2.70	20–300	Dec.–May	Irrigation, land preparation
			Rol	65	1.01	300–800	Dec.–May	
			Riada	65	–	>800	–	
Groundwater*	Regulated	Wells	Several systems	?	12.70	1.5–20	Mar.–Dec.	Irrigation, land recovery Irrigation, crop washing, human consumption, animal watering

Source: adapted from del Callejo and Vásquez (2007)
*Not included in this paper

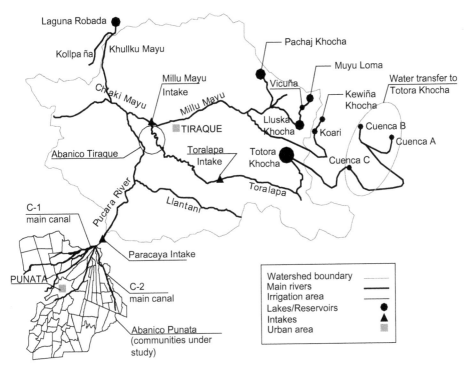

Figure 2.   Hydrological network of the study area.

Source: adapted from Rojas and Montenegro (2007), cited in Saldías (2009).

the different systems and one community might have access to several irrigation systems (Figures 4 and 5). A community with access to more systems can secure water throughout the year. The current access pattern is the result of the struggles in the past by the communities to access the different water sources to secure their livelihood. This is explained in the next section.

How did the irrigation systems evolve?

Reservoir systems

These reservoirs originated before the Agrarian Reform, when landlords built dams and canals to supply their plots in Abanico Punata with water during the dry season. During the Agrarian Reform, the land was given to the people working at the *haciendas* and they organized themselves in peasant communities.

The LR reservoir was built in 1929. After the Agrarian Reform, a group of communities of Punata claimed rights over this source (Montaño 1995). Years later, the dam was rebuilt and for this the communities involved needed an extra labour force. Therefore, they invited neighbouring communities to join the system in exchange for labour (Tuijtelaars *et al.* 1994). The choice to limit the access to nearby communities (connected through infrastructure) was made to avoid problems during water delivery (e.g. water theft, losses) and to maintain control (personal communication, community leaders). The system was opened to individuals, but linked to community membership (Tuijtelaars *et al.* 1994). After construction, the entry for new members was closed because water rights were linked

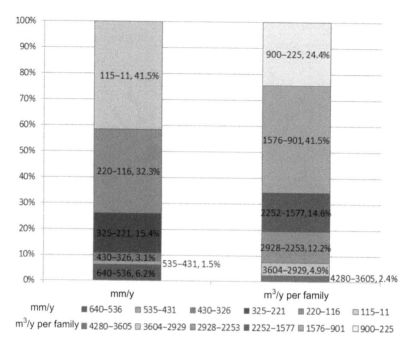

Figure 3. Water allocation in relation to area and population.

Source: adapted from Saldías (2009).

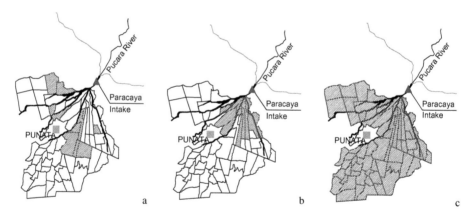

Figure 4. Communities with access to a) Laguna Robada; b) Lluska Khocha–Muyu Loma; c) Totora Khocha.

Source: adapted from Ampuero (2007).

to labour investment. Moreover, including more users would imply less water available per user. Later, an agreement was achieved with the Servicio Nacional de Desarrollo de Comunidades (Communities Development National Service - CDNS) and the German cooperation to increase the dam's capacity and improve irrigation infrastructure. In 1986, a new dam with a capacity of 2.2 hm^3 was constructed as part of the Punata project (Programa de Riego Altiplano Valles - PRAV), Phase 1.[5]

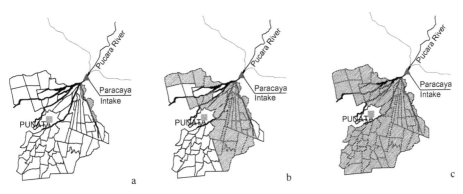

Figure 5.   Communities with access to a) Pilayacu; b) Mit'a; c) Rol and Riada.

Source: adapted from Ampuero (2007).

The LK-ML reservoirs have a similar background. Another group of communities[6] of Punata organized themselves to build dams for the lakes Lluska Khocha, Muyu Loma and Wiskana Khocha, claiming old rights over these sources (Saravia 1998). Supported by the CDNS, they started the construction of the Lluska Khocha dam and later the Muyu Loma dam. By the time they finished the construction, filtration problems occurred and additional labour was needed. By then PRAV was already effective and the repairs were included in the project. The Muyu Loma dam, with a capacity of 1 hm³, was finished in 1987 and one year later the Lluska Khocha dam with a capacity of 1.25 hm³. These reservoirs are interconnected and therefore known as LK-ML (Montaño 1995). The system was opened to members of the organizing communities; after conclusion of the works, access was closed for new users.

The first construction phase of the TK dam took place before the Agrarian Reform with the participation of surrounding communities. The second phase, in the 1960s, was supported by the Ministry of Peasant and Agricultural Affairs and the Alliance for Progress Program with the participation of communities from Punata and Tiraque. The initial capacity was 0.80 hm³. A third phase was part of Punata Project, Phase 2, and aimed to increase the dam's capacity to 22 hm³ with water transfers from neighbouring watersheds Cuenca A (Kewiñal), Cuenca B (Condoraño) and Cuenca C (Lagunillas) (Figure 2). The project was initially designed to cover water demand in Punata, including that of communities served by the LR and LK-ML systems and of new downstream communities, which up to then did not benefit from any regulated flow system. It did not include the communities of Tiraque; they achieved their inclusion after negotiations under an agreement known as "60–40". It established the sharing of the water and the rights and responsibilities over the infrastructure. Communities in Punata get 60% of the yearly volume stored and communities in Tiraque, 40%. Moreover, 0.80 hm³ is reserved for the communities that participated in the first phase (personal communication, former project staff and community leaders).

The inclusion of communities in the project area was subject to a balance between water availability and irrigation demand, influenced also by aspects such as topography that constrain gravity systems. The initial design focused mainly on technical aspects (efficiency, productivity), neglecting social aspects. It was conceived to operate combining the three reservoirs with one single organization to supply water according to demand and rights. Yet, this approach was heavily opposed by the users of LR and LK-ML. For them, the sources were different in terms of volumes and a sense of ownership was created due to the work done (personal communication, former project staff and community leaders). Thus, a process of "homologation of water rights"[7] was introduced which would respect local perceptions of systems' distinctiveness and history, and find correlations between water rights of the different systems (see e.g. Gandarillas *et al.* 1994).

Currently every reservoir system has an irrigation committee responsible for the management and the use of water. The three committees constitute the Punata Irrigation and Services Association (ARSP), which brings together all users. Its purpose is to provide support to the committees in terms of operation and maintenance, water rights management and organizational strengthening, as well as to promote infrastructure improvement and liaisons with institutions at national and international levels for cooperation (Delgadillo and Lazarte 2007). This association is also part of the peasant irrigators' movement, a group that has strongly influenced Bolivia's water governance policies in the last decades (see e.g. Perreault 2005, 2008).

Scheduled systems of the river

Four modalities of scheduled systems of the river are found. They are called Pilayacu, Mit'a, Rol and Riada. The operation of these modalities is linked to the water availability in the river. They are operational during the wet season. The Pilayacu modality operates up to a river discharge of 20 L/s. When the river discharge is larger than this but smaller than 300 L/s, the modality set in place is Mit'a. Next, the Rol modality operates when the discharge is larger than 300 L/s. The last modality, Riada (flood) operates when the river discharge exceeds 800 L/s (Centro AGUA 2009).

Two Pilayacu systems are found in the area: Pilayacu Pucara and Pilayacu La Villa. The access to each system is exclusively for the communities Pucara and La Villa, respectively. These communities have the permanent and exclusive right to use water from these systems (Bleumink and Sijbrandij 1990). Here the focus is on Pilayacu Pucara. The Pucara community gained access to this system as a legacy of the *hacienda,* which was the 'original' right holder. During this time, *colonos* (landlords' workers) used the water under agreements with the landlord. After the Agrarian Reform, the community became the right holder. Water rights were distributed among former *colonos* according to a categorization of the services rendered. For instance, full-time service entitled one to 6 hours of irrigation, part-time service to 3 hours and so on. The water flow is continuous, but subject to water availability in the river. The flow agreed as Pilayacu discharge is respected when other systems are in use. The community is responsible for management (Saravia 1998).

The Mit'a system also dates back to the *haciendas.* Landlords used water from the river to irrigate. Following the Agrarian Reform, water rights passed to former *colonos* as community property. Water allocation was initially linked to land (Gerbrandy 1998) but that is no longer the case. During Mit'a, communities are entitled to a time allotment of 12 to 24 hours. Nowadays not all community members have water rights to this system; however, they can be inherited. One can find right holders entitled to 20 minutes, or 10 minutes. Thirty-three communities upstream use this system. Again, the community is responsible for management (Saravia 1998).

The Rol system was set up for downstream communities without rights to the Mit'a system. During this modality, the time allocated per community varies from 12 to 24 hours depending on the number of users (Montaño 1995). Community members can access this system as long as they comply with the requirements: union membership, maintenance work and monetary contributions (if necessary), and attendance at meetings. Water is allocated to irrigator families when the river discharge is above 300 l/s. The organization responsible is the *Central Campesina.* Lastly, the Riada is accessible to all communities and their members. There is no special organization.

Sharing the infrastructure

The systems described are gravity systems, i.e. water is conveyed through open canals. The reason is mainly economic because investments for this type of infrastructure are affordable by either the State or the users themselves. Infrastructure plays a major role. On the one hand, it has conditioned the access to the various water sources; on the other hand, it is the link among communities in reference to irrigation. Moreover, investing in infrastructure is linked to water rights acquisition, a phenomenon known as hydraulic property creation (see e.g. Coward 1990, Gerbrandy and Hoogendam 2002). Maintenance of the infrastructure is a way to keep water rights and a requirement for access (Gutierrez 2005).

Users from the different systems are related through the irrigation infrastructure. For reservoir systems, water is delivered at different times (there is no mixing of water), so people can distinguish among sources (Gutierrez and Gerbrandy 1998a). There is a whole organization around this, involving the different committees and the communities (see e.g. Saravia 1998). For river scheduled systems, either flumes or marks along the river are used to establish the flow rate at each moment in time (personal communication, former project engineer).

Legal context and water rights

The irrigation systems are community-managed, based on customary laws, mainly because of their historical construction as 'local hybrids' and because of the lack of a formal legal framework regarding water management (Perreault 2005). In this respect, the Irrigation Law, promulgated in 2004, is slowly introducing changes because it provides some sort of formal legality through the process of accreditation of the use of water for irrigation.[8] *Registration* and *authorization* constitute two ways of recognizing and granting water rights.[9] The law stipulates the registration of water rights for indigenous people, peasant communities, associations and peasant unions. In this way, the State permanently guarantees the right to use the water sources according to "customary practices".Moreover, it incorporates some key aspects: (1) recognizing the National Irrigation Service (SENARI) as a participatory decision-making body; (2) responding to the lack of a legal framework; and (3) raising issues of *culturality* – the differentiation between indigenous and non-indigenous, peasant and non-peasant – and how to address this in the normative framework (Salazar 2005).

Certainly, this law is the result of the social struggles that followed the 'water war' in Bolivia (see e.g. Perreault 2008). As stated in the preamble, it seeks to provide "legal security" in favour of the communities based on "customary practices". Yet, water management practices "are inherently communal, place-based, and variable in time and space" (Perreault 2008, p. 835). Thus, it is questionable whether the law captures the complexity of water systems in the Bolivian Andes, particularly in cases of water transfers. The question arises from the fact that conflicts have emerged between upstream and downstream communities in the course of registration of water sources. This is a sensitive issue for downstream communities, which on the one hand are entitled to the water sources by "customary practices", but on the other hand are spatially disadvantaged because they are located away from the source. Communities upstream see this as an opportunity to claim water sources or at least get some sort of compensation. According to SENARI's official website, as of January 2012, only 42 registrations had been completed and 11 were in process.[10] Nothing is registered for the case study area. In this sense, the law opens new spaces for renegotiating water rights. As this suggests, water policies in Andean countries are often introduced without taking into account the historical background and the lessons of local water management systems and the different ecological and social context (Vos *et al.* 2006, p. 38).

The article now turns to the development of water rights in the irrigation systems in question to show the embedded complexity and to explain the need for contextualized water management frameworks. One can distinguish two moments in time for water rights acquisition in the different systems: (1) when the system was created; and (2) when the system started operating and entrance of new members was stopped. This is important because in most systems the member communities and water rights were then fixed. Three mechanisms of water rights acquisition are distinguished: (1) customary law (traditional use of the river, e.g. the Mit'a and Pilayacu systems); (2) labour investment in the construction of the infrastructure (reservoir systems, e.g. the LR, LK-ML and TK) (Gerbrandy and Hoogendam 2002); and (3) de facto rights: no member can be deprived of water (Gutierrez 2005) (e.g. Rol). After the 'closure' of the system, the mechanisms of acquisition include transfers, inheritances and sales (personal communication, former community leader). These mechanisms are possible in all systems, depending on the local rules.

Nowadays, water rights are linked to individuals and not to land (personal communication, former project staff). It is implicit that the right holder does not lose the right in case of land sale and is authorized to use the water outside the community. This adds flexibility (Gutierrez 2005). It responds to the fact that farmers own land in different areas, sometimes even in different communities, and they need to bring water to their plots. The exception is the Pilayacu system, where right holders can use water only within the community boundary (Blanco 1997).

Water rights, also called shares, are expressed in units of time. In LR, LK-ML and TK one share is equivalent to 30 minutes; in the Mit'a modality it is equivalent to 20 minutes. The Rol modality does not have a specific time set; turns are defined at the moment water is distributed according to availability in the river. This way to express water rights incorporates transparency during distribution (Bleumink and Sijbrandij 1990), as well as notions of equity, since "everyone receives a water share in time proportional to his contribution to the water system" (Gutierrez and Gerbrandy 1998b, p. 244). Water rights are associated with communities because that is the space for dialogue, solutions and agreements. Water agreements are also directly connected to the entwined canal networks, giving location-particular shape to the water-territorial boundaries (personal communication, former project staff).

Discussion and conclusions

Community-managed irrigation systems in the interandean valleys of Bolivia are complex, multi-faceted structures embedded in hydro-social networks. Water transfers and usage of multiple water sources with compound, overlapping flows are widespread practices, primarily because of a mismatch between water demand and availability for agricultural production.

The water allocation shows that some communities access more irrigation systems (sources) than others. By accessing different sources, communities aim to guarantee agricultural production and consequently secure their livelihood. This is a way to cope with water scarcity adapted to (or resulting from) local reality. Water allocation also reflects the strategic use of their resources and their organizational capacity for collective action. After exploring the historical development of the different systems it was found that water access is influenced by factors such as: (1) historical claims; (2) organizational capacity; (3) resource availability; and (4) geographical position and infrastructure. The first factor refers to the traditional use of the water source. The second factor deals with the organizational capacity to achieve common goals, which depends heavily on the leadership and the willingness and support of community members. The third factor refers to availability of resources to invest in water access (labour, money, knowledge, technology and access to it). The fourth factor influences physical water control and conditions and inclusion in (and exclusion from) the system. The LR and LK-ML systems are used as examples.

It is noteworthy that the criteria used in this paper to evaluate water allocation are based on 'theoretical' water rights. The volumes of water were estimated based on time allotment and the average discharges (volume/time) of the different irrigation systems. These are useful for an overview; however, they fail to include local arrangements such as lending or selling of water turns. Thus, 'theoretically', some communities have access to more water sources than others, but this does not mean that it is static or that at the user level people cannot benefit from other sources. Again, this shows the complexity of the studied irrigation systems.

The IWRM paradigm emphasizes that sustainable water management, guaranteeing the welfare of future generations, can best be achieved at the watershed level. However, it is questionable whether this approach is applicable to regions where water management goes beyond the watershed boundaries. As this paper shows, community-managed irrigation systems in the interandean valleys have a different conceptualization of water control, space, and the required institutions and networks from the ones promoted by 'engineered' irrigation systems. The local definitions of socio-territorial and political boundaries and scales play a crucial role – definitions that are seldom taken into account in water system and policy design. Moreover, in the realm of water control, crucially, it is the bottom-up conceptualization, control and consolidation of space and place that determine the possibilities for communities and user families to construct their hydro-social networks and waterscapes, gain and maintain legitimacy, materialize water rights and secure water access. The case study shows the overlap of different systems, the water transfers from different sources and the transboundary management of the different usage systems; this makes subscribing them to the boundaries of a physical unit for water management entirely problematic. It also shows that the recently introduced changes in policy still fail to capture this complexity. Consequently, new approaches are needed for water resources management in complex environments, approaches that start with the realization that water management is an intrinsically territory-based resource, and that its management and governance are embedded in historically and culturally specific patterns of resource use. A historical understanding of local water control and the rationale behind current practices regarding water management in particular contexts is vitally important.

Acknowledgements

Field research for this paper has been executed within the framework of the international research alliance Justicia Hídrica/Water Justice (see www.justiciahidrica.org). The authors thank the staff of Centro AGUA. Special thanks to Alfredo Duran, Rigel Rocha and Vladimir Cossio for their support during fieldwork. Additionally, Dr Stijn Speelman would like to thank the Research Foundation Flanders for funding his Post-doc fellowship.

Notes

1. For water allocation among/within communities in terms of *quantity* and *opportunity*, see Saldías *et al.* (2011).
2. Elaborated by Centro AGUA; information extracted includes communities' names, number of families, time allotment (water rights) and discharges. This information was incomplete and has therefore been complemented with other sources.

3. In total, 14 interviews were conducted in 2009 with former staff of the project (3) and community members and leaders (11). Names remain confidential.
4. Bread, alcoholic beverages and dairy products.
5. In the 1970s, the German cooperation supported (technically and financially) irrigation projects in Bolivia for agricultural development. The PRAV project included two projects: Huarina in La Paz and Punata in Cochabamba. The latter had two phases: Phase 1 involved the improvement of existing infrastructure (storage capacity of LR, LK-ML and Koari) and Phase 2 involved the construction of the TK reservoir for new users from both Punata and Tiraque. The Huarina project stopped and the Punata project (Proyecto de Riego Inter-Valles - PRIV) continued, focusing on Punata-Tiraque (personal communication, former project staff).
6. Discrepancies exist regarding the number of communities and users. It is not possible to verify this in official minutes.
7. The idea to operate the three reservoirs jointly required a process of "homologation of water rights"; this would find correlations between water rights from the different reservoirs. Assuming there would be sufficient water in the TK reservoir, a volume of 3200 m^3/y was established as one share. New irrigators could acquire shares in TK directly, while those bene-fitting from LR and/or LK-ML had to complete their shares with water rights in TK reservoir. Consequently, one irrigator could have a time allotment in either LR and/or LK-ML, and in TK, to complete one or two shares.
8. Ley de Promoción y Apoyo al Sector Riego para la Producción Agropecuaria y Forestal. No. 2878. 08 de octubre de 2004. Retrieved from: http://www.sederi-ch.gob.bo/sites/all/themes/ubiquity/publicaciones/Ley%20de%20Riego%202878.pdf
9. *Registration*: administrative act by means of which the State, through the National Irrigation Service (SENARI), recognises and grants the right to the use of water sources for irrigation to indigenous and first nations people, indigenous and peasant communities, associations, organizations and peasant unions, guaranteeing juridically and permanently water resources according to uses and customs.
   Authorization: administrative act by means of which the State, through the National Irrigation Service (SENARI), grants the right to the use of water for irrigation in the agricultural and forestry sector to legal or individual persons that are not contemplated as subjects of Registration (Ley de Promoción y Apoyo al Sector Riego para la Producción Agropecuaria y Forestal, Ley N° 2878, Ley de 8 de octubre de 2004. Retrieved from: http://www.sederi-ch.gob .bo/sites/all/themes/ubiquity/publicaciones/Ley%20de%20Riego%202878.pdf, authors' trans-lation).
10. For the 42 authorizations completed, see http://www.senari.gob.bo/regulacion.asp, and for the 11 in process, see http://www.senari.gob.bo/regulacion2.asp

References

Ampuero, R., 2007. *Análisis de actores y marco institucional de la gestión de agua en Punata*. Cochabamba: Centro AGUA.

Beccar, L., Boelens, R., and Hoogendam, P., 2002. Water rights and collective action in commu-nity irrigation. *In*: R. Boelens and P. Hoogendam, eds. *Water rights and empowerment*. The Netherlands: Koninklijke van Gorcum.

Blanco, A., 1997. *Proceso de Transferencia y Autogestión en los Sistemas de Riego Punata. Proceso de Autogestión en la represa de Totora Khocha*. Cochabamba: Facultad de Ciencias Agrícolas, Pecuarias, Forestales y Veterinarias "Martín Cárdenas" – UMSS.

Bleumink, H., and Sijbrandij, P., 1990. *De monoflujo a multiflujo. Organización de riego en el Valle Alto de Cochabamba*. Irrigation and Water Engineering Group. Wageningen University, The Netherlands.

Boelens, R., 2009. The politics of disciplining water rights. *Development and Change*, 40 (2), 307–331.

Boelens, R., 2010. Water rights politics. *In*: K. Wegerich and J. Warner, eds. *The politics of water: a survey*. London: Routledge.

Boelens, R., and Zwarteveen, M., 2005. Prices and politics in Andean water reforms. *Development and Change*, 36 (4), 735–758.

Centro AGUA, 2009. *Base de Datos Pucara*. Cochabamba: Centro AGUA.

Centro AGUA, 2012. *Proyecto GIRH: construccion de una estrategia para promover la ges-tion integral de recursos hidricos cuenca hidrosocial Pucara (Tiraque-Punata, Cochabamba)*.

Available from: http://www.centro-agua.org/index.php?option=com_content&view=category&layout=blog&id=36&Itemid=64 [Accessed 17 January 2012].

Coward, E.W., 1990. Property rights and network order: the case of irrigation works in the Western Himalayas. *Human Organization*, 49 (1), 78–88.

Cruz, R., 2009. *Estudio hidrológico de la micro-región Tiraque Valle. Compitiendo por el agua: entendiendo el conflicto y la cooperación en la gobernanza local del agua. Valoración de los recursos hídricos del municipio de Tiraque.* Cochabamba: Centro AGUA.

del Callejo, I., and Vásquez, S., 2007. *Caracterización y cambios en el uso del agua en Punata.* Cochabamba: Centro AGUA.

Delgadillo, O., and Lazarte, N., 2007. *Gestión de los sistemas de aprovechamiento de agua en el municipio de Punata.* Cochabamba: Centro AGUA.

FAO-CISTEL, 1998. *Zonificación Agroecológica del Valle Alto – Cochabamba, Bolivia.* Proyecto: FAO GCP/RLA/126/JPN. Cochabamba: FAO.

FM Bolivia. 2010. Reducción de agua en represas del valle alto de Cochabamba afecta producción agrícola y lechera, 6 October. Available from: http://www.fmbolivia.com.bo/noticia37656-reduccion-de-agua-en-represas-del-valle-alto-de-cochabamba-afecta-produccion-agricola-y-lechera.html

Gandarillas, H., et al., 1994. *Dios da el agua ¿Qué hacen los proyectos? Manejo de agua y organización campesina.* La Paz: PRIV-HISBOL.

Gerbrandy, G., 1998. Reparto de Agua en un Río en los Valles Interandinos. *In*: R. Boelens and G. Davila, eds. *Buscando la Equidad.* The Netherlands: Van Gorcum.

Gerbrandy, G., and Hoogendam, P., 2002. Materializing rights: hydraulic property in the extension and rehabilitation of two irrigation systems in Bolivia. *In*: R. Boelens and P. Hoogendam, eds. *Water rights and empowerment.* The Netherlands: Van Gorcum.

Grigg, N.S., 2008. Integrated water resources management: balancing views and improving practice. *Water International*, 33 (3), 279–292.

Gupta, J., and Van der Zaag, P., 2008. Interbasin water transfers and integrated water resources management: where engineering, science and politics interlock. *Physics and Chemistry of the Earth*, 33 (1–2), 28–40.

Gutierrez, Z., 2005. *Appropriate designs and appropriating irrigation systems: irrigation infrastructure development and users' management capability in Bolivia.* The Netherlands: Wageningen University.

Gutierrez, Z., and Gerbrandy, G., 1998a. Multiple-source irrigation systems and transparency: the case of Punata, Bolivia. *In*: R. Boelens and G. Davila, eds. *Searching for equity: conceptions of justice and equity in peasant irrigation.* The Netherlands: Van Gorcum.

Gutierrez, Z., and Gerbrandy, G., 1998b. Water distribution, social organisation and equity in the Andean vision. *In*: R. Boelens and G. Davila, eds. *Searching for equity: conceptions of justice and equity in peasant irrigation.* The Netherlands: Van Gorcum.

Hendriks, J., 2006. Legislación de aguas y gestión de sistemas hídricos en países de la región andina. *In*: P. Urteaga and R. Boelens, eds. *Derechos Colectivos y Políticas Hídricas en la Región Andina.* Lima: IEP.

Molle, F., 2009. River-basin planning and management: the social life of a concept. *Geoforum*, 40 (3), 484–494.

Molle, F., Wester, P., and Hirsch, P., 2010. River basin closure: processes, implications and responses. *Agricultural Water Management*, 97 (4), 569–577.

Mollinga, P.P., 1997. *On the water front.* The Netherlands: Wageningen University.

Montaño, H., 1995. *Sostenibilidad de Sistemas de Riego Manejado por Campesinos en la Asociación de Riego y Servicios Punata.* Cochabamba: Facultad de Ciencias Agrícolas, Pecuarias, Forestales y Veterinarias "Martín Cárdenas" – UMSS.

Moss, T., 2004. The governance of land use in river basins: prospects for overcoming problems of institutional interplay with the EU Water Framework Directive. *Land Use Policy*, 21 (1), 85–94.

Newson, M., 1997. *Land, water and development: sustainable management of river basin systems.* London: Routledge.

Perreault, T., 2005. State restructuring and the scale politics of rural water governance in Bolivia. *Environment and Planning A*, 37 (2), 263–284.

Perreault, T., 2008. Custom and contradiction: rural water governance and the politics of *usos y costumbres* in Bolivia's irrigator movement. *Annals of the Association of American Geographers*, 98 (4), 834–854.

Prudencio, A., 1998. *Manejo Integral del Agua en el Valle de Cochabamba.* Foro del Agua de Cochabamba: Documento base. Asociación de Investigación y Desarrollo Andino – Amazónico. Cochabamba: CONDESAN. Retrieved from http://www.condesan.org/memoria/Docscochabamba.htm

Rojas, F., and Montenegro, E., 2007. *Potencial hídrico superficial y subterráneo del abanico de Punata.* Cochabamba: Centro AGUA.

Salazar, L., 2005. *Concentración Social para la elaboración de leyes: El Caso de la Ley de Riego No.2878 en Bolivia – Testimonios.* Cochabamba: COSUDE, Inter Cooperation, Agua Tierra Gente.

Saldías, C., 2009. *Revelando la distribución del agua. Abanico Punata, área de influencia de la cuenca Pucara, Bolivia.* MSc thesis, Justicia Hídrica research programme, Wageningen University, the Netherlands.

Saldías, C., Speelman, S., and Van Huylenbroeck., G., 2011. A source of conflict? Distribution of water rights in Abanico Punata, Bolivia. Proceedings of *VI water resources management conference*, 23–25 May, Riverside, California.

Saravia, R.I., 1998. *Gestión de agua de las diferentes fuentes en la comunidad de Pucara.* Cochabamba: Facultad de Ciencias Agrícolas, Pecuarias, Forestales y Veterinarias "Martín Cárdenas" – UMSS.

Shah, T., Makin, I., and Sakthivadivel, R., 2000. Limits to leapfrogging: issues in transposing successful river basin management institutions in the developing world. *In*: C. Abernethy, ed. *Intersectoral management of river basins. Proceedings of the international workshop, Integrated water management in water-stressed river basins in developing countries: strategies for poverty alleviation and agricultural growth*, 16–21 October 2000, Loskop Dam, South Africa.

Swyngedouw, E., 1997. Neither global nor local: 'glocalization' and the politics of scale. *In*: K.R. Cox, ed. *Spaces of globalization: reasserting the power of the local.* New York: Guilford Press.

Tuijtelaars, C., *et al.*, 1994. *Mujer y Riego en Punata. Aspectos de género. Situación de uso, acceso y control sobre el agua para riego en Punata.* Cochabamba: PEIRAV.

Uphoff, N., 1986. *Improving international irrigation management with farmer participation: getting the process right.* Studies in Water Policy and Management, No. 11. Boulder: Westview.

Vos, H., Boelens, R., and Bustamante, R., 2006. Local water control in the Andean region: a fiercely contested field. *Water Resources Development*, 22 (1), 37–48.

Wegerich, K., 2010. The Afghan water law: "a legal solution foreign to reality"? *Water International*, 35 (3), 298–312.

Warner, J., Wester, P., and Bolding, A., 2008. Going with the flow: river basins as the natural units for water management? *Water Policy*, 10 (S2), 121–138.

Wester, P., 2008. *Shedding the waters: institutional change and water control in the Lerma-Chapala Basin, Mexico.* The Netherlands: Wageningen University.

Examining the emerging role of groundwater in water inequity in India

Veena Srinivasan[a] and Seema Kulkarni[b]

[a]Ashoka Trust for Research in Ecology and the Environment, Bangalore, India; [b]Society for Promoting Participative Ecosystem Management (SOPPECOM), Pune, India

This article addresses a gap in the water equity literature arising from the simultaneous use of surface water and groundwater in India. Using two diverse case studies – one agricultural (Kukdi) and one urban (Chennai) – we demonstrate how gaps in planning, design and policy exacerbate inequity. Groundwater abstraction from user wells allows wealthier users to both free-ride and capture a greater share of the resource. By converting a public resource to a private one, it worsens inequity and jeopardizes the sustainability of water projects. The article suggests that better monitoring, inter-agency coordination and rethinking water entitlements and norms are needed for going forward.

Introduction: equity in water issues in India

There is a rich database of case-study research on water conflicts and equity issues in India. These cover a wide range of issues, ranging from equitable allocation of inter-state rivers to debates over large dams, where civil engineers' claims of 'efficiency' of large dams have been shown to mask the fact that there are winners and losers in the beneficiary and displaced communities (Duflo & Pande, 2007; Mehta, 2007; Thakkar, 2004), as well as inequities between head-end and tail-end users in surface-water canal-irrigated projects (Kulkarni, Sinha, Belsare, & Tejawat, 2011; Mollinga, 2008; Patil, 1995). Reviews of watershed development projects, promoted as a benign alternative to large dams, have also argued that farmers are not passive observers of watershed projects; they wield their power to influence the location and size of watershed structures and they may generate inequities between upstream and downstream users of water (Batchelor, Rama Mohan Rao, & Manohar Rao, 2003; Glendenning, van Ogtrop, Mishra, & Vervoort, 2012; Kerr, 2002; Narayana, 1987). Similarly, within the groundwater literature there is ample evidence that richer farmers are able to drill deeper (Shah, 1988) and thus capture much of the available groundwater resource. Most previous analyses of equity in the water sector are confined to specific scales and contexts. The problem is that unlike other natural resources like land or forests, water is a mobile resource that exists in multiple interconnected forms – e.g. groundwater and surface water – often with differential systems of rights and regulations. Few studies have analyzed the equity dimensions arising from the multi-dimensional, multi-scalar nature of water.

This article addresses a particular gap in the water equity literature arising from the simultaneous use of multiple forms of the resource: surface water and groundwater. The existence of water as a visible surface-water component and an invisible groundwater component produces unique challenges in the management of water systems. Water policies in India continue to be divided along groundwater–surface water lines, resulting in "hydroschizophrenia" (Llamas & Martínez-Santos, 2005). While hydrologists have long pointed out interconnections between surface water and groundwater, for example that groundwater pumping often occurs at the expense of surface-water flows (Bredehoeft, 2002; Sophocleous, 2000), such analyses are largely missing from Indian case-study research (Ranade, 2005). Pumping technologies are becoming cheaper, and electricity more accessible; this has led to an explosive growth of groundwater use in India. While the resulting declines in groundwater levels in India have been widely discussed (Moench, 1992; Rodell, Velicogna, & Famiglietti, 2009; Shah, Roy, Qureshi, & Wang, 2003), studies that link groundwater pumping to declining flows in rivers due to loss of baseflow remain scarce. To our knowledge, the social dimensions of conjunctive use of ground-water and surface-water resources have not received any attention.

This article is organized as follows. First we explore two case studies – one involving rural agricultural water use, and the other urban water use. In the next section, we discuss the lessons learned from the two case studies. In the following section we discuss the implications for policy and in the final section we propose a way forward to examine water governance issues in the context of the current expansion in groundwater use.

Case studies

Agricultural groundwater use: the Kukdi Major Irrigation Project

Public-sector irrigation is rapidly changing in India, and the state of Maharashtra is no exception to this. Surface projects that were designed independently of groundwater are increasingly being used to harness groundwater. This has been a gradual ongoing process which has shifted benefits from public to private. There is enough evidence across the country that shows expansion of groundwater irrigation and reduction in the area under surface irrigation (Janakarajan & Moench, 2006). Public water sourced through gravity flows is increasingly being sourced through pump irrigation, either through lifts on canals or through groundwater in command areas (Shah 2011).

Poor cost recovery; centralized management and political interference; weak admin-istration; overemphasis on repair and rehabilitation rather than reforming irrigation bureaucracy; lack of participation of users, etc., have contributed to poor performance of public-sector irrigation. However, what is insufficiently discussed is the burgeoning groundwater economy in the canal commands, which has changed the face of public-sector irrigation. Increasing use of groundwater in canal commands has huge implications for equity, both in terms of direct access to water and decision making in its use and allocation, but also indirectly in terms of changing land and labour relations. Figure 1 shows the increasing penetration of groundwater into canal commands in Maharashtra. Of the total irrigated area in canal commands, that irrigated by wells increased from 26% in 2000–01 to 37% in 2010–11, while that irrigated by canals decreased from 73% to 62% in the same period.

This section discusses the case of the Kukdi Major Irrigation Project in Maharashtra to show how the benefits of canal commands are increasingly accumulated in the hands of the few, with an increasing number of wells in command areas, and potentially serious implications for equity. The Society for Promoting Participative Ecosystem Management (SOPPECOM, 2012) carried out a study to understand whether the water entitlement

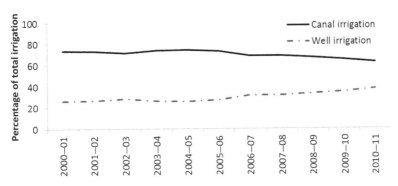

Figure 1. Surface-water and groundwater irrigation in canal command areas in Maharashtra show a decline in canal irrigation and an increase in well irrigation. Source: Irrigation Status Report 2010–11, Water Resources Department, Government of Maharashtra.

programme – a part of the reform process in the irrigation sector in Maharashtra – was in fact promoting the much-promised equity in water distribution. A water entitlement is defined as an authorization to use water and is issued by the water regulatory authority or the river-basin agency. It is relevant for surface storages and evolved through a set of parameters that include designed utilization in three seasons, land within command areas, river losses and gains, and evaporation. Water entitlements for agriculture are worked out after allocations for the other two categories, industry and domestic use, are worked out. Importantly, use of groundwater in canal commands is not taken into account while working out entitlements. Thus, the key finding was that the programme was completely misplaced in the current context of canal irrigation in Maharashtra, which was character-ized by the infiltration of groundwater abstraction. The study was done in two irrigation projects in Maharashtra; one of these, the Kukdi Major Irrigation Project, is discussed here.

The Kukdi Major Irrigation Project

The Kukdi project is a multi-river, multi-reservoir and multi-canal project. Its 156,000 ha command area cuts across three districts of north (Ahmednagar) and west (Pune and Solapur) Maharashtra. It has five dams and a dense network of right- and left-bank canals, distributaries, and finally the command areas at the minor level. In short, it has several levels at which water is distributed, and these levels cut across the political and physical boundaries of the region, making it a complex project to govern.

This discussion draws from a study by SOPPECOM (2012) that looked at the command areas of several water user association (WUAs), including insights from two WUAs which are located on a single distributary of the Kukdi project in Ahmednagar District. They are located at the middle and the tail end of the distributary and give a fair understanding of how water distribution is affected by location and how people at different locations cope with their locational advantages and disadvantages.

The Sahakari Pani Vapar Sanstha (Sahakari Co-operative WUA) is a tail-end WUA and currently does not receive any water from the Kukdi project despite being in its command. The WUA was formed 10 years ago, and although it does not receive any water from the system, it continues to function as a co-operative. Its command area is 200 ha, and its membership is 100–150 farmers. Despite the lack of surface water, the command area was lush green. Further probing revealed that the command area of this WUA has a large number of wells, which benefit about 90% of the farmers in the winter season. However, summer crops are grown by only the 20% of farmers who are able to pump water from the adjacent river. Drip irrigation is practised in about 50% of the area; the major crops are onion and pomegranate.

The Dharmanath Pani Vapar Sanstha (Dharmanath Co-operative WUA) is a middle-reach WUA with a command area of 465 ha and a membership of 324 farmers. This middle-reach WUA is better off than the tail-end one. However, not more than 100 ha of it is covered by flow irrigation. Only about 15–20 of the 324 farmers benefit from flow irrigation. Like the tail-end WUA, this one had several wells in its command area: bore-wells that were below 300 ft., and numerous dug wells. There were also a series of pipelines from the streams to the farms. The command areas were green, despite the low demand recorded for flow irrigation. Bananas, pomegranates, onions and sugar-cane were grown, mostly irrigated through a drip network. Low water demand for flow irrigation also meant low recovery of water charges. The complaint from the farmers was that in the last 10 years or so the command area has received water for not more than 12–15 days each year, which amounts to about 2–3 rotations in the entire year, where ideally there should be a minimum of 4–6 in each of the seasons of winter and summer.

In both the WUA commands there were similar responses to the uncertainties associated with flow irrigation, although there were some variations due to the locational advantage in the middle-reach WUA. In the tail-end WUA, there was no hope that the flow-irrigation system would ever work for them. The head and the middle-reach WUAs were drawing most of the water, depriving their own tail-end users in the process. As a result of this, many farmers opted out of the system and started setting up their own systems, either building river lifts, digging wells or going for bore-wells. This provided them some amount of assured water supply and the much-needed control over water.

The middle-reach WUA benefitted some of the head-end farmers and thus did not opt out of the system completely but innovated and improvised by shifting to pump irrigation either through ground or the closest surface source like the river or streams. Most of the wells in the command of this WUA are located in these streams, where water is freely lifted from canals or delivered through channels or pipelines made by the landowners. Consider the case of Sandeep Salke, a prosperous farmer who harnesses canal water from a stream nearby. He owns 40 acres of land, and is a member of two WUAs, but pays no water bill to either since he does not directly benefit from flow irrigation. Instead, he chose to lay a pipeline from a stream 4–5 km away to irrigate his 40 acres. He depends on the canal flows to recharge the stream from which he draws water. This is how he ensures his water security and control – the public resource becomes a private one for those who can afford to harness it. This is largely the story of the few prosperous farmers on this system. Others are direct beneficiaries in the sense that they take the canal water either into their wells directly or soak their land with water to recharge their wells. Both the WUAs – one better located in the distributory and the other very unfavourably located – have shifted over to groundwater and/or lift irrigation, both of which are in the private domain, offering better water control for the farmers. However, we have seen in the cases of both WUAs that the shift to the private domain was made possible by a large public investment in the form of a surface irrigation project.

The reasons for these shifts are manifold, and the most evident one expressed by the farmers is the poor functioning of surface systems, which do not offer any water security to most farmers, thus forcing some farmers – mostly small farmers with no wells in the command areas – to move out of the system, or those with wells and better resources to innovate and improvise, converting the public resource into a private one. There has been a general discontent about the number of rotations and about water not released per the demands of the WUA. The canal inspectors and section officers we spoke to in Kukdi agreed with this observation and said that expansion of the command area without due attention paid to the increase in river, canal and reservoir lifts means that command-area entitlements are low and frequent rotations are not feasible.

Many of the older farmers narrated how the situation was far better 20 years ago, with more than 70–80% of the commands being irrigated by surface flows. Then, well irrigation was seen as a more expensive option, and people preferred the gravity flows. However, according to an irrigation officer, demands on the water gradually increased from other non-irrigation uses, particularly domestic water for the growing urban areas and also industries. Political pressure in the head reaches of the Kukdi project also led to construction of a series of weirs on the river where dam waters were released and later lifted by the farmers in those locations. In another section, the officer informed us that 50% of the command under his jurisdiction draws water from river lifts. Thus, the landscapes of the command areas were changing rapidly.

Cropping patterns changed with better markets, infrastructural supports, and subsidies offered to horticulture, floriculture, etc. Cereals were thus replaced by crops that promised better 'value' and required very high control over water. Drip irrigation thus became a favoured form of irrigation in these areas. Most farmers within the commands have a well or a bore-well.

We see that the whole nature of flow irrigation is changing. These systems were designed to operate independently of groundwater interactions. Today, groundwater has become an important element in the system, with potentially far-reaching implications. Earlier, only a small number of farmers had wells. Now almost all farmers in the command have wells. More and more farmers are relying on wells and looking at flow irrigation as a supplement, mainly as a means of recharging their wells. The uncertainty created by lack of information, and inability to provide sufficient number of rotations due to over-extended commands, has exacerbated these trends in Kukdi. The turnover to WUAs has in fact led to an increased informalization of the system, which has also led to concentration of irrigation and increasing exclusion of farmers.

Insights from the Kukdi case study

The study provides insights not only about the concentration of water but also about the deep-seated changes that are underway in command-area irrigation. These changes are a result of changing irrigation policies but also of shifts in agricultural and land-use policies. Crop patterns have been changing, and more efficient methods of irrigation such as drip and sprinkler have also meant that farmers prefer to receive much of their water as groundwater recharge, which is recharge as a result of canal flows, rather than direct flow application. This has made for sharpening inequalities between head and tail, within the system at all levels. The study shows that the pump-irrigation economy has in fact taken over the surface-irrigation systems, with serious implications for equity within the command areas of canals but also across sectors: domestic, agriculture, and industrial users. The head reaches of canals are able to extract both surface water and groundwater, forcing the tail-enders to gradually opt out of the public system. Flow canals are increasingly being used as recharge canals by those who have the ability to combine groundwater and surface water, thereby increasing their water security and control. Tail-enders are increasingly being thrown out of the system. Small landholders within commands with no wells are increasingly moving out of agriculture, and those at the tail end with proximity to rivers are going in for lift irrigation, for which the recurrent costs of electricity are very high. Apart from the inequities that are directly linked to access to water, shifts in cropping patterns, or alternatively reallocations in water from subsistence crops to commodity crops, alter the arrangement in land, water and labour relations.

Urban groundwater use: Chennai

As India urbanizes, one of the major challenges will be that of supplying water to its burgeoning urban population. Traditionally, urban water supply has been conceived to be about centralized piped water schemes. Most water supplied to large metropolitan areas is sourced from surface-water reservoirs (though some smaller towns have piped schemes sourced from bore-wells). Self-supplied groundwater sourced from private wells has hitherto played a very small role in formal government policy on urban water supply, although this is changing. For the purposes of this article, the main distinction drawn is between centralized, public, surface-water-based supply and decentralized, private, groundwater abstraction.

Official norms on how much water urban dwellers *ought* to get are typically developed from the perspective of designing and financing piped supply infrastructure. Several organizations and expert committees, such as the Central Public Health and Environmental Engineering Organisation, National Master Plan India – International Drinking Water Supply and Sanitation Decade, and the National Institute of Urban Affairs, have proposed norms for basic infrastructure and services. These norms vary by size and class of urban settlement and approach to urban water supply. The norms are based on a per capita estimate of the water needs of a 'typical' urban dweller, with provisions to account for industry and commercial water use and for pipeline leakage.

Recent surveys of water users, however, reveal a very different picture. Groundwater is playing an increasingly important role in urban water supply. Several studies of Indian megacities (Shaban & Sharma, 2007; Zérah, 2000) have shown that between 25% and 80% of households rely on private wells for some portion of their water needs. The implications of this expansion of urban groundwater use and piped supply–groundwater linkage and consequent implications for revision of urban supply norms have yet to be understood.

Most government documents do not formally acknowledge the role of groundwater in meeting urban water demand, although recently there have been some indirect references. The limits to meeting urban water demand through water imported through massive inter-basin transfer schemes are increasingly being recognized. The recent 12th Five Year Plan calls for a greater reliance on "local sources" (Shah, 2013), including groundwater. The high-level Zakaria Committee (1963) recognizes that urban supply comes at a cost and makes provisions for differential abilities to finance urban infrastructure as well as the feasibility of augmenting supply with local groundwater. The committee notes that in small towns it would be possible to meet certain water uses such as gardening or washing of clothes from local sources such as wells, lakes and rivers, which would not be possible in bigger towns. The committee acknowledges the role of self-supplied groundwater mainly to justify differences in infrastructure investments in small towns versus large cities. Beyond this, there is little mention of the role of groundwater in providing urban water security. Importantly, none of the recent policy documents examine the implications of the increased reliance on local sources or the institutional changes needed.

Similarly, the urban rainwater harvesting movement in India is about recharging local aquifers. Many cities such as Chennai and Bangalore now have rainwater harvesting laws which require households and commercial and institutional establishments to capture rainwater collected on the roof and direct it to infiltration pits on the property. In most cases, the urban water utility or the urban development authority is charged with enforcement of the law. The enthusiasm for rainwater harvesting is an implicit acknowledgement that a portion of urban water needs will be met from groundwater via private wells. However, despite the passage of such laws there are no noticeable changes in design of urban water norms or institutions monitoring groundwater. Urban water utilities continue to plan and build projects to import, treat and distribute 135 litres per capita per day to households with piped connections.

Case study of Chennai, India

The role of groundwater in increasing inequity is analyzed in an urban context using a case study of Chennai (formerly Madras), India's fourth-largest city. According to the 2011 census, about 7 million people reside in the urban agglomeration, which includes peri-urban areas, towns and villages. The water utility, Metrowater, supplies mainly the municipal area; supply is currently being expanded to outlying suburbs. The water utility supplies water from rain-fed reservoirs and well-fields outside the city (Metrowater website, http://chennaimetrowater.gov.in/departments/operation/developwss.htm) as well as from inter-basin transfer projects – the inter-state Telugu Ganga Project and the intra-state Veeranam project – and more recently from desalination plants.

Although only 5–10% of the city's centralized piped water is sourced from groundwater (from well fields to the north), private well supply is much larger. A large fraction of households have private wells. Two large household surveys, conducted in 2004 and 2006, respectively, suggest that 60–70% of Chennai's households use private wells to supplement water. Similarly, ward-level 2001 Housing Census data from Chennai show 46% of households depending on groundwater as their primary source of drinking water, although this fraction varied spatially throughout the city. Households access groundwater directly through their own private wells as well as indirectly through the private tanker market. The conflicts between tankers and peri-urban villages are well documented and well addressed elsewhere (Janakarajan, 1999; Ruet, Gambiez, & Lacour, 2007). However, because tanker water is expensive, it typically accounts for only a small fraction of the urban water demand. Therefore, this article focuses only on the component of groundwater use arising from direct abstraction of groundwater via private wells.

Groundwater plays a critical role in Chennai's overall water supply. It is an important supplementary source of water in dry months and drought years. For instance, when Chennai suffered from a severe multi-year drought in 2003–2005, Chennai's reservoir system dried up completely. This occurred both because of the lack of inflows due to the failed monsoon as well as the failure of the inter-state Telugu Ganga project to deliver water. As reservoir levels dropped, piped supply was curtailed and then completely halted for a period of almost 12 months (Srinivasan, Gorelick, & Goulder, 2010a). Households turned to their own wells to augment supply. This curtailment of piped supply had a dual effect on the aquifer: extraction increased, and recharge from leaky pipelines dropped; the water table fell approximately 8–10 m. Many wells went dry. As households lost access to both of their lowest-cost sources of water, wells and piped supply, they were forced to purchase expensive water from private tankers and suffered losses in well-being.

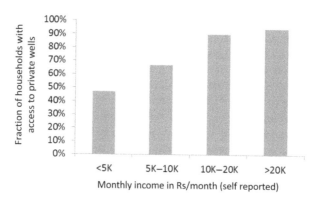

Figure 2.   Private well ownership correlated with income per a household survey of ~1500 house-holds in Chennai conducted in 2006 (Srinivasan, 2008).
Note: 1 Rupee ~ 0.022 USD in 2005.

Everyone lost access to piped supply during the 2003–2004 drought, but the households that suffered the most were the ones whose wells also went dry (Srinivasan et al., 2010a).

The Chennai case study demonstrates the welfare implications of this massive exploi-tation of groundwater. But these welfare implications were not uniform: middle- and upper-income households were more able to invest in private water systems of increasing complexity, particularly deep bore-wells. The Chennai household survey data (Figure 2) corroborate this: unsurprisingly, well ownership is correlated with household income. A significant fraction of household water was sourced from groundwater, and richer house-holds were able to consume more because they had better access to *both* groundwater and piped water. Indeed, even in a wet year like 2006, household survey data show that people with private wells used significantly more water than people without wells (Table 1). A previous household survey conducted during the severe 2003–2004 drought showed that the poorer households lacking piped connections were the hardest hit as water tables fell and they lost access to shallow bore-wells (accessed via hand pumps), which went dry (Vaidyanathan & Saravanan, 2004). The only households that were able to access water cheaply were those with deep, productive bore-wells. Overall, wealthier households were more able to diversify and sustain consumption levels in all periods, but especially during droughts.

Although Chennai Metrowater was proactive in supplying water via tankers to slum areas, slum households were completely dependent on the vagaries of tanker supply. Interviews with local residents suggest they developed mechanisms of collective action to lobby Metrowater officials to secure a lifeline supply of tanker water, but at considerable expense of time and money. As a result, water use tended to be inequitable across income classes. One insight from this is that private well abstraction is not something that a water

Table 1.   Differential consumption by well-owning versus non-well-owning households in 2006 (a wet year).

	Water use (litres per capita per day)
Households with piped supply and wells	133
Households with no piped supply, only private wells	78
Households with no piped supply or wells	35

Source: Srinivasan et al. (2013).

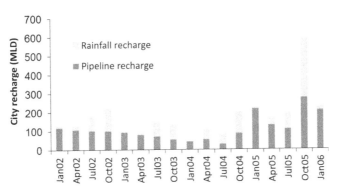

Figure 3. Leakages from the public delivery system constitute about half of the total recharge into the aquifer. Based on a groundwater model; model development and calibration are described in Srinivasan (2008).
Note: MLD, milliion litres per day.

utility can control for. It cannot be addressed easily by 'managing' the resource differently, since the utility cannot control who abstracts groundwater, when, or how much.

Furthermore, analysis shows that access to groundwater worsened existing inequities in public supply. Overall, groundwater played an important role in 'buffering' consumption during dry seasons and multi-year droughts, but groundwater availability was biased by piped supply availability. A groundwater model of Chennai (Srinivasan et al., 2010a) showed that leakage from pipelines was a significant contributor to groundwater recharge and availability, contributing almost 50% of the urban recharge (Figure 3). As a result, the neighbourhoods that receive plentiful piped supply are also the ones with shallow groundwater levels. In other words, the use of groundwater amplifies existing inequities in piped water distribution.

The case study showed that the groundwater aquifer essentially functioned as a substitute for surface-water storage, but with very different legal and institutional controls. Chennai has not been able to increase its surface-water storage capacity to keep pace with population growth. Analysis of Chennai's water reservoir system indicates that reservoir storage is extremely constrained. The city does not really have the capacity to store either the peak monsoon flows or the Telugu Ganga project imports. The result is that in wet periods, the Metrowater utility expands its hours of supply; much of this water leaks out from the ageing piped network and recharges the aquifer. In dry periods, piped supply is cut back, and users turn to their private wells to meet their needs. Thus, the water is abstracted back via private wells. In effect, the aquifer plays the function of a balancing surface-water reservoir. The net result of this is an unintentional switch from public to private supply during droughts. The water which was meant to be supplied to all consumers through the public supply system is now only available to users who have invested in bore-wells.

The Chennai case study also raises questions about the long-term sustainability of public water utilities. Private wells constrain how much a water utility can charge for water, with implications for its financial sustainability. The marginal cost of extracting water from a bore-well works out to INR 5 to 7 per kL. In contrast, the long-run marginal cost of water from new sources could be as high as INR 45/kL (e.g. in the case of Chennai's desalination plants). Metering water and charging more to recover a portion of the cost of piped water delivery has long been suggested to control demand and to ensure the long-term financial sustainability of the public water supply. However, if the water utility were to greatly increase water charges, the wealthier households with bore-wells could opt out of the system altogether, choosing to rely partly or entirely on privately abstracted and treated groundwater. Only the poorer users would be dependent on public supply, severely straining the financial viability of the water utility. It should be pointed out that this is not intended to imply that private wells are undesirable. Indeed, previous studies show that rainwater harvesting and dependence on local water sources is desirable. However, private groundwater abstraction needs to be accounted for in planning, designing and managing urban water systems (Srinivasan, Gorelick, & Goulder, 2010b).

Insights from the Chennai case study

The Chennai case study highlights the critical role that groundwater plays in buffering urban consumers against drought and dry seasons. There is an increasing reliance on private wells as public supply remains unreliable. The problem is that this expansion in groundwater use via private wells has exacerbated existing inequities in access to water for domestic needs – partly through differential well ownership rates and partly because leaky pipelines recharge groundwater more in wealthier neighbourhoods, which also get more hours of piped supply. However, existing government regulations take a rather contradictory approach to self-supplied groundwater. On the one hand, rainwater-harvesting mandates implicitly assume that self-supplied groundwater will play a major role in achieving urban water security. On the other hand, design standards and urban water supply norms largely ignore this. To ensure equitable and sustainable access to urban water, it is critical that policy makers acknowledge the role of self-supply.

Discussion

The two case studies presented in this article highlight a trend that is being observed throughout India: the substitution of publicly supplied surface water by privately abstracted groundwater. The case studies bring to light problems with current policies arising from this.

First, groundwater expansion *complicates water entitlements*. Current policies on water entitlements and norms on water allocation – both within and across sectors – fail to consider conjunctive water use. In the agricultural sector, many states are considering creating water entitlements based on surface-water storages, assuming a particular pattern of surface-water use as independent of groundwater. This is problematic because rights to groundwater and surface water are treated very differently. While entitlements are created to surface-water resources, groundwater remains open access and unregulated. Landowners are entitled to abstract as much groundwater as they can – the abstraction is unmetered – and at the same time, electricity is free for farmers. The equity implications arising from this expansion of groundwater use apply not only to other groundwater users but to surface-water entitlements as well. Similarly, in urban areas, current urban supply norms continue to focus exclusively on public piped water delivery, completely ignoring the large component of urban water use that is groundwater-based.

Second, current policies fail to acknowledge the interconnectivity of groundwater and surface water that *exacerbates inequitable allocation of water*. Both case studies show how losses from the public delivery system result in recharging wells. In the Kukdi case study, private wells allow head-end farmers to further consolidate their capture of the resource. In the Chennai case, wealthy households can augment unreliable piped supply with self-supply from their own wells, while poorer users cannot. Drawdowns by users with deep bore-wells cause the water table to drop, depriving the nearby slums that are dependent on shallow bore-wells and hand pumps.

Third, groundwater expansion necessitates *different forms of monitoring and regulation*. Much of the formal policy addressing equity – both in urban and rural areas – is focused on centralized surface-water schemes, which are easier to track, monitor and regulate. Whereas centralized piped supply schemes are planned, financed and controlled by government agencies, private wells are planned, financed and controlled by individual households. The problem is that groundwater is difficult to quantify because it is extracted by the users based on their ability to invest in extraction mechanisms. It is virtually impossible to track or regulate, and current institutions are poorly equipped for this task.

Finally, however, none of these policies have looked at the present use of surface water and its substitution by groundwater and how this is impacting the *financial sustainability of these systems*. Recent reforms in the water sector have included regulation, repair and maintenance and participatory management of public water. Emphasis on policies around improving performance of the public surface systems has changed over time. In the initial period, the bulk of the investment went into infrastructure improvement, and repair and maintenance of the canal systems. This was followed by policies geared towards institutional restructuring and economic reforms. Since groundwater use is unregulated, it does not generate any revenue for the state. In fact, it leads to creation of a perverse subsidy for those with adequate means to abstract water without paying for it. This is a kind of free riding that does not promote efficiency, revenue generation or equity. The Kukdi study clearly brings out the need to bring the wells in the commands under the jurisdiction of the WUAs. The Chennai study brings out the need to bring groundwater use from urban wells under the jurisdiction of the water utility.

Implications for contemporary water policy

In recent years, there have been several new legislative and policy initiatives at the state and national levels. In India, water is a state subject; national framework laws must be passed or ratified by the states. The central government has jurisdiction only over the sharing of inter-state rivers. A few of these initiatives, as well as the recent draft national water framework law, were re-examined based on how they approach the issue of conjunctive use of groundwater and surface water.

First, many of the recent initiatives acknowledge in principle the need to account for the interconnected nature of surface water and groundwater. For instance, the Maharashtra State Water Policy (GOM, 2003) states that the "isolated and fragmented approach to surface and groundwater development and management" has resulted in deteriorating quality of surface and groundwater and calls for "integrated and coordinated development of surface and ground water and their conjunctive use" at all stages of project planning and development. Similarly, the draft national water framework bill by the Alagh Committee (MoWR, 2013) advocates river-basin management and requires states to "manage groundwater conjunctively with surface water of any basin of which it is a part, taking into account any interconnections between aquifers or between an aquifer and a body of surface water". However, in both cases, the prescribed approach to water entitlements and WUAs focuses only on surface-water deliveries to the offtake point; command-area wells are not mentioned.

Similarly, inter-state water-dispute tribunals are either inconsistent or silent on groundwater extraction and the possible impacts on streamflows. The Krishna and Godavari

Water Disputes Tribunal conceded the interconnectivity between surface and groundwater but noted that groundwater flow is not fully calculable from the technical point of view, and therefore "not fully cognizable" from the legal point of view (as reported in Cauvery Water Disputes Tribunal, 2007). In the Narmada Water Disputes Tribunal, while it is not categorically stated that the allocation refers only to surface water, this can be inferred based on the inter-sectoral allocations of water proposed by the states (Ranade, 2005).

Overall, while the principle of integrated management of surface water and groundwater in the design and planning of projects is mentioned in many contemporary policies, there is very little concrete guidance on how to operationalize this. Often, the implementation rules contradict the principle of conjunctive management. Integration of surface water and groundwater cannot occur without a major realignment of how groundwater and surface-water organizations function on the ground. As written, existing policies would merely perpetuate the status quo.

The way forward

Several researchers have investigated case studies of groundwater governance across the world (Faysse, Petit, Bouarfa, & Kuper, 2012) and attempted to explain why groundwater is effectively governed in some places. The studies show that participation of local groundwater users in framing the rules is critical. For groundwater governance to be effective, top-down rules must be enforced by bottom-up local institutions, because local users have a "comparative advantage" over government in having the relevant information and potentially being able to monitor resource use, provided they have an incentive to manage the resource sustainably, especially when their long-term livelihoods depend on groundwater (Rica, López-Gunn, Llamas, Bouarfa, & Kuper, 2012). However, the likelihood and success of collective action around groundwater have been found to be influenced by several factors, including a shared understanding of the problem (Faysse et al., 2012), the nature and extent of the aquifer (Ostrom, 2009), the level of dependence on the aquifer for livelihoods (Mukherji & Shah, 2005), the transaction costs of regulating a large number of small rural users (Shah, 2009), the presence or absence of farmer lobbies (Mukherji, 2007), the homogeneity of groundwater users (Shah, 2009), and the ability to enforce restrictions effectively (Rica et al., 2012). However, the present authors are not aware of any successful cases of groundwater governance in India. The widely cited Andhra Pradesh Farmer Managed Groundwater Systems experiment was an informational, not regulatory intervention (FAO & BIRDS, 2010). The objective was to give farmers information on the level of recharge in a given monsoon so they could reduce the risk of second crop failure if they knew the rains had failed. Thus, the project was designed to protect farmers from incurring losses, not necessarily to control groundwater depletion.

This article has used two very different case studies to illustrate equity issues arising from the massive groundwater exploitation underway in India. The access and equity implications of this development have been poorly studied, and current policies have failed to take account of them. For instance, the recent 12th Five Year Plan of the Planning Commission of India rightly emphasizes local sources but does not quite address the institutional structure needed to ensure equitable access to water resources – particularly in light of the conjunctive use of surface water and groundwater. Based on these case studies, a series of actions that recognize water as a multi-dimensional, multi-scalar resource are recommended.

Recognize water as a common-pool resource

The interconnectivity between surface water and groundwater implies that the present "hydroschizophrenia" of creating entitlements to surface water while allowing groundwater to remain open access cannot continue. In reality, there is only one, interconnected resource, which is a common-pool resource. Any reforms of water rights must take cognizance of this. Creation of entitlements to just surface water, leaving groundwater as an open-access resource, is likely to exacerbate inequity.

Create institutions to manage the water resources of the region

At the moment, surface-water and groundwater monitoring agencies at the national and state levels are separate (e.g. the Central Water Commission versus the Central Ground Water Board). Very little coordination or exchange of data occurs between them. Indeed, the very conceptual framework for quantifying and allocating water resources allows for 'double counting'. If the long-term equity and sustainability of public water delivery are to be ensured, there is an urgent need for better inter-agency coordination. Moreover, scientific research on how surface water and groundwater are connected will be needed. Although most water professionals verbally acknowledge the interconnectivity, there is very little guidance on how to operationalize this knowledge in ways that can be used to manage a specific project or water utility.

Track overall water use, not just water delivered by centralized supply schemes

There is a need for monitoring mechanisms to track total water use, not just water delivered by centralized piped or canal schemes. At present, to our knowledge, groundwater use is not being formally tracked by any monitoring agency in any canal irrigation project or in any urban area. Although the number of wells may be tracked, water use is not. It is true that monitoring groundwater is not easy, because groundwater use is controlled by the end user, who has little incentive to disclose how much is abstracted. However, modern technologies – feeder-level electricity data in agricultural areas; tracking divisional sewage collection in the case of urban settlements – suggest the possibility of tracking groundwater use, at least by groups of users, if not individual users. Only if all water use in all seasons is considered can equitable access to water be ensured.

Create norms based on overall water use

Finally, to ensure equitable access to water, water entitlement reforms and water delivery norms must consider conjunctive use of surface-water and groundwater resources. Norms and entitlements dealing with only centralized water supply are inherently flawed and open to exploitation by the privileged few who can afford deep bore-wells and opt out of (or free ride on) the public delivery system.

Acknowledgements

We thank the Center for Global, International and Regional Studies at the University of California, Santa Cruz, for hosting the workshop, Multi-Scalar and Cross-Disciplinary Approaches towards Equitable Water Governance (21–23 February 2013), which made this collaborative piece possible. Funding for our travel and stay was provided by this US National Science Foundation Workshop in Cultural Anthropology. We thank the meeting organizers, Flora Lu, Ben Crow, Constanza Ocampo-Raeder, Sarah T. Romano, and the workshop attendees for sparking the ideas presented herein. We acknowledge funding for previous work on the Chennai case study from the Stanford Woods

Institute for the Environment and on the Kukdi case study from the IWMI-TATA Water Policy Research Program.

References

Batchelor, C., Rama Mohan Rao, M., & Manohar Rao, S. (2003). Watershed development: A solution to water shortages in semi-arid India or part of the problem. *Land Use and Water Resources Research, 3*(1), 1–10.

Bredehoeft, J. D. (2002). The water budget myth revisited: Why hydrogeologists model. *Ground Water, 40*(4), 340–345. doi:10.1111/j.1745-6584.2002.tb02511.x

Cauvery Water Disputes Tribunal. (2007). *Report of the Cauvery Water Disputes Tribunal with the decision*, volume 3. New Delhi: Cauvery Water Disputes Tribunal. Retrieved February 22, 2014 from http://wrmin.nic.in/writereaddata/linkimages/FINALDECISIONOFCAUVERYWATERTRIBUNAL4612814121.pdf

Duflo, E., & Pande, R. (2007). Dams. *The Quarterly Journal of Economics, 122*(2), 601–646. doi:10.1162/qjec.122.2.601

FAO, & BIRDS. (2010). Andhra Pradesh Farmer Managed Groundwater Systems Project (APFAMGS Project) (Terminal Report No. GCP/IND/175/NET) (p. 79). Secundarabad, India: Food and Agriculture Organization (FAO) and Bharithi Integrated Rural Development Society (BIRDS).

Faysse, N., & Petit, O. (2012). Convergent readings of groundwater governance? Engaging exchanges between different research perspectives. *Irrigation and Drainage, 61*, 106–114. doi:10.1002/ird.1654

Faysse, N., EL Amrani, M., EL Aydi, S., & Lahlou, A. (2012). Formulation and Implementation of policies to deal with groundwater overuse in Morocco: Which supporting coalitions? *Irrigation and Drainage, 61*, 126–134. doi:10.1002/ird.1652

Glendenning, C., van Ogtrop, F., Mishra, A., & Vervoort, R. (2012). Balancing watershed and local scale impacts of rain water harvesting in India—A review. *Agricultural Water Management, 107*, 1–13. doi:10.1016/j.agwat.2012.01.011

GOM. (2003). Maharashtra State Water Policy. Government of Maharashtra. Retrieved from http://www.google.co.in/url?sa=t&rct=j&q=&esrc=s&source=web&cd=1&cad=rja&ved=0CCgQFjAA&url=http%3A%2F%2Fwww.cseindia.org%2Fuserfiles%2FmaharashtraSWP.pdf&ei=yPjbUreyBcO_rge5sYH4AQ&usg=AFQjCNGnBAEtdw8CXfNbrQKKQxHWXLT1u-g&sig2=DJnbHSSP2-UDk7mMqUIRzQ&bvm=bv.59568121,d.bmk

Janakarajan, S. (1999). Conflicts over the invisible resource in Tamil Nadu: Is there a way out? In: M. Moench, E. Caspari, & A. Dixit (Eds.), *Rethinking the mosaic: Investigations into local water management*. Kathmandu: Institute for Social and Environmental Transition and Nepal Water Conservation Foundation.

Janakarajan, S., & Moench, M. (2006). Are wells a potential threat to farmers' well-being? Case of deteriorating groundwater irrigation in Tamil Nadu. *Economic and Political Weekly, 41*(37), 3977–87.

Kerr, J. (2002). Watershed development, environmental services, and poverty alleviation in India. *World Development, 30*(8), 1387–1400. doi:10.1016/S0305-750X(02)00042-6

Kulkarni, S., Sinha, P., Belsare, S., & Tejawat, C. (2011). Participatory irrigation management in India: Achievements, threats and opportunities. *Water and Eenrgy International, 68*(6), 28–35.

Llamas, M. R., & Martínez-Santos, P. (2005). Intensive groundwater use: Silent revolution and potential source of social conflicts. *Journal of Water Resources Planning and Management, 131*(5), 337–341. doi:10.1061/(ASCE)0733-9496(2005)131:5(337)

Mehta, L. (2007). Whose scarcity? Whose property? The case of water in western India. *Exploring New Understandings of Resource Tenure and Reform in the Context of Globalisation, 24*(4), 654–663. doi:10.1016/j.landusepol.2006.05.009

Moench, M. (1992). Drawing down the buffer: Science and politics of ground water management in India. *Economic and Political Weekly, 27*(13), A7–A14.

Mollinga, P. P. (2008). Water, politics and development: Framing a political sociology of water resources management. *Water Alternatives, 1*(1), 7–23.

MoWR. (2013). Report of the Committee for Drafting of National Water Framework Law. Ministry of Water Resources, Government of India. Retrieved from http://www.google.co.in/url?sa=t&rct=j&q=&esrc=s&source=web&cd=1&cad=rja&ved=0CCoQFjAA&url=http%3A%2F%2Fmowr.gov.in%2Fwritereaddata%2Flinkimages%2Fnwfl1268291020.pdf&ei=offbUtu_N8bDrAej7oCoCw&usg=AFQjCNGIIytTxioG1nlL049AkCPHrPtP6g&sig2=wMMFfvphmNdB4-Oi87KKXmQ&bvm=bv.59568121,d.bmk

Mukherji, A. (2007). The energy-irrigation nexus and its impact on groundwater markets in eastern Indo-Gangetic basin: Evidence from West Bengal, India. *Energy Policy, 35*(12), 6413–6430. doi:10.1016/j.enpol.2007.08.019

Mukherji, A., & Shah, T. (2005). Groundwater socio-ecology and governance: A review of institutions and policies in selected countries. *Hydrogeology Journal, 13*(1), 328–45.

Narayana, V. D. (1987). Downstream impacts of soil conservation in the Himalayan region. *Mountain Research and Development*, 287–298. doi:10.2307/3673207

Ostrom, E. (2009). A general framework for analyzing sustainability of social-ecological systems. *Science, 325*(5939), 419–22. doi:10.1126/science.1172133.

Patil, D. R. (1995). *Water users' association in Minor 7, Mula Project: Farmers' experience.* Ahmedabad, India and Colombo, Sri Lanka: International Irrigation Management Institute.

Ranade, R. (2005). "Out of sight, out of mind": Absence of groundwater in water allocation of narmada basin. *Economic and Political Weekly, 40*(21), 2172–2175. doi:10.2307/4416673

Rica, M., López-Gunn, E., & Llamas, R. (2012). Analysis of the emergence and evolution of collective action: An empirical case of Spanish groundwater user associations. *Irrigation and Drainage, 61*, 115–125. doi:10.1002/ird.1663

Rodell, M., Velicogna, I., & Famiglietti, J. S. (2009). Satellite-based estimates of groundwater depletion in India. *Nature, 460*(7258), 999–1002. doi:10.1038/nature08238

Ruet, J., Gambiez, M., & Lacour, E. (2007). Private appropriation of resource: Impact of peri-urban farmers selling water to Chennai Metropolitan Water Board. *Cities, 24*(2), 110–121. doi:10.1016/j.cities.2006.10.001

Shaban, A., & Sharma, R. (2007). Water consumption patterns in domestic households in major cities. *Economic and Political Weekly, 42* (23), 2190–2197. doi:10.2307/4419690

Shah, M. (2013). Water: Towards a paradigm shift in the twelfth plan. *Economic & Political Weekly, 48*(3), 41.

Shah, T. (1988). Externality and equity implications of private exploitation of ground-water resources. *Agricultural Systems, 28*(2), 119–139. doi:10.1016/0308-521X(88)90031-5

Shah, T. (2009). *Taming the anarchy: groundwater governance in South Asia.* Washington, DC: Resources for the Future; and Colombo, Sri Lanka: International Water Management Institute.

Shah, T. (2011). Past, present, and the future of canal irrigation in India. *India infrastructure report*, 2011, 70–87.

Shah, T., Roy, A. D., Qureshi, A. S., & Wang, J. (2003). Sustaining Asia's groundwater boom: An overview of issues and evidence. *Natural Resources Forum, 27*(2), 130–141. doi: 10.1111/1477-8947.00048

Sophocleous, M. (2000). From safe yield to sustainable development of water resources—the Kansas experience. *Journal of Hydrology, 235*(1), 27–43. doi:10.1016/S0022-1694(00)00263-8

SOPPECOM. (2012). Maharashtra Water Resources Regulatory Authority: An Assessment. Presented at the IMWI-Tata Annual Partner's Meet, Ahmedabad.

Srinivasan, V. (2008). *An integrated framework for analysis of water supply strategies in a developing city: Chennai, India.* PhD Dissertation. Stanford University, Stanford, CA.

Srinivasan, V., Gorelick, S. M., & Goulder, L. (2010a). A hydrologic-economic modeling approach for analysis of urban water supply dynamics in Chennai, India. *Water Resources Research, 46* (7). doi:10.1029/2009WR008693

Srinivasan, V., Gorelick, S. M., & Goulder, L. (2010b). Sustainable urban water supply in south India: Desalination, efficiency improvement, or rainwater harvesting? *Water Resources Research, 46*. doi:10.1029/2009WR008698

Srinivasan, V., Seto, K. C., Emerson, R., & Gorelick, S. M. (2013). The impact of urbanization on water vulnerability: A coupled human–environment system approach for Chennai, India. *Global Environmental Change, 23*(1), 229–39.

Thakkar, H. (2004). Big dams in North East India: For whose benefits? For what benefits?. *Rounglevaisuo Dams Update, 1*(1), 8.

Vaidyanathan, A., & Saravanan, J. (2004). *Household water consumption in Chennai City: A sample survey.* Report by Centre for Science and Environment, New Delhi.

Zakaria Committee. (1963). *Augmentation of financial resources of urban local bodies.* Report of the Committee of Ministers, Constituted by the Central Council of Local Self Government.

Zérah, M. -H. (2000). Household strategies for coping with unreliable water supplies: The case of Delhi. *Habitat International, 24*(3), 295–307. doi:10.1016/S0197-3975(99)00045-4

The colonial roots of inequality: access to water in urban East Africa

Brian Dill[a] and Ben Crow[b]

[a]Department of Sociology, University of Illinois at Urbana-Champaign, USA; [b]Department of Sociology, University of California, Santa Cruz, USA

While water access is a major concern for all residents in Dar es Salaam and Nairobi, the difficulty of hauling water is particularly pronounced in the informal settlements that are significant portions of both cities. This is an inequality that has only recently begun to be recognized as an injustice between rich and poor. Rooted in the segregation of colonial rule, it is sustained by the continuing injustice of land policies and the multiple complications involved with upgrading urban settlements.

Introduction

The Millennium Development Goals (MDGs) were established at a UN Summit in 2000 (UN, 2014). They set a series of targets for poverty reduction, achieving universal primary education and gender equality, reduction of infant mortality, and combating AIDS, malaria and other diseases. Under the goal of ensuring environmental sustainability was a target of halving the proportion of the population without sustainable access to safe drinking water and basic sanitation. Although the world as a whole reached this target five years ahead of schedule, Sub-Saharan Africa is not on track to do so by 2015 (UN MDG, 2012). Gains have been made in the countries south of the Sahara, and yet the region still accounts for more than 40% of people worldwide without access to safe drinking water (UN MDG, 2013). Those who struggle to meet their daily water needs are often concentrated spatially and economically. Although urban dwellers tend to have better access to water than their rural counterparts, the most vulnerable residents of Africa's burgeoning urban informal settlements have fared particularly poorly. They are liable to be triply disadvantaged in terms of the local availability of water, the effort required to secure it, and the price paid per litre.

The MDGs are just one example of the attention that development actors and scholars have paid to the plight of the nearly 1 billion people without access to water. In December 2003, the United Nations General Assembly took the additional step of proclaiming 2005–2015 the International Decade for Action 'Water for Life' in its resolution A/RES/58/217. The heightened interest in water that officially started on World Water Day (22 March 2005) will continue well beyond the conclusion of this designated decade. In May 2013, the UN's High-Level Panel on the Post-2015 Development Agenda released its much-anticipated report, outlining a vision and a framework for future global development efforts (UN, 2013). The 6th goal, of the 12 presented in the report, commits the

112

global community to the ambitious objective of achieving universal access to water and sanitation by 2030.

The vast applied and academic literatures have engaged with the water issue in two primary ways. The first pertains to the negative health outcomes that are a direct result of inadequate access to water and sanitation. Numerous studies have observed that under-served urban settlements are regularly subjected to outbreaks of cholera, dysentery and diarrhoea, and as a result have extremely high rates of infant mortality (Bartram & Cairncross, 2010; Fewtrell et al., 2005; Gundry, Wright, & Conroy, 2004; Jalan & Ravallion, 2003; Mara, 2003). The other approach is concerned with the institutional reforms required to improve the quality and availability of water, arguing that the adequate management of finite freshwater resources requires appropriate governance arrangements (Bakker, Kooy, Shofiani, & Martijn, 2008; Franks & Cleaver, 2007; World Bank, 2010). A decade ago, these arrangements were limited to a global push for the privatization of public water utilities (Goldman, 2005). The many subsequent and documented shortcomings of privatization have led scholars and practitioners to inter-rogate and advocate for more comprehensive approaches to resolving the water crisis, such as by integrating water into the development of other sectors, exploring the connec-tions between resource use and service delivery, and improving the performance of public utilities and user associations (World Bank, 2010).

This article engages with the water issue on quite different terms. We take as our point of departure the position advanced in the Santa Cruz Declaration on the Global Water Crisis that is included in this special issue of *Water International*: that the water crisis is fundamentally one of injustice and inequality. That is, the crisis is not primarily driven by water scarcity, nor is it most productively characterized as a lack of appropriate institu-tions. Instead, the global water crisis is best understood as many overlapping injustices in terms of access to, the allocation of, and the quality of water. The declaration, which developed out of a workshop at the University of California, Santa Cruz, in February 2013, draws much-needed attention to the fact that access to water resources is not evenly distributed across societies. Poor and affluent societies alike can be riven by inequalities when it comes to water, and these inequalities are embedded in particular histories.

This article examines the water crisis in two African cities: Dar es Salaam, Tanzania, and Nairobi, Kenya. While there are significant differences between the two cities, the similarities are more compelling. In Tanzania, for example, safe drinking water is recog-nized as a basic human right in the Water Resources Management Act of 2009. Kenya undertook the reform of its water institutions in 2002 and included a human right to water in its 2010 constitution. But in spite of these propitious policies, both countries' primary cities have large, underserved and rapidly growing populations; the majority of their poor residents lack household connections. Moreover, both of these cities are broadly illustra-tive, in terms of origins and outcomes, of a crisis that has proved to be particularly recalcitrant in Sub-Saharan Africa. For example, access to water is marred by profound inequality. The ability of households in each of these cities to meet their basic water needs is not commensurate, in terms of cost and ease of access, across residential areas. By focusing on the inequalities and injustices that are inherent to this crisis in these cities, we endeavour to explain their historical antecedents, as well as to evaluate contemporary efforts to improve access to water. We note also that debates about slum upgrading in Sub-Saharan Africa have recognized that improvements in infrastructure can also address a key underlying source of insecurity. The provision of formal water access can give de facto security of land tenure (Gulyani & Bassett, 2007).

The first section of this article summarizes the current water situation in Dar es Salaam and Nairobi. What proportion of each city's population has access to water? How is access distributed spatially and economically? The discussion highlights the inequities associated with access to water in terms of cost, geographic location, and ease of acquisition. This is followed by a discussion of the historical origins of contemporary impediments to access. How did decisions taken during the colonial era create the conditions for current inequities? The final section of the article draws attention to some of the ongoing efforts to improve access in both cities. Although these efforts are indeed steps in the right direction, they fall far short of what is necessary, largely because they neither acknowledge nor have the ability to reverse the historical production of inequality.

Unequal access in contemporary Dar es Salaam and Nairobi

Africa is the world's least urbanized continent, with a majority of the population expected to become urban only in 2030. Eastern Africa remains the continent's least urbanized subregion. The situation, however, is changing rapidly. Two of the subregion's largest cities, Dar es Salaam and Nairobi, experienced annual growth rates of 4% between 2000 and 2005. Such growth has been associated with a deteriorating quality of life for the poorest residents, who struggle to meet their basic needs. A recent UN report on African cities, for example, observed that "access to social services and infrastructures is dependent on income rather than population density (although there is a clear correlation between income and residential densities), with excellent standards of provision in well-off areas and next to none in high-density, low-income areas" (UN-HABITAT, 2010). This section shows how residents of both cities' sprawling informal settlements struggle to meet their daily requirements for water.

Dar es Salaam

Among the many challenges confronting Dar es Salaam's residential areas, lack of access to improved water sources is arguably the most enduring, problematic and important. As a result of the water system's low production capacity, high rates of leakage, and limited coverage, the vast majority of the city's households regularly struggle to meet their daily water needs. The scale of the problem is extraordinary. While Dar es Salaam is by no means a mega-city, its population has grown steadily over the past several decades, increasing from 850,000 residents in 1978 to more than 4 million today. Approximately 75% of the city's current residents live in informal settlements. Although the lack of access to water is most acute in informal settlements, the city-wide lack of distribution networks means that residents of surveyed areas are by no means assured delivery of this essential resource.

Two studies of Dar's chronic water woes over the past decade give a good indication of both the scope and the enduring nature of the problem. The first is a formal report that was produced by ActionAid (2004), a prominent international NGO:

The water system [has] failed to keep up with population growth in the city, and by 2003 only 98,000 households in a city of 2.5 million people [approximately 16% of the population] had a direct water connection. Only 26% of water was being billed, 60% was lost through leaks, and a further 13% through unauthorized use, illegal taps and non-payers. Even those with connections only received water irregularly, and the water quality was poor. In low-income areas, the vast majority of households had no water connection at all, relying instead on buying water from kiosks, water vendors or their neighbors, at more than three times the price.

A more recent assessment, produced by Twaweza, a Tanzanian NGO, is even less favourable. In the summer of 2012, the organization undertook another iteration of its Listening to Dar project, an ongoing survey conducted via mobile phones of city residents and their views of major public services (Twaweza, 2012). The study found that more than 76.9% of the city's households did not have access to the water mains via a tap inside either their building or their plot. While the majority of the city's households do claim to have access to an improved source of water, nearly 7% of respondents reported drawing their water from an uncovered well. Studies have shown that using water from these sources triples the risk of diarrhoea.

To make matters worse, many of those who are (un)fortunate enough to have a proper connection with the Dar es Salaam Urban Water and Sewerage Corporation (DAWASCO), the public parastatal company that operates the city's water-supply system, receive water infrequently. Fewer than one in seven households with access to piped water indicated on the Twaweza survey that water had flowed from the tap on all days of the previous week. More than half claimed that they received water no more than three days per week. Part of the problem is that many of these residents are connected to the main transmission pipes with the thin, one-inch 'spaghetti' pipes that are commonly found in the city. These pipes, which are above ground along the side of the road, are often damaged with time or cut as other residents seek out water. The resulting leakage reduces the pressure in the system.

The official price of water in Dar es Salaam is TZS 20 for a 20 litre container, or the equivalent of TZS 1 per litre. (At the current exchange rate of TZS 1600 per USD 1, this works out to just over US cent per container.) This tariff was formally established in June 2010, when the Energy and Water Utilities Regulatory Authority set the price at which water may be sold from public standpipes and water kiosks in Dar es Salaam. Unfortunately, only 13.6% of respondents claimed in a recent survey that a public kiosk or tap was available in their neighbourhood, and a mere 1.2% claimed to purchase water from community taps (Uwazi, 2010). As a result, households in the city without direct access to piped water pay a median price of TZS 100 for 20 litres of water, five times the official rate.

The burden of paying a higher price for water falls disproportionately on the poorest residents of Dar es Salaam. This is because they are more likely not only to reside in unplanned areas that lack access to piped water but also to rent rather than own their houses. Thus, even if they happen to live in a neighbourhood where the distribution network is available, their status as renters provides them with little incentive to pay for a formal connection if the owner has not already done so. The one-time connection fee is simply too costly for many of the city's residents.

Nairobi

The situation in Nairobi is similarly problematic. The majority of the city's residents rate improvement in water provision as their top development priority (Citizen's Report Card, 2007; Gulyani, Talukdar, & Kariuki, 2005). Most of the urban poor lack a private water connection. Even in those estates where there is a household connection, competitive pumping by neighbours exacerbates frequent prolonged scarcities to deny them water

(Citizens, 2007). As a result, "both the poor and the non-poor are consuming little water and incurring high costs for it" (Gulyani et al., 2005).

Income inequality in Nairobi is amongst the highest in African cities, and living conditions in the more than 200 informal settlements are amongst the worst in Africa (UN Habitat, 2010). Infant mortality in these heterogeneous and dynamic informal settlements, where some 60–70% of the population of Nairobi lives, is estimated to be four times the rate of Nairobi as a whole (Zulu et al., 2011). Housing is dense, with families of five to seven people living in small, tin-roofed houses measuring 10 feet by 10 feet. Few residences in informal settlements have toilets. In some areas, neighbourhood groups or NGOs have built small pay-for-use toilet and shower blocks. Services in general are limited. Electricity is available to some through illicit connections to nearby overhead lines. Policing is constrained, and there is generally no garbage collection. The population of Nairobi grew from 120,000 in 1948 to 3.1 million in 2009. Recent population growth rates for the city averaged 5% between 1989 and 1999, and 4% between 1999 and 2009 (Zulu et al., 2011). Most of this growth occurred in the expansion of informal settlements. Until the late 1980s, informal settlements were threatened with demolition. The city has only since 2004 begun to focus on water provision for this population, and is still experimenting with alternative technical and social approaches to provisioning.

In informal settlements, most people buy water from a tap owned by a structure holder. (Land may be notionally owned by the government, but those who built structures there first have use rights and have become de facto landlords.) The water comes from illicit connections to one or more mains, preferably those with 24-hour water pressure, supplying high-prestige areas. There are a few wells and open springs. The unit price of water varies seasonally and in response to local interruptions but is routinely high. Some estimates suggest that households spend 20% of their income on water (UNDP, 2006). There is, in addition, the significant time and work taken to collect water. Multiple 10-to-20-minute trips carrying 44 lb. jerry cans of water are required for a household to do laundry, for example.

In Kibera, the largest informal settlement in Nairobi, women have regular routines for responding to frequent water shortages. These routines include the cessation of informal enterprise activities (sale of snacks, hairdressing, tailoring, laundry and child-care), post-poning laundry and bathing, and reducing the number of meals in the household to one per day (Crow & Odaba, 2010).

In informal settlements, the role of municipal government is limited. This began to change with Kenyan democratic reforms, which resulted in the Water Act of 2002. The Nairobi water utility separated from the city council in 2003 (discussed further below). This enabled a new focus on provision of water to the majority of the population, living in informal settlements. Efforts are being made to improve water supply, also discussed below. But there are setbacks, as when in the summer of 2007 the police enforced the disconnection of water and electricity services as part of a drive against organized crime (Crow & Odaba, 2010), and progress is slow.

Access to basic water resources is very unequal in both Dar es Salaam and Nairobi. Inequality has, unfortunately, kept pace with each city's rapid population growth. The poorest residents of these cities are triply disadvantaged with respect to the availability of water, the effort required to acquire it, and the price paid per litre. These inequalities did not manifest themselves overnight, however; they are the product of each city's particular history. The next section examines the historical origins of these contemporary inequalities. It looks, in particular, at how decisions taken during the colonial era set the stage for each city's chronic water crisis.

Historical origins of contemporary inequalities

Like many African cities, Dar es Salaam and Nairobi came into existence during the colonial era as overseas extensions of metropolitan powers. The former was an important port on the East African coast; the latter, a critical railway centre and beachhead of the white-settler economy in the Kenyan hinterland. Although there are significant differences between these cities, as there are among all African urban centres, they have had to struggle with many of the same issues: overcoming significant poverty; dealing with informal sectors and settlements; managing crime and violence; governing justly; and addressing severe underdevelopment and socio-spatial inequality (Myers, 2011). The lack of equitable access to water resources is yet another issue of common concern, one that can also be traced back to each city's origins as a colonial creation.

Dar es Salaam

The contemporary widespread lack of access to the formal water-supply system in Dar es Salaam is undeniably a consequence of choices made during Tanzania's colonial period. The overarching principle of colonial urban policy in Dar es Salaam was segregation. This was true for the Germans, who were the first European colonizers and the first to define discrete zones of residential settlement for Europeans, Asians (primarily Indians) and Africans in 1891. It was also the case for the British, who administered the territory of Tanganyika and hence its capital, Dar es Salaam, as a League of Nations Mandate at the conclusion of World War I. Europeans primarily resided in Zone I, a large area of premium land which included the original German quarter that extended to the north-east of the city centre, as well as nascent settlements along the coast to the north. The Asian population was concentrated in Zone II, which consisted of the land adjacent to the harbour and the commercial area that was essentially the city centre. Africans were relegated to Zone III, a parcel of land which did not have direct access to either the ocean or the harbour and which was separated from the rest of the town by an empty, sanitary corridor.

Zone III had the least stringent building and sanitation codes. Africans were allowed to construct buildings of any type and with traditional materials, such as mangrove poles and palm fronds. And while the colonial state organized the zone on a grid pattern, it did not regulate either prices or density. As a result, the slow expansion of Zone III did not keep pace with population growth; many Africans were priced out of the area and pushed into the adjacent countryside. Although services in the expanded Zone III were meagre, particularly in terms of water, sanitation, refuse collection and street lighting, at least there was something. There were no services at all in the rural areas beyond the city's boundaries. Thus, the seeds for uncontrolled urbanization were sown early, and it was only a matter of time before the official town and its peripheral communities merged into the sprawling city of today.

The growth of Dar es Salaam's urban population and unplanned settlements did not slow but rather accelerated when Tanzania achieved political independence in December 1961. The postcolonial government abolished the colonial laws that had restricted the flow of Africans into urban areas and dictated their settlement patterns. The results were both predictable and unfortunate: the city grew at unprecedented rates (14% per year between the 1948 and 1967 censuses) and settlement occurred in an ad hoc manner (Lugalla, 1995). The vast majority of the city's new residents found themselves in areas without basic services.

One consequence of Dar es Salaam's long trajectory of uncontrolled growth is that access to water is not *entirely* correlated with wealth. To be sure, those living in the elite neighbourhoods along the coast – formerly known as Zone I – have long had access to piped water. But because most contemporary residential areas have developed far from the limited trunk infrastructure, residents with the economic means to connect to the water-supply system and to pay for regular water services often have no more opportunity to do so than their less fortunate neighbours. That said, they are certainly better positioned to pay the inflated water prices borne by customers not served by the city infrastructure.

Nairobi

Inequalities in both income and in water and sanitation in Nairobi also have their roots in the redistribution of land, and the city planning, of the colonial period. The settler origins and racial zoning of Nairobi are succinctly captured in this paragraph from Amis (1984):

> Nairobi is a settler city which owes its existence to its position on the Uganda Railway and its function as a service center to the white settler economy of Kenya. It was originally conceived as a European city where Africans were 'tolerated' only for their labour power. To achieve this with minimum of public expenditure and a disease-free urban environment, Nairobi was systematically racially zoned in the major plans of 1905, 1927 and 1948. The result of this was an extremely unequal land distribution which means that for the impover-ished African majority the availability of urban land has been and still is severely restricted.

Nilsson and Nyanchaga (2008) argue that racial segregation under colonial rule was gradually transformed, after it was made illegal in 1923, into segregation through the use of strict building codes and sanitary regulations. Housing and services for Africans were limited, partly to restrict migration. The Nairobi City Council allocated only 1–2% of its revenue between 1932 and 1947 to services to Africans. To the extent that facilities were designed and constructed, design criteria around World War II allowed for domestic water consumption of 50 UK gallons (220 litres) per person per day for "non-natives" and 10 gallons (45 litres) per person per day for Africans. After independence, restrictions on migration were reduced, and there was rapid growth of informal settlements.

The race-based segregation of colonial rule was gradually transformed after indepen-dence into residential zones separated first by class and then, within that, by ethnicity. A new African ruling class moved into fully serviced areas of the city as settlers moved out. The unserved African settlements of colonial times turned into residential areas for migrants and the poor:

> The post-colonial governments of Kenyatta and Moi sustained the colonial land policies and did not reverse the huge land inequalities that were created by the colonial government. Instead the post-colonial governments allocated land in favour of a new emerging group of political and economic elites. (Syagga & Mwenda, 2010)

The continuing division of the city between informal, unplanned settlements with limited services and fully serviced residential areas is illuminated by Furedi's (1973) analysis of the "African crowd" in Nairobi. Furedi suggests that the educated African middle class under British rule became the ruling elite after independence. Mass protests of the African crowd, founded in the needs of the non-industrial working class, were suppressed in the final years of colonial rule and immediately after independence. In Nairobi, the repre-sentation of popular needs and the rights of the majority living in informal settlements are only gradually gaining recognition. The families of these petty traders, domestic workers, and private and public employees have only limited voice.

Periodic efforts have been made to extend water and sanitation to cover the whole city, but they have so far been largely ineffectual. Nilsson and Nyanchaga (2008) suggest that institutional change in Kenyan urban water supply generally occurs slowly, over long periods. And no serious attempt has been made to address the insecurity of tenure in much of Nairobi. Such endeavours would be subject to significant resistance from those who have benefited from the accumulation of land.

In both cities, the colonial segregation of land has continued to shape water access and service provision. Racial segregation has generally been transformed into segregation by income and occupation, though there are parts of both cities where households with decent income have inadequate access to water. The historical context of contemporary injustice is infrequently recognized. Open recognition of the colonial roots of deprivation might possibly provide an uncontroversial foundation for reconsidering and addressing these inequities. The next section examines some recent efforts to improve access to water resources in Dar es Salaam and Nairobi. To date, these efforts have had mixed results.

Contemporary initiatives to improve access

Two changes in institutional focus influence contemporary initiatives to provide access to water in both Dar es Salaam and Nairobi: privatization and the emergence of community-based approaches. The privatization of urban water systems occurred with dizzying speed across the global South as the last millennium came to a close. But the multinational corporations that had eagerly jumped at the opportunity to control the sale of an essential resource found it difficult to make a profit or to meet the terms of their contracts. The transfer of water resources back to municipal authorities has occurred in many cities in recent years (cf. Pigeon, McDonald, Hoedeman, & Kishimoto, 2012). Privatization took different forms in Dar es Salaam and Nairobi, but in both cities water companies reformed by the influence of privatization initiated new types of relationship between the utility and the residents of informal settlements.

Dar es Salaam

In the span of just two years, Dar es Salaam privatized the state-owned urban water utility and then returned it to municipal control. Privatization had been an initial condition of World Bank funding for a city-wide project to rehabilitate and extend the water-supply system and sewerage services. The consortium of foreign and domestic investors that was granted the lease proved unable to abide by the terms of their contact, particularly in terms of connecting new customers, and did little to advance the project. This led to the termination of the contract and the creation of the parastatal company, DAWASCO, that is now solely responsible for the management, operation and maintenance of the water-supply system in the city (Dill, 2013).

Although the project is primarily oriented towards large-scale infrastructure improvement, funders and the water authority have recognized that the majority of the city's residents live in unplanned settlements that are not proximate to the main transmission network and thus are not likely to be reached with services for quite some time. In an effort to ensure that the underserved population living in informal settlements received at least stopgap services, community-based actors were mobilized by a range of development actors.

The manner in which the residents of informal settlements are induced to take responsibility for their own development has evolved over the past decade. Initially, there was a coherent effort to produce a particular type of local development actor: the community-based organization (CBO) (Dill, 2009). Typically formed by and for individuals residing in a geographically bounded and administratively defined area, usually a municipal sub-ward, CBOs, which sit at the nexus between urban neighbourhoods and a wide range of local, national and international development actors, were the primary vehicles for popular participation in development efforts from the mid-1990s until the early part of the new millennium.

The Hanna Nassif Community-Based Upgrading Project was the first to facilitate the formation of CBOs as a means to improve infrastructure and services in Dar es Salaam's unserved settlements. Located just 4 km from the city centre, Hanna Nassif is an unplanned residential area that, at the time of the project, had a population of approximately 20,000. Residents experienced seasonal flooding and chronic lack of access to water. The project is noteworthy not only because of its many enduring accomplishments but also because it was the first to be based on the assumption "that successful community involvement requires a Community-Based Organization (CBO) that would be the entry point to the community and would facilitate community mobilization" (Lupala, Malombe, & Konye, 1997 p. 10). The community-based approach to infrastructure improvement that was first put into practice in Hanna Nassif has become fully institutionalized in Dar es Salaam; it has, however, also proved to have some shortcomings. Although improving access to water was typically important for these organizations, they tended to have diverse agendas and a level of autonomy that made them difficult to supervise and direct by external actors. As a consequence, the more recent approach has been to establish water user associations (WUAs) that can be more easily monitored and have limited responsibilities and capacities.

WUAs were first identified in the 2009 Water Supply and Sanitation Act. In contrast to the previous policy that had simply encouraged CBOs to finance and develop local water-supply systems, the new act required each WUA to "meet the costs of operation and maintenance of its water supply system and make a contribution, in cash or kind, to its capital costs" (URT, 2009, part IV, section 37(4)). This means that urban residents are now obligated, at the outset, to pay a share of the costs associated with water-supply and sanitation projects. This requirement may be particularly onerous for the most marginalized members of a community. That said, municipal officials are not merely demanding more from residents; they have also recognized that they need to be more systematic in their efforts to prepare residents, through their WUAs, to carry out their responsibilities. This is yet another logical step in the institutionalization of the community-based approach to infrastructure development and service delivery. To date, there are no published evaluations of WUAs in Dar es Salaam, so the jury is still out on whether their impacts will be equitable and sustainable.

Nairobi

In Nairobi there was a different pattern of change in water institutions and in government relation to informal settlements. In contrast to Dar es Salaam, there has been only periodic reliance by the water utility on community-based groups, and only in some informal settlements. Nevertheless, changes in water institutions and services to informal settlements reflect the gradual (and incomplete) reversal of the city and national governments' approach toward these settlements.

In Kenya (and in many cities of the global South), the approach of city and national government toward informal settlements (slums) has evolved through three phases:

(1) Demolition and eviction, and the long continuing shadow of demolition and illegality. In this phase, water and other services were rarely provided.
(2) Reluctant recognition of the needs of informal settlements, even though they are excluded by the insecurity of land tenure, informality of institutions, and inattention by city government.
(3) Current attempts to upgrade and improve slums.

Syagga (2012) describes how the demolition-and-eviction phase in Kenya (1895–1970s) expanded the area of slums because there were no alternatives. The second phase (1980s), encouraged by civil rights groups and international pressure, recognized the need for slum improvement but was curtailed in the 1990s by the reduction of state funding and changes of government roles required by structural adjustment loan conditions. Syagga notes that in Nairobi at this time, slum upgrading was "stalled but slum development [expansion] continued". The third and current phase, starting in the late 1990s, seeks to integrate slum upgrading into wider development concerns.

In this third phase, slum upgrading in Nairobi has often, with some exceptions (Otiso, 2003), taken the form of demolition of sections of informal settlements in order to replace them with high-quality tenements. Informal-settlement residents are mostly unable to take advantage of the new housing (Nilsson & Nyanchaga, 2008). This new real estate is desirable, and it is taken by the better-off. Such 'gentrification' and renewed eviction of the poor has been widely noted in Sub-Saharan Africa (Gulyani and Bassett 2007).

The water department of the Nairobi City Council was given autonomy, but in contrast to the situation in Dar es Salaam, private water companies from Europe or America were not put in charge. In 2003, the Nairobi Water and Sewerage Company was established. This change was in the context both of the wave of privatization across Africa and, more importantly, of a set of reforms associated with the establishment of a new constitution for Kenya, which included a human right to water (Government of Kenya & Ministry of Water and Irrigation, 2007). The new Nairobi Water and Sewerage Company (NWSC) was still owned by the Nairobi City Council, but with separate revenue and responsibilities. This meant that water revenues were no longer available to the municipality for other purposes than water and sanitation. There have been advances in water and sanitation, but the water agency continues to be racked by allegations of corruption. The NWSC board of directors was forced to resign in 2009 (IRC, 2009).

Soon after its establishment, the NWSC started a separate Informal Settlement Department with a distinct mandate. The organization as a whole was to be focused on revenue, but this new department had a social responsibility to extend water and sanitation access.

Notwithstanding the utility's recognition of the needs of informal settlements, the social context of water supply has made recent initiatives largely unsuccessful. National political upheavals, the influence of traders providing illicit water supply within settlements, and the internal politics of the Nairobi water company, slow or undermine the realization of initiatives.

Two recent schemes (Crow & Odaba, 2010) have involved disparate experiments in new relations between the water company and the residents of particular informal settlements. The first scheme tried to engage with the prevailing structure of illicit water trade. The second sought to build community-based organizations around new pipe and meter systems. These pilot schemes were initiated in three large informal settlements in Nairobi: Kibera, Mukuru and Mathare.

The first scheme, Maji Bora Kibera (Better Water for Kibera) emerged over several years, from 2003 to 2007. This scheme sought to regularize relations between the Nairobi water company and the existing illicit and sometimes powerful water traders who provided water in Kibera. It was built on the extensive discussions of Shagun Mehrotra, a junior professional with the World Bank Water and Sanitation Program (Brocklehurst, Mehrotra, & Morel, 2005). Nearly 80% of the traders in Kibera joined the association. It planned to negotiate unpaid bills with the water company, regularize existing illicit connections, forgo the 'bypassing' of water meters, report leaks and end corrupt connections with utility plumbers. The association made initial headway in its meetings with the water company. But the negotiations broke down in a stormy meeting in spring 2007. And a few months later, in the wake of conflict over the 2007 elections, the Nairobi police directed the disconnection of illegal water and electricity connections in informal settlements across Nairobi, including Kibera. After that, water traders in Kibera were distrustful of the company, and Maji Bora Kibera went into decline. This innovative approach, albeit depending on often-powerful traders, many of whom are also structure holders or de facto landlords, is now largely forgotten.

A second approach in Nairobi tried to connect new pipes and meters to 'community organizations'. It followed the model of a modestly successful water-access project in an informal settlement in Kisumu in western Kenya (Nilsson, 2011). In Kisumu, a new network of pipes had been laid, reaching from the Kisumu water company trunk mains into the informal settlement. The management and maintenance of these pipes was left to community groups and traders. They made repairs, organized connections, and collected bills. In this way, the scheme established 'delegated management'. The Nairobi water company tried to follow suit with a similar delegated-management scheme in the city's Mukuru informal settlement. Here they ran into trouble, because powerful water traders running the pre-existing Mukuru water system took over control of the 'community groups' managing the water (Crow & Odaba, 2010).

Another iteration of this approach was tried in another large informal settlement, Mathare. This time there was prior engagement with community-based organizations. The outcome of this ongoing scheme is not yet apparent. But by 2012, the approach of delegated management to community-based organizations was out of favour. The engineer newly taking charge of the Informal Settlements Department told one author that he preferred direct ownership and management of all pipelines by the Nairobi water company.

In contrast to Dar es Salaam, community-based organizations in Nairobi have taken a less prominent role in the provision of water and sanitation, but their role has been increasing since the water reforms passed in 2002.

Contemporary water-access initiatives in Dar es Salaam and Nairobi have focused on the social organization of water supply at least as much as on the technological. They begin to recognize the key role of community-based actors and their relation to water utilities. This could be the beginning of an intermittent and, in the case of Nairobi, uncertain democratization of water access. In Dar es Salaam, the community-based approach has created new development actors, and the approach has been institutionalized. In Nairobi, an initiative to build on the existing structure of water trade in one settlement has been lost, and the experiment in delegating the management of local pipes and bill collection to community groups appears out of favour.

Conclusion

The widespread lack of access to water in Africa's informal settlements is much more than an unfulfilled Millennium Development Goal. It is fundamentally a crisis of injustice and inequality. Regular and affordable access to water is indispensible for people to have the opportunities to live lives that they have reason to value and for societies to ensure a modicum of equality. Access to water is essential for many of the informal enterprises that guarantee the livelihoods of the majority of Africa's urban dwellers. Easy access to water within a house or compound also offers the possibility of freeing women and girls from the physical and temporal demands of water collection. While water access is a major concern for all residents in Dar es Salaam and Nairobi, the difficulty of hauling water is particularly pronounced in the informal settlements that are significant portions of both cities. This is an inequality that has only recently begun to be recognized as an injustice between rich and poor. Rooted in the segregation of colonial rule, it is sustained by the continuing injustice of land policies, the lack of political will, and the multiple complications involved with upgrading urban settlements. Community-based urban upgrading projects have had multiple successes, particularly in Dar es Salaam, where they are more established. But they have also left large numbers of urban residents without access to a basic necessity.

Nevertheless, we wish to reiterate that the current lack of access is inequitable and unjust, not merely in terms of the quantity of water consumed or its cost but in terms of denying all residents the same opportunities to live lives that they value. In addition, we wish to stress that although ongoing efforts to rectify the situation may be a step in the right direction, they fall far short of what is necessary, largely because they neither acknowledge nor have the ability to reverse the historical production of inequality.

Access to water is a recognized priority for the residents of informal settlements in both cities. Focus on this as an injustice with roots in colonial rule could provide a plane above current social divisions that a range of actors, including CBOs and city government, could use to push city-wide initiatives. Steps toward democratization of water access have been made in both cities. Beyond these steps, a political focus on remedying water inequities could enable change that both addresses immediate community needs and might in some cases provide some element of residential security.

References

ActionAid. (2004). *Turning off the taps: Donor conditionality and water privatization in Dar es Salaam, Tanzania*. London: ActionAid International.

Amis, P. (1984). Squatters or tenants: The commercialization of unauthorized housing in Nairobi, Kenya. *World Development, 12*(1), 87–96. doi:10.1016/0305-750X(84)90037-8

Bakker, K., Kooy, M., Shofiani, N. E., & Martijn, E. (2008). Governance failure: Rethinking the institutional dimensions of urban water supply to poor households. *World Development, 36*(10), 1891–1915. doi:10.1016/j.worlddev.2007.09.015

Bartram, J., & Cairncross, S. (2010). Hygiene, sanitation, and water: Forgotten foundations of health. *PLoS medicine, 7*(11), e1000367. doi:10.1371/journal.pmed.1000367

Brocklehurst, C., Mehrotra, S., & Morel, A. (2005). *Field note: Rogues no more? Water Kiosk operators achieve credibility in Kibera*. Nairobi, Kenya: WSP.

Citizens'-Report-Card. (2007). Citizens' Report Card on Urban Water, Sanitation and Solid Waste Services in Kenya: Summary of Results from Nairobi, Kisumu and Mombasa. Nairobi, Supported by Water and Sanitation Program, Ministry of Water and Irrigation, SANA, Kara, Ilishi Trust.

Crow, B., & Odaba, E. (2010). Access to water in a Nairobi slum: Women's work and institutional learning. *Water International, 35*(6), 733–747. doi:10.1080/02508060.2010.533344

Dill, B. (2009) The paradoxes of community-based participation in Dar es Salaam. *Development and Change, 40*(4), 717–743. doi:10.1111/j.1467-7660.2009.01569.x

Dill, B. (2013). *Fixing the African state: Recognition, politics, and community-based development in Tanzania*. New York, NY: Palgrave.

Fewtrell, L., Kaufmann, R. B., Kay, D., Enanoria, W., Haller, L., & Colford, J. M. Jr (2005). Water, sanitation, and hygiene interventions to reduce diarrhoea in less developed countries: a systematic review and meta-analysis. *The Lancet Infectious Diseases, 5*, 42–52. doi:10.1016/S1473-3099(04)01253-8

Franks, T., & Cleaver, F. (2007). Water governance and poverty: A framework for analysis. *Progress in Development Studies, 7*(4), 291–306. doi:10.1177/146499340700700402

Furedi, F. (1973). The African crowd in Nairobi: Popular movements and élite politics. *The Journal of African History, 14*(2), 275–290. doi:10.1017/S0021853700012561

Goldman, M. (2005). *Imperial nature*. New Haven, CT: Yale University Press.

Government of Kenya, Ministry of Water and Irrigation. (2007). *Water sector reform in Kenya and the human right to water*. Nairobi: Ministry of Water and Irrigation.

Gulyani, S., & Bassett, E. M. (2007). Retrieving the baby from the bathwater: Slum upgrading in Sub-Saharan Africa. *Environment and Planning C: Government and Policy, 25*(4), 486–515. doi:10.1068/c4p

Gulyani, S., Talukdar, D., & Kariuki, R. M. (2005). Universal (Non)service? Water Markets, Household Demand and the Poor in Urban Kenya. *Urban Studies, 42*(8), 1247–1274. doi:10.1080/00420980500150557

Gundry, S., Wright, J., & Conroy, R. (2004). A systematic review of the health outcomes related to household water quality in developing countries. *Journal of Water and Health, 02*(1), 1–13.

IRC. 2009. Kenya: Nairobi water board sent packing following reports on malpractices. Retrieved from http://www.source.irc.nl/page/49480.

Jalan, J., & Ravallion, M. (2003). Does piped water reduce diarrhea for children in rural India?. *Journal of Econometrics, 112*, 153–173. doi:10.1016/S0304-4076(02)00158-6

Lugalla, J. (1995). *Crisis, urbanization and urban poverty in Tanzania: A study of urban poverty and survival politics*. Lanham, MD: University Press of America.

Lupala, J., Malombe, J., & Konye, A. (1997). Evaluation of Hanna Nassif Community Based Urban Upgrading Project, Phase I. Technical Report. International Labor Organization.

Mara, D. D. (2003). Water, sanitation and hygiene for the health of developing nations. *Public Health, 117*, 452–456. doi:10.1016/S0033-3506(03)00143-4

Myers, G. (2011). *African cities: Alternative visions of urban theory and practice*. London and New York: Zed Books.

Nilsson, D., & Nyanchaga, E. N. (2008). Pipes and politics : A century of change and continuity in Kenyan urban water supply. *The Journal of Modern African Studies, 46*(1), 133–158. doi:10.1017/S0022278X07003102

Nilsson, D. (2011). Pipes, Progress, and Poverty Social and Technological Change in Urban Water Provision in Kenya and Uganda 1895–2010. Stockholm Papers in the History and Philosophy of Technology.

Otiso, K. M. (2003). State, voluntary and private sector partnerships for slum upgrading and basic service delivery in Nairobi City, Kenya. *Cities, 20*(4), 221–229. doi:10.1016/S0264-2751(03)00035-0

Pigeon, M., McDonald, D., Hoedeman, O., & Kishimoto, S. (Eds.) (2012). *Remunicipalisation: Putting water back into public hands*. Amsterdam: Transnational Institute.

Syagga, P. M., & Mwenda, A. K. (2010). Political economy and governance issues surrounding policy interventions in the land sector in Kenya. Final Report prepared for the World Bank. Retrieved from http://s3.marsgroupkenya.org/media/documents/2011/02/5d3fa1a0637024c34db8ab899d7c6242.pdf

Syagga, P. M. (2012). Land tenure in slum upgrading projects. In *Slum upgrading programmes in Nairobi: challenges in implementation*. Nairobi: French Institute for Research in Africa (IFRA); 2011.

Twaweza. (2012). "Listening to Dar, Report 15: Access to Water." Retrived from http://dl.dropboxusercontent.com/u/6504020/listeningtodar/R15 ENG.pdf

UNDP. (2006). *Human development report 2006: Beyond scarcity: power, poverty and the global water crisis*. New York: UNDP.

UN Habitat. (2010). The State of African Cities 2010: Governance, Inequality and Urban Land Markets Nairobi, November.

UN MDG. (2012). *The millennium development goals report 2012*. New York: United Nations.

UN MDG. (2013). *The millennium development goals report 2013*. New York: United Nations.

UN. (2013). *A new global partnership: Eradicate poverty and transform economies through sustainable development*. New York: United Nations.

UN. (2014). *Millennium development goals and beyond 2015*. Retrieved from http://www.un.org/millenniumgoals/

United Republic of Tanzania (URT). (2009). *The water supply and sanitation act, 2009*. Dar es Salaam: Government Printer.

Uwazi. (2010). Water prices in Dar es Salaam: Do water kiosks comply with official tariffs? Policy brief TZ.09/2010E. Technical report, Uwazi at Twaweza, Retrieved from http://www.uwazi.org/uploads/files/Water kiosks in DSM Englsih.pdf

World Bank. (2010). *An evaluation of World Bank Support, 1997-2007: Water and development, Volume 1*. Washington, DC: The World Bank.

Zulu, E. M., Beguy, D., Ezeh, A. C., Bocquier, P., Madise, N. J., Cleland, J., & Falkingham, J. (2011). Overview of migration, poverty and health dynamics in Nairobi City's slum settlements. *Journal of Urban Health, 88*(Supplement 2), 185–199.

Popular participation, equity, and co-production of water and sanitation services in Caracas, Venezuela

Rebecca McMillan, Susan Spronk and Calais Caswell

School of International Development and Global Studies, University of Ottawa, Ottawa, Canada

This article argues that the technical water committees in Venezuela are an example of co-production of public service delivery between state and citizen. In practical terms, the committees help to reduce information asymmetries between service providers and citizen-users and improve accountability. Unlike depoliticized notions of co-production that have been celebrated in the mainstream development literature, however, this experiment in urban planning promotes participation as empowerment, because the committees are part of a wider political agenda, engage citizens in a broader process of social change, promote rethinking of the concept of citizenship, and have thus far avoided elite capture.

Introduction

The World Health Organization (2012) estimates that 780 million people still lack access to improved drinking water and 2.5 billion people have no access to improved sanitation. The UN Millennium Development Goals (MDGs) (United Nations, 2013) include an ambitious target for water and sanitation: to cut in half by 2015 the proportion of people without sustainable access to safe drinking water and basic sanitation. Public water utilities in developing countries are routinely criticized for failing to provide adequate water services to the poor. Indeed, the perceived failure of governments to provide 'water for all' is one of the reasons for the dramatic changes in urban water governance over the past two decades. Often large and inefficient, public water utilities struggle as institutions to remedy failed state-led planning models.

To date, much of the debate on improving equity in water-sector reforms has been polarized between assessing the merits of 'public' versus 'private' forms of delivery. This debate, however, has tended to obscure the principal problem in developing-country contexts: the systematic failure of water companies to connect the poorest of the poor, no matter who owns and operates them. Indeed, the barriers that limit poor people's access to water – such as poverty and political powerlessness – are likely to persist whether the provider is publicly or privately owned and operated. The formation of participatory institutions that include key stakeholders in the service-delivery process has therefore been promoted as a way to promote good governance and equity in the water and sanitation sector.

This article discusses the impact of citizen participation in service delivery in the public water and sanitation utility in Caracas, Venezuela. In Caracas, 'technical water committees' (*mesas técnicas de agua*, or MTAs) emerged in marginalized neighbourhoods in the early 1990s to meet the needs of poor populations. Since the election of Hugo Chávez in 1998, the process of establishing MTAs has been institutionalized and promoted at the national level. Despite the apparent success of the model, and the opportunities it presents for learning about 'best practices' in participation, the MTAs and other participatory initiatives in Venezuela have received little attention in the international development literature (Buxton, 2011).

This article argues that the MTAs in Venezuela represent a form of co-production that has sought to improve equity by empowering poor citizens. In practical terms, the MTAs help to reduce information asymmetries between service providers and citizen-users, and improve accountability. Unlike depoliticized notions of co-production that have been celebrated in the mainstream development literature, this article argues that citizens' participation in the MTAs creates the possibility of empowerment, because the committees engage citizens in a wider process of social change and promote a radical rethinking of the concept of citizenship.

The first part of the article reviews the mainstream literature on co-production in development sociology, outlining an alternative framework based on the insights of Hickey and Mohan (2005), who argue that empowerment initiatives must be part of a broader political project, change power relations, expand the notion of citizenship, and avoid elite capture. The second part of the article recounts the history of the MTAs and discusses their achievements based on Hickey and Mohan's criteria for empowerment. The conclusion discusses the productive tensions that emerge given the dialectical relationship between 'invited' (officially sanctioned) and 'invented' (autonomous, grassroots) spaces.

Co-production, participation and empowerment

Much of the mainstream development literature on co-production is built on the notion that civic action creates synergistic relationships between the state and civil society, in a process known as 'building social capital'. In a path-breaking symposium published in *World Development*, eminent development sociologist Evans (1996) claims that state–society synergy can be a catalyst for development in conditions where there is complementarity, simply defined as the "mutually supportive relations between public and private actors", and embeddedness, or "the ties that connect citizens and public officials across the public-private divide" (p. 1120).

Similar to 'social capital', the notion of co-production is conceptually ambiguous and used to refer to a wide range of institutional arrangements (Bovaird, 2007; Mitlin, 2008). Part of the theoretical ambiguity around the concept of co-production relates to the fact that it tends to be simply defined as an organizational form into which any ideological content can be poured. Ostrom's (1996) definition of co-production, for example – "the process through which inputs used to provide a good or service are contributed by individuals who are not in the same organization" (p. 1073) – includes anything from public–private partnerships between the state and multinational corporations to community wells managed by a community and a local non-governmental organization.

In the contemporary context of neoliberal austerity, the concept of co-production has been mobilized to justify shrinking public spending and the withdrawal of the state from guaranteeing the conditions of social citizenship. Neoliberal reforms have eroded peoples' livelihoods and access to the most essential services, at the same time that they have opened up certain public realms of participation from which 'the poor' had been previously excluded (Miraftab, 2004; Molyneux, 2008). As Jaglin (2002) and Spronk (2009) observe, when coupled with the privatization of public services, participation is mobilized as a way to stabilize commercial relationships, as poor people find themselves volunteering to shovel ditches to shore up the profits of multinational water companies. Such co-production arrangements work to legitimate unequal power relations, not to change them.

The above analysis suggests that participation within co-production arrangements will not always lead to democratization and material improvements for the poor. Based on their critical review of participatory development initiatives, Hickey and Mohan (2005) identify four factors that increase the likelihood that participation will promote social transformation rather than maintain the status quo. First, participation must be part of a broader political project that seeks to challenge existing power relations, "rather than simply work around them for more technically efficient service delivery" (p. 250). Second, they argue that participatory approaches have the greatest potential to achieve social transformation where they aim not only to change power relations within the scope of development interventions, e.g. by changing power dynamics between planners and citizens, but also to address broader structures of oppression and envision more inclusive and sustainable development models. The third criterion, the pursuit of citizenship, involves not just bringing people into the political process but also democratizing these processes "in ways that progressively alter the 'immanent' processes of inclusion and exclusion that operate within particular political communities, and which govern the opportunities for individuals and groups to claim their rights to participation and resources" (p. 251). Lastly, attention needs to be paid to local power dynamics, particularly how participation may simply conceal the persistence of patronage relations that reinforce elite privilege. In short, participatory arrangements must avoid elite capture.

The authors clarify that not all of these conditions must be in place to make participatory initiatives worthwhile. Indeed, they suggest that political learning can take place in any form of participation. The following analysis, based on 19 semi-structured interviews and participant observation conducted over 4 months in Antímano, a poor parish (a subdivision of the municipality) in western Caracas, suggests that the MTAs represent a case of empowerment, because they meet all 4 criteria.

The case study of Antímano has been chosen because the parish is home to a number of Venezuela's longest-running MTAs and because current and former staff of Hidrocapital, the state water utility for Caracas, identified Antimano's MTAs as an example of 'best practice' due to their high and sustained levels of participation.[1] Antímano has also been classified as the poorest parish in Caracas (Goldfrank, 2011, p. 113). A single case study was chosen to assess changes in service quality, relationships between the communities and state representatives, and community development. However, findings from this case may not be representative of all MTAs, and further comparative research across cases is clearly needed to shed more light on this innovative reform process.

The technical water committees in Caracas, Venezuela

Water service problems have long plagued Caracas, particularly the city's populous hillside *barrios* (low-income, informal settlements). Prior to 1999, government water service policy was highly discriminatory. It prioritized building networks in the formal neighbourhoods and within the formal city, and it privileged the high-income areas over the middle-income areas (Cariola & Lacabana, n.d., p. 6). The result is the present situation of 'water apartheid': the upper- and middle-class areas, where the majority of the residents self-identify as 'white' according to the most recent census, benefit from high-quality services, while the lower-income areas, where most residents identify as 'mixed-race', develop informally in the absence of attention from the state.

The *barrios* grew due to progressive waves of rural–urban migration in the twentieth century. With the dawn of the petro-economy in the 1920s, wealth and employment opportunities became concentrated in urban areas, and the agricultural economy declined, which fueled migration to Venezuela's cities, particularly Caracas. New migrants settled all corners of the capital, often illegally building makeshift homes on public or privately owned land in marginal areas without access to urban services. Once their houses were established, residents mobilized to build clandestine connections to the urban water system, or eventually secured limited infrastructural improvements through public demonstrations or clientelistic ties with the party in power. The resultant piecemeal development of services constitutes what Bakker (2010, p. 22) describes as an 'archipelago': incomplete, fractured water and sanitation networks, and highly uneven service access within neighbourhoods.

The high altitude of the *barrios* makes delivering services in these areas particularly difficult. The city's average neighbourhood is at 800–1000 m above sea level – above its principal water reservoirs – and many *barrio* neighbourhoods are at much higher elevations. Considerable energy and an elaborate infrastructure are needed to pump water to households. For this reason, the Caracas aqueduct system is considered one of the most complex in the world.[2]

Neglected by water-utility and state officials, *barrio* residents' only recourse for poor services was public protest. To demand improvements, residents would deliver petitions, block main thoroughfares, occupy the offices of Hidrocapital, and even temporarily kidnap utility officials until they agreed to improve service delivery (Arconada, 2005a; Francisco, 2005; McCarthy, 2009). These water-related protests reached their peak during the early 1990s, prompting the newly elected, left-leaning mayor (1993–1996) of Caracas's Libertador Municipality, Aristóbulo Istúriz, and his team to seek innovative solutions to the water crisis. The water situation in the city had gone from bad to worse in the late 1980s and early 1990s. Neoliberal reforms introduced during that period reduced public spending on services, which revealed the inequity of Venezuela's political system. From 1958 to 1993, the country had a 'pacted democracy' referred to as Punto Fijo, under which the electorate accepted the sharing of power between the three dominant political parties[3] with the promise of limited redistribution. Under neoliberalism, the political spoils-sharing continued; only social redistribution declined dramatically (Buxton, 2004).

The water sector was deeply affected by the corruption of Punto Fijo period. From 1974 onwards, the Instituto Nacional de Obras Sanitarias (INOS), at the time responsible for service planning, delivery, regulation and infrastructure, was plagued by clientelism and corruption, severely undermining the institute's ability to provide services. There were allegedly people on the payroll of the utility that did not actually work there but were rather political allies of the dominant parties (Francisco, 2005). This inefficiency, combined with decreased investment in services in the 1980s, led to a marked deterioration of urban services, and ultimately to the dismantling of INOS in 1989. In its place, the *hidrológicas* were created: a national water company, Hidroven, and its regional subsidiaries, including Hidrocapital, which provides water services for the capital region. That same year, there were over 500 recorded water protests in Venezuela, more than one a day (Francisco, 2005).

Santiago Arconada, a long-time labour organizer who also worked in the Istúriz and later the Chávez administration, reflects on the situation in his home parish of Antímano, a popular area in the west of the Libertador municipality: "When Aristóbulo's administration began in 1993, the water crisis was dire. We had water only once every two months! And during the times when we didn't have piped water, we had to wait in line to get water from tanker trucks. It was horrible."[4] Sewage ran freely in the streets.

Widespread discontent with the neoliberal reforms and the corruption of the Punto Fijo system eventually led to Hugo Chávez's electoral victory. However, before the left turn at the national level, the cracks in the Punto Fijo system were already manifest at the level of municipal politics. Reforms introduced in 1984 and 1989 mandated a degree of fiscal decentralization,[5] and in 1992 legislation passed that replaced centrally appointed regional officials with directly elected ones. This allowed for the election of mayors and governors not affiliated with the three dominant parties, including members of Aristóbulo Istúriz's left-leaning La Causa Radical party (Fernandes, 2010, p. 59). It also opened up political space for progressive local development initiatives. However, given their highly localized nature and limited financing, they were ill equipped to offset the devastating negative impacts of neoliberal reforms.[6]

The MTAs were among these local experiments. When Mayor Istúriz came to office in 1993, he assembled a "change team" of progressive reformers within his administration to address the water and sanitation problem and other pressing urban issues. Many members of the team had long histories of organizing in the *barrios*, student movements, unions, and other progressive organizations, and shared a strong commitment to people's participation. Together, they piloted a new model of participatory local governance called the 'parish government', which brought together local civil society, community members, and city councillors.

At a now famous parish meeting in Antímano on 6 March 1993, the idea for the MTAs was first conceived (Goldfrank, 2011). Together with progressive reformers within Hidrocapital, including Jacqueline Faría, then manager of the Caracas metropolitan water system, the municipality overcame national-government resistance and gained approval for the proposal.[7] With few resources, but a strong commitment to the project, the team implemented pilot water committees in two of Caracas's populous parishes: Antímano and El Valle.

The early initiatives were extremely successful in improving local services and encouraged unprecedented levels of community organization in the *barrios*, but the experiment was short-lived. In the 1995 elections, Istúriz lost to the right-wing candidate Antonio Ledezma. City officials, threatened by the reforms, quickly disbanded the MTAs and parish governments (Goldfrank, 2011). However, the early experience with the MTAs provided valuable lessons that would be taken up again after the election of Chávez in 1998. Indeed, Istúriz and several social movement activists who were part of his original change team would go on to play key roles in the Chávez administration.

When Hugo Chávez assumed office in 1999, the water crisis was one of his mandate's first political challenges. The electorate had huge expectations of the political outsider who had won a landslide victory on an anti-neoliberal social democratic platform. According to Santiago Arconada, "He rose to the occasion in such a forceful, convincing way that in the first few years the water experience was really the face of the Bolivarian Process."[8]

In March and April of his first year in office, Chávez appointed a new leadership team for the water sector. Jacqueline Faría[9] was appointed president of Hidrocapital. Faría, "an engineer by education and an activist by vocation", had first-hand experience with the

MTAs from her tenure in Hidrocapital during the Istúriz administration (McCarthy, 2009, pp. 10–11). Fellow progressives Cristobal Francisco Ortiz and Alejandro Hitcher Marvaldi were appointed vice-president of Hidrocapital and president of Hidroven, respectively (Arconada, 2005b). All three had sharpened their political teeth in student movements in the 1980s (Perfil biográfico de Alejandro Hitcher, 2010).

From 15 to 30 May 1999, the new leadership convoked a workshop of long-time social activists to discuss what became known as the 'communal management' of Hidrocapital. Delegates came from a variety of backgrounds and included trade unionists, students, environmentalists, cooperative activists, academics, and members of different neighbourhood and cultural groups. The MTA experience under Istúriz figured centrally in the workshop discussions, and many elements of that experience were ultimately adopted. In 1999–2000, Hidrocapital implemented a variation on the MTA model throughout the capital region (Arconada, 2005b).

The first step in creating a people-centred water service was to establish Hidrocapital's community management office, which serves as the main point of contact between communities and the utility (McCarthy, 2009, p. 11). In its first two years, the MTA model proved itself unequivocally in Caracas. When two disasters hit the capital – the Vargas landslide[10] and the collapse of the El Guapo River Dam, which left parts of the capital without water for days – the organizations helped ensure that people could access water during the crises. And they were gaining national recognition for their achievements. By 2001, thanks to success at the local level, MTAs became national public policy (Lacabana et al., 2007). Today, there are an estimated 9000 MTAs nationwide;[11] as of 2011, the MTAs had initiated 1500 community-managed infrastructure projects (Mesas Tecnicas de Agua, 2011).

MTAs as part of a broader political project

There is little doubt that Chavez's Bolivarian Revolution (or *el proceso* as it is known in local terms) is one of the most radical experiments of the New Left in Latin America, and that as such the MTAs are part of a broader process of political change. What is less well known is that the MTAs and many of the other participatory initiatives promoted by the Chávez government, such as the communal councils,[12] were actually inspired by earlier models that predated Chávez by at least a decade. In this sense, the Bolivarian process is not a simple 'top-down' project but rather a dialectical process in which the state responds to local struggles and initiatives that emerge outside of the official channels of the state or at local levels of government. These grass-roots initiatives are then 'scaled up' and institutionalized at the state level.

Since the beginning of the process, Venezuela has seen many concrete gains in terms of advances in health, education and housing, thanks to a massive increase in social investments funded by the country's vast oil wealth. From 1998 to 2006, social spending as a percentage of GDP increased from 8.2% to 20.8% (Griffiths, 2010, p. 614). As a result of social spending and the state provision of welfare through the various missions of the Bolivarian Revolution, numerous quality-of-life indicators have also improved. According to Wilpert (2005), "infant mortality … dropped from 18.8 per thousand to 17.2 per thousand between 1998 and 2002, and life expectancy … increased from 72.8 to 73.7 years in the same period" (p. 20). Between 2002 and 2005, poverty and extreme-poverty rates in Venezuela decreased by 18.4 and 12.3 per cent, respectively, which was "the second sharpest decline in the continent" (Ellner, 2010, p. 90). More recent figures indicate that the percentage of Venezuelans living below the poverty line fell from 49.4%

in 1999 to 27.6% in 2008 and that inequality seems to have been reduced by as much as 18% in the same period (Grugel & Riggirozzi, 2011, p. 10).

These impressive accomplishments can be attributed partly to participatory local projects in which thousands of citizens have participated. In the aftermath of the devastating bosses' strike in 2002–2003, Chávez announced seven 'missions' aimed at delivering health, education and food to ordinary people, with the active participation of the citizenry. This represents a radical change from the way that social programmes have traditionally been delivered in the country (Hawkins et al., 2011; Mahmood et al., 2012).

Like these other programmes, the MTAs have been accompanied by a dramatic increase in state spending, which distinguishes them from neoliberal forms of co-production in which citizen participation compensates for declining public investment in water and sanitation. According to Ministry of Environment statistics, the government has invested an average of USD 600 million per year in improving services since Chávez's election in 1998 ("Estado venezolano", 2011), meaning that Venezuela has one of the highest per capita investment levels in water and sanitation in Latin America. Between 2001 and 2011, it is estimated that the government invested a total of USD 7.518 billion in water and sanitation, compared with only USD 2.4 billion between 1989 and 1998 (Castillo, 2011) – a threefold increase. Between 2010 and 2012, special funds invested in the water and sanitation sector totalled close to USD 2.718 billion, of which USD 1.18 billion was related to sanitation (Interamerican Development Bank, 2012, p. 4).

Changing power relations between state and citizen

The MTAs represent a co-production arrangement between the state and citizens that is part of a broader process of changing state–society relations. A key element of the model is its attempts to bridge the divide between 'development experts' and community members by mobilizing knowledge for both technical and political ends. This approach to grass-roots co-production draws on local expertise, not just as a means for collecting technical information but also as a way to raise the political capacity of the poor to make claims on the state. As Diana Mitlin (2008) observes, more attention needs to be "given to co-production as a political process that citizens engage with to secure changes in their relations with government and state agencies, in addition to improvements in basic services" (p. 352). In her work on 'bottom-up', grass-roots co-production arrangements amongst shack dwellers in Pakistan, Namibia, South Africa and India, Mitlin (2008) explores how social movements have used co-production as a means to advance access to services and goods that meet basic needs, and also to change the role of citizens in relation to the state programmes. Mitlin goes beyond the instrumental focus of the co-production literature to emphasize how these experiences have helped build the political capacity of poor people's movements to make claims on the state, but also to go beyond it. In what she calls "grassroots co-production", local social-movement organizations involved in advocacy over housing rights challenge the notion that the only legitimate agents for the planning and construction of infrastructure are state agencies.

As in the housing projects analyzed by Mitlin, citizen participation in the MTAs serves both an instrumental and political function. Since the water system is so complex, Hidrocapital does not have the capacity to provide adequate services in the *barrios* without information from users. Inspired by the 'pedagogy of the oppressed' approach of Brazilian educator Paulo Freire, the MTA methodology is also a political strategy for breaking down the intellectual division of labour between those who plan and make decisions (the bureaucrats and technocrats) and those who carry out community work.

To form an MTA, a community must follow a participatory three-step methodology with support from the utility: (1) census; (2) plan or sketch; and (3) diagnosis. Through the census, community members collect data on who lives in their neighbourhood and the status of their services. The plan or sketch is a map of the community. The map is important because the utility needs to know where the communities have laid their pipes; however, the exercise is also a symbolic way for communities to assert their right to the city. Arconada recalls that under the administration of Istúriz, the *barrios* of Antímano (home to over 150,000 people) did not even appear on Caracas city maps.[13] The map is a way of putting communities in 'the plans', assigning themselves importance in political priorities. The mapping process also serves as a way of building a collective memory of the community and its history. The third and final step is a diagnosis of the service deficiencies. In cooperation with the utility, the communities identify problems and plan solutions, often through community-managed infrastructure projects.

The most significant formal channel through which MTAs may influence government policy and planning is the community water council (*consejo comunitario de agua,* or CCA). At regularly scheduled CCA meetings, representatives from the technical water committees that share the same water distribution system (or from the same parish, depending on its size) come together with water utility staff to discuss service issues. The council has three main functions: to prioritize issues from identified needs; to organize solutions and create work plans; and to follow up on work. The monitoring function is crucial because the council acts as a form of "social control" over the water company (Arconada, 2005a, p. 134). According to Arconada, the oversight led to immediate changes in the way the water utility worked, because it was required for the first time to document its work in a way that was presentable to the public (2005b, p. 195).

The biggest achievement of information sharing through the MTAs has been in managing water distribution in areas where piped water is delivered to households in rotation. In elevated parishes such as Antímano, there is insufficient pressure to pump water to all sectors (neighbourhoods) simultaneously, so most sectors receive water only periodically. Before, communities rarely knew when the water would arrive. It would often come in the middle of the night, and residents would have to sacrifice sleep to fill up their water tanks.[14] They would also risk missing the opportunity to store water if they were not vigilant (McCarthy, 2010). Now, the engineers work with the community to deliver water according to a predetermined schedule, so that communities can better plan their water storage. The utility directs water according to an elaborate system of valves, and there is continuous follow-up through CCA meetings, phone calls and household visits to verify that water is arriving on time. Given the complexity of the system, the utility would never be able to know whether water was arriving at certain blocks without this cooperation. The government eventually envisions transferring the entire management of the local water cycle to the communities themselves, including the operation of the valves.[15]

Through these practices, the MTAs are challenging the distinction between community and expert knowledge. Indeed, the term 'technical' was deliberately added to the name of the water committees to build the communities' confidence in their ability to make decisions about water service, rather than deferring to 'specialists' (McCarthy, 2009, p. 12). When engineers visit neighbourhoods, residents emerge from their homes to explain to the engineers how their water system works. In CCA meetings, it is not uncommon for community members to contradict an engineer's proposals, explaining to them why a certain proposed solution would not work.

This close cooperation represents a radical departure from Hidrocapital's previous practices. As Victor Díaz, a community coordinator for Hidrocapital, explains, the utility staff previously never set foot in the *barrios* but "planned everything from air-conditioned offices".[16] Manuel González, another member of the early reform team in Hidrocapital, refers to the shift as a transition from a 'technical' logic to a 'social' logic (cited in Lacabana et al., 2007). This new logic means using creativity to find solutions that do not always conform to conventional engineering practices. Initially, these changes were met with considerable resistance within the utility. As Arconada explains, engineers would frequently complain that they didn't study engineering to "talk to poor people in the *barrios* at 8 o'clock at night".[17] Today, the model has been institutionalized, and most engineers recognize the importance of participation for improving services. However, questions remain about how far the MTA participants have been empowered to advance their broader strategic objectives.

Promoting a new form of citizenship

Florencia Gutiérrez, a long-time MTA *vocera* (spokesperson)[18] and a 67-year-old grand-mother from Antímano's largest *barrio*, Santa Ana, swells with pride as she points out improvements in the parish of Antímano. From sewer pipes and storm drains to sub-sidized food distributors and community health centres, community members themselves led all of the work, through the communal councils and communes. "We are building an *urbanismo*", she explains.[19] By *urbanismo*, she means a dignified urban space, in contrast with a *barrio*. While *barrio* is often used as a derogatory term, an *urbanismo* is a source of pride. It also implies self-sufficiency: that goods and services are produced and distributed locally. Through the MTAs, communal councils, and communes, community activists are actively shaping the urban development process and redefining what it means to be a *barrio* resident. In doing so, they are advancing a vision of citizenship based on the notion of 'popular power'.

The form of co-production expressed in the MTAs challenges the liberal notion of citizenship in which citizens enter into a 'social contract' with the state, under which they agree to be ruled in exchange for certain privileges (or 'rights') and protections (Rousseau, 1987, cited in Purcell, 2003, p. 565). Under this vision, citizens' input into state decisions is usually institutionalized through an electoral system, which establishes delegation as the organizing principle of democracy (Motta, 2011, p. 35). The liberal view of universal citizenship overlooks the conditions of 'partial citizenship' that have long been the reality for the majority of citizens in Venezuela, as formal political equality masks underlying inequalities in the economic realm.

The notion of *poder popular* ('popular' or 'public' power) advanced through the MTAs represents a reinterpretation of citizenship that goes beyond 'inclusion' towards addressing processes of uneven development. This vision of popular power views social exclusion as a structural rather than individual problem and seeks to redress exclusion through the redistribution of power and resources. The Venezuelan government has therefore granted greater weight to marginal areas in its promotion of social programmes and participatory organizations. According to Smilde (2011), this reorientation has resulted in a "change in citizenship as formerly marginalized sectors of society become the central focus of the government and are receiving full benefits of modern citizenship" (p. 22).

Women have been particular beneficiaries of the participatory initiatives at the local level, and the main protagonists in the MTAs. Allen et al. (2006) argue that women's leadership within the MTAs has been responsible for substantial achievements, and that their participation has "created a change in the way the peri-urban poor perceive their reality, creating a positive attitude towards new forms of social inclusion and hope for improvements in livelihoods" (p. 77). On average, 75% of all MTA participants are women (Allen et al., 2006; Lacabana, 2008; McCarthy, 2012), and in Antímano the rate of female participation is slightly higher, at 80%.[20] The vast majority of these participants are mothers and grandmothers from various socio-economic backgrounds, although the youngest active member interviewed for this study was 16 years old.

MTAs and elite capture

Finally, it is important to recognize that communities are not always harmonious spaces. Community organizing is inevitably fraught with interpersonal conflict and struggles for power (see also Bakker, 2008, 2010), which means that participatory arrangements are subject to elite capture. MTA *voceras* lamented that in some sectors, a small group of people take over the communal council for their own gain. For example, they cited cases where lead *voceras/voceros* were hoarding funds and distributing resources such as housing and scholarships to their own families and friends, without consulting the rest of the community through neighbourhood assemblies (as required by law). These abuses were allegedly facilitated because members of the same family controlled the communal councils' Community Administration and Finance and Social Accountability Units, in contravention of the Organic Law on Communal Councils. In some cases, one *vocera* would take control of official documents or the communal council seal, which is necessary for all correspondence with state institutions, and withhold them from rest of the community members.[21]

These types of abuses differ from more conventional forms of elite capture, because these community organizers are not political or economic elites. However, there is concern that some *barrio* residents may be better equipped than others to assume leadership positions in the communal organizations and may therefore be able to manipulate the processes for their own ends. Acting as a spokesperson is a major time commitment, which may be out of reach for people who have full-time employment or young children. Moreover, it requires levels of literacy and skills in financial planning, which not all residents have. However, the capacity building and literacy efforts implemented by the education and health missions may help mitigate the risks of exclusion and elite capture in the long term.

Elite capture may also be less of a problem in the MTAs compared with housing projects or scholarship programmes because water infrastructure improvements are generally designed for an entire neighbourhood. Moreover, all community projects must be approved by majority vote, in an assembly open to all neighbourhood residents over 15 years of age, with a minimum quorum of 20% of all residents (Ley Orgánica de los Consejos Comunales, 2006, p. 20), and the communal councils are held to strict reporting requirements by government agencies. In reality, given the prevalence of internal community conflict and jealousies, it is also possible that many of the claims of corruption are exaggerated (a point also raised by Lacabana et al., 2007).

Nonetheless, longer-term ethnographic research is needed to ascertain whether participation has facilitated an inequitable distribution of resources within neighbourhoods. This is especially true given that pre-existing asset distributions and related power inequities may be less visible in urban areas due to people's involvement in the informal economy and their control over 'hidden' resources (Marcus & Asmorowati, 2006).

A related question is the extent to which the MTAs depart from traditional forms of political clientelism. For liberal analysts of the process, the Chávez government's mobilizing and channelling resources to its base represents a mere continuation of the clientelist relations of Punto Fijo. They understand the revolutionary parties' electoral victories through the lens of clientelism, arguing that the government buys votes through the social missions (see e.g. Corrales & Penfold-Becerra, 2007). While it is true that mission and MTA participants – like the majority of Venezuela's poor and marginalized citizens – are predominantly supporters of the process, explaining government support in terms of 'buying votes' is far too simplistic.

People's commitment to the Bolivarian Process and their participation in the popular organizations is motivated not just by material rewards but also by an increased sense of dignity and self-esteem and a strong identification with Chávez and the broader political project (Fernandes, 2010). In addition, the concerted efforts to promote participation, community organization, and political consciousness-raising discussed above distinguish the MTAs and other social organizations from past examples of political gift-giving meant to guarantee political subordination (Ciccariello-Maher, 2013, p. 130). The demands on community activists' time alone belie claims that the Chávez government is giving 'handouts'.

Moreover, even observers (such as Michael McCarthy) who are critical of the politicization of the MTAs acknowledge that although there may be isolated instances of favouritism in the distribution of funds to communal councils and other organizations, "membership in the PSUV [Partido Socialista Unido de Venezuela, Chávez's party] is not a widely applied prerequisite for participating in or forming a council or for a constituted council to generate effective participation in its relationship with the state" (McCarthy, 2012, p. 139). Hawkins et al. (2011, p. 211) confirm this finding in their study of the government's social missions.

Nonetheless, it is possible that the increasingly overt partisan politicization of the MTAs will lead some low-income opposition members to self-exclude (Hawkins et al., 2011, p. 211; interview with Santiago Arconada, 24 August 2012). More systematic comparative research on the MTAs as they continue to evolve will help determine whether the MTAs continue to deepen democracy or revert to past patterns of political clientelism.

Conclusion: Tensions that emerge between 'invited' and 'invented' spaces

This article has argued that while the MTAs in Venezuela have not resolved the water problems in Caracas, they have significantly improved equity in service delivery. More importantly, rather than narrowly harnessing the participation of the poor in one-off, isolated initiatives, the MTAs represent a form of co-production that not only helps to deliver the 'material emblems of citizenship' but has enhanced the political and organizational capacity of citizens to push for citizenship rights, thus transforming power relations.

Co-production initiatives such as the ones in Caracas rest on a productive tension that emerges between 'invited' and 'invented' spaces for building popular power, highlighting the dialectical nature of the process of transformation in Venezuela. Faranak Miraftab (2004) emphasizes the importance of maintaining 'invited' spaces – defined as spaces "occupied by those grassroots and their allied non-governmental organizations that are legitimized by donors and government interventions" (p. 1) – in order to meet the practical needs of poor citizens for food, shelter and basic sanitation. However, the 'invented spaces', defined as those that are "occupied by the grassroots and claimed by their collective action" (p. 1), are equally important for maintaining the independent capacity of the community to push for structural change, for it is in these spaces that citizens directly confront the authorities and the status quo.

The creation of new institutionalized relationships between the utility and the communities represents a major improvement in one key respect: communities no longer have to protest for their water services. But many participants and social movement activists in Venezuela raise concerns that within these 'invited spaces', the community organizations are becoming increasingly bureaucratized, which limits experimentation and creativity by forcing community organizations to organize within only official, state-sanctioned channels. MTA participants fear that increased rigidity may ultimately slow the transformative process.

Action in autonomous spaces is also important because it is unclear whether the MTAs would outlast a change in political administration, particularly at the national level. Victor Díaz raised this point at a CCA meeting in Antímano just days after Chávez's electoral victory in the 7 October 2012 presidential elections. While Chávez's victory would have been considered a landslide in the US or Canada, it was his closest margin – nearly 11%. The election of Nicolás Maduro, following Chávez's death in March 2013, was even narrower (just 1.5%). Many attribute the increasingly close margins to the problems of bureaucracy and inefficiency described above. At the meeting, Díaz, himself an Antímano resident and long-time social activist, acknowledged that the work on water and sanitation had not been fast enough, and admonished participants to continue to put pressure on the government. "I want to pressure the water utility further, but I can't do it alone." He also encouraged the committees to be more autonomous. "I wouldn't be here if the opposition had won; they would send new civil servants." Carmen Rojas summed up the central dilemma created by 'invited spaces' for fostering popular power: "No one is just going to give us popular power. We need to take it ourselves."

Notes

1. Interviews with Victor Díaz, current coordinator of Hidrocapital's Community Management Office and community promoter for Antímano, 20 August 2012; and Santiago Arconada, long-time water activist and first coordinator of Hidrocapital's Community Management Office, 24 August 2012.
2. Interview with Hidrocapital community promoters, 28 August 2012. (Community promoters are utility staff people who act as liaisons between the neighbourhood MTAs and the water company.)
3. The parties to the Punto Fijo pact were: Acción Demócratica (AD, Democratic Action); the Christian-democratic Comité de Organización Política Electoral Independiente (COPEI, Committee of Independent Political Organization); and the small, leftist Unión Republicana Demócrática (URD, Democratic Republican Union). The primary goal of the agreement was to share power and resources among these three parties while excluding challengers, primarily the Communist Party of Venezuela (Wilpert, 2006, p. 12).
4. Interview with Santiago Arconada, 24 August 2012.
5. As Fernandes (2010, pp. 58–59) describes, the outcomes of decentralization were contradictory. On the one hand, decentralization was part of broader neoliberal measures promoting fiscal austerity and also led to the concentration of resources in wealthier municipalities, exacerbating uneven urban development. On the other hand, decentralization allowed for the emergence of new power bases outside of the country's traditional corporatist structures, which promoted greater diversity in political-party activity and also encouraged social-movement organizing outside of clientelist networks.
6. By the late 1990s, poverty had reached astronomical levels. At the end of 1996, 86% of the Venezuelan population was poor, and 65% lived in extreme poverty (Buxton, 2004, p. 122).
7. Interview with Victor Díaz, Hidrocapital community coordinator, 20 August 2012.
8. Interview with Santiago Arconada, 24 August 2012.

9. At the time, Faría was a member of the Patria Para Todos party, a splinter group of La Causa Radical, which also promoted participatory forms of democracy.
10. The tragedy struck December 16, 1999, in the coastal state of Vargas. Over the course of three days, torrential rains, floods and landslides killed tens of thousands of people, destroyed thousands of homes and completely disrupted the state's infrastructure, including the water service.
11. Interview with Victor Diaz, 20 August 2012.
12. Communal councils are elected neighbourhood planning bodies that identify and prioritize community needs and execute community development projects. With the adoption of the 2006 Organic Law on Communal Councils, the MTAs were subsumed under the communal councils as a working group. It is the communal councils that receive and manage the finances for MTA-led water and sanitation projects. Enshrined in law in 2009, the communes bring together all of the communal councils in a given geographic area to better coordinate projects and initiatives, including the development of the social economy.
13. Interview with Santiago Arconada, 24 August 2012.
14. Focus group with Antímano MTA spokeswomen, 22 November 2012. Lacabana and Cariola (2005) also make similar observations.
15. Interview with Hidrocapital community promoters, 28 August 2012. According to the promoters, in some cases the community members already change the valves; however, in Antímano and most other parishes the operations engineers manage the pumps and valves. These engineers work for a private company, Empresa Carrillo, under contract with Hidrocapital. According to Hidrocapital promoter Dircia García, having the communities manage the valves would be a source of local employment (since the utility would pay them), and safer. Security is a perennial concern in the *barrios*, and outsiders are often at greatest risk.
16. Interview with Victor Díaz, 20 August 2012.
17. Interview with Santiago Arconada, 24 August 2012.
18. Spokespeople (*voceros/voceras*) are elected representatives of neighbourhood MTAs. The term 'spokesperson' is used instead of terms such as 'president' or 'leader' in view of the preference for more horizontal organizational structures.
19. Personal communication with MTA spokesperson Florencia Gutiérrez, 28 October 2012.
20. Interview with Victor Díaz, 4 December 2012.
21. Personal communication with *voceras* during a tour of Carapita, 2 October 2012; interview with MTA *vocera* Sulay Morales, 11 November 2012.

References

Allen, A., Davila, J., & Hofmann, P. (2006). *Governance of water and sanitation services for the peri-urban poor*. London: Development Planning Unit, University of London.

Arconada, S. (2005). The Venezuelan experience in the struggle for people-centred drinking water and sanitation services. In B. Balanyá, B. Brennan, O. Hoedeman, S. Kishimoto, & P. Terhorst (Eds.), *Reclaiming public water: Achievements, struggles, and visions from around the world* (pp. 131–137). Porto Alegre, Brazil: Transnational Institute.

Arconada, S. (2005b). Seis años después: Mesas técnicas y consejos comunitarios de aguas. *Revista Venezolana de Economía y Ciencias Sociales, 12*(2), 187–203.

Bakker, K. (2008). The ambiguity of community: Debating alternatives to private-sector provision of urban water supply. *Water Alternatives, 1*(2), 236–252.

Bakker, K. (2010). *Privatizing water: Governance failure and the world's urban water crisis*. Ithaca, NY: Cornell University Press.

Bovaird, T. (2007). Beyond engagement and participation: User and community coproduction of public services. *Public Administration Review, 67*, 846–860.

Buxton, J. (2004). Chapter 6: Economic policy and the rise of Hugo Chávez. In S. Ellner & D. Hellinger (Eds.), *Venezuelan politics in the Chávez Era: Class, polarization & conflict* (pp. 113–130). Boulder, CO: Lynne Rienner Publishers.

Buxton, J. (2011). Forward: Venezuela's Bolivarian democracy. In D. Smilde & D. Hellinger (Eds.), *Venezuela's Bolivarian democracy* (pp. x-xxii). Durham, NC: Duke University Press.

Cariola, C., & Lacbana, M. (n.d.). *WSS practices and living conditions in the peri-urban interface of metropolitan Caracas: The cases of Bachaquero and Paso Real*. (Report for Service Provision

Governance in the Peri-urban Interface of Metropolitan Areas Research Project). United Kingdom: Development Planning Unit, University College London.

Castillo, A. (2011, February 19). Hidrológicas asegurarán suministro de agua potable a proyectos de la Misión Vivienda. *Correo del Orinoco*. Retrieved from www.correodelorinoco.gob.ve

Ciccariello-Maher, G. (2013). Constituent movements, constitutional processes: Social movements and the new left in Latin America. *Latin American Perspectives*, *40*(3), 126–145.

Corrales, J., & Penfold-Becerra, M. (2007). Venezuela: Crowding out the opposition. *Journal of Democracy*, *18*(2), 99–113.

Ellner, S. (2010). Hugo Chávez's first decade in office: Breakthroughs and shortcomings. *Latin American Perspectives*, *37*(1), 77–96.

Estado venezolano ha invertido $600 millones para mejorar servicio de agua potable. (2011, March 22). *Agencia Venezolana de Noticias*. Retrieved from www.avn.info.ve

Evans, P. (1996). Government action, social capital and development: Reviewing the evidence on synergy. *World Development*, *24*, 1119–1132.

Fernandes, S. (2010). *Who can stop the drums? Urban social movements in Chávez's Venezuela*. Durham, NC: Duke University Press.

Francisco, C. (2005). Cambio y equidad del servicio del agua en Venezuela. *Cuadernos del Cendes*, *22*(59).

Goldfrank, B. (2011). *Deepening local democracy in Latin America: Participation, decentralization, and the left*. University Park, PA: Pennsylvania State University.

Griffiths, T. (2010). Schooling for twenty-first-century socialism: Venezuela's Bolivarian Project. *Compare*, *40*(5), 607–622.

Grugel, J., & Riggirozzi, P. (2011). Post-neoliberalism in Latin America: Rebuilding and reclaiming the state after crisis. *Development and Change*, *43*(1), 1–21.

Hawkins, K. A., Rosas, G., & Johnson, M. E. (2011). The Misiones of the Chávez government. In D. Smilde and D. Hellinger (Eds.), *Venezuela's Bolivarian democracy: Participation, politics, and culture under Chávez*. Durham, NC: Duke University Press.

Hickey, S., & Mohan, G. (2005). Relocating participation within a radical politics of development. *Development and Change*, *36*(2), 237–262.

Interamerican Development Bank. (2012). *Bolivarian Republic of Venezuela Río Guaire sanitation loan proposal* (VE-1037). Retrieved from IADB website: http://www.iadb.org/en/projects/project-description-title,1303.html?id=VE-L1037

Jaglin, S. (2002). The right to water versus cost recovery: Participation, urban water supply and the poor in Sub-Saharan Africa. *Environment and Urbanization*, *14*(1), 231–245.

Lacabana, M. et al. (2007).*Las mesas técnicas de agua en el contexto de los cambios institucionales, la democracia participativa y la participación popular en Venezuela*. CENDES-HIDROVEN.

Lacabana, M. (2008). *An overview of the water supply and sanitation system at metropolitan and peri-urban level: The case of Caracas. Service provision governance in the peri-urban interface of metropolitan areas research project*. London: The Development Planning Unit.

Lacabana, M. (2010). *Confrontando la indefensión social: Mesas técnicas de agua en Venezuela*. Unpublished manuscript.

Lacabana, M., & Cariola, C. (2005). Construyendo una nueva cultural de agua en Venezuela. *Cuadernos del Cendes*, *22*(59), 111–133.

Ley Orgánica de los Consejos Comunales. República Bolivariana de Venezuela. Asamblea Nacional No. 751. (2006).

Mahmood, Q., Muntaner, C., del Valle Mata Leon, R., & Perdomo, R. E. (2012). Popular participation in Venezuela's barrio adentro health reform. *Globalizations*, *9*(6), 815–833.

Marcus, A., & Asmorowati, S. (2006). Urban poverty and the rural development bias: Some notes from Indonesia. *Journal of Developing Societies*, *22*(2), 145–168.

McCarthy, M. (2009). Herramientas de la revolución? (Tools of the revolution?): The case of the Mesas Tecnicas de Agua and politicized participatory water policy in Venezuela. Paper presented at the 2009 Congress of the Latin American Studies Association, Rio de Janeiro, Brazil, June 11–14.

McCarthy, M. (2010). The Bolivarian or "Chavista" way to reach the UN Millennium Development Goals for water and sanitation. *John Hopkins University Global Water Magazine*. Retrieved from http://globalwater.jhu.edu/magazine/article/the_bolivarian_or_chavista_way_to_reach_the_un_millenium_development_goals_/

McCarthy, M. (2012). The possibilities and limits of politicized participation: Communal councils, coproduction, and poder popular in Chávez's Venezuela. In M. A. Cameron, E. Hershberg, & K.

E. Sharpe (Eds.), *New institutions for participatory democracy in Latin America: Voice and consequence* (pp. 123–147). New York, NY: Palgrave-Macmillan.

Mesas Técnicas de Agua han recibido más de Bs. 500 milliones. (2011, August 11). *Correo del Orinoco*. Retrieved from http://www.correodelorinoco.gob.ve/ambiente-ecologia/mesas-tecnicas-agua-han-recibido-mas-bs-500-milliones/

Miraftab, F. (2004). Invited and invented spaces of participation: Neoliberal citizenship and feminists' expanded notion of politics. *Wagadu, 1*, 1–7.

Mitlin, D. (2008). With and beyond the state – Co-production as a route to political influence, power and transformation for grassroots organizations. *Environment & Urbanization, 20*(2), 339–360.

Molyneux, M. (2008). The 'neoliberal turn' and the new social policy in Latin America: How neoliberal, how new? *Development and Change, 39*(5), 775–797.

Motta, S. C. (2011). Populism's Achilles' heel: Popular democracy beyond the liberal state and the market economy in Venezuela. *Latin American Perspectives, 38*(1), 28–46.

Ostrom, E. (1996). Crossing the great divide: Coproduction, synergy, and development. *World Development, 24*, 1073–1087.

Perfil biográfico de Alejandro Hitcher, nuevo Ministro de Ambiente y Recursos Naturales. (2010, January 27). Retrieved from http://www.aporrea.org/poderpopular/n149693.html

Purcell, M. (2003). Citizenship and the right to the global city: Reimagining the capitalist world order. *International Journal of Urban and Regional Research, 27*(3), 564–590.

Smilde, D. (2011). Introduction: Participation, politics, and culture – Emerging fragments of Venezuela's Bolivarian democracy. In D. Smilde & D. Hellinger (Eds.), *Venezuela's Bolivarian democracy: Participation, politics, and culture under Chávez* (pp. 1–27). Durham, N.C: Duke University Press.

Spronk, S. J. (2009). Making the poor work for their services: Neo-liberalism and 'pro poor' privatization in El Alto, Bolivia. *Canadian Journal of Development Studies, 28*(3–4), L1–L17.

United Nations. (2013). *Millennium development goals and beyond 2015 Factsheet, Goal 7: Ensure environmental sustainability.* Retrieved from http://www.un.org/millenniumgoals/environ.shtml.

Wilpert, G. (2005). Venezuela: Participatory democracy or government as usual? *Socialism and Democracy, 19*(1), 7–32.

Wilpert, G. (2006). *Changing Venezuela by taking power: The history and policies of the Chávez government.* London: Verso.

World Health Organization. (2012). *UN-Water global annual assessment of sanitation and drinking water (GLAAS) 2012 report: The challenge of extending and sustaining services.* Geneva, Switzerland: World Health Organization.

Creating equitable water institutions on disputed land: a Honduran case study

Catherine M. Tucker

Department of Anthropology, Indiana University, Bloomington, IN USA

This article explores the decade-long process by which village-level water committees established a reserve in 2002 to protect communal mountain springs in the Montaña Camapara region of Honduras. In so doing, it considers the conditions under which shared dependence on water resources may motivate cooperation and foster equitable access to water in the face of difficult challenges posed by conflicts over land and water rights claims and degradation of the resource.

Introduction

Struggles over land rights and access to natural resources are associated with resource degradation, institutional failures, and human suffering. In particular, conflicts over water resources threaten to become more frequent as demands increase (Amery, 2002; Homer-Dixon, 2001), and water scarcity has been described as the major environmental challenge of the twenty-first century (Marks, 2009). Moreover, bodies of fresh water are often common-pool resources, which pose challenges for governance because restricting access (excludability) proves difficult, and deterioration occurs with overuse (McKean, 2000). In addition, rights to water are often inequitably distributed, which poses challenges for human health, poverty alleviation, and economic development (e.g. Bakker, 2001; Carmichael, Schafer, & Mazumdar, 2008; Sterling, 2007). Challenges for managing watersheds and distributing water equitably can multiply in contexts where multiple stakeholders have competing legal or traditional claims, or poorly defined property rights. Ineffective or inappropriate institutions (rules-in-use), including unclear or inequitable property rights, have been viewed as root problems of poor natural resource management and associated social tensions. While progress has been made in discovering the contexts associated with effective institutions (Acheson, 2006; Ostrom, 2005), more research is needed to understand why competition for scarce resources sometimes evolves towards sustainable and equitable resource management, but in other instances escalates to entrenched conflict.

In this context, the case of the Montaña Camapara Reserve in western Honduras presents an interesting opportunity for study. The reserve was created to protect mountain springs that provide water to surrounding settlements, through collective action carried out by the settlements' water committees. The circumstances of the reserve's creation (detailed below) presented a number of obstacles to the grass-roots effort to achieve sustainable and equitable access to water. First, the reserve exists on a mountain subject to ongoing land disputes. The mountain is divided among three *municipios* (similar to counties in the United States). Each owns communal land titles to its land, and they have been struggling over their boundaries on the mountain for generations. Two of the *municipios* are embroiled in a legal battle over boundaries drawn between them by the national government, which ignored historic land titles. Boundary disputes indicate insecure property rights, which are frequently associated with resource degradation (Blaikie & Brookfield, 1987; Hecht & Cockburn, 1990; White & Martin, 2005). Second, the reserve was formed despite tensions and discord between water users who supported the reserve and water users who opposed the reserve because they farmed on the mountain. Discord among resource users appears antithetical to shared trust, which is recognized as an integral component for the successful management of communally owned resources (Dietz, Ostrom, & Stern, 2003; Ostrom, 2005). Third, a number of farmers working on the mountain had to give up land for the reserve to be created. Protected areas that deny prior land claims and displace residents tend to experience strong resistance, weakening of local institutional arrangements, and resource over-exploitation (Nagendra, Karna, & Karmacharya, 2005; Tucker, 2004; West, Igoe, & Brockington, 2006). Fourth, at the reserve's inception, the *municipios* lacked monitoring and enforcement arrangements, which are associated with effective resource conservation (Gibson, Williams, & Ostrom, 2005). Finally, the region has been experiencing rapid change during the past two decades with the introduction of cell phones, electricity, Internet, development programmes, new roads, and dramatic expansion of export coffee production. Rapid change tends to weaken local institutions and may lead to environmental degradation (Dietz et al., 2003; Gunderson & Holling, 2002; Janssen, 2000).

In light of the initial circumstances, the probabilities for conservation and equitable management of water resources in the Montaña Camapara Reserve seemed slim. I have followed the process that led to the emergence of the reserve and its maintenance across 20 years, beginning in 1993, when I arrived to conduct a year of doctoral research, and through an additional 15 fieldwork visits. Given the apparent contradictions to theoretical expectations, I first doubted that the reserve would be created, and then expected it to fail. Instead, I witnessed local water committees undertaking well-organized collective action to create and manage the reserve, protect their water sources, and assure provision of potable water to their populations and that of other settlements who joined the process. Curiosity led me to explore the following questions:

- How did the Montaña Camapara Reserve become established despite property-rights disputes and discord among resource users?
- How have water committees dependent on Montaña Camapara worked towards equitable access to water? Specifically, what kinds of institutions did they create to support equitable distribution of water?

Theoretical perspectives

To analyze the emergence of the Montaña Camapara Reserve, this article draws on two broad theoretical approaches: conflict resolution and institutional analyses of common-pool resources governance. These approaches explore different sets of questions but offer complementary insights for evaluating the circumstances in which disputes over common-pool resources may be negotiated to move towards peaceful interactions and effective management.

Conflict resolution theories explore questions about the nature of disputes and how tensions may be mitigated or resolved (Burton & Dukes, 1990). Environmental conflict resolution has emerged as a distinct field of endeavour (Dukes, 2004). The goal is to formulate the best possible, mutually advantageous agreements that mitigate or resolve conflict over environmental problems. The process generally begins with an assessment of the problem's suitability for mediation (Bean, Fisher, & Eng, 2007). The assessment considers characteristics known to influence the potential for productive interactions among disputing parties, including participants' willingness to collaborate, the number of participants, and the degree of difficulty of the case. As may be expected, mutually acceptable agreements prove harder to reach where discord is entrenched, the parties involved resist mediation or negotiation, and the number of participants is high (Burton & Dukes, 1990). Successful outcomes are contingent upon effective engagement of all parties, integration of reliable information, and participants' available time, skills and resources. A mediator can be useful to facilitate the process (see also Bush & Folger, 2005; Emerson, Orr, Keyes, & McKnight, 2009); however, mediators are not always necessary (Dukes, 2004). In a comparative study of 52 cases in the United States mediated by environmental conflict resolution processes, agreements were reached in 82% of the cases. A majority of participants in these processes reported improvements in the level of trust and ability to work collaboratively with the other participants, although they sometimes expressed reservations (Emerson et al., 2009).

Institutional analysis explores how formal and informal rules-in-use influence human behaviour, along with their implications, interactions and outcomes. Following North (1990), this study distinguishes institutions from organizations. Institutions are rules that guide people as to what they may do, must do or must not do in a given situation. They range from legally enforced laws to shared understandings about correct behaviour (Ostrom et al., 2002). Organizations, by contrast, are groups of individuals brought together for a common purpose, including political parties, councils, schools, corporations, government units, and cooperatives (North, 1990). Research on environmental change has been working to understand the characteristics of institutions that support or undermine sustainable use of natural resources (Dietz et al., 2003; Young, 2002). Studies of common-pool resource management tend to focus on local-level institutions, given that local users directly affect the resource base. Management regimes that govern common-pool resources sustainably typically incorporate well-defined boundaries, clear recognition of the rightful users, locally appropriate institutions, participation by a majority of the users, and accessible conflict-resolution mechanisms (Ostrom, 2005). Effective regimes incorporate monitoring and enforcement, often in association with graduated sanctions for rule breakers (Ostrom, 2005). Research indicates that face-to-face interaction, shared experiences and understandings, and trust contribute to the evolution of successful institutional arrangements (Chhatre & Agrawal, 2008; Gibson et al., 2005). Studies of conflict resolution similarly point to the importance of face-to-face negotiations and the building of relationships as integral factors in reaching mutually acceptable agreements (Susskind & Ozawa, 1983).

In the case of Montaña Camapara, the various disputants in the process of reserve creation shared a set of experiences related to their place of residence, dependence on the same natural resources, and options for livelihood. These experiences provided a foundation for building support for the reserve across a decade of collective action by the water committees.

As the following discussion will explore, the water committees developed many of the dimensions associated with successful environmental conflict resolution and sustainable common-pool resource management. The article first considers the contexts in the research site, the history of the property-rights conflicts, and the contrasting as well as shared experiences across the three *municipios*. It then addresses the research questions by examining (1) the process of reserve creation in the contexts of conflict, (2) the institutions that evolved to support equitable access to water, and (3) the lessons learned.

Methods

The data for this study were collected primarily between 2008 and 2012, but benefited from my knowledge of the research site gained through longitudinal, ethnographic fieldwork that began in 1993. I conducted fieldwork in each of the three *municipios* that share Montaña Camapara, and visited all but one of the settlements that currently draw water from it (Table 1). I carried out interviews with municipal authorities, knowledgeable residents, forest guards, and representatives of government and non-government agencies and development programmes. I tracked down and interviewed 8 of the farmers who agreed to abandon their mountain land, 34 current and former water committee leaders, and the 5 living mayors (over two electoral periods) involved in reserve creation (1 had passed away). I also attended water committee meetings in 10 settlements that depend on Camapara's water, and talked with participants afterwards. The study also benefitted from archival research and review of the documents and records kept by the water committees, municipal governments, the Honduran Institute for Forest Conservation and Development and the Honduran Coffee Institute.

Description of the site

The three *municipios* that share Montaña Camapara—La Campa, San Marcos de Caiquín (hereafter, Caiquín) and Santa Cruz—are located in the Department (similar to a state) of Lempira (Figure 1). This department is the poorest in Honduras. Nearly one-fifth of the population is unlikely to live to age 40; 37.2% of the population over the age of 15 are unable to read or write; and 71.8% of children under the age of 5 suffer from malnutrition (PNUD, 2006). Lempira is named after a leader of the indigenous Lenca people, who led a region-wide uprising against the Spanish invaders in the sixteenth century. To this day, the department retains its association with indigenous Lenca culture and traditions, even though the Lenca language disappeared in the early 1900s.

The three *municipios* have maintained certain aspects of their Lenca heritage, and they have distinct traditions of syncretic agricultural and religious rituals that combine Catholic and indigenous symbols (Chapman, 1985). Although the observance of traditional rituals has been declining, each of the *municipios* maintains its own syncretic expressions of culture, including dances and music performed at their patron saints' festivals. La Campa is known for traditional Lenca pottery, which today has found popularity with tourists and regional collectors (Castegnaro de Foletti, 1989; Tucker, 2010).

In each of the three *municipios*, residents assert community identity based on shared territory, traditions, communal projects and a sense of place. This identity dates at least back to the Spanish colonial period; each of them appears in 16th-century censuses as a *comunidad indigena* (indigenous community). The people of La Campa call themselves *Campeños*, those of Santa Cruz are *Cruzeños*, and the people of Caiquín are *gente de Caiquín*. Hereafter, I use 'community' to refer to the people of each *municipio*, given that their identities as indigenous communities predate their political designations as separate

Table 1. Contexts and constraints affecting municipal-level engagement with the reserve.

	San Marcos de Caiquín	La Campa	Santa Cruz
Dependence on Montaña Camapara's water in 2002	~50% of the population[c]	~85% of the population	none
Dependence on Montaña Camapara's water in 2010	~60% of the population	~98% of the population	none
Completion of first potable water project on Montaña Camapara	1990	1986	–
Number of settlements dependent on potable water projects on Montaña Camapara in 1992	2	4[e]	none
Number of settlements dependent on potable water projects on Montaña Camapara in 2002	2	10[e]	–
Number of settlements dependent on potable water projects on Montaña Camapara in 2012[a]	8[d]	16[e]	–
Estimated number of households depending on water from Montaña Camapara in 2012[b]	574[d]	1209[e]	none
Major obstacles or threats to the reserve	13 families settled on land inside its reserve area	Insecure property rights to its side of Montaña Camapara	Farmers using its side of Montaña Camapara for subsistence agriculture
Forest conditions on Montaña Campara	Mature cloud forest with reforestation in degraded areas within reserve boundaries	Mature cloud forest with reforestation in abandoned fields within reserve boundaries	Nearly 100% deforested

[a] Includes small settlements and hamlets that do not appear on maps of the region.
[b] Estimates are based on water committee reports of participating households.
[c] The people drew water from streams running off Montaña Camapara and nearby springs prior to construction of water projects.
[d] Includes settlements in the neighbouring *municipio* of San Manuel Colohete that draw water from Caiquín's side of Montaña Camapara.
[e] Includes settlements in the neighboring *municipio* of Gracias that draw water from La Campa's side of Montaña Camapara.

municipios. Their distinctions as well as common experiences have been reinforced through a long history of economic and political marginalization under the colonial government and the subsequent Honduran nation. Santa Cruz has most stubbornly defended itself from outside interventions that might undermine its traditional ways of life; it ranks as the poorest of Lempira's *municipios.* Most residents of Santa Cruz depend on subsistence agriculture and migrate seasonally to pick coffee. In Caiquín and

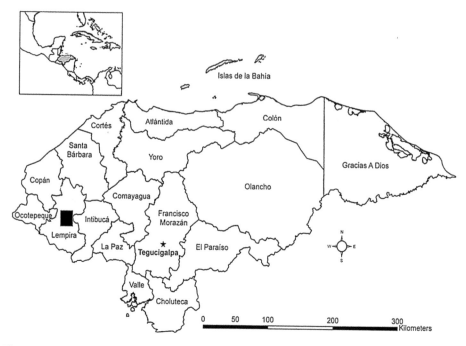

Figure 1. Map of Honduras, showing study area.

La Campa, dependence on subsistence corn and beans has declined as farmers have turned to coffee production and wage labour.

Each *municipio* has about a dozen settlements scattered across its territory, varying in population from several dozen to over 1000 people. Occasional rivalries occur between settlements and within and between families. Some of the deepest enmities revolve around land disputes and inheritance rights. Most residents of each *municipio* see themselves as having something in common that exceeds their differences (Tucker, 2010). Although Campeños, Cruzeños and gente de Caiquín may cast aspersions on each other, social ties and kin networks cross municipal boundaries. *Compadrazgo* (fictive kin relations) links families in ritual obligations that tend to reinforce shared economic or political interests. Families in Caiquín and La Campa are often linked by marriage, and some farmers own land on both sides of the municipal borders. It is rarer, however, for Cruzeños to marry outside their *municipio*.

The *municipios* share a strong tradition of local governance. For many years, they received minimal government support or assistance. As a result, residents had to provide labour and resources for municipal projects, including the construction of roads, schools and municipal buildings. While the *municipios* no longer require communal labour, they continue to exhort their people to work together on public projects. Participation in local government and traditions of communal labour provided a background for collective action for water governance.

Each *municipio* traditionally owns forests, pastures, clay and sand beds, and gravel deposits as common property, dating back to Spanish colonial land titles granted to indigenous communities. In the late twentieth century, privatization of common property occurred as population growth, a national land-titling programme, and transitions to coffee production coincided (see Tucker, 2008, for a more detailed discussion). The land-titling programme fell short because most residents could not afford the costs of titling the lands to which they had usufructuary rights. At first, mountain slopes with marginal soils remained as common property. However, residents in need of cash sold their fields and forests to outsiders, who easily paid for private land titles and established extensive coffee plantations. As outsiders acquired land, many residents experienced land shortages that drove them to clear marginal mountain slopes, including Montaña Camapara, for agriculture.

Origins of conflict and border disputes

The people of Santa Cruz, La Campa and Caiquín have been disputing the boundaries between their lands since the colonial period. Although they obtained their first land titles in the sixteenth and seventeenth centuries, the exact borders between their lands remained uncertain for generations, because the earliest titles included few permanent landmarks (Tucker, 2008).

Following independence from Spain in 1821, the emergent nation of Honduras began dividing its territory into subordinate regional and municipal units. La Campa and Caiquín were placed within the *municipio* of Gracias, while Santa Cruz became part of Erandique. Each retained a sense of identity, aspiring to autonomous governance and augmentation of their land titles. In 1864, Campeños and gente de Caiquín hired a surveyor to officially demarcate their lands and provide definitive land titles. The land title's text, accompanied by a hand-drawn map, indicates that the surveyor attempted to walk along the proposed border-lines but was blocked by deep canyons and cliffs. Therefore, he declared that the border ran in a straight line sighted from peak to peak across the highest three mountains between the two communities and ending on the peak of Montaña Camapara, where the boundaries of La Campa and Caiquín meet those of Santa Cruz. Residents of Santa Cruz, La Campa and Caiquín accompanied the surveyor to mark their shared borders. Together they built stone cairns to mark the borders; but the sightlines between La Campa and Caiquín proved controversial, as did the precise location of the peak on Montaña Camapara. It was next to impossible to ascertain the exact location of the border-line on the complex terrain, and forest cover obscured the topography of the mountain peaks. Farmers on all sides moved the cairns at will.

The disputes over land contributed to the three communities' desire to gain independent municipal status. Santa Cruz achieved it in 1926, following the 1864 borders. In 1921, La Campa succeeded in submitting all of the documents required for municipal status (Fiallos, 1991). Caiquín, preferring to be subordinate to La Campa rather than to the more distant and demanding Gracias government, asked to be included in La Campa's bid for municipal status. As a result, Caiquín nominally became part of La Campa, but its residents still considered themselves to be autonomous. Caiquín soon began to petition the national government for independence from La Campa, but it lacked the population and infrastructure required by Honduran law to become a *municipio*.

Across decades, anger and frustration mounted over the mobile border markers, in association with property damage related to land conflicts along the borders. In the early 1980s, the three communities agreed to construct concrete border markers reinforced with steel rebar sunk into the bedrock. Together they hired a surveyor to confirm adherence to the 1864 titles. The three communities argued over several marker positions, but cooperated to build concrete markers at key points. Even so, new debates over land emerged

Figure 2. Map of the Department of Lempira and study *municipios* (counties).

continually as landowners expanded their fields and grazed their livestock, surreptitiously or accidentally, on neighbouring *municipios'* lands.

In 1995, Caiquín won its battle to become an independent *municipio* (Tucker, 2008). A map of Caiquín's boundaries was drafted in Tegucigalpa and approved by the national legislature. Neither La Campa nor Santa Cruz was consulted when Caiquín's borders were mapped. Although Honduran law mandates that official delineation of municipal border-lines must be done on location by an accredited surveyor in the company of witnesses from each of the bordering *municipios*, no surveyor ever came. When La Campa's municipal authorities saw the map later that same year, they discovered that almost one-fifth of their land title had been allocated to Caiquín, including two of La Campa's settlements and its entire portion of Montaña Camapara. (Figure 2) La Campa promptly began an expensive legal battle to rectify the error. Caiquín's authorities acknowledged that a mistake had been made, but declined to support La Campa's litigation. The situation places Montaña Camapara, especially that part claimed by La Campa, in unclear legal status. Santa Cruz, for its part, maintains that La Campa and Caiquín have made incursions on its land to the north and south of the mountain.

Although competition for land constitutes the foundations of conflict between Caiquín, La Campa and Santa Cruz, their enmities also trace to differences in Lenca identity, traditions, subsistence, ritual practices and relationships with outside powers. These have been further confounded by socio-economic differences that have become accentuated with the expansion of export coffee production since the 1990s.

The emergence of water committees

In the study site, expanding populations and increasing concern for water quality compelled the formation of settlement-based water committees. The first water committee was formed in the late 1970s by two neighbouring settlements. They cooperated to build the first water project on Montaña Camapara, which they completed in 1986. At that time, the national government held responsibility for managing potable water, but in reality, rural populations received minimal assistance and had to provide water for themselves. Through the 1990s, eight more water committees formed with the goal of drawing water from Montaña Camapara. They negotiated arrangements to share responsibilities for building and maintaining two reservoirs and associated aqueducts. Households that wanted potable water joined their settlement's water committee; participation typically reached 100%. To accomplish their goals, water committees designed and enforced rules that required member households to assist in the construction of the water system, cooperate to maintain the project, pay modest water fees, and attend committee meetings. The obligations to maintain and monitor the water project remind people of the real cost of water, even though their fees stay relatively low (currently around USD 0.75 per month). The experience of building and managing water projects, and instituting rules, provided the basis for the more complex organizational processes of creating the reserve.

In 1992, the national government passed the Agricultural Modernization Act (Decree 31–92) and began to decentralize natural resource management. Honduran water law underwent restructuring, and the Honduras Secretary of Health and the National Autonomous Aqueduct and Sewage Service (SANAA) developed a potable water and sanitation framework that became formalized in 2003 as Decree 118–2003 (Phumpiu, 2008; Republic of Honduras, 2003). The new law placed responsibility for rural water provision with municipal governments, which may delegate the responsibility to settlement (town and hamlet) water committees. The law established principles for water committee governance, such as election of a board of directors and fee collection. The 2003 law accommodated Honduras' rural reality of pre-existing water provision systems, while providing guidelines that in many ways support the principles associated with sustainable and effective common-pool resource management. Since 2003, the study site's water committees have generally conformed to the national law.

Establishing the reserve despite property-rights conflicts

In the early 1990s, Campeño farmers seeking land for coffee plantations and subsistence crops began clearing forest for agricultural fields on Montaña Camapara. They did not have permits to do this, although by municipal law, agricultural clearing of communal land required municipal approval. At this time, four water committees in La Campa had completed water projects on the mountain, and another project had been completed in Caiquín. Each project had respective municipal approval, and additional committees had projects in the planning stage. Water committee members discovered the clearings and observed that sediment from runoff was polluting the springs. Within weeks, one of the settlements organized to purchase barbed wire, and the people went to the mountain to build a sturdy fence around their reservoir. They continued upslope to the border with Santa Cruz, and south to the line with Caiquín. When they finished, about 40 hectares of mature cloud forest had been fenced off. From the border with Santa Cruz, they could see that Cruzeños had already cleared land for agriculture up to the La Campa border. On the

Caiquín side, several new clearings had appeared, created by subsistence farmers who had been displaced from lower, more fertile land by coffee growers.

Representatives of La Campa's water committees immediately complained to their municipal authorities and requested that Montaña Camapara be protected. The mayor at the time agreed that the water sources needed protection, but the farmers on Montaña Camapara did not want to give up the land they had cleared, even though it had been done without permission. By local custom and agrarian law in Honduras, agricultural land uses have priority over forest, so the farmers had legal precedent on their side. Meanwhile, the water had turned grey from soot and soil from slash-and-burn fields that washed with runoff into the reservoirs, and water flow was declining as the cloud forest was cut. While cutting most forests does not reduce water flow because rainfall remains unchanged, cutting down cloud forests reduces water availability because atmospheric moisture condenses on the dense canopy and drips to the ground. When trees are cut, that water is lost (Hamilton, 1995).

During the 1990s, water committee leaders and people in favour of the reserve pursued a multipronged strategy to build support for the reserve. Interviews that I conducted with key actors in La Campa and Caiquín mentioned the following strategies. (1) Reserve supporters, who were farmers themselves, talked with the farmers who had fields on Montaña Camapara to change their minds. One person characterized their efforts as educational; others said that they aimed to *conscientizar* (develop awareness and sense of responsibility) among their neighbours and farmers on the mountain who did not understand the need for the reserve. (2) Water committee meetings were used as venues for members who supported the reserve to express concerns over threats to water. These meetings require attendance of every head of household who uses water from the settlement water project; thus, farmers with land on the mountain had to attend, and they faced neighbours who criticized their lack of concern for water quality. (3) Water committee presidents met with La Campa's mayor and searched for strategies to promote the reserve's creation. (4) Reserve supporters and the mayor made declarations in municipal council meetings in favour of the reserve. (5) Water committees invited international development organizations – including the Lempira Sur Project of the Food and Agriculture Organization of the United Nations (FAO) and Spain's Solidaridad – to provide training and orientation, and mapping of the mountain's watersheds (Navarro, 2002). The development organizations also provided transportation for face-to-face meetings among water committee representatives and municipal leaders, which helped build shared understanding about the need for the reserve. (6) La Campa's mayor met several times with his counterpart in Caiquín. First, in the late 1990s, they came to an agreement to respect the 1864 boundary lines between La Campa and Caiquín on Montaña Camapara. This oral contract is reconfirmed each time new mayors in Caiquín and La Campa are elected. Subsequently, the two mayors considered the problem of protecting Montaña Camapara.

It was clear to Campeños and gente de Caiquín that protecting their parts of the mountain would not be enough; therefore, the two mayors approached the mayor of Santa Cruz. Santa Cruz owns the peak and the highest points of Montaña Camapara, and Cruzeños use it for agricultural activities that affect the watersheds of La Campa and Caiquín that lie below. The Cruzeño mayor faced a populace that vehemently opposed any restrictions on their traditional land rights and practices; he received two death threats just for discussing the possibility of creating the reserve. In Caiquín, some families had settled on Montaña Camapara out of desperation. They were barely eking out a living, and Caiquín's mayor saw no possibility of relocating them. Even so, the three mayors carried out face-to-face conversations and agreed in principle that Montaña Camapara should be protected.

In 2001, the emerging accord was set back when national elections took place and new mayors were elected in the three *municipios*. The new La Campa mayor, Don Juan, had returned home after establishing a career as a lawyer and representative in the national legislature. Creating a reserve on Montaña Camapara became one of his first priorities. By contrast, Caiquín's new mayor encountered depleted municipal coffers and pressing financial obligations, and the new mayor of Santa Cruz entered his position with no intention of supporting a reserve that his electorate bitterly opposed. La Campa had widespread support for the reserve, and its municipal council proceeded to declare a reserve on its side of Montaña Camapara in 2002. The reserve encompassed 337 hectares and La Campa's 7 water reservoirs that existed at that time.

Through collaboration among the water committees, water beneficiaries, the FAO, and Solidaridad, the process of demarcating the reserve began, even though several major challenges remained: (1) Farmers still worked fields and coffee plantations within the proposed boundaries, and demanded compensation to move off the mountain. (2) Coffee farmers had begun building a road through the reserve to help them transport their harvest. (3) Hunters still viewed Montaña Camapara as a communal hunting ground. (4) Caiquín and Santa Cruz had not committed to protect the cloud forest and watersheds on their sides.

For over a year, municipal authorities and water committees continued talking to the Campeño farmers whose land now fell inside the declared reserve. Two farmers departed voluntarily; they had adequate landholdings and coffee plantations in other parts of the municipio. One man died, and his family relinquished its claim to land in the reserve. Others asked for land to replace what they would lose by leaving Montaña Camapara. La Campa's mountain farmers, however, had not built their homes within the reserve, and losing land did not mean that they would lose their houses. On 26 July 2004, Don Juan met with 12 of the Campeño farmers who had resisted leaving their land. The farmers, most of whom I interviewed individually in 2010,[1] recalled the meeting with vivid, highly correlated details.

"Don Juan told us that we had no choice, we would have to leave our land."

"We felt that we would have to go along with the water committees, in the interests of keeping peace."

"By that time we had realized that it was only a matter of time, the public was against us, but we did not want to give up the land without some compensation."

"Some of us had no other land. We had little hope of finding similar, good land. We asked to be given something for all that we would lose."

"The mayor listened to us. He said that there was no money in the municipal coffers, but he would find a way to give us something, although it would have to be paid in small portions over a period of time."

"After a long talk, we agreed that we would move off the mountain if we each received 4000 lempiras [about USD 200] in compensation."

The municipal secretary recorded the meeting minutes, and at the end drew up the agreement, which affirmed that the farmers would depart their land in the reserve in return for the promised monetary compensation. Everyone present at the meeting signed the agreement with their names or thumb prints. Additional farmers came forward after this meeting, and four received partial compensation for relinquishing land claims within the reserve. Altogether, 19 farmers left La Campa's reserve. Each of these farmers removed their fences, left their land, and then fulfilled their duty as community residents to complete the fencing of La Campa's part of the reserve. The water committees accepted responsibility for covering the promised compensation, which they did by levying an additional fee on all of the households who used water. Over the course of two years, almost all signatories received the HNL 4000 as agreed. Don Juan paid the small amount still pending from his own pocket.

As I interviewed farmers who had left the reserve, I asked how they felt about the process. Each one expressed a degree of regret, and the few who had not found suitable land to purchase in replacement revealed lingering anger. Nonetheless, every one of them expressed support for the reserve, and they believed that their decision protected the water upon which their families depended as much as everyone else. Several stated that as members of the community, individuals must sometimes recognize a common good, even if it is personally difficult. Most of them mentioned that the forest had returned to the land they had relinquished.

Other challenges have proven more difficult to address. Municipal authorities stopped the road that was being built through Montaña Camapara, and water committees built concrete posts to prevent vehicles from entering the reserve where the road breached the boundaries. But a few coffee farmers still argue that the road should be built, and several of the concrete posts have been damaged. A continuing educational campaign discourages hunting in the reserve; even so, one of my guides brings along a rifle "just in case we see a dangerous animal" while we collect botanical samples.

Farmers in Caiquín and Santa Cruz who continue to work on the mountain pose the greatest challenges to the reserve. Unlike La Campa's, Caiquín's mayor and water committees have been unable to raise funds to compensate farmers living on Camapara, and 13 families continue to live and farm on the mountain. These farmers include some of Caiquín's poorest families. Compensation is more costly and difficult in Caiquín. The land is more expensive than in La Campa due to the influx of absentee landlords who acquired coffee plantations on much of Caiquín's best land. Caiquín's authorities have not figured out a way to collect the taxes due from these outsiders, who neither live in nor invest in Caiquín, and the municipality teeters on the edge of bankruptcy. (La Campa, by contrast, has been relatively successful in compelling absentee landlords to pay taxes and water-use fees by enforcing municipal and water committee by-laws.) Nevertheless, Caiquín has succeeded in fencing off a reserve that encompasses its major water sources, and secondary succession has begun to grow in some of the deforested sections.

Santa Cruz's municipal authorities agree in theory to the importance of the reserve, but they are shackled by limited resources and resistance among the families who hold land on its side of the mountain. Santa Cruz does not benefit from Montaña Camapara's water, because all of the springs belong to Caiquín and La Campa, and water does not flow uphill. Therefore, Santa Cruz does not prioritize the reserve's protection (Table 1).

Reserve creation as a process of environmental conflict resolution

The processes that led to the creation of this reserve contained many aspects of successful environmental conflict-resolution processes, even though there was neither conscious effort nor knowledge on the part of the participants to follow conflict resolution theory. The water users and municipal authorities of La Campa and Caiquín included all of the interested parties in negotiations, and they sought outside assistance as well as relevant information. They benefitted from informative seminars on forest conservation and technical assistance made available through international development programmes. The mayors, while motivated to defend their own *municipios'* interests, were willing to meet face to face to discuss their situation, and they found common ground on which to consider the reserve. All of the participants, including the Campeño farmers who eventually departed the reserve, engaged actively in the process. Participant engagement appears to be a crucial key to conflict resolution (Emerson et al., 2009; Orr, Emerson, & Keyes, 2008). Environmental conflict resolution often benefits from an outside mediator, and in this case, the actors who came close to playing that role were representatives of the FAO and Spanish development programmes in the region. In providing transportation, information and technical assistance, the representatives of the programmes supported the process of creating the reserve. Throughout the process, members of the local water committees worked patiently to convince their neighbours, municipal authorities and the farmers on Montaña Camapara that creating a reserve and conserving the cloud forest was the best way to protect their water supply.

Environmental conflict resolution recognizes that willingness to participate interacts with stakeholders' abilities to engage (including skills, resources and capacity). These factors shape eventual outcomes (Emerson et al., 2009). In the process of creating the reserve, each *municipio* had its own contexts and constraints, which translated into their willingness to participate. La Campa's residents and authorities had compelling circumstances that proved conducive to reaching an agreement with farmers on the mountain. By 2002, when the reserve was first created, a majority of the people in La Campa depended on Montaña Camapara for potable water, including the farmers whose clearings on the mountain had negatively impacted the water quality of springs and reservoirs. Their shared dependence on water provided a context in which they were highly motivated to find a workable solution to disagreements among themselves and with neighbouring *municipios*.

Caiquín had less dependence than La Campa on Montaña Camapara's water. Nearly half of Caiquín's settlements have other sources of water available to them, and the settlements that depend on Montaña Camapara's water had completed only one water project on the mountain when La Campa started pressing to create a reserve. As more of Caiquín's settlements formed water committees and started building water projects, they became committed to protecting their investment, and formally declared the reserve on their side of the mountain. Unfortunately, Caiquín faces enduring difficulties to relocate farmers who would be destitute if they were to leave their land and houses without compensation. Moreover, a wealthy and powerful absentee landowner owns a large parcel of cleared land in Caiquín's reserve. He has no interest in conservationist arguments and has not been swayed by the social pressure to 'do the right thing' that influenced residents of La Campa.

Santa Cruz presents the most difficult obstacles to conservation of Montaña Camapara's forests and watersheds. A number of residents depend on the land to plant subsistence crops and graze livestock. Due to their historical relationships with La Campa and Caiquín, which include perceptions that they have already lost some of their land, many Cruzeños view the reserve as an excuse by their neighbors to expropriate more land.

Institutional arrangements that promote equity

Throughout the process of reserve creation and maintenance, the water committees have developed rules that promote equitable responsibilities and benefits among water users. The beneficiaries of water projects, which today include nearly 100% of La Campa's households and around 60% of Caiquín's, provide labour to maintain the fences that surround their respective reserve areas, as well as cleaning and repair of water tanks and reservoirs. This usually means that each head of household works a turn on water-project maintenance once or twice a year; residents of smaller settlements serve more frequently. If a household is unable to send a worker, a substitute can be sent or hired, or a fee equivalent to one day's labour can be paid. Because every household sends at least one adult member to work periodically on Montaña Camapara, the people understand the conditions on the mountain and the vulnerability of their common source of water to incursions.

Every household that fulfils its responsibilities to the water committee has the right to water. The committees make special arrangements for water users experiencing illness or economic difficulties, but enforce fines for shirkers who fail to pay or skip labour obligations without an explanation. Thanks to the democratic arrangements of the committees, along with peer pressure facilitated by relatively small sizes of the settlements, it is difficult for an individual to avoid responsibilities or use a disproportionate amount of the water. All of the 20 water committees currently depending on Montaña Camapara's water (some of which serve more than one settlement) are highly successful in providing water regularly to the majority of their populations. They do have some technical difficulties associated with seasonal low water pressure, which disproportionately affects households near the ends of the water lines. Several committees have faced financial management and leadership problems, but water users tend to address problems as soon as they perceive a risk to their water supply.

Since the reserve's creation, La Campa and Caiquín have developed monitoring mechanisms for the reserve. At the municipal level, they have municipal forest-protection units staffed by one or two natural resource guards who patrol the reserve's borders, among other responsibilities. Caiquín's water users pay an extra monthly fee to support two guards hired specifically to monitor their fences and water infrastructure. In addition, water committee representatives check the fences as part of their maintenance activities. With consistent monitoring, the reserve area has had only a few incursions. In one case, a desperately poor Campeño cleared a maize field to feed his family, and it crossed into the reserve. Municipal authorities called the man to a hearing, and arranged a compromise that allowed him to keep the land. In return, the farmer promised to make no further inroads, maintain the reserve's fence in the vicinity of his land, and guard that edge of the reserve against incursions. The decision to let him keep the land converted him into a defender of the reserve who constantly monitors a vulnerable edge.

Lessons and conclusions

Studies have argued that negotiation and participatory processes can lead to effective institutions for resource management and potentially mitigate conflict (Petts, 1988; Rocheleau, 1994; Tang & Tang, 2001; Uitto & Duda, 2002; Xu & Jim, 2003). In the case of the Montaña Camapara Reserve, the land disputes have been kept in check through negotiation and participatory experiences, as well as social networks that link Caiquín, La Campa, and to a lesser extent Santa Cruz. La Campa's reserve remains in legal limbo, due to the errors made when Caiquín's borders were drawn, yet agreements made between the mayors provide local confidence that La Campa's rights will be respected. Cloud-forest protection and recuperation have been constrained because Caiquín has not yet been able to remove farmers living on its side of the mountain, and Santa Cruz continues to oppose any reserve on its part of the mountain. Even so, the *municipios* have kept the lines of communication open, and educational efforts have helped people to learn about the benefits from sustainable water management and reserve protection. Moreover, the associated water committee rules for reserve maintenance build awareness of the natural resource base, provide opportunities for information exchange and sociality, and reinforce democratic procedures that support water equity.

The more interesting elements in the creation and governance of the Montaña Camapara Reserve lie in the grass-roots efforts that make it possible and the results which show support for equitable access to potable water. The mountain's springs and streams are protected because water users made a commitment to negotiate with the people who posed threats to the water sources. The users also found support from municipal authorities, and have today developed rules and practices that better conserve the watershed. The current institutions largely fulfil the principles associated with sustainable management of common-pool resources, including monitoring and enforcement with graduated sanctions, accessible mechanisms for conflict resolution, and shared trust established through face-to-face interactions (Chhatre & Agrawal, 2008; Gibson et al., 2005; Ostrom, 2005). Although La Campa is still engaged in legal efforts to reverse the national decision that ignored its land rights to Montaña Camapara, the reserve has clearly demarcated boundaries and well-defined user groups.

The creation and ongoing protection of the Montaña Camapara Reserve provides evidence that it is possible, albeit difficult, for local populations to work together to find solutions to manage complex environmental problems. As in other studies, this research shows that partnerships between local and external entities can facilitate productive outcomes (Berkes, 2007; Chambers 1997; Ostrom, 1996). From an institutional perspective, the rules have evolved to fit the local circumstances and to meet people's needs. From the perspective of environmental conflict resolution, it appears that calibrating conflict-resolution processes and subsequent agreements to fit local constraints can lead to mutually acceptable outcomes for participants. At the same time, some of the participants made more sacrifices than others, and the goals of protecting the entire mountain have not been attained. The case of Montaña Camapara Reserve indicates that shared dependence on a critical natural resource such as water can provide an incentive for cooperation, collective action and community building even in contexts of conflict and contested property rights.

Acknowledgements

Support and funding for this research came from the Wenner-Gren Foundation for Anthropological Research (Grant No. 7748), the Inter-American Institute for Global Change Research (CRN-2060), the Center for the Study of Institutions, Population, and Environmental Change, and Indiana University. I am deeply grateful to the residents, water committees, and municipal authorities of San Marcos de Caiquín, La Campa and Santa Cruz in Honduras; they responded openly to questions, welcomed me to their meetings, and offered frank assessments of the historical and current challenges they continually confront in managing Montaña Camapara and its resources. Staff at the Honduran Institute for Forest Conservation and Development and former employees of the Food and Agriculture Organization of the United Nations and Spain's Solidaridad development projects in Lempira kindly provided useful documents and information. Throughout my research, the participants revealed their commitment to the ongoing demands of water resource management, discussed the challenges they faced and continue to address, and imparted the lessons they have learned. I appreciate their thoughtful and modest reflections, and the optimism they imparted.

Note

1. Of these 12 farmers who left the mountain, 8 were still farming land in La Campa in 2010. Two had departed to find work in an urban area, one had died, and another was incapacitated due to a stroke.

References

Acheson, J. M. (2006). Institutional failure in resource management. *Annual Review of Anthropology*, 35, 117–134.

Amery, H. A. (2002). Water wars in the Middle East: A looming threat. *The Geographical Journal,* 168(4), 313–323.

Bakker, K. J. (2001). Paying for water: Water pricing and equity in England and Wales. *Transactions of the Institute of British Geographers,* 26(2), 143–164.

Bean, M., Fisher, L., & Eng, M. (2007). Assessment in environmental and public policy conflict resolution: Emerging theory, patterns of practice, and a conceptual framework. *Conflict Resolution Quarterly,* 24(4), 447–468.

Berkes, F. (2007). Community-based conservation in a globalized world. *Proceedings of the National Academy of Sciences of the United States of America,* 104(39), 15188–15193.

Blaikie, P., & Brookfield, H. (1987). *Land degradation and society.* New York, NY: Routledge.

Burton, J., & Dukes, F. (1990). *Conflict: Practices in management, settlement and resolution.* New York, NY: St. Martins Press.

Bush, R. A. B., & Folger, J. P. (2005). *The promise of mediation: The transformative approach to conflict.* San Francisco, CA: Jossey-Bass.

Carmichael, M., Schafer, S., & Mazumdar, S. (2008). Troubled waters. In P. McCaffrey (Ed.), *Water supply* (pp. 101–104). New York, NY: H. W. Wilson.

Castegnaro de Foletti, A. (1989). *Alfarería Lenca contemporanea de Honduras.* Tegucigalpa: Editorial Guaymuras, SA.

Chambers, R. (1997). *Whose reality counts? Putting the first last.* London: Intermediate Technology Publications.

Chapman, A. (1985). *Los hijos del copal y la candela: Ritos agrarios y tradición oral de los Lencas de Honduras.* México, DF: Universidad Nacional Autónoma de México.

Chhatre, A., & Agrawal, A. (2008). Forest commons and local enforcement. *Proceedings of the National Academy of Sciences of the United States of America,* 105(36), 13286–13291.

Dietz, T., Ostrom, E., & Stern, P. C. (2003). The struggle to govern the commons. *Science,* 302, 1907–1912.

Dukes, E. F. (2004). What we know about environmental conflict resolution: An analysis based on research. *Conflict Resolution Quarterly,* 22(1–2), 191–220.

Emerson, K., Orr, P. J., Keyes, D. L., & McKnight, K. M. (2009). Environmental conflict resolution: Evaluating performance outcomes and contributing factors. *Conflict Resolution Quarterly,* 27 (1), 27–64.

Fiallos, C. (1991). *Los municipios de Honduras.* Tegucigalpa: Editorial Universitaria.

Gunderson, L., & Holling, C. S. (Eds.). (2002). *Panarchy: Understanding transformations in human and natural systems.* Washington, DC: Island Press.

Gibson, C. C., Williams, J. T., & Ostrom, E. (2005). Local enforcement and better forests. *World Development,* 33(2), 273–284.

Hamilton, L. S. (1995). Mountain cloud forest conservation and research: A synopsis. *Mountain Research and Development,* 15(3), 259–266.

Hecht, S., & Cockburn, A. (1990). The fate of the forest: Developers, destroyers and defenders of the Amazon. Chicago: University of Chicago Press.

Homer-Dixon, T. (2001). Environment, scarcity, and violence. Princeton, NJ: Princeton University Press.

Janssen, M. A. (Ed.). (2000). *Complexity and ecosystem management: The theory and practice of multi-agent systems.* Cheltenham: Edward Elgar.

Marks, S. J. (2009). *Aqua shock: The water crisis in America.* New York, NY: Bloomberg Press.

McKean, M. A. (2000). Common property: What is it, what is it good for, and what makes it work? In C. C. Gibson, M. A. McKean, & E. Ostrom (Eds.), *People and forests: Communities, institutions and governance* (pp. 27–56). Cambridge, MA: MIT Press.

Nagendra, H., Karna, B., & Karmacharya, M. (2005). Examining institutional change: Social conflict in Nepal's leasehold forestry programme. *Conservation & Society,* 3 (1), 72–91.

Navarro, E. (2002). Agua para más de 500 años. Socializando la esperanza series. Tegucigalpa: Litografía López S. de R. L. for FAO, Secretaria de Agricultura y Ganadería and Gobierno de Holanda.

North, D. C. (1990). Institutions, institutional change and economic performance. Cambridge: Cambridge University Press.

Orr, P. J., Emerson, K., & Keyes, D. L. (2008). Environmental conflict resolution practice and performance: An evaluation framework. *Conflict Resolution Quarterly,* 25(3), 283–301.

Ostrom, E. (1996). Crossing the great divide: Coproduction, synergy, and development. *World Development,* 24(6), 1073–1087.

Ostrom, E., Dietz, T., Dolšak, N., Stern, P. C., Stonich, S., & Weber, E. U. (Eds.). (2002). *The drama of the commons.* Washington, DC: National Academies Press.

Ostrom, E. (2005). *Understanding institutional diversity.* Princeton, NJ: Princeton University Press.

Petts, G. E. (1988). Water management: The case of Lake Biwa, Japan. *The Geographical Journal,* 154(3), 367–376.

Phumpiu, P. (2008). *The politics of Honduras water institutional reform.* (TRITA-LWR Report 3020). Stockholm: Royal Institute of Technology, Department of Land and Water Resources Engineering.

PNUD (Programa de las Naciones Unidas para el Desarrollo). (2006). *Informe sobre desarrollo humano Honduras 2006: Hacia la expansión de la ciudadanía.* San José, Costa Rica: Litografía e Imprenta Lil, S.A.

Republic of Honduras. (2003). Decreto 118–2003, Ley marco del sector agua potable y saneamiento. *Diario Oficial La Gaceta,* 8 (de Octobre).

Rocheleau, D. E. (1994). Participatory research and the race to save the planet: Questions, critique, and lessons from the field. *Agriculture and Human Values,* 11(2), 4–25.

Sterling, E. J. (2007). Blue planet blues. *Natural History,* 116(9), 29–31.

Susskind, L., & Ozawa, C. (1983). Mediated negotiation in the public sector: Mediator accountability and the public intec rest problem. *American Behavioral Scientist,* 27, 255–279.

Tang, C. -P., & Tang, S. -Y. (2001). Negotiated autonomy: Transforming self-governing institutions for local common-pool resources in two tribal villages in Taiwan. *Human Ecology,* 29(1), 49–67.

Tucker, C. M. (2004). Community institutions and forest management in Mexico's Monarch Butterfly Reserve. *Society and Natural Resources,* 17(7), 569–587.

Tucker, C. M. (2008). *Changing forests: Collective action, common property and coffee in Honduras.* Dordrecht: Springer.

Tucker, C. M. (2010). Private goods and common property: Pottery production in a Honduran Lenca community. *Human Organization,* 69(1), 43–53.

Uitto, J. I., & Duda, A. M. (2002). Management of transboundary water resources: Lessons from international cooperation for conflict prevention. *The Geographical Journal,* 168(4), 365–378.

West, P., Igoe, J., & Brockington, D. (2006). Parks and peoples: The social impact of protected areas. *Annual Review of Anthropology,* 35, 251–277.

White, A., & Martin, A. (2005). Who owns the worlds forests? Forest tenure and public forests in transition. In J. Sayer (Ed.), *The Earthscan reader in forestry and development* (pp. 72–103). London: Earthscan.

Xu, S. S. W., & Jim, C. Y. (2003). Using upland forest in Shimentai Nature Reserve, China. *Geographical Review,* 93(3), 308–327.

Young, O. (2002). *The institutional dimensions of environmental change: Fit, interplay and scale.* Cambridge, MA: MIT Press.

Democratizing discourses: conceptions of ownership, autonomy and 'the state' in Nicaragua's rural water governance

Sarah T. Romano

Political Science and International Affairs, University of Northern Colorado, Greeley, CO, USA

ABSTRACT
The interconnected discourses of ownership, autonomy, and state roles and responsibilities in the water sector are a strategic feature of the mobilization of water committees in Nicaragua. In particular, this paper argues that the effectiveness of these discourses in supporting water committees' goals of political inclusion and legal recognition owes to how they reflect the day-to-day, historical and contemporary experience of water management at the grassroots, including how this work implicates the state. Ultimately, this case demonstrates how discourses 'from below' can have a democratizing effect on water governance by helping to carve out space for marginalized actors' policy interventions.

Introduction

We're the volunteer army of the community. (Luis Adolfo Vargas Ríos, CAPS San Lorenzo, Boaco, 17 February 2010)

[The CAPS] are nothing more than groups of voluntarily organised people [...] taking actions and make efforts that correspond to the state to assure the population's access to the vital liquid in the urban–rural zones in the country. (Rodríguez et al., 2008)

In the early 2000s, an anti-water privatization social movement and legal–institutional restructurings of the water sector in Nicaragua set the stage for the political mobilization of hundreds of rural, community-based water committees. Exercising de facto control over thousands of small-scale water systems nationally, the Comités de Agua Potable y Saneamiento (Potable Water and Sanitation Committees – CAPS) constitute a large-scale resource management scheme in the country's rural areas, uniting upwards of 30,000–40,000 rural residents in the day-to-day work of securing access to water for domestic use.[1] The challenge of improving access to clean water in Nicaragua is stark: 52% of rural residents lack access to drinking water (compared with 25% for urban) (Herrera, 2007), and ongoing deforestation and contamination of water sources threaten access to this critical resource for immediate and future consumption and use. In this context, the environmental and social implications of community-based water management are significant. CAPS-managed water systems, including wells and gravity-fed systems, enable water access to over an estimated 1 million rural residents nationwide – a little over half the country's rural population (Medrano et al., 2007).

Notably, despite their crucial role in enabling water access in rural areas, CAPS have confronted a political and legal landscape that has effectively 'hid' them for over 30 years. Importantly, the Nicaraguan state has, since the early 1970s, facilitated the emergence of community-based arrangements for water provision in rural areas by coordinating the investments of domestic and international agencies in the water sector and, in certain regions, playing a direct role in water system construction and technical training.[2] Nevertheless, the work of managing and maintaining water infrastructure over time has fallen to residents. Surprisingly, until very recently, CAPS have not benefitted from a comprehensive legal framework acknowledging, delineating and legally backing their organizations.[3] This political exclusion and legal invisibility were starkly highlighted in the early 2000s when CAPS were left out of primarily urban-based, 'public' debates on water privatization and government- and non-governmental organization (NGO)-sponsored consultations on an impending water law. CAPS' exclusion from the process of designing the General Water Law (Law 620), published in 2007, was compounded by the law's glaring omission of the CAPS. Emphasizing urban-focused state agencies, the law reflected a skewed landscape of water governance and provided a strong impetus for the mobilization of rural water managers.

Since the mid-2000s, mobilized water committees have pursued a dual objective: to protect and promote local autonomy and ownership over rural water systems as well as to demand state recognition and increased support of rural water management. This pursuit of multi-scalar political goals has required collective action linking CAPS across communities for the first time. Specifically, dozens of new 'CAPS networks' at the municipal and national levels have allowed rural water managers to 'scale up' their locally focused, territorially based collective action (Romano, 2012a, 2012b; see also Fox, 1996; Hoogesteger & Verzijl, 2015; Perreault, 2003, 2005). These multi-scalar networks are also multi-sectoral in nature; alliances with domestic NGOs and international development agencies have been of crucial importance to the ability of CAPS as rural social actors to mobilize to transcend their localities, engage with state authorities and integrate themselves into policy-making processes. Indeed, the organizational, financial and technical support of extra-local allies (like the Netherlands Development Organization and the Swiss Agency for Cooperation and Development) has enabled a trans-community and scaled-up mobilization strategy – hence facilitating the development of collective political agency amongst CAPS (Romano, 2012a, 2012b).[4]

As the principal spaces and means of collaboration amongst CAPS from different regions, CAPS networks have facilitated the development of collective political objectives and served as platforms for creating public awareness of CAPS' rural water management. Representing upwards of 2000 water committees, these networks have also facilitated engagement with state actors, in part evident through the 2010 passing of the Special CAPS Law (Law 722). While the law resulted from and reflected CAPS' political interventions, it is proving to be a double-edged sword, as water committees confront burdensome legal and financial responsibilities alongside what are perceived as new, beneficial prerogatives for water committees.[5] Law 722's continued implementation and outcomes merit further study; indeed, it is likely that inequities with regard to the distribution of the law's heralded benefits will be stark, as Seemann (2016) found with regard to state decentralized water policies in Bolivia. However, this paper focuses on the political dynamics *preceding* the law's passing to demonstrate how political assertions on the part of rural water committees helped to engender novel forms of political inclusion among these rural stakeholders.

Specifically, the paper identifies how local configurations of 'water, power, identity, and cultural politics' (Boelens, 2014, p. 3) – or grassroots hydrosocial territories – have been publically projected in a way that has proved instrumental towards promoting the political inclusion and legal recognition of water committees vis-à-vis the state. It argues, in particular, that the interconnected discourses of autonomy, ownership, and state roles and responsibilities in the water sector reflect the day-to-day, historical and contemporary, experience of water management at the grassroots. While the CAPS' water management continues to encompass diverse ecological landscapes and membership regimes – and hence a diversity of infrastructural and organizational technologies – the CAPS as common property regimes (CPRs) share similarly constructed socionatural 'water territories' across Nicaragua.[6] That is, as discussed further below, CAPS interviewed and observed for this research reflect shared senses of community belonging, commitments to enabling water access, and morally guided decision-making in local water management. New networks, importantly, gave an opportunity to construct collective *framings* of shared experiences and values in rural water governance across multiple water committees. These discourses, in turn, have informed the political and legal reshaping of the water sector in Nicaragua, evident in the carving out of political space for marginalized actors' policy interventions and the passing of Law 722. Ultimately, this case generates insight into how discourses generated 'from below' can have a democratizing effect on water governance by promoting more inclusive public policies and policy formation processes.

Research for this project was conducted primarily in several northern highland departments in Nicaragua, including Estelí, Jinotega, Madrid and Matagalpa, across 14 months of field research (2004–10 and 2014). These regions were selected as they were known nationally for their vibrant and well-organized CAPS networks, making them apt sites to examine how CAPS have mobilized and engaged with state actors. In addition to extensive participant observation of local-, municipal- and national-level CAPS meetings, semi-structured and informal interviews were conducted with dozens of key informants including CAPS members; staff at domestic and international NGOs, multilateral organizations, and state agencies and ministries; as well as local and national elected officials. Review and analysis of primary and secondary data sources, including NGO and multilateral organizations' websites and written materials, newspaper articles, as well as national legislation, also have significantly informed the contextualization of interview and observation data gathered for this paper.

Hydrosocial territories 'from below' and shifting regimes of representation in water governance

In rural areas across the global South, thousands of local, non-state groups like the CAPS operate 'below the radar' of formal decentralization initiatives or state policies to compensate for inadequate state reach or capacity to manage or distribute resources necessary for livelihoods (Perreault, 2006, 2008; Pretty & Ward, 2001; Romano, 2012b; see also the literature on CPRs). Although the organizational and financial contributions of domestic and international NGOs and multilateral organizations have been of crucial importance for the ability of grassroots organizations to function locally as well as to stretch across political scales, operating outside of formal policy or decentralization initiatives can allow community-based resource managers to work largely independently from state actors at the grassroots. In fact, many community-based regimes have what are called here an 'organic empowerment' to manage resources like forests and water sources. In these circumstances, empowerment emerges from the day-to-day labours associated with resource management, rather than as a result of official state policy bestowing legal responsibility upon a particular group or level of government (Romano, 2012b). Notably, CAPS' empowerment to manage resources – as well as to provide water service – has emerged from the imperative to manage water infrastructure after financing agencies (like UNICEF) leave, or 'abandon', communities once water systems are completed.

While tenuously and unevenly connected to official state realms of water governance, community-based water managers are embedded in local 'water territories' (Boelens, Hoogesteger, Swyngedouw, Vos, & Wester, 2016). Integral to what Boelens (2014) refers to as 'hydrosocial cycles', these grassroots territories reflect socially constructed and dynamic social, political and ecological landscapes, and constitute a

> scheme of mutual belonging that enables the rebirth of collective imaginations. [... Water] territories involve socio-natural webs with landscapes and waterscapes in which people live and make livelihoods and identities, for which people feel responsible, in which they are morally involved. (Boelens, Getches, & Guevara-Gil, 2010, p. 19)[7]

Hydrosocial territories thus encompass both material and ideological dimensions of water management, as well as influence processes of identity formation for those groups intimately involved in water extraction, use and management at the grassroots (Boelens, 2008, 2009, 2014; Boelens et al., 2010). As socio-natural configurations, hydrosocial territories reflect the value-laden and ideological dimensions of community-based water management, including shared commitments to ensuring water access and protecting local water systems and access regimes from potential threats. In the Nicaraguan case, community norms and commitments manifest in CAPS' value-laden decision-making, such as their selective implementation of by-laws – like water shut-offs – that may harm residents.

As scholars have noted, struggles over water are not solely material. Grassroots conceptions and experiences of water governance intersect, and often come into conflict with, extra-local, including 'top down', representations of water governance – including those outlined in state policy (Boelens, Hoogesteger, & Baud, 2015; Perreault, 2006; Romano, 2012b; Roth, Boelens, & Zwarteveen, 2005). Indeed, as legal frameworks are implemented or restructured – as they have been across the global South in recent decades – a salient concern arises over how to reconcile longstanding, yet 'unofficial', prerogatives to control local water systems and sources with state policies that in practice undermine local regimes (Boelens et al., 2010). States' adoption of 'one size fits all' legal characterizations of grassroots organizations, for example, can effectively deprive communities of their water management prerogatives or responsibilities via the imposition of culturally or politically inappropriate rules and/or decision-making bodies (e.g., Baer, 2008; Bakker, 2010; Boelens, 2008; Boelens et al., 2010; Shiva, 2002; Swyngedouw, 2004). As the legitimate authorities within grassroots water territories – and from which they 'draw strength' their organizing efforts (Spronk & Webber, 2007) – grassroots organizations for water management may be inclined to confront state-constructed hydro-social territories perceived as threatening to their local regimes, thus setting the stage for a 'battle [...] over the right to culturally define, politically organise and discursively shape [water use systems]' (Boelens, 2008, p. 50).

As this paper will argue, discursively projecting grassroots hydrosocial territories offer politically marginalized water users a tool to confront dominant political and legal arrangements.[8] Water users and managers may leverage representations of their organizations and socio-environmental landscapes to achieve certain objectives like (re) shaping public opinion, persuading decision-makers (including elected officials) and/or mobilizing a base of supporters. Importantly, a legal 'representation by omission' (as opposed to an *inaccurate* representation) may constitute a threat in the eyes of water committees. In Nicaragua, CAPS operated for over 30 years without explicit inclusion in formal policy frameworks, rendering rural water committees overlooked in political and legal terms beyond their communities. CAPS' mobilization ensued in the context of the development of Law 620, a comprehensive legal framework for the use, administration, conservation and regulation of the country's freshwater resources. Upon its passing, Law 620 did not reference CAPS except for a brief and underdeveloped casting of rural water management in its *Reglamento*.[9] CAPS' mobilization via networks constitutes a response to a top-down representation of national water governance in which rural water committees were largely absent as resource managers and political actors.

In important respects, discourses of water governance that are used in mobilization efforts – including reflections of local hydrosocial territories – are akin to collective action 'frames' as explored in the vast literature on social movements (Goodwin & Jasper, 2004; McAdam, McCarthy, & Zald, 1996; Oliver & Johnston, 2000; Snow, 2004; Westby, 2002). Framing processes, or 'the collective processes of interpretation, attribution, and social construction' (McAdam et al., 1996, p. 2), have been cast as a 'dynamic, negotiated, and often contested processes' (Benford & Snow, 2000, p. 3). The collective action frames that result encompass 'shared meanings' and 'mediate between opportunity and action' (McAdam et al., 1996, pp. 2, 5). While framing as a dynamic and contested process of meaning-making has been well documented within the social movement scholarship, there is still much to learn about how frames influence social movement trajectories and affect the achievement of movement goals (Benford & Snow, 2000).

In part because they reflect the conveying of material experiences of grassroots water management, CAPS' discourses do not always serve as 'strategic' frames in a social movement sense. That is, CAPS' discourses do not appear to constitute the outcome of collective *strategizing*, but reflect a process of 'mak[ing] sense of both daily life and the grievances that [they] confront' (Oliver & Johnston, 2000, p. 42). Hence, while framing is oftentimes qualified as 'strategic' in the social movement literature, CAPS' discursive framing, it is argued here, can best be understood as producing instrumental, yet not self-consciously strategic, 'emergent' frames (McAdam et al., 1996; Oliver & Johnston, 2000).

Interestingly, CAPS' framing processes and resulting frames have not been overtly contentious in the way described in much of the social movement literature. In fact, what is perhaps most surprising about CAPS' discourses in light of theories of collective framing is how effective they were in garnering an apparent 'consensus' among CAPS participating in networks (Klandermans, 1984, as cited in Benford & Snow, 2000). Media outlets and state actors followed suit, with the buy-in of the latter being all the more counterintuitive as the principal target of CAPS' policy interventions. These outcomes may owe to how a case of 'representation by omission' means grassroots groups can benefit from a relatively 'blank' discursive terrain as they promote their organizations and agendas in the public sphere. Importantly, the Nicaraguan state's failure legally to delineate and politically include CAPS as materially and socially consequential actors in water governance gave these water committees leverage as they began to advocate for recognition of their organizations.

This paper bridges the concept of hydrosocial territories with the social movement scholarship's work on framing in order to account for the way in which collective projections of the former can – much like movement frames – play a distinct and crucial role in promoting movement objectives. It aligns with Boelens (2014) in that this research has not principally sought to 'verify' divergent 'truth contents" in the discourses examined, but rather to '[understand] how such claims to truth are being used in practice, how they shape perceptions of reality and also define socionatural reality itself: how they form part of particular hydrosocial cycle constructs and truth–knowledge–power triangles' (p. 240). That is, it aims to document and analyse grassroots discourses, drawing attention to how these reflect collective perceptions of locally rooted territorial and political claims.[10] Moreover, this research suggests that discourses referencing the material, territorially based work of rural water committees may be particularly instrumental towards advancing movement objectives because of their 'empirical credibility', or their 'fit between the framings and events in the world' (Benford & Snow, 2000, p. 620). CAPS' discourses reflect rural residents' everyday, historical and contemporary, experience as resource managers, including their morally guided work at the grassroots. These discourses have resonated with broader publics, and importantly with state actors, including national agency staff and elected officials. In so doing, these discourses have contributed to shaping 'realities' of rural water management now codified in national policy.

The Nicaraguan Potable Water and Sanitation Committees (CAPS): transcending exclusion with discourse

Since the 1970s, when state and private investments in rural water projects began in Nicaragua, the invisibility of the CAPS in legal and political terms beyond their communities has belied their geographic scale and material consequences as resource managers. Arguably, the infrastructural and membership constraints of their community-based regimes has supported remaining geographically and politically isolated from one another as water committees.

Yet, CAPS' confrontation of multi-scalar issues engendered trans-community collective action. Water committees' visibility on a national stage and collective assertions with regard to water policy owe in large part to their formation of over 30 municipal level 'CAPS networks' and a National CAPS Network – both multi-CAPS organizations formed in response to the threat of water privatization and exclusion from Law 620 in the early 2000s.[11] The coalescing of individual CAPS' experiences in new network spaces has allowed for the construction of collective representations of grassroots hydrosocial territories: networks have served as forums for CAPS members to discuss shared grievances, receive training on and analyse national legislation, and develop a collective action platform. Networks have thus enabled the honing of a collective voice and sense of agency that is embedded within shared experiences as 'CAPS' at the grassroots.

The discourses honed within CAPS networks have made public appearance by several means: a collective action platform, political pronouncements, participation in public events and use of media outlets – including billboards and radio spots. CAPS have also engaged directly with state actors in network spaces and in government spaces of representation.

The following sections examine the interconnected discourses of autonomy, ownership, and state roles and responsibilities in the sector, drawing attention to the ways these reflect and leverage the CAPS' embeddedness in grassroots hydrosocial territories vis-à-vis top-down, exclusive water territories, towards achieving movement aims.

Discourse of autonomy and ownership: 'aquí somos los dueños'[12]

The CAPS' discourses of autonomy and ownership reflect their identity as resource managers as well as their particular position within the political economy of water governance in Nicaragua. As local water management authorities, CAPS exercise a great degree of control over rural water flows and 'markets', despite not always having official legal ownership over water systems and sources. Notably, CAPS' discourses of independence and ownership reflect deeply held commitments to their communities' well-being. Identifying strongly as volunteers, most CAPS are not compensated in economic terms for their work. Some CAPS explained their work in terms of an emotional connection to one's community. As a CAPS coordinator from Boaco expressed:

> We as community leaders, who work with love for our community, for our municipality, the majority [of us] live in poverty. Why, because we don't have ability as others have to get out of it. If in a year we work one month, two months, that's a lot. [...] At times when someone has a job, a little bit of work, [he's] not going to work in service of the community, if [he's] working to have his own money, his own life. Meanwhile we [the CAPS] don't do that. We're the volunteer army of the community. (17 February 2010)

This CAPS member characterizes water management as emerging from the 'love' of one's community, and implicitly, of one's family and neighbours within the community. Other CAPS similarly framed the commitment reflected in being a part of a water committee: 'Hay que enamorarse del proyecto' ('You have to fall in love with the project'), and 'We have love [for the water system ...] as if it belongs to us' (5 November 2009).[13] These statements indicate a perception that this work 'requires', as well as infuses, an emotional connection to the water project and a commitment to one's community; in short, a commitment grounded in local moral economies that does not lend itself to replication or substitution by external actors (Arnold, 2008; Bakker, 2010; Edelman, 2005; Scott, 1976; Trawick, 2001).

Importantly, CAPS' discourses reflect a hydrosocial territory in which the limited and geographically uneven contributions of the state over time to rural water provision have produced an *imperative* for rural residents to develop autonomous water management capacities. In interviews and network meetings, CAPS members spoke openly and emphatically about the little support they receive from government. Referencing local CAPS' experience in Muy Muy prior to the formation of the municipal network in 2005, and its subsequent representation in the Municipal Development Council (CDM), CAPS coordinator Esperanza explains: 'The truth is that before [participating in the CDM] we never had the support of any mayor. And we have been around for a long time' (8 August 2008). A Sébaco CAPS member likewise emphasized her community's independent work: '[W]e've crawled along by ourselves, with our own money, for our own work. We have never, to this day, had help from a mayor' (11 December 2009). Rural residents, including CAPS members, invest time, labour and, in some cases, monetary resources into constructing rural water systems that they must sustain in the wake of system construction. This investment and set of responsibilities fosters a strong sense of ownership over local water territories and the technical, organizational and financial systems they encompass.

Despite not always having formalized legal ownership over water systems and sources, CAPS' acute sense of ownership has fostered a willingness and ostensible eagerness to articulate claims to water systems. One CAPS coordinator in Boaco recounted a direct interaction with his municipality's mayor, objecting to his attempt to put local water systems in the hands of the Citizen Power Councils (CPCs)[14]: 'I said to the mayor, "One moment, here the community we are the owners and señores of this water system. We understand that the [last] mayor donated the water source to us, because [he] bought it"' (17 February 2010). A similar assertion of local authority and ownership from the mouths of CAPS appeared in a special issue of the popular publication *Enlace* in 2008. CAPS coordinator María asserted: 'We are the owners of these resources, and we are the ones who have worked to have water in [our] homes.' In reference to a public water post built by the Nicaraguan Aqueduct and Sewerage Enterprise (ENACAL, the state water company) that preceded the community's construction of its 'own', and ultimately less costly, water system, she continues: 'why would someone from outside [of the community] come wanting to control us or put in a [water] meter?' (Enlace, 2008, p. 7). These statements reflect a strong sense of ownership over and responsibility for local water systems, engendered by a tenuous relationship to extra-local, including state, actors historically.

CAPS' interactions with government officials – who may demonstrate a lack of awareness of CAPS and the water provision they facilitate – have also constituted opportunities for CAPS to assert an autonomous role in rural water governance. One such example of this occurred after the passing of Law 620 in 2007: CAPS member José Francisco of San Dionisio participated in a workshop led by the Nitlapán Research Center entitled 'Water Management in Nicaragua: Implications of the Law [620] and Water Regulation'. As the only CAPS member to participate in the event, José Francisco found himself in the company of representatives of state institutions and domestic and international NGOs. In an interchange with a government official from the Ministry of Natural Resources and the Environment (MARENA), he explained CAPS' extensive water provision, one that the government had yet to recognize:

> I was the one there from the CAPS. [...] So I spoke as a CAPS. Someone from MARENA comes and she tells me that the CAPS aren't anything, they don't know anything. 'What?' I tell her. 'One moment. *The CAPS, we're five thousand* [nationwide], *and we serve 200,000 people across the country. And no one gives us anything. If you don't know us, now, yes, you're going to begin to know us.*' (19 March 2010, emphasis added)

José Francisco hence counters the MARENA representative's discounting of his organization's work by touting CAPS' extensive contributions to water management at a national scale.

A parallel experience unfolded as CAPS members negotiated with state officials how their name would be represented in the 'Special CAPS Law'. The draft law referred to the Comités de Agua Potable, or CAP, omitting the 'S' as had been done in the Regulations of Law 620. As CAPS' negotiations with state actors revealed, the 'S', referencing 'sanitation', holds not only symbolic importance but also material and financial significance. The name represents a collective identity and speaks to tangible dimensions of grassroots hydrosocial territories, including what CAPS assert to be their rightful prerogatives within the water and sanitation sector. In November 2009, CAPS leaders from the National CAPS Network met with representatives of ENACAL and the National Assembly's Commission of Environment and Natural Resources to discuss the law, including CAPS' capacity to provide for basic sanitation in their communities. Jessenia from Ocotal recalled an ENACAL representative questioning the capacity of CAPS, yet following with the acknowledgment that their company had not had sufficient coverage in rural areas. She recounted her subsequent intervention:

> So that's when I asked to speak and I said: 'You're telling us that we're incompetent. Excuse me but we – the water committees – are more competent than ENACAL. Because, yes, we have entered into rural communities and you [ENACAL] haven't been able to. And there we are. We're getting water [*el vital líquido*] to the people. You think that we're not capable of providing sanitation but we're already doing so. We chlorinate water, we do sanitation campaigns. If we haven't constructed drainage systems, latrines, or hygienic services, it's because we haven't had the legality [to do so]'. (5 November 2009)

Via their national network, CAPS members fiercely contested losing the prerogative of local sanitation – here represented by having their name altered within impending national legislation. Importantly, 'winning' the struggle over the S, at once discursive and material, depended upon CAPS' articulations of their contributions to sanitation at the community level. National deputies' ultimate siding with the CAPS – and in opposition to ENACAL – demonstrates the credibility and, arguably, the legitimacy they ascribed to the CAPS when confronted with water committees' self-representations with regard to rural water governance.

Even though CAPS have, perhaps frustratingly, encountered a public and public officials with little awareness of their water management work, they have benefitted in ways from their extra-legality, or representation by omission. That is, rather than challenge a *pre-existing* legal delineation of their organizations and work, CAPS have sought to shape their integration into a state-sanctioned political and legal water territory in which they were previously invisible.

Discourse of stateness: 'un trabajo que corresponde al estado'[15]

As demonstrated, CAPS members frequently invoke the state in defending their local water management. They do so in part to establish that they work with very few state resources and little support, which lends sympathy to their cause. Like water users in Ecuador discussed in this special issue (Hoogesteger, Boelens, & Baud, 2016; see also Boelens, 2008), CAPS also encourage and promote an *augmented* role of the state in rural water management at the same time that they aggressively assert independence.[16] In the CAPS' case, discursively juxtaposing their state-like work with the state's shirking of its legal responsibilities – and those that are a part of popular imaginaries of the state – has bolstered claims to autonomy and calls for political recognition.

For example, CAPS' discourses of water governance call attention to the state's failures at supporting water access, including state actors' neglect of environmental conservation responsibilities. The 1987 Nicaraguan Constitution makes the state responsible for basic services, citing 'an obligation of the State to promote, facilitate and regulate provision of basic public services', including water, and affirming that 'access to these is an inalienable right of the population' (Article 105). Law 620 confirmed the state's role in securing water access: 'water is a finite and vulnerable resource essential for life and development [...] whose access is a right associated with life and human health that should be guaranteed by the State' (Preamble). CAPS members interviewed and observed in meetings showed awareness of the state's legally ascribed responsibility for water provision. As Sébaco CAPS member Marbelly expressed:

> Since 2007 to the present [...] I've been in the CAPS' struggle in which we're aspiring to a law that supports us. Now we've realized, from Article 150 in Law 620, that it's an obligation of the local government to look after [*velar por*] the water systems. And they've never done it. Of course, as we didn't know the law, we never went to knock on their door. The [NGOs], yes, because we would go to them to ask for backing. But I tell you…the only thing [the government] has done is give a signature so the [NGOs] could work in the community. (11 December 2009)

Across interviews, CAPS members told similar tales of lack of involvement on the part of state actors. Local governments have not been the only entity of significance: many CAPS members reported that the state health agency MINSA did not take regular water samples from rural water systems. In effect, many CAPS were unaware of the actual quality and potability of the water flowing through their 'potable water' systems. CAPS' local water management and environmental stewardship may be actively undermined by state actors as well. Interviews with CAPS and local government officials suggested that the results of state agency water tests may not be shared with communities, at times as a way to avoid dealing with serious issues of water quality. Some CAPS spoke of the state as complicit in the destabilization of local conservation efforts through a lack of sufficient regulation, such as not penalizing companies or landowners for legal breaches. One CAPS member pointed out during a 2009 National CAPS Network meeting that as *campesinos* they do not receive 'payments for environmental services' from the government, but in practice they are fulfilling a role as stewards of the forests and water sources (5 November 2009).

National media outlets' taking up the work of sharing experiences emanating from the CAPS' networks demonstrates the legitimacy and credibility being ascribed publically to rural water committees. Published articles from both major newspapers, *El Nuevo Diario* and *La Prensa*, have emphasized the state's legal duties in the sector while projecting an image of water committees as the de facto guarantors of potable water in the country's rural areas. According to one article:

> The CAPS are practically the only ones who enable the provision of potable water in rural areas, given that the Nicaraguan Aqueducts and Sewerage Enterprise (ENACAL) only covers urban areas, although not all. *Where ENACAL is not* [present], *there you'll find the CAPS.* (Pérez, 2010, emphasis added)

The CAPS are portrayed – much as they have sought to portray themselves – as 'filling in the gaps' left by the state in rural areas. The other leading newspaper similarly asserted that water committees were working where ENACAL 'doesn't offer public service' (García, 2008b), and another article cited Law 620's provision requiring local governments to prioritize water projects – rural and urban – above all others (García, 2008a). Journalist Mendoza quoted a CAPS member from Muy Muy to draw attention to Article 66, which specifies that the government must prioritize allocation of water for human consumption before all other uses: 'This article provides an important tool so that the municipal authorities prioritize water for the population, be [the people] where they are' (Mendoza, 2008).

Reformers in the state advocating for the CAPS' recognition have likewise adopted a discourse recognizing the state's shirking of its water and sanitation responsibilities. In 2008, the bipartisan Environment and Natural Resources Commission submitted to the President of the National Assembly an introductory report entitled *The Potable Water and Sanitation Committee Law Project* (Rodríguez et al., 2008). According to the report, the CAPS

> are nothing more than groups of voluntarily organised people at the community level, in charge of the maintenance and sustainability of potable water and sanitation projects. In few words, *taking actions and making efforts* [gestiones] *that correspond to the state* to assure the population's access to the vital liquid in the urban–rural zones in the country. (emphasis added)

Thus, characterizing the CAPS' work as legally assigned to the state, the commission members demonstrate the bottom-up influence of water committees in national policy discourses. Notably, the deputies further sanctioned demands for legal recognition by paraphrasing CAPS' voices heard during consultations on the proposed law:

> [An] expression gathered during the consultations was the following: 'In passing this law, it would be hardly a cancellation of one of the many debts that the government and the deputies have with us, given that we've been doing the work of the government without receiving anything in return; better said we've had to sacrifice with the paying of high prices for the materials that we buy. In general we have been abandoned by the government and the municipalities.

The unanimous passing of the Special Potable Water and Sanitation Committee Law (Law 722) in 2010 served to recognize CAPS as self-governing, independent grassroots organizations in Nicaragua. The law – which outlines a multi-step process for CAPS to register their organizations and obtain legal status – cites one of CAPS' governing principles to be 'respect and defense of their autonomy and independence' and assigns the state the 'obligation' of 'guaranteeing and strengthening [CAPS'] promotion and development'. The law thus reflects not only the official recognition water committees had pursued since the formation of the first CAPS network in 2005 but also codifies a commitment on the part of the national state to support the CAPS.

The question perhaps arises as to why state actors would pass such a law, especially one that delineates an expansion of state responsibilities and accountability to the rural sector after decades of neglect. In line with this paper's argument, the state's contesting of CAPS' assertions of autonomy, ownership and fulfilling of state roles in the sector would have meant contesting the empirical credibility of CAPS as frame articulators, and the 'truthfulness' of their discourses reflecting decades of experience in the sector. In many regards, not acting upon these from-below pressures would have been politically disadvantageous. This in part has to do with issues regarding the state's own legitimacy. As Boelens (2009) argues, when confronting conflicts over water 'the State and its legal system face the need to incorporate local fairness constructs and solve normative conflicts in order not to lose legitimacy and discursive power in the eyes of its citizens' (p. 315). The 2006 election of President Ortega, representing the Sandinistas' return to power at the national level, lends credence to this interpretation when considering the FSLN's citizen participation and empowerment agenda; that is, recognizing CAPS dovetailed with national agendas seeking to reshape the state's relationship to organized sectors. State actors may have deemed the public and political value of this recognition all the more important given the crisis of legitimacy Ortega has faced, given claims of electoral fraud and undermining of representative institutions (Chamorro, Jarquín, & Bendaña, 2009; Ética y Transparencia, 2008; Prado, 2010).

Conclusions: expanding democratic representation in water governance

As one facet of CAPS' trans-community mobilization strategy, collective discourses of autonomy, ownership, and state roles and responsibilities in the water sector have been instrumental in promoting water committees' goals of legal recognition and political inclusion. Notably, the collective claims emanating from the CAPS networks reveal a conception of grassroots hydrosocial territories that stands in stark contrast to the states' underdeveloped legal representation of rural water governance. CAPS' discourses reflect water committees as the legitimately empowered, autonomous agents of rural water management, while simultaneously labelling and leveraging this work as legally assigned to the state. CAPS discourses are thus both at odds with and *strengthened by* the state's own top-down discursive representations of water governance: the state's legal 'representation by omission' gave CAPS a strategic advantage as they worked to fill the void in public knowledge about their work.

The longer-term effects on grassroots hydrosocial territories of CAPS' mobilization and their formalization in accordance with state law require further research. Preliminary follow-up research in 2014 raised the question of how Law 722, as it is implemented, will create and/or deepen cleavages among CAPS within and across regions. As Boelens (2009) insightfully asserts, 'The legalisation of some is accompanied by the illegalisation of others' (p. 318). Indeed, some state officials have indicated that 'legalized' CAPS will be prioritized for state financial support and technical assistance.[17]

Yet, perhaps counterintuitively, *legalized* CAPS may be disadvantaged financially and legally because of the state's burdensome accounting requirements. The issue of CAPS' local autonomy, and whether this will be strengthened or inhibited by the new law, also merits further examination.

Importantly, CAPS' novel inclusion in policy processes, including their role in shaping Law 722, not only demonstrates the buy-in of state actors to a 'reality' of rural, grassroots water governance that committees promoted via their discourses. Water committees' engagement in policy formation also reveals a shifting landscape of water governance towards greater inclusiveness of democratic policy processes and decision-making in Nicaragua. Indeed, the state's inclusion of grassroots actors previously 'below the radar' of formal water governance matters as a reconfiguration of actors, decision-making and spatial scales. This reconfiguration has integrated CAPS and state actors alike into a reshaped, national-scale hydrosocial territory in which grassroots actors have carved out space – new space owing largely to the bottom-up interventions of rural water committees in which discourse played a consequential role. Given the dynamism of the state–society interface in Nicaragua, it remains to be seen how such a reconfigured hydrosocial territory, and its related governance structures, will be both sustained as well as transformed over time.

Notes

1. These estimates are based upon an estimated CAPS membership of five to 10 residents and reflect documented water systems (through part of 2004 only) in the National Information System of Rural Water and Sanitation (SINAS), a database created by the Nicaraguan government and international donors in the mid-1990s.
2. A 1973 USAID grant entitled 'Rural Community Health Services Grant 542-15-530-110', which two years later created the Programa Rural de Acción Comunitaria en Salud (Rural Community Health Program), provides evidence of the early state role in fomenting community-based water provision in rural areas (Donahue, 1983).
3. Several reports have detailed the legal framework for the water and sanitation sector in Nicaragua (e.g., FANCA, 2006; Government of Nicaragua and PAHO, 2004; Gómez, Ravnborg, & Rivas, 2007; Romano, 2012b).
4. Domestic NGOs have also played a crucial facilitative role with regard to CAPS networks. In Matagalpa, these include the Organization for Economic and Social Development in Urban and Rural Areas (ODESAR) and the Association for the Development of Northern Municipalities (ADEMNORTE), although the latter has since dissolved due to insufficient funding.
5. This preliminary finding is based upon field research conducted in June 2014.
6. McKean (2000) defines CPRs 'as institutional arrangements for the cooperative (shared, joint, collective) use, management, and sometimes ownership of natural resources' (p. 1).
7. Certainly, the social 'webs' that embed community-based resource management are oftentimes characterized by gendered, socio-economic and ethnicity-based inequities that call upon scholars not to romanticize community-based resource management, nor to presume homogeneity along these dimensions in examining 'communities' (Boelens, 2009; see also Meinzen-Dick & Zwarteveen, 1998).
8. This argument parallels Boelens's (2014) characterization of Andean water users' 'meta-physical arguments' as 'weapons to counteract hegemonic water policies' (p. 245).
9. 'In communities where the service provider [ENACAL] does not have coverage, systems will be administered by the community, specifically the Potable Water Committees, who will guarantee service to the community, all below the supervision and control of ENACAL' (Article 75).

10. The author favours reference to discourses and discursive framings over 'frames' as the CAPS' language can be seen as constituting discursive representations of their values as well as morally guided work and roles at the community level. 'Discourse' better captures the extent to which ideology and language are mutually constitutive, and thus allows 'do [ing] justice to the ideational complexity of a social movement' (Oliver & Johnston, 2000, p. 38; see also; Munson, 1999; Jasper, 1999; Westby, 2002).
11. CAPS from over 80 municipalities have participated as representatives in the National CAPS Network, which, as of 2014, had garnered the participation of an estimated 1900 CAPS. At the time of research, departmental-level networks were also under discussion in Jinotega, Leon and Nueva Segovia.
12. 'Here we are the owners.'
13. 'Tenemos amor [... lo] tenemos como propio de nosotros.'
14. In 2007, President Daniel Ortega decreed the creation of the Consejos del Poder Ciudadano (Citizen Power Councils – CPCs) as decision-making bodies at multiple levels of governance. The CPCs received harsh criticism as an alleged party tool of the National Sandinista Liberation Front (FSLN) and have accompanied a broader narrowing of space for autonomous civil society since Ortega's election (Anderson & Dodd, 2009; Chamorro et al., 2009; Prado, 2010).
15. 'Work that corresponds to the state.'
16. As Boelens (2008) has argued, 'it would be mistaken to suggest that local user organisations try to avoid interaction with the state or development institutions to defend their autonomy [...] both the state and the users try to achieve the most favourable ratio of investment versus control for their purposes, where local user groups try to gain more access to state resources and international funding without handing over local normative power' (p. 62).
17. As of June 2014 there were 1200 registered CAPS nationally (interview with INAA [Instituto Nicaragüense de Acueductos y Alcantarillados, Nicaraguan Institute of Aqueducts and Sewerage], 13 June 2014).

Acknowledgements

The author thanks Kent Eaton, Jonathan Fox and Hector Perla for guidance in developing and honing an earlier version of this paper. Thanks are also due to the guest editors for useful comments and suggestions, as well as to two anonymous reviewers. The author also extends considerable gratitude to members of the CAPS networks and non-governmental organizations in Nicaragua who made this research possible.

Funding

This work was supported by grants from Fulbright, the UC Pacific Rim Research Program, the UC Santa Cruz Department of Politics, and the UC Santa Cruz Chicano/Latino Research Center.

References

Anderson, L. E., & Dodd, L. C. (2009). Nicaragua: Progress Amid Regress? *Journal of Democracy*, *20*(3), 153–167.

Arnold, T. C. (2008). *The San Luis valley and the moral economy of water*. Cambridge, MA: MIT Press.

Baer, M. (2008). The global water crisis, privatisation, and the Bolivian water war. In J. M. Whiteley, H. Ingram, & R. W. Perry (Eds.), *Water, Place, and Equity* (pp. 195–224). Cambridge, MA: MIT Press.

Bakker, K. (2010). *Privatizing Water: Governance Failure and the World's Urban Water Crisis.* Ithaca, NY: Cornell University Press.

Benford, R. D., & Snow, D. A. (2000). Framing processes and social movements: An overview and assessment. *Annual Review of Sociology, 26* (1): 611–639. doi:10.1146/annurev.soc.26.1.611

Boelens, R. (2008). Water rights arenas in the Andes: Upscaling networks to strengthen local water control. *Water Alternatives, 1*(1): 48–65.

Boelens, R. (2009). The politics of disciplining water rights. *Development and Change, 40*(2): 307–331. doi:10.1111/dech.2009.40.issue-2

Boelens, R. (2014). Cultural politics and the hydrosocial cycle: Water, power and identity in the Andean highlands, *Geoforum, 57,* 234–247. doi:10.1016/j.geoforum.2013.02.008

Boelens, R., Getches, D., & Guevara-Gil, A. (2010). Water struggles and the politics of identity. In R. Boelens, D. Getches, & A. Guevara-Gil (Eds.), *Out of the mainstream: Water rights, politics and identity* (pp. 1–25). London: Earthscan.

Boelens, R., Hoogesteger, J., & Baud, M. (2015). 'Water reform governmentality in Ecuador: Neoliberalism, centralization and the restraining of polycentric authority and community rule-making. *Geoforum 64,* 281–291. doi:10.1016/j.geoforum.2013.07.005

Boelens, R., Hoogesteger, J., Swyngedouw, E., Vos, J., & Wester, P. (2015). Hydrosocial territories: A political ecology perspective. *Water International, 41*(1), 1–14. doi:10.1080/02508060.2016.1134898

Chamorro, C., Jarquín, E., & Bendaña, A. (2009). Understanding populism and political participation: The case of Nicaragua. *Woodrow Wilson Center Update on the Americas.*

Donahue, J. M. (1983). The politics of health care in Nicaragua before and after the revolution of 1979. *Human Organization, 42*(3), 264–272. doi:10.17730/humo.42.3.x737h47hqw3r2785

Edelman, M. (2005). Bringing the moral economy back in to the study of 21st-century transnational peasant movements. *American Anthropologist, 107*(3), 331–345. doi:10.1525/aa.2005.107.issue-3

Enlace. (2008). Comités de Agua Potable: La Población Organizada para Resolver Su Problema de Agua. Enlace. Managua, Nicaragua: EDISA.

Ética y Transparencia. 2008. *Informe Final Elecciones Municipales.* Managua, Fundación Grupo Cívico Ética y Transparencia.

FANCA. (2006). *Las Juntas de Agua en CentroAmérica: Valoración de la Gestión Local del Recurso Hídrico.* Retrieved from http://www.fanca.co.cr/?cat=1006&title=Publicaciones&lang=es

Fox, J. (1996). How does civil society thicken? The political construction of social capital in rural Mexico. *World Development, 24*(6), 1089–1103. doi:10.1016/0305-750X(96)00025-3

García, N. (2008a, May 29). Aguadoras Comunitarias Piden Reconocimiento y Apoyo Oficial. El Nuevo Diario. Retrieved from http://www.elnuevodiario.com.ni/contactoend/17082.

García, N. (2008b, October 3). Iniciativa De Ley De CAPS Sale De Gaveta y Llega a Secretaría. El Nuevo Diario. Retrieved from http://www.elnuevodiario.com.ni/contactoend/28648.

Government of Nicaragua and PAHO. 2004. *Análisis Sectorial De Agua Potable y Saneamiento De Nicaragua.* Managua.

Gómez, L. I., Ravnborg, H. M., & Rivas, R. H. (2007). *Institucionalidad Para La Gestión Del Agua En Nicaragua.* Managua: Institución de Investigación y Desarollo Nitlapán.

Goodwin, J., & Jasper, J. M. (2004). *Rethinking social movements: Structure, meaning, and emotion.* Oxford: Roman & Littlefield, Inc.

Herrera, R. S. (2007). *ABC del agua y su situación en Nicaragua.* ENACAL, Managua.

Hoogesteger, J., & Verzijl, A. (2015). Grassroots scalar politics: Insights from peasant water struggles in the Ecuadorian and Peruvian Andes. *Geoforum 62,* 13–23. doi:10.1016/j.geoforum.2015.03.013

Hoogesteger, J., Boelens, R., & Baud, M. (2016). Territorial pluralism: water users' multi-scalar struggles against state ordering in Ecuador's highlands. *Water International, 41*(1), 91–106. doi:10.1080/02508060.2016.1130910

Jasper, J. M. (1999). *The art of moral protest: Culture, biography, and creativity in social movements.* Chicago, IL: University of Chicago Press.

Klandermans, B. (1984). Mobilization and participation: Social-psychological expansions of resource mobilization theory. *American Sociological Review, 49,* 583–600.

McAdam, D., McCarthy, J. D., & Zald, M. N. (1996). *Comparative perspectives on social movements: Political opportunities, mobilizing structures, and cultural framings*. Cambridge: Cambridge University Press.

McKean, M. A. (2000). Common property: What is it, what is it good for, and what makes it work. *People and Forests: Communities, Institutions, and Governance*, Working Paper No. 3, Food and Agriculture Organization, 27–55.

Medrano, E., Tablada, O. K., Baltodano, A., Medina, F., Swagemakers, N., & Obando, W. (2007). *22 años de experiencia recopilada sobre el trabajo de acueductos rurales*. Managua.

Meinzen-Dick, R., & Zwarteveen, M. (1998). Gendered participation in water management: Issues and illustrations from water users' associations in South Asia. *Agriculture and Human Values*, 15(4), 337–345. doi:10.1023/A:1007533018254

Mendoza, F. (2008, August 21). Se constituye red defensora del agua y del medio ambiente. *El Nuevo Diario*. Retrieved from http://www.elnuevodiario.com.ni/departamentales/24759

Munson, Z. (1999) *Ideological production of the Christian right: The case of the Christian coalition*. Department of Sociology, Harvard University. Unpublished manuscript.

Oliver, P. E., & Johnston, H. (2000). What a good idea! Ideologies and frames in social movement research. *Mobilization: An International Quarterly*, 5(1), 37–54.

Perreault, T. (2003). Changing places: Transnational networks, ethnic politics, and community development in the Ecuadorian Amazon. *Political Geography*, 22, 61–88. doi:10.1016/S0962-6298(02)00058-6

Perreault, T. (2005). State restructuring and the scale politics of rural water governance in Bolivia. *Environment and Planning A*, 37(2), 263–284. doi:10.1068/a36188

Perreault, T. (2006). From the guerra del agua to the guerra del gas: Resource governance, neoliberalism and popular protest in Bolivia. *Antipode*, 38(1), 150–172. doi:10.1111/anti.2006.38.issue-1

Perreault, T. (2008). Custom and contradiction: Rural water governance and the politics of usos y costumbres in Bolivia's irrigators' movement. *Annals of the Association of American Geographers*, 98(4), 834–854. doi:10.1080/00045600802013502

Pérez, W. (2010, February 11). Comités De Agua Con Menos Trabajo Por Sequía. *La Prensa*. Retrieved from http://www.laprensa.com.ni/2010/02/11/nacionales/341781-comites-de-agua-con-menos-trabajo-por-sequia.

Prado, S. (2010, August). Municipal autonomy is more threatened than ever. *Revista Envío*, 349. Retrieved from http://www.envio.org.ni/articulo/4226.

Pretty, J., & Ward, H. (2001). Social capital and the environment. *World Development*, 29(2), 209–227. doi:10.1016/S0305-750X(00)00098-X

Rodríguez, F. J., Martínez, J. A., Martínez, J. A., Silwany, N. S., Incer, O. A., Zeledón, S., & González, J. M. (2008). Informe de Consulta y Dictamen: Ley Especial de Comités de Agua Potable y Saneamiento. Nicaraguan National Assembly. Retrieved from http://legislacion.asamblea.gob.ni/SILEG/Iniciativas.nsf/0/d5fa5a48b5802a05062574ef00610db0?OpenDocument#_Section4.

Romano, S. T. (2012a). From protest to proposal: The contentious politics of the Nicaraguan anti-water privatisation social movement. *Bulletin of Latin American Research*, 31(4), 499–514. doi:10.1111/blar.2012.31.issue-4

Romano, S. T. (2012b). *From resource management to political activism: Civil society participation in Nicaragua's rural water governance* (Unpublished doctoral dissertation). University of California, Santa Cruz.

Roth, D., Boelens, R., & Zwarteveen, M. (2005). *Liquid relations: Contested water rights and legal complexity*. Piscataway, NJ: Rutgers University Press.

Scott, J. C. (1976). *The moral economy of the peasant: Rebellion and subsistence in Southeast Asia*. New Haven, CT: Yale University Press.

Seemann, M. (2016). Inclusive recognition politics and the struggle over hydrosocial territories in two Bolivian highland communities. *Water International*, 41(1), 157–172. doi:10.1080/02508060.2016.1108384

Shiva, V. (2002). *Water wars: Privatization, pollution and profit*. London: Pluto Press.

Snow, D. A. (2004). Framing processes, ideology, and discursive fields. In D. A. Snow, S. A. Soule, & H. Kriesi (Eds.), *The Blackwell Companion to Social Movements* (pp. 380–412). Maldon, MA: Blackwell Publishing.

Spronk, S., & Webber, J. R. (2007). Struggles against accumulation by dispossession in Bolivia: The political economy of natural resource contention. *Latin American Perspectives, 34*(2), 31–47. doi:10.1177/0094582X06298748

Swyngedouw, E. (2004). *Social power and the urbanization of water: Flows of power.* Oxford: Oxford University Press.

Trawick, P. (2001). The moral economy of water: Equity and antiquity in the Andean commons. *American Anthropologist, 103*(2), 361–379. doi:10.1525/aa.2001.103.issue-2

Westby, D. L. (2002). Strategic imperative, ideology, and frame. *Mobilization: An International Quarterly, 7*(3), 287–304.

Adjudicating hydrosocial territory in New Mexico

Eric P. Perramond

Southwest Studies & Environmental Programs, Colorado College, Colorado Springs, CO, USA

ABSTRACT

The US state of New Mexico shifted its management and legal treatment of water in the 20th century to a private property access right, weakening communal notions of water. This article explains how New Mexico has redefined and territorialized water rights as private property through the adjudication process and administrative governance rules. State adjudication of water rights disrupts horizontal social relations. The process also results in territorialization – not of fluid water per se – but of water users themselves. As water users have adjusted to this rescaling of governance, the state has found new ways to govern users vertically through water-crisis measures.

Introduction: territorializing waters

Territorialization is both a material (spatial) and a conceptual (theoretical) process of transformation, as is amply demonstrated by past scholarship and in this special issue (Hannah, 2000; Murphy, 2013; Sack, 1986; Sluyter, 2002). The concept of territorialization is also vital to understanding and explaining changing notions of sovereignty (Agnew & Muscarà, 2012; Hannah, 2000). Here, however, this paper discusses how the state territorialization, through water rights adjudications in the US state of New Mexico, intersects with reshaping local and regional hydrosocial cycles and relationships (following Boelens, Hoogesteger, Swyngedouw, Vos, & Wester, 2016; Linton & Budds, 2014; Swyngedouw, 2009). The state's adjudication process has inserted new vertical relationships between local users and state experts and has often disrupted horizontal social relationships on the ground between water users (Saurí, 1990). The paper then discusses recent administrative attempts by the state to manage water, even before adjudication is completed. In this way, it reframes Wainwright and Robertson's (2003) arguments regarding how *science* territorializes people within the state to argue that administrative *law* produces similar effects of reterritorialization on subject peoples. Law, in fact, presumes much more about capturing existing social fabrics than does science (Latour, 2010). Yet, both work in similar ways, and together law and science territorialize and recruit water users so that the state can manage its own hydrosocial relationship with New Mexicans (see also Lane, 2011). The crux of the argument here is that the process of territorialization is not about water itself, but about the water users. It is an attempt legally and politically to capture and metabolize water

175

users into a single framework that is at cultural odds with how water is perceived, used, valued and reproduced at the community level.

The examples of state reterritorializations of hydrosocial communities are drawn from the author's long-term fieldwork on local ditches known as *acequias*. The practices of *acequias* – such as the customary sharing of waters – are local and adaptive, since users recognized the fluctuating quantity of water in semi-arid New Mexico (Meyer, 1984). First, the paper explains adjudication, as inscribed into state water code to map state legal spaces in the socio-natural relationships between New Mexicans. Second, it turns to ground-level responses to adjudication by irrigators. Third, it details how *acequias* and their proponents were reterritorialized because of their response to the adjudication efforts. Finally, it describes how 'water crises', namely a drought in 2002, led to new powers for the state engineer to manage water users even before adjudication is accomplished. Collectively, these episodes and examples illustrate how extant and emergent hydrosocial territories are actively produced.

The production of territory through adjudication

The 1907 state water code of New Mexico directed the Office of the State Engineer (OSE) to adjudicate all water rights in the state. The process continues to this day, with no definite end date in sight. Adjudication, as the length of the process suggests, is complex even as it attempts to simplify and formalize existing past users and legal entities into the new state space (Meehan & Moore, 2014; Perramond, 2013). But adjudication goes beyond simplification and formalization. Were it only about these two procedures, adjudication would not be as controversial, contested or slow-going. Adjudication imposes a completely new, often alien, form of hydrosocial relationship as the state inserts itself as the authoritative body for water management. The state has asserted its right to dole out 'publicly owned water' to private water users, bypassing notions of communal governance and pre-existing hydrosocial society (such as the *acequias*). In the end, adjudication produces a more vertical, hierarchical form of hydrosocial relations than past horizontal and more equitable systems that pre-existed the agency directed to manage water for the state. This territorialization of water users, then, is a more three-dimensional process than simple bird's-eye mapping projects – the vertical integration between water rights holders and the state introduces the hierarchical relationship between the state and individuals on the ground.

Grossly simplified, and specific to the state of New Mexico, adjudication begins with a hydrographic mapping survey of all properties with water rights, or claims to them, and goes through several administrative and legal stages. After OSE staff complete their field and archival research, letters presenting 'Offers of Judgments' are sent to the documented, legal water rights holders, who may either accept or dispute the offers. An Offer of Judgment documents the location, priority in time and quantity of water used by the water rights holder. In the next phase, these individual water rights offers are made public and other neighbouring water users can contest the offer (the *inter se* process of adjudications).

Thus, claimants for water are given repeated opportunities to document their water rights and may also dispute the rights of others in the same basin. What was meant to be a process for a simple accounting of water has resulted in disputes along ditches, between ditches and between users. This new atomization of water rights to the individuals on streams and ditches, with rights attached to the body of the householder, respatializes the water users in ways that were not common before. Furthermore, the cartographic process made the spatial rearrangement of water territory more legible by and for the state, reconfiguring social relationships. Logically, as a response to adjudication, people have altered their behaviours during and after adjudications. Social relations have been reshaped, starting with the individual user and cascading through the entire network at ditch, stream, basin and inter-basin levels. Thus, as adjudication legally maps watersheds, it also assigns a new importance to individual water rights. This alienates older, customary notions that considered water as a basis for village and communal life.

The new level of adversarial resource tension has close ties to past resource adjudications. In the infamous land adjudications of the 19th and 20th centuries, many Hispano heirs to Spanish or Mexican land grants lost access to those lands, which were transformed into national and state forests or simple-fee private holdings (Correia, 2009; Ebright, 1994; Kosek, 2006). The evolving Anglo-American legal framework did not recognize 'commons', and therefore did not recognize the de facto community-held land grants (Correia, 2013; Montoya, 2002). Land adjudication was rapid, in contrast to the slow plod of water adjudication. 'Water adjudication', however, is a term loaded with negative connotations for many New Mexicans, for whom history is not in the past. Rather, it is a lived and oral remembrance of past injustices. At the ditch level, informants expressed concern that water adjudication would alienate more resources away from the villages and the ditch associations (*acequias*) that long-managed local waters. These irrigators recognized that water rights under the state code are about the state's power in doling out and disciplining (at the individual level) through the assignment of water rights (Boelens, 2009). In addition, *acequia* members perceived a creeping administrative effort to bureaucratize and sterilize social relationships that have long been informal on ditches.

US legal jurisprudence only indirectly addresses communal waters (Meyer, 1984; Worster, 1985). Water, like land, can be either privately used or publicly held. Publicly held water is not communal water. This is a critical distinction: *acequias* assume a communal norm and separate level of controlled access by users to the *acequia* ditch. *Acequia* use rights are considered separate and distinct from a 'private right to water' as the state would consider it. The state, with the 1907 water code, considered the water itself as a public, state property to which citizens could have use rights based on historical first use. One water user, Alberto, was critically philosophical about the water code:

> Sure, the code is used by judges, lawyers, they love it. For us, though, it was just a piece of paper that didn't mean anything to us until the State Engineer started to use it to poke his nose into our local *acequia* management. It changed us, my neighbours and all, we don't get along like we used to. I think part of it is the adjudication process for sure [...].(A. Norberto, interview with the author, Mimbres, NM, 8 August 2011).

In communal notions of water justice, doling out water to individuals puts at risk the entire *acequia* system. Thus, perceptions and experiences with water adjudication are not only influenced by the past land adjudications, but also by conflicting norms of whether water should stay with the land (*acequias*) as a communal resource or be severable as a private 'resource' that can be sold away from the land (under New Mexico state law). These shifting and clashing norms of water as movable property or as part of a commons are at the core of changes in social relationships in New Mexico.

Acequias as hydrosocial landscapes

Acequias are deceptively simple, gravity-fed canals that begin with a diversion from natural stream courses, using rocks and brush or hardened concrete (Figure 1). *Acequia* canals, once diverted, follow the land's contour along a valley wall to bring water to a village, and small distribution outlet canals (often called *sangrias*) allow that water to be applied to relatively flat, arable land. The vast majority of these earthen canals were built by their users, or *parciantes* (in Spanish), before modern surveying equipment was available, and, remarkably, follow a downward slope of 0.5–2.0° their entire length (Rivera, 1998). No firm numbers exist, but estimates of somewhere between 600 and 800 *acequias* in the state of New Mexico are largely accepted. The majority are clustered around the main tributaries that feed the Rio Grande. Some of the individual ditches

Figure 1. Example of an earthen *acequia*, with minor flow, near the Rio Chupadero just north of Santa Fe. Photo: author, 2009.

are up to 10 km (approximately 6.2 miles) long, running from an initial point of diversion off a stream, through an agricultural village to the end of the ditch as it drains back into the natural stream channel (through a tail-water ditch called a *desagüe*). These features, however, are merely the biophysical 'hardware' of the social cycle of *acequias*. The ditches and institutions also share many commonalities with other local irrigation systems found throughout the Americas (Boelens, Getches, & Guevara-Gil, 2010; Perreault, 2008). During the Spanish and Mexican periods of rule in New Mexico, *acequias* became shared, joint or at least adjacent between Pueblo Indian settlements and Spanish *mestizo* village settlements (DuMars, O'Leary, & Utton, 1984).

Acequias are not only physical ditches, they are also institutions. Ditch governance rules are overseen by a *mayordomo* (ditch boss), annually elected, and three ditch commissioners. Commissioners then designate a records-keeper, a treasurer for dues and a president for the commission. These assignments are now less flexible and interchangeable, yet in the past it was not unusual for the *mayordomo* to handle all of the irrigation and administrative work involved (Crawford, 1988, 2003; Rivera, 1998; Rodriguez, 2006). Entire basins have been transformed by this local and communal form of hydrosocial territory: once arid valleys are now amplified to a wider agro-riparian environment. While the dependence on food crops has dropped dramatically over the last century, the canals and institutions still support important garden crops for household consumption, farmers markets and livestock production. An important percentage of all cropland is now put to alfalfa for livestock and as a way to prove the full use of water rights. Even with these changes, *acequias* created an entirely new environment. This historical hydrosocial institution shapes behaviour and rules, as well as the customary act of water sharing, known as the *reparto* or *repartimiento* (Baxter, 1997). The *reparto* was a *mayordomo*'s responsibility of dividing the available water in any irrigation season to ensure a just distribution of water for all users.

Today's modern water adjudication process reveals how water rights and notions of water justice are differentially treated, whether through property rights or ditch rights rhetoric. Adjudication also makes dates of first use of water a rigid marker for allocation. Water law in the western US is based on the doctrine of prior appropriation. Accordingly, the longer one has been using water, the more senior the water right user is considered. This is colloquially referred to as the 'first in time, first in right' aspect of prior appropriation. A year 1703 water right, for example, is considered 'senior' to one from, say, 1898. The 1703 user then has the legal right to claim all of their water rights before the 1898 user gets a single drop. This system, while seemingly clear between only two dates, gets far more complicated when hundreds of water users share the same water course. It also creates an immediate adversarial framework for recrafting the ditches as state water territory, as users are pitted against each other to prove earlier dates. As one user, quipped: 'New Mexicans always fight about history, but now we fight about specific dates, too, and that's just twisted, man' (A. Norberto, interview with the author, Mimbres, NM, 8 August 2011).' Conversely, *acequias* share water by quantity, or rather by the amount of time for irrigation, almost never allocating full water rights based on historical first use. More commonly, *acequias* have attempted to assert ditch-wide dates to avoid in-fighting between irrigators. This push for a ditch-wide single historical priority date has not always been successful. Even when success-ful, ditches sharing a stream with another *acequia* with different priority dates are then pitted against each other to claim these prized 'first in time' rights for seniority. Yet these new social relationships are crafted not only through the adjudication of *acequias*. As discussed below, the latest drought management measures adopted by the state engineer have also reshaped territorial norms for ditch governance.

Adjudication is an example par excellence of creating new hydraulic and social relationships, although the process was putatively created simply to understand what water was in use and what water might be available for future use. The process has changed what it purported to map and understand. It also depended on local participation and knowledge, as state technicians were sometimes 'blind' to local water relationships and uses, as described below. The territorialization of water users was fundamental to crafting a state-visible society of water users, even if local water users refused to be completely legible during the adjudication mapping process.

Mapping state territory over *acequias*

The arrival and presence of field mapping technicians is still remembered by water users who underwent the process. The legally mandated technical mapping process conducted by the OSE took months if not years in some basins.

> We never thought they [state technicians] would leave us alone', one elderly farmer from Taos said, 'and the mapping part of it was OK I guess, but it really put everyone in a strange position you know [...] watching each other for water use. Watching each other wasn't new, we'd always done that [chuckles], but watching and not trusting each other, that was disturbing. It created tension, you know?' (A. Martinez, interview with the author, Taos, NM, 8 November, 2009)

Technicians who mapped fields in the late 1960s remember that it was not easy work. As one of the surveyors sent to Taos to ground-truth aerial photos put it, 'sometimes we'd be out there for 6–8 hours at a time, everyone staring, sun blazing, just checking air photo baselines in the field' (F. Thompson, interview with the author, Santa Fe, NM, 5 January 2010). As mentioned above, the mapping process started to change existing social relationships between neighbours and water users. Mapping was not a one-way process, and discussions on the ground, during surveys, were frequent. Over time, changes in mapping technology clearly shaped the ease or difficulty of mapping. Remote mapping techniques removed much of the human contact between OSE technicians and local irrigators, except for occasional field checks. Advances in technology, however, did not dispel discord and mistrust. As one of those adjudicated, Martinez (interview with the author, Taos, NM, 8 November, 2009) reflected on this at some length:

> If anything, having those technicians around led to a lot of water use, sure, but people were also trying to behave, not steal water out of turn. It was like not running a red light when the cops are around, you know? No one wants to be called out or get a ticket. So on the surface, we were friendly, and didn't steal water. But definitely it was tense; people were nervous, it changed things. We just didn't trust our neighbours anymore. I don't think some people ever got over that [...] and now of course they don't even need to send anybody, they just look at satellite stuff. Wouldn't surprise me if they started using drones to watch us [chuckles darkly].

Although scant historic water gauge data or water use data are available for ditches, dozens of informants asserted, as Martinez (2009) did, that 'my neighbours started using a lot more water when the field mappers [from OSE] were here doing the maps and stuff'. These ethnographic statements regarding increased water use are borne out in the quantitative summaries from interviews, where some 93% of all respondents (n = 211 as of 2012) claimed that water use changed during the adjudication process (Perramond & Lane, 2014, table 1). No one claimed or witnessed less water use in that period. Additionally, an irrigator from the Mimbres basin opined that 'that whole process [of adjudication] changed how we deal with each other, it added a lot of suspicion back then [1970s] that hasn't really disappeared. It's only made things worse in a dry area' (J. Holman, interview with the author, Mimbres, NM, 10 August 2011). Such unease about adjudication continues to this day.

Part of that unease is the perception and reality of being monitored in new ways: in adjudication, once an entire basin has been mapped, served and all claims settled, a final decree for the basin is recorded by a judge in state courts. That basin is then considered adjudicated and falls under priority administration of waters by the OSE (note that this is a highly simplified description). Some legal scholars, like Tarlock (1989), remain sceptical of the finality in most general stream adjudications. The entire process is about quantifying water, made explicit on mapped terrain, and legal due process. The result, however, produces a rather fixed notion of water, a snapshot quantification of 'who is using what water' and a static assignment of water duties per acre. The adjudication process, then, is a momentary glimpse of water use at the time of the mapping and research. It is not dynamic and does not account for crop or irrigation changes over time. The data are, by their nature, obsolete once produced. That said, the data do provide estimates for maximum water usage in a basin, a fact not lost on both water users and the OSE.

Mapping and the legal process of adjudication made external monitoring and a 'sense of the state' evident. This was the visible, surface aspect of water governance and social territorialization. But territory is not simply a physical space; it also depends on the internalization of those visible changes on the landscape. Just as adjudication transformed water use among and between ditches physically, it also reshaped the institutional operations of *acequias* themselves, eroding the social aspect of these hydrosocial local societies.

Even before adjudication unfolded, *acequias* were beholden to a set of state-level governance rules, which were increasingly bolstered in the 20th century as the state began the process of quantifying surface waters. However, adjudication made it even more important to formalize the long-informal and oral agreements on ditches and in *acequia* governance rules. Over the course of a century, the *acequias* were given, or recognized as having, particular properties and powers. The *acequias* have the power of eminent domain. For example, they can access other property to maintain their irrigation ditches, or use private property for their access points. The *acequia* also has an easement right, a spatial buffer along the ditch to maintain the irrigation ditch borders (*bordos*) that are used for walking along and maintaining the ditch. *Acequias* are authorized to borrow money and enter into contracts for maintenance and improvements. There is, for example, an *acequia* community ditch fund, and the *acequias* are eligible to apply for capital improvement funds through the state legislature, although the latter is difficult to accomplish. Yet *acequia* associations do not have the power formally to tax, so any expenses of maintenance and improvements are shouldered by the individual *parciantes* served by their irrigation ditch. In other words, *acequia* associations can collect dues for maintenance of the ditch and to fund the operations of the commission and the *mayordomo*'s modest stipend.

Thus, even as the state recognized *acequias* as vital village-level governance units in New Mexico, it also demanded new administrative rules of *acequias*, embedding these water districts into the larger bureaucratic state. The state allowed and simultaneously circumscribed the purview of these local institutions. For many *parciantes*, dealing with the technicians, attorneys and water experts of the OSE is their most intimate contact and experience with state officials and bureaucracy.

Not all the statutes have been constraining to these ditch institutions, and much of the push-back from *acequias* against state encroachment of their governance rules has been led and organized by the New Mexico Acequia Association (NMAA), based in Santa Fe (New Mexico's capital). NMAA is now almost 25 years old as a non-profit and has worked to help *acequias* organize not only for adjudication but also for internal governance and helping individual *acequias* parse out better by-laws so that they are recognized by the OSE and remain eligible for state funding and grants. NMAA has especially been effective at preparing ditch associations for pending or ongoing adjudi- cation procedures. Yet these actions by NMAA have also formalized the largely informal rules, which used to govern *acequias* and encourage more legalistic paper by-laws, for all *acequias*. By-laws across *acequias* now strongly resemble each other, and these new rules make *acequias* more visible to state governance.

As part of these social changes wrought by the state, even activist non-profit groups like NMAA have been recruited into the process of territorializing water. NMAA helps organize individual *acequia* ditch organizations into regional ones to accommodate and prepare them for adjudication. The organization thus reterritorializes the *acequias* before the state finalizes the new spatial culture of water that results from adjudication. But NMAA also conceptually adjusts local water rules and users to the state norms that will eventually govern and manage them. The formalization of by-laws is not necessarily a disadvantage for ditches, but is an example of territorial conformity so that by-laws can resemble one another, whereas in the past these local by-laws were full of eccentric and place-specific details typically written in elegiac Spanish.

To be sure, some localized territorial powers remain for the *acequia* officers. *Acequia* commissioners, for example, can deny ditch (access) rights to an individual who has not paid his or her dues to be part of the *acequia*. These are annual dues, and debts are recorded and accumulated. As a result of the distinction, delinquent *parciantes* may retain their state 'water rights' but cannot access the water until they have paid back into the ditch organization to use those water rights. So the right to use water, recognized by the state of New Mexico, depends on the ability to access the water through the *acequia*'s infrastructure and governance structure. The spati- ality of access, then, is governed at two levels: the state hypothetically assigns rights to individuals for the *possibility* of water use, while the *acequia* controls the access point to the ditch itself for the *actual* use of water. Since New Mexico is governed by prior appropriation law, it is up to the state to determine when a water right is no longer in use. Non-use of a water right is grounds for water right forfeiture, yet few *acequias* report this to the OSE, since the total volume of water in the ditch is vital for water to be available to the entire community. The state has the ability to file a water-right forfeiture, but this measure is rarely used for a single user, and the OSE prefers that the *acequia* sort out any delinquent *parciantes* before attempting to move water away from an *acequia*.

The story of hydrosocial state relations, then, is not one of totalizing state hegemony in which *parciantes* are crushed by the weight of the state and economic development imperatives. *Acequia* activists have certainly pushed back; they now have an effective non-profit organization at their disposal (the NMAA), and champions in the state legislature who represent the traditionally Hispano *acequia* districts (especially in northern New Mexico). For example, since 2004, the *acequias* have gained the ability to create water banks and to oversee the approvals or denials of water transfers that would affect the irrigation users on a shared ditch. These new measures are necessary for internal and external factors that may affect *acequias* in the near future. First, water banking (as a concept) allows *parciantes* to 'trade' use rights in any given year. Thus, the non-use of water by one *parciante* can be fully utilized by another *parciante* and the total amount of water used by the entire ditch remains the same. This is another clever way to make 'full beneficial use' of the duty of water on the ditch level so that local water stays on the land at the local territorial level. Water banking alleviates some concerns about the non-use of individual water rights, and is a forward-thinking mechanism should the state install stream gauges above and below the ditch for long-term water monitoring.

The second measure, allowing *acequias* to change their by-laws to have oversight on any water transfers by *parciantes*, is largely to prevent net water loss from the local community. Since individual water rights are severable from the land to which they are appurtenant, this measure is an attempt to counter the transfer of water. *Parciantes* are often tempted to sell water rights to another user. The concern is less about sales to other users on the ditch. Rather, it is more about the departure of water from the ditch, such as a water rights sale to a local municipality or a domestic mutual water supply association. If a *parciante*, with rights to 20 acre-feet (roughly 25 mega-litres) of water per year were to sell his or her rights to a downstream city completely, in theory the ditch would have to allow that water to move down the ditch or stream completely. If that water is an important percentage of the total water in the ditch, water levels in the ditch would decline over time, affecting irrigators.

Water banking and transfer denial powers were created to prevent water from moving away from ditches. These measures may help the ditch cope with drought and water shortages and reassert local, territorial notions of water. Both measures, however, further embed *acequia* institutional responses into the state's framework of legal visibility because they have been accorded these powers. And the measures may not be enough to ward off state intervention when it comes to water shortages or drought. The state, in response to the drought of 2002, created different strategies for water scarcities, as described below. These new administrative powers allowed the state engineer to begin actively managing water even prior to a full adjudication in any given basin. Below, the paper focuses on two small basins in New Mexico, one fully adjudicated (the Rio Mimbres) and the other unadjudicated (the Rio Gallinas), to discuss the implications of the new rules.

Territorial active water management

Because of the slow pace inherent in adjudication, the OSE created new administrative rules so that it could begin to manage water even without a full accounting and certification of legal water rights. In 2003, after the record-breaking 2002 drought year in New Mexico and in the Southwestern US in general, the OSE was able to convince the state legislature to enact a new form of 'active' water resource management. These new rules, known under the acronym AWRM, for active water resource management, gave broad and flexible powers to the state agency (and the state engineer).

The OSE claimed it needed these new powers to enforce proportional water sharing, to decrease conflicts between water users and also to designate particular water basins as needing intervention. These measures were approved by the New Mexico state legislature as official statutes to the water code. They also gave broad powers to monitor (i.e., measure) stream flows as part of the arrangement. The measures actively reshape two spatial dimensions of the water territorialization by the state. First, the legislation inserts a new level of water agency above *mayordomos*, a social re-engineering of decision-making in a basin. This underlines the notion that adjudication and AWRM are meant to territorialize water users, not necessarily water itself. Second, the new measures allow for intrusive technological surveillance by way of water metering. The latter has created difficulties not solely because of the notion of being monitored, but also because the concrete water meters can actually alter the flow of water in the ditches, as discussed in more detail below. On the former aspect of state social intrusion, informants likened the AWRM powers to having created 'a new water pharaoh in New Mexico, the state engineer or the special master can just turn on the river or turn it off' (J. A. Medina, personal communication with the author by email, 17 January 2010). One of the early basins to ask for state assistance in resolving water disputes between a city and *acequias* was the Rio Gallinas, where the small city of Las Vegas is located (Figure 2). Little did local water users know the consequences of asking for assistance.

The territorial dimensions to AWRM were immediately apparent to the water users who decided, or had decided for them, that OSE active management was necessary on their basin. The new powers allowed the engineer to install water or river masters for particularly problematic water basins, or where there were chronic water shortages and conflicts. As one new river master put it to a stunned audience of *parciantes* in 2003 in Las Vegas (NM), 'there's a new sheriff in town'. Locals took this as might be expected, that this new river master was going to be in charge. Or, as Hector, a resident of the Gallinas River basin stated back in 2010 (H. Gurule, interview with the author, Las Vegas, NM, 8 February 2010), the new river master was to be a

> kind of *mayordomo* for the *mayordomos* that are already trying to manage water. We don't really need an extra level of management you know? Sure, there was disagreement in the basin and we asked the State Engineer for some help in managing conflict, but we had no idea that our invitation would bring in this level of control; now they have complete control over the head-gates and we have to ask permission from this jerk to get water downstream. It's terrible.

This new infrastructure control via metering has not gone uncontested. One can, for example, find all manner of vandalized water meters and head-gates in basins being 'actively managed' by the state. There is pragmatism to these measures, as water users

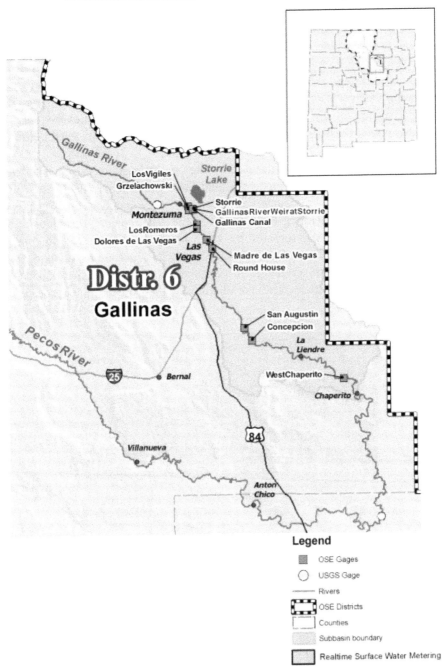

Figure 2. The Rio Gallinas basin, where the City of Las Vegas, New Mexico, and local *acequia* associations have long disputed water uses, quantities and priorities. This was one of the first basins that featured active water resource management (AWRM) rules and a special master for the basin.

recognize. The state is trying to manage conflict in drought-stricken basins. But the state's techniques and implementation of AWRM have been problematic for social *and* biophysical reasons, underscoring recent scholarship that watershed management approaches are not a magical cure for water management (Cohen & Davidson, 2011). First, there is the clear conflict over hierarchical governance by appointing a state-level *mayordomo* to control locally elected *mayordomos*. Thus, the perception that the 'water master' rules local *mayordomos* is problematic since the special master controls head-gate keys and locks and does not allow for actually active *mayordomos* to manage and allocate water when it is available. Second, as the social relationships change, so do the hydrologic balances of this calculus. Canal and stream-flow dynamics have physically changed because of these new rules and the insertion of certain managerial technologies such as meters (Figure 3).

For example, the water meters installed by the OSE typically constrain water flow to create a measurable level of cubic feet per second past a gauge point. The problem pointed out by local irrigators is that the water meters slow water flow and make the irrigation canal back-up, typically overwhelming the capacity of the canal channel and flooding lands upstream of the meter. Local response has often been one of contesting the use, or at least the type, of these water meters combined with the questioning of state expertise (following Birkenholtz, 2008). When formal channels produce no change, vandalism has also occurred on ditches with meters. Like adjudication then, the technology meant simply

Figure 3. One of the meter gauges on the Rio Gallinas near Las Vegas, New Mexico. The meter constrains and backs up water flow in the ditch, as it purports to measure flow. Photo: author, 2010.

to measure water has changed the flow, and changed the relationships to water up- and downstream. Metering, as a form of scientific state technology, is another example where 'ecological science and state-making are interwoven [...] and the state requires a certain kind of nature that will confirm the state's territorial integrity' (Wainwright & Robertson, 2003, p. 213). Meters are the latest physical evidence of the state's efforts to territorialize water through the control of water users.

Towards hydraulic territory

In its effort to territorialize its 'own' water, New Mexico reconceived, implemented and now governs its hydrosocial territory as a state entity even though overlapping water values (and local hydrosocial territories) still exist across the state and, indeed, in other parts of the Southwest (Espeland, 1998). This process reflects both theoretical concerns and quite pragmatic ones. First, the state territorialization of water users illustrates that the closing of a water frontier, like the closing of the 19th-century land settlement frontier, is a long and tortuous process and remains incomplete (Blomley, 2003; Perramond, 2013). Second, on a more practical level, the codification process of water rights often takes so long it is nearly impossible to gauge whether any governance failure might be evident (Bakker, 2010). Combined with the notion of hydrosocial cycles, territorialization of water and water users illuminates how laws, scientific metering, canals, streams and the latest state rules on drought measures are reflected in these reciprocal and cyclical relationships and then hardened in state boundary formation.

The lack of explicit recognition for communal access to resources or particular local customs in water sharing is notable in the New Mexico water code. Yet more implicit notions of a shared water territory survive and are being negotiated. *Acequias* long divided and shared water so that all *parciantes* would have access. When scarcity occurred, all users were allocated proportionally less water. When the snowpack and water flows were good, people were afforded their usual watering times and volumes, especially true in times of surplus. More recent water statutes have attempted to correct the plodding pace of water adjudication. The state's AWRM drought measures are starting to duplicate historical practices. Yet this is not the same hydrosocial cycle or practice, as special masters have 'hardened' the decision-making process (vertically) and have often barred timely access to head-gates by shutting out *mayordomos* from their traditional practices.

Emel and Brooks (1988) noted how internal changes to property rights within the law itself are possible as the notion of scarcity becomes a concern. Scarcity, again, was the trigger for a new level of active state intervention at the local, ditch level. AWRM allows the state to regulate water users prior to a full accounting of water rights. It allows the state to measure stream flow and, more importantly, to monitor water users. It also allows 'special masters' of these arrangements to assert new social and political powers among those who had long decided how to allocate waters. AWRM, thus, has created a new version of a social network that cements the state's notions of territorial water even if it remains incomplete. One of the notable arrangements under AWRM is for 'alternate management plans' for water in a basin. These plans are remarkably close to what *acequias* have long done at the local level: sharing water, agreeing to not place 'calls' on the river by priority date. Yet in these alternate plans there is no single acknowledgment that the historic and still existing *acequias* were already practising shortage-sharing agreements.

Just as adjudication identifies, personalizes and legalizes individual water rights as a private good and not as a common good, the state recognizes the need for some 'communal' understanding and management of water even if the state engineer refuses to use the term. Although prior appropriation did not initially create a space for communal or common resources, some statutes on water banking and transfers do allow for governance of these local commons (Perramond, 2013). The state has internalized some of this reality in the new AWRM measures and allowed for alternate water plans to share water in a basin when water is scarce, as *acequias* themselves have long done.

Conclusions

Water is unstable territory for state territorialization processes in semi-arid New Mexico. As adjudication failed to make water citizens quickly legible to the state, New Mexico responded with new tactics of governance to be able to read and govern its territorial water users in more efficient ways, such as the new AWRM rules now in place. Thus, in New Mexico adjudicatory delays led to the AWRM mechanisms in basins where water conflict was on the rise. Drought became the catalyst through which the state actively inserted itself to affirm a new level of state–local governance. Scarcity was the pretext as the state agency vertically embedded itself into and between (horizontal) regional *acequia* systems, reshaping notions of institutional and individual reciprocity. In the AWRM example, metering of water by the state itself reinforces this new vertical relationship to the state itself, breaking some (horizontal) social ties at the landscape level.

The hydrosocial cycle framework joined with our latest understandings of territory (Dawson, Zanotti, & Vaccaro, 2014; Murphy, 2013; Wainwright, 2008) can illuminate many of the socio-ecological aspects of water (Linton & Budds, 2014) that go well beyond a spatial politics of water. Because New Mexico remains, in many ways, a postcolonial society embedded in a larger nation-state, the notion of water territory is one imposed by the state itself (following Wainwright, 2008). How the state territorializes water, however, is through water users as objects, as mapped entities and products (of the state). In other words, the state engineer's task is more about water *user* governance than water governance.

Water complicates notions of assumed state territory and straightforward narratives about how territorialization creates new forms of social relations. Adjudication, however, treats water as a fixed and stable object, as the OSE maps 'cement' hypothetical and maximum water use figures to particular users, fields, and crops during hydrographic surveys. Although this grand accounting of water remains incomplete in New Mexico, the notion of water as part of a state's territory endures. In both adjudication and AWRM, the state has struggled to deal with water as a mobile governance challenge and has tried to impose more static and permanent yet vertically integrated solutions on streams that are rarely perennial or fixed. Irrigators, *parciantes* and *mayordomos* have long understood that the fixity of water in this dry region is an illusion itself, and that water remains remarkably uncooperative (Bakker, 2010).

Acknowledgments

The author gratefully thanks all 243 New Mexicans who generously shared their time and insights. The author thanks the two anonymous reviewers for their comments that led to substantive and theoretical improvements in this paper, as well as the editors of this special issue for their guidance.

Funding

This work was supported between 2006 and 2015 by the Colorado College Hulbert Center for the Southwest Jackson Fellowship Fund, the Dean's Office, Social Sciences Executive Committee and several Keller Venture Faculty–Student Collaborative Research Grants.

References

Agnew, J., & Muscarà, L. (2012). *Making political geography*. Lanham, MD: Rowman & Littlefield.

Bakker, K. (2010). *Privatizing water: Governance failure and the world's urban water crisis*. Cornell, NY: Cornell University Press.

Baxter, J. (1997). *Dividing the waters: 1700–1912*. Albuquerque, NM: University of New Mexico Press.

Birkenholtz, T. (2008). Contesting expertise: The politics of environmental knowledge in northern Indian groundwater practices. *Geoforum, 39*, 466–482. doi:10.1016/j.geoforum.2007.09.008

Blomley, N. (2003). Law, property, and the geography of violence: The frontier, the survey, and the grid. *Annals of the Association of American Geographers, 93*(1), 121–141. doi:10.1111/1467-8306.93109

Boelens, R. (2009). The politics of disciplining water rights. *Development and Change, 40*(2), 307–331. doi:10.1111/dech.2009.40.issue-2

Boelens, R., Getches, D., & Guevara-Gil, J. A. (Eds.). (2010). *Out of the mainstream: Water rights, politics and identity*. London: Earthscan.

Boelens, R., Hoogesteger, J., Swyngedouw, E., Vos, J., & Wester, P. (2016). Hydrosocial territories: A political ecology perspective. *Water International, 41*(1), 1–14. doi:10.1080/02508060.2016.1134898

Cohen, A., & Davidson, A. (2011). The watershed approach: Challenges, antecedents, and the transition from technical tool to governance unit. *Water Alternatives, 4*(1), 1–14.

Correia, D. (2009). Making destiny manifest: United States territorial expansion and the dispossession of two Mexican property claims in New Mexico, 1824–1899. *Journal of Historical Geography, 35*(1), 87–103. doi:10.1016/j.jhg.2008.02.002

Correia, D. (2013). *Properties of violence: Law and land grant struggle in northern New Mexico*. Athens, GA: University of Georgia Press.

Crawford, S. (1988). *Mayordomo: Chronicle of an acequia in Northern New Mexico*. Albuquerque, NM: University of New Mexico Press.

Crawford, S. (2003). *The river in winter: New and selected essays*. Albuquerque, NM: University of New Mexico Press.

Dawson, A., Zanotti, L., & Vaccaro, I. (Eds.). (2014). *Negotiating territoriality: Spatial dialogues between state and tradition*. (Routledge Studies in Anthropology). London: Routledge.

DuMars, C. T., O'Leary, M., & Utton, A. E. (1984). *Pueblo Indian water rights: Struggle for a precious resource*. Tucson, AZ: University of Arizona Press.

Ebright, M. (1994). *Land grants and lawsuits in northern New Mexico* (1st ed.). *New Mexico land grant series* Albuquerque, NM: University of New Mexico Press.

Emel, J., & Brooks, E. (1988). Changes in form and function of property rights institutions under threatened resource scarcity. *Annals of the Association of American Geographers, 78*(2), 241–252. doi:10.1111/j.1467-8306.1988.tb00205.x

Espeland, W. N. (1998). *The struggle for water: Politics, rationality, and identity in the American Southwest*. Chicago, IL: University of Chicago Press.

Hannah, M. (2000). *Governmentality and the mastery of territory in nineteenth-century America*. Cambridge: Cambridge University Press.

Kosek, J. (2006). *Understories: The political life of forests in New Mexico*. Durham, NC: Duke University Press.

Lane, K. M. D. (2011). Water, technology, and the courtroom: Negotiating reclamation policy in territorial New Mexico. *Journal of Historical Geography*, *37*, 300–311. doi:10.1016/j.jhg.2011.01.004

Latour, B. (2010). *The making of law: An ethnography of the Conseil d'Etat*. Malden, MA: Polity.

Linton, J., & Budds, J. (2014). The hydrosocial cycle: Defining and mobilizing a relational-dialectical approach to water. *Geoforum*, *57*, 170–180. doi:10.1016/j.geoforum.2013.10.008

Meehan, K., & Moore, A. W. (2014). Downspout politics, upstream conflict: Formalizing rain-water harvesting in the United States. *Water International*, *39*, 417–430. doi:10.1080/02508060.2014.921849

Meyer, M. C. (1984). *Water in the Hispanic Southwest: A social and legal history, 1550-1850*. Tucson, AZ: University of Arizona Press.

Montoya, M. (2002). *Translating property: The Maxwell land grant and the conflict over land in the American West, 1840-1900*. Berkeley, CA: University of California Press.

Murphy, A. (2013). Territory's continuing allure. *Annals of the Association of American Geographers*, *103*(5), 1212–1226. doi:10.1080/00045608.2012.696232

Perramond, E. (2013). Water governance in New Mexico: Adjudication, law, and geography. *Geoforum*, *45*, 83–93. doi:10.1016/j.geoforum.2012.10.004

Perramond, E., & Lane, M. (2014). Territory to state: Law, power, and water in New Mexico. In A. Dawson, L. Zanotti, & I. Vaccaro (Eds.), *Negotiating territoriality: Spatial dialogues between state and tradition* (pp. 142–159TBA). New York, NY: Routledge.

Perreault, T. (2008). Custom and contradiction: Rural water governance and the politics of 'usos y costumbres' in Bolivia's irrigators' movement. *Annals of the Association of American Geographers*, *98*(4), 834–854. doi:10.1080/00045600802013502

Rivera, J. (1998). *Acequia culture*. Albuquerque, NM: University of New Mexico Press.

Rodriguez, S. (2006). *Acequia: Water-sharing, sanctity and place*. Santa Fe, NM: School of American Research Press.

Sack, R. (1986). *Human territoriality: Its theory and history*. Cambridge: Cambridge University Press.

Saurí, D. (1990). *From mayordomos to state engineers: Historical change in New Mexico water rights* (Unpublished Ph.D. dissertation). Graduate School of Geography, Clark University, Worcester.

Sluyter, A. (2002). *Colonialism and landscape: Post-colonial theory and application*. Lanham, MD: Rowman & Littlefield.

Swyngedouw, E. (2009). The political economy and political ecology of the hydrosocial cycle. *Journal of Contemporary Water Research and Education*, *142*, 56–60. doi:10.1111/j.1936-704X.2009.00054.x

Tarlock, D. (1989). The illusion of finality in general water rights adjudications. *Idaho Law Review*, *25*, 271–289.

Wainwright, J. (2008). *Decolonizing development: Colonial power and the Maya*. London: Blackwell.

Wainwright, J., & Robertson, M. (2003). Territorialization, science and the colonial state: The case of Highway 55 in Minnesota. *Cultural Geographies*, *10*, 196–217. doi:10.1191/1474474003eu269oa

Worster, D. (1985). *Rivers of empire: Water, aridity, and the growth of the American West*. New York, NY: Pantheon Books.

Downspout politics, upstream conflict: formalizing rainwater harvesting in the United States

Katie M. Meehan and Anna W. Moore

Department of Geography, University of Oregon, Eugene, USA

This article examines the formalization of rainwater harvesting (RWH) and the implications of new policy trends for water governance. Analysis of 96 RWH policies across the United States indicates three trends: (1) the 'codification' of water through administrative rather than public law; (2) the institutionalization of RWH through market-based tools; and (3) the rise of policies at different spatial scales, resulting in greater institutional complexity, new bureaucratic actors, and potential points of friction. Drawing on the cases of Colorado and Texas, the article argues that states with diverse legal traditions of water enable more successful regulatory environments for downspout alternatives.

"Who owns the rain? Turns out, not you."
Reporter, KSL News, Salt Lake City, Utah

(Hollenhorst, 2008)

Introduction

In August 2008, Utah officials informed Mark Miller, a Salt Lake City car dealer, that his cistern – an underground tank for storing rain – violated state law. In an effort to be more 'green', Miller had been collecting rooftop rain at his dealership, which he used to wash cars. Boyd Clayton, deputy state engineer at the time, explained that such diversions violated the principle of prior appropriation, the backbone of water law in much of the American West. "Obviously if you use the water upstream," Clayton said, "it won't be there for the person to use it downstream." Responding to a media firestorm, Utah's Division of Water Rights (DWR) clarified its take on water rights and wrongs. As long as rain is merely controlled or directed – for example, with gutters – a water right is unnecessary. A water right is needed, however, if rain is stored, then later used for a purpose other than release back into the basin (Gittins, 2009). Local journalists lampooned the possibility of enforcement: "Homeowner projects, although technically illegal, are likely to stay off the state radar screen" (Hollenhorst, 2008). Miller expressed surprise at the decision. "Utah is the second-driest state in the nation," he said. "Our water laws ought to catch up with that" (Hollenhorst, 2008).[1]

Who owns the rain? In spotlighting the upstream conflict spawned by downspout technologies, this story points to the broader challenges associated with formalizing and governing small-scale institutions and technologies. In recent years, governments in the United States have promoted rainwater harvesting as 'sustainable' and 'green': as a technique to reduce stormwater runoff and conserve municipal supply (Cummings, 2012; Pandey, Gupta, & Anderson, 2003).[2] Worldwide, rain catchment is increasingly considered a vital strategy in adaptation to climate change: harvesting is widespread in Australia, India, Japan and Mexico, and has the potential to meet between 48% and 100% of residential water demand in Brazil (Ghisi, 2006). Despite its popularity, surprisingly little is understood about the process of formalizing rain: how *de jure* institutions render *de facto* practices 'official'. As scholars in law and social science have shown, formalization is a process in which society, nature and space are organized through legal techniques and policy tools such as city ordinances, judicial opinions, zoning, and property rights (Blomley, 2004, 2008; Boelens, 2009b; Jepson, 2012; Perramond, 2013). Conflict often arises when local norms contradict or defy formal rules and rights (Boelens & Doornbos, 2001), or when local organizations contest state efforts to 'territorialize' or take control over resources like public space or water flows (Boelens, 2011; Perreault, 2008). Indeed, local users mobilize 'customary' water norms and institutions to articulate livelihood claims and gain legitimacy, particularly as states restructure water governance (Perreault, 2008). What remains less clear, however, is why and how formalization succeeds, particularly in places – like the United States – that offer many different institutional avenues to territorialize water.

This article examines *how* rain comes to fall under state jurisdiction: through which policy mechanisms, at what spatial scale, and with what effect. Analysis of rainwater harvesting (RWH) policies in the United States identifies three major trends: (1) the 'codification' of rainwater through administrative rather than public law; (2) the institutionalization of rain catchment through market-based tools and mechanisms; and (3) the rise of policies at different spatial scales, resulting in greater institutional complexity, new bureaucratic actors, and potential points of friction. Collectively, these trends mark a shift in the US tradition of water governance – from a state-level to a 'shotgun' approach – and further suggest that socio-legal complexity may *encourage* rainwater harvesting. Drawing on a closer analysis of Colorado and Texas, we argue that states with diverse legal traditions regarding water enable more successful regulatory environments for downspout alternatives.

Rainwater harvesting is important not merely as water 'savings' for municipalities but also as a site of water *production* in the household, generating new forms of institutional decision making and social power. In the United States, research indicates that a single residential rain barrel (190 L) is able to provide approximately 50% water-savings efficiency of non-potable indoor water demand in water-rich cities in the East, Midwest and Pacific Northwest regions, and between 25% and 30% in water-strapped cities in the Mountain West, Southwest and most of California (Steffen, Herriman, Pomeroy, & Burian, 2013). Whether sanctioned or not, such technological systems are supported by institutional norms and practices – even in the United States, where "modern" water provision systems are often assumed to have replaced or eclipsed informal, non-networked, or traditional water management (Perramond, 2013). The present analysis is therefore developed with an eye on how the findings might compare with other countries, albeit under different legal, regulatory, and socio-economic settings.

The article continues with a short description of the research methods and database. Key findings are then presented based on the analysis. To elicit more insight into the success and failure of formalization, two case studies (Texas and Colorado) are contrasted. The article concludes with a consideration of the broader implications of this research.

Methods and analysis

In collecting data, we aggregated information on all policies related to rainwater harvesting in the United States, including 50 states and several territories. We sought to include all possible sites and spatial scales of governance: from municipal ordinances to state laws. Two existing data-sets provided useful starting points: the websites of the American Rainwater Catchment Systems Association (ARCSA) and the National Conference of State Legislatures (NCSL).[3]

Predominantly, policies were tracked that address active rainwater harvesting – as opposed to policies aimed strictly at passive harvesting, greywater, or stormwater – to emphasize legal inroads with regard to the direct appropriation of water supply. Active rainwater harvesting is defined as the constructed collection and storage of rain (e.g. with roof catchment and cisterns); passive harvesting involves landscape design that encourages direct infiltration into the soil (Cummings, 2012; Lancaster, 2006). Compared to greywater or stormwater, active rainwater systems yield water of higher quality: in Mexico, for example, household systems with affordable filtration mechanisms can produce near-potable-grade water from rooftop runoff (Adler, Campos, & Hudson-Edwards, 2011).

Overall, 96 policies were compiled and tracked. Five of these were high-profile state laws that have been declared 'dead in the water' since initial inclusion by legislative trackers. As Figure 1 shows, harvesting policies are found in areas with and without substantial rainfall: in 15 states and territories, 12 cities and 3 counties (Yavapai and Pima Counties in Arizona and Bernalillo County in New Mexico). Four states that use the Colorado water law doctrine – Arizona, Colorado, New Mexico and Utah – feature harvesting policies, a point discussed in greater detail in a later section. While harvesting is prevalent in the arid Southwest, several rain-laden areas in the East have enacted

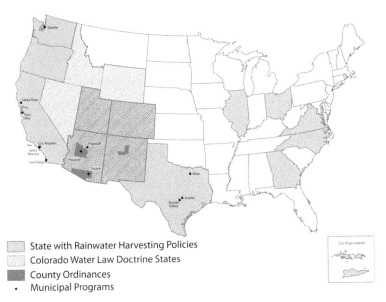

State with Rainwater Harvesting Policies
Colorado Water Law Doctrine States
County Ordinances
• Municipal Programs

Figure 1. Map of rainwater harvesting legislation in the United States.

Note. All geographically outlying states (e.g. Alaska, Hawaii) and territories (e.g. American Samoa, Guam, the Northern Mariana Islands, Puerto Rico) were included in our initial search for rainwater legislation. Those without existing laws or policies on the books are not included in this map.

innovative laws and regulations. For example, Ohio allows rain catchment for *potable* use, in contrast to most legislation, which restricts harvested water to non-potable use.[4]

The policies span two decades, from 1992 to 2013. For the most part, RWH legislation is fairly recent, implemented between 2008 and 2012. Exceptions tend to involve rather straightforward amendments to existing legislation. For example, in 2008 the US Virgin Islands updated building and landscaping codes from 1964 to incorporate rainwater harvesting, a long-standing practice on the islands. Revision of greywater codes often preluded rain catchment policies. For example, Colorado water rights statutes originally published in 1969 were revised in 2002 to reflect legal uses of greywater; this predated 2009 legislation to allow limited forms of rainwater harvesting. Similarly, California adopted greywater changes to the state water code as early as 1992, and Arizona implemented greywater permits and regulations in the late 1990s.

Texas was also an early adopter. The state legislature passed laws in the early 2000s – along with publication of the highly regarded *Texas Manual on Rainwater Harvesting* (TWDB, 2005; the first edition was published in 2002) – that paved the way for municipal-level legislation and provided a model for other states. Oregon, Colorado and Utah introduced comprehensive statewide policies in 2008, 2009 and 2013, respectively. California, by contrast, saw early attempts at comprehensive RWH legislation fail in 2012. In its place, municipal governments in California have taken the lead, implementing rebate programmes in Palo Alto, San Diego, Santa Monica and Santa Rosa. Further, the city of Los Angeles' landmark Low Impact Development ordinance, effective May 2012, requires new and redevelopment projects greater than 500 square feet (46.45 m^2) to capture the first 0.75 inches (1.9 cm) of a storm event. Compliance mechanisms include the installation of RWH barrels, permeable pavement, planter boxes, dry wells, and rain gardens (LA Stormwater, n.d.). The following sections tease out several patterns in the data-set, including the institutional pathways of formalization, the broader role of rebates and incentives, and the implementation of policies across different spatial scales.

The codification of rain

Three main trends emerged during analysis. First, unlike 'big' water in rivers or aquifers, rainwater is governed mostly through the minutiae of bureaucracy: building and municipal codes. Within the data-set, 40% of the policies modified rules for residential design, construction or plumbing, the clear tool of choice for regulation (Table 1). All together, 34 instances were found of revised state statutes or codes in 12 states and territories, including the US Virgin Islands. That number does not include changes to state codes and statutes mandated by other legislative processes listed in Table 1. Indeed, most of the RWH policies identified in Table 1 involve revisions to code and statutes in some way; however, the 34 instances of revisions noted here represent stand-alone legislation.

Most policies regulate micro-level harvesting systems, those ideally suited to single-family homes. Some policies, such as in Illinois, Ohio and Oregon, go so far as to define 'plumbing' and make specific provisions for acceptable design, installation and water use in rainwater systems. Take for example HB 1585 in Illinois. This bill, first proposed in 2011 and now sitting in the Rules Committee (as of 2013), aims to (1) define 'plumbing' within Illinois law; (2) enrol the state Department of Health to establish new standards for harvesting systems; (3) set limits to water use (non-potable only); and (4) ultimately place the state responsibility of water with plumbers.

Table 1. Regulation of rainwater harvesting in the United States.

Type	Number	Scale	Location
Revised building, planning and landscaping codes	34 state, 4 municipal	State and municipal	Arizona, California, Colorado, Georgia, Illinois, North Carolina, Oregon, Texas, Utah, Virginia, Washington, Virgin Islands
Rebates	15	Municipality, utility district, county	Arizona, California, Colorado, Texas, New Mexico
Laws	10	State	Arizona, Colorado, Illinois, Oklahoma, Oregon, Texas, Utah
Non-binding guidelines	6	County, state	Arizona, Georgia, Oregon, Texas, Virginia, Washington
Committees	2	State	Arizona, Ohio
Revised tax codes	2	State	Arizona, Texas
Ordinances	2	County, City	County of Santa Fe, NM; Los Angeles
Income tax credits, corporate and individual	1	State	Virginia, Arizona (expired)

Source: ARCSA, NCSL, authors.
Note. Policies were clustered according to general type of institutional mechanism. Non-binding guidelines include all formally produced guidelines, reports and planning documents, perhaps indicative of future policy directions but otherwise free of legal obligations. Likewise, committees are supported by state legislation but do not formally regulate RWH.

Also uncovered were several policies encouraging or mandating macro-level harvesting systems at industrial sites, government complexes, agricultural operations, and other large-scale developments. Macro-level systems typically incorporate stormwater and rainwater catchment into structural designs and can store thousands of litres. Given their cost, the widespread use of macro-level systems is unlikely unless required by regulation or rewarded with incentives (Cummings, 2012).

While such codes seem tiny and technical, and remain consistent with state law, collectively they represent a paradigm shift in water supply governance. In the United States, water has historically been considered a public resource and governed by two main systems of water law: riparian and prior appropriation (Dobkins, 1959; Wilkinson, 1985).[5] While the bulk of surface and groundwater is governed through these legal doctrines, rainwater – itself only a small percentage of the total water supply – falls in another domain of US water governance: administrative law. Between 1880 and 1920, a period of dynamic urban growth in the United States, cities implemented strict plumbing and construction codes designed to protect urban dwellers from epidemic diseases and public health threats, the result of poorly engineered sanitary systems (Melosi, 2000). In effect, these codes opened the door for government regulation of water in the home. Such progressive legislation, however, later restricted options for households looking to creatively re-engineer water use. For example, in California, until the passage of Water Code §14875-14877.3 in 1992, the reuse of greywater – domestic wastewater without toilet refuse or 'blackwater' – was effectively illegal (Meehan, 2012). Yet, because rain is directly appropriated through downspouts and cisterns – and treated by and large as private property by individual homeowners – harvesting is institutionalized through plumbing codes, not water rights.

Formalization through financial incentives

The analysis also identified the process of formalization through market-based tools. The second most popular type of regulation, nearly 15% of policies are remunerative, in the form of rebates, income tax credits, and other financial incentives within the boundaries of state law (Table 2). Policies in this category mostly target single-family residences. Rebates range from $25 in Allen, Texas, to $2000 in Tucson, Arizona, and Sunset Valley, Texas, to $4400 in the Seattle (Washington) Public Utilities District. Common justifications for rebates and incentives are to promote municipal conservation through augmenting alternative supply sources (e.g. Prescott, Arizona) or to reduce stormwater runoff and pollution in local water bodies (e.g. Santa Monica, California).

Rebates have been adopted by municipal and city governments in Arizona, California, New Mexico and Texas. For example, the city of Santa Monica, California, offers three types of rebates: for downspouts and gutters, the infrastructural conduits; for rain barrels, the storage-filtration mechanisms; and for cisterns, the workhorses of storage. Virginia and, until recently, Arizona offered state tax credits for individuals and corporations who install harvesting systems.[6] With state House Bill 7070, Rhode Island proposed a tax credit for installing cisterns, but the legislation was held in committee in 2012.

At first glance, rebates and other financial incentives reflect the broader global trend toward water *marketization*, or what Ken Conca (2006, p. 215) refers to as the "process of creating the economic and policy infrastructure" – such as rebates and tax credits – "for treating water as a marketed commodity". While related to commodification and privatization, marketization is distinct, referring to "a broader set of linked transformations related to prices, property rights, and the boundary between the public and private spheres" (p. 216). In theory, rebates and incentives encourage greater individual (or household) involvement and introduce market-based principles into the management of water, reflecting the broader shift toward the 'neoliberalization' of water management worldwide (Bakker, 2010; Harris & Roa-García, 2013). Within these shifts, policies become key fulcrums for reconfiguring governance.

Among the incentives listed in Table 2, the San Diego (California) programme comes closest to a direct valuation of water. Effective 1 March 2013, San Diego residents can receive a $1.00 rebate for every gallon (3.59 L) of stored rain, up to 400 gallons (1,514 L). Systems must be connected to a gutter, and harvested water may be distributed only through a hose or bucket – in effect, only for non-potable backyard uses (City of San Diego Public Utilities, no date). In contrast to other programmes, the San Diego rebate signals a move toward quantitative valuation of rain itself, rather than equipment or infrastructure. While the San Diego programme does not equate to a full commodification of water (a process that necessarily involves exchange), it nonetheless introduces the pricing and valuation of rain into the norms of household water use.

While the rise of rebates and incentives signals a clear trend in RWH policies, we recognize that RWH muddles any straightforward understandings of marketization and neoliberal water management. First, in contrast to neoliberal reforms that 'roll back' state involvement to free the individual consumer, RWH rebates and incentives necessitate a greater –not a declining – role for the state, enrolling new bureaucratic agencies and authorities (e.g. public departments of health, construction and housing) not normally involved in water supply. Second, while harvesting does mark a shift towards individual responsibility and management, the specific roles of private-sector actors depend on their function and form. Harvesting households, for example, may be motivated by environmental or political values, a distinct situation from entrepreneurs motivated by profit.

Table 2. Rebates and financial incentives for rainwater harvesting.

Target	Location	Amount	Notes
Water	San Diego, CA	$1.00 per gallon (3.79 L) of water stored	Established in 2013; 400-gallon (1514 L) limit
Water	Santa Rosa, CA	$0.25 per gallon (3.79 L) of water stored	Limited to estimated peak month of water use per site; cannot exceed cost of materials
Water	Kitsap County, CO	50% rate reduction for surfaces from which rain is harvested	For new or remodelled properties; rate reduction determined by total parcel acreage or measured impervious surface area
Infrastructure	Santa Monica, CA	Up to $200 each (up to 8)	For rain barrel
Infrastructure	Palo Alto, CA	$50	For purchase and installation of rain barrel
Infrastructure	Prescott, AZ	$50–300	For passive and active systems; established 2009
Infrastructure	Flagstaff, AZ	$100	For new systems with at least 1000-gallon (3,785 L) capacity; established 2012
Infrastructure	Tucson, AZ (Tucson Water)	$2000	Covers half of materials and installation
Infrastructure	Tucson, AZ (AZ MDWI District)	$50	Covers rainwater-harvesting and greywater systems
Infrastructure	Allen, TX	$25 per barrel (up to 2)	Tied to participation in a specific programme
Infrastructure	Austin, TX (residential)	$30	For rain barrel
Infrastructure	Austin, TX (commercial)	$500	For installation of system over 300 gallons (1,136 L); prior approval necessary
Infrastructure	Sunset Valley, TX	Up to $2000	Covers 20% of system cost
Landscape	Albuquerque, Bernalillo County, NM (Water Utility Authority)	$1.50 per square foot (0.09 m^2)	Landscape area must be composed of native vegetation and served by rainwater system

Spatial scales of regulation

Finally, the analysis reveals a proliferation of policies at different spatial scales. While cities and counties have taken active roles in institutionalizing harvesting, seven states – Arizona, Colorado, Illinois, Oklahoma, Oregon, Texas and Utah – have adopted 10 broad-sweeping RWH laws at the state level. Notable highlights of these legislative activities include the following.

- California and Washington attempted but failed to pass comprehensive statewide RWH legislation. As a result, California has seen increasing activity at the municipal level, while Washington developed policy guidance to provide an alternative interpretation of the state's water law.
- Arizona and Oregon passed legislation that allows municipalities to implement RWH in accordance with state law. Both sets of laws include permit procedures and guidelines for system design, materials, fees, installers and acceptable water applications.
- In a reversal of its previous position, which prohibited rainwater harvesting, Colorado now allows RWH, but only in very restricted circumstances.
- Texas has passed the most comprehensive state-level legislation, actively encouraging rainwater harvesting via mandates, incentives, funding, training programmes and demonstration projects.

Even as these laws represent huge steps forward, the politics of regulating rainwater harvesting across different scales remains complex. We use the term 'scalar friction' to call attention to the potential for tension between state law and municipal governments as a result of the increasing codification of RWH in cities. California serves as an excellent example. During the 2012 legislative cycle, two state bills with harvesting components failed to be implemented into law (AB 1750, the Rainwater Capture Act, and AB 2398, the Water Recycling Act). Meanwhile, several city governments – including major metropolitan areas in Southern California – have changed their municipal ordinances and launched successful rebate programmes. In Washington state, after nine unsuccessful attempts at state legislation, in 2009 the Department of Ecology issued an interpretive policy statement declaring that a water right is not required for rooftop harvesting – a more direct but still temporary approach to reconcile local actions (such as the Seattle Public Utilities District's RainWise programme) with state laws.

It is important here to clarify that these local policies, while adopted in part to provide RWH opportunities where none existed at the state level, *are* consistent with overall state water law. Yet, with the rise of RWH policies at different spatial scales – for example, the city of Los Angeles – new bureaucratic actors and tensions with state water authorities may emerge, particularly as local policies grow in significance, popularity and scope. Since World War II, US federal laws, regulations and judicial decisions have made "inroads" into the traditional domain of state water law, consequently "eroding" the grip of state-level power over water (Wilkinson, 1985). Yet, the emergence of rainwater harvesting policies in the past decade represents an entirely different inroad: through the codification of rain catchment, municipalities have established new avenues of decision making over water supply. As the next section demonstrates, this splintered mode of water governance presents both opportunities and challenges.

Downspout politics

Why does formalization work in some areas, but not others? To provide a more in-depth explanation, this section compares two states with radically different approaches to rainwater governance: Texas and Colorado. While each state features distinct climate regimes and population pressures, overall limited water availability has prompted dramatic divergences in water policy and regulation. Colorado is a mid-latitude interior continental state, with complex topography that results in dramatic climate and precipitation differences. Overall, the state is characterized as semi-arid, with a statewide annual precipitation average of 43.2 cm, and precipitation gradually increases as one travels eastward, toward Kansas (Doesken, Pielke, & Bliss, 2003). Snowmelt is the primary source of water for much of Colorado, and the state occupies the headwaters of the Colorado River – the 'spigot' for large portions of the American West.

Colorado has done very little to encourage rainwater harvesting across all scales of governance. Colorado Senate Bill 80, passed in 2009, represents a small step forward. It allows landowners with existing well permits to capture and store rain for beneficial use on their property, but only if networked water is unavailable. In the wake of national media controversy following the bill's passage, in which the *New York Times* declared, "It's now legal to catch a raindrop in Colorado" (Johnson, 2009), the Colorado Department of Water Resources issued a press release, noting:

> The passage of Senate Bill 09–080 that allows people to capture precipitation has led to some confusion. The bill is very limited in scope and does not enable everyone to start catching and using precipitation to supplement their existing water supplies. SB-80 allows limited collection and use of precipitation for landowners, only if ALL of the criteria below (note especially the last one) are met: (1) the property on which the collection takes place is residential property; (2) the landowner uses a well, or is legally entitled to a well, for the water supply; (3) the well is permitted for domestic uses according to specific statutes; (4) there is no water supply available in the area from a municipality or water district; (5) the rainwater is collected only from the roof and; (6) the water is used only for those uses that are allowed by, and identified on, the well permit. (Haynes 2010)

In effect, SB 80 restricts rain catchment to a handful of Colorado residents. Beyond SB 80, we uncovered only one additional RWH policy. House Bill 1129, also passed in 2009, authorized 10 pilot projects to capture precipitation in new real estate developments. The bill has been characterized as a policy experiment to determine the amount of rain that can be captured without having an effect on existing water rights (Cummings, 2012, p. 549). Two additional policies – the Individual Sewage Disposal Systems Act (changes to Colorado revised statutes) and the Water Rights Determination and Administration Act – address greywater more fully but make no mention of rainwater harvesting for potable or non-potable use.

Texas, in contrast, sits at the cutting edge of rainwater policy. The Texas climate ranges from semi-arid, in the western portion of the state, to subtropical-subhumid in central, eastern, and coastal areas. Potential evapotranspiration rates exceed rainfall in nearly every month of the year, contributing to a prominent moisture deficit and periods of prolonged drought, such as the 2011 water year, where 97% of Texas suffered from extreme to exceptional drought (Winters, 2013).

Governments in Texas have passed a variety of policies, including revisions to state water, property, health and tax codes, and various city rebates (in Allen, Austin and Sunset Valley). The centrepiece of national harvesting legislation, Texas House Bill 3391 (State of Texas, 2011), was signed into law on 17 June 2011 and is widely considered "one of the most far-reaching and comprehensive pieces of legislation regarding rainwater harvesting in recent years" (National Conference of State Legislatures [NCSL], 2012). Among its notable provisions, HB 3391 includes legal protections and sanctions a wide range of RWH practices:

- It allows potable and non-potable uses of rainwater.
- Financial institutions may consider loans to projects using harvested rainwater as their sole source of water. Furthermore, building permits cannot be denied solely because the facility will implement RWH.
- New state facilities with roofs of at least 50,000 square feet (4,645 m^2), and in areas in which the average rainfall is at least 20 inches (50.8 cm) must include RWH technology for potable and non-potable indoor and landscaping use in their design and construction.
- Rules must be developed on the installation and maintenance of RWH systems for indoor potable purposes and connected to a public water supply system (e.g. the two conduits must not come into contact).
- All municipalities are encouraged to promote RWH at residential, commercial and industrial facilities – particularly school districts – through the use of discounts and rebates.
- The Texas Water Development Board will provide training on rainwater harvesting for staff in municipalities and counties.

What explains such policy differences? The dearth of RWH-friendly policies in Colorado largely stems from its strict interpretation of the prior-appropriation doctrine. The US Congress recognized the doctrine in 1866 and in 1870, and Colorado adopted the legal system almost immediately (Wilkinson, 1985). Even as other Western states have modified the rigidity of prior appropriation by accepting tenets of other legal traditions, most notably riparian rights, Colorado still adheres closely to prior appropriation – leading to some calling it the Colorado Doctrine. Water law in Colorado is strictly defined within the state constitution and managed centrally through the Department of Water Resources.

Texas marches to a different beat. It helps that the state has "an extremely diverse hydrological picture" (Bath, 1999, p. 121), as illustrated by the "I-35 problem". The I-35 interstate freeway marks the separation between the well-hydrated eastern half of the state and the arid west, which sees as little as 18 cm of annual rain. Under Spanish and Mexican rule, water rights in Texas were allocated through the Hispanic model of water law (with conceptual roots in Spain and Northern Africa). Indeed, all lands granted before 20 January 1840 – some 10.5 million hectares – follow civil law from Spain and Mexico (Dobkins, 1959). Since 1840, Texas water law has incorporated both Hispanic civil law and English common law (the riparian doctrine). By the late 1800s, Texas adopted the prior-appropriation doctrine for its arid region, causing a century of confusion over water rights (Bath, 1999). The 1967 Water Rights Adjudication Act finally addressed the confusion between appropriated and riparian claims; riparian rights became subject to a beneficial-use provision, and all unappropriated water required a use permit granted by state authorities. By the late 1980s and early 1990s, legislation incorporated water conservation and the preservation of riparian zones. With its complex legal traditions, Texas has facilitated both the devolution of harvesting policy to municipal and county governments (e.g. HB 3391), and the grass-roots rise of policies and programmes legally consistent yet independent of state processes (e.g. rebates in Austin).

Legal and institutional complexity – between formal and informal systems, and among legal traditions – is often seen as a disadvantage in water policy, particularly in cases where overlapping or contested property rights stoke conflict (Jepson, 2012). Yet, as Bryan Randolph Bruns and Ruth Meinzen-Dick (2005, p. 12) argue, such complexity "also create[s] space for maneuver, formation of coalitions, and institutional innovation to solve problems". Indeed, legal pluralists have demonstrated "how various legal orders, such as customary law, religious law, and project law, continue to evolve and interact" within and between "local" and "formal" systems (p. 12). In the case of Texas, years of negotiation among different 'bundles of rights' and water users has cultivated institutional innovation at multiple scales of governance, leading to greater acceptance of rainwater harvesting. As water supply alternatives are increasingly formalized – not only in 'developing' countries but also in 'developed' settings like the United States – lessons from Texas and other US sites of institutional flexibility offer important lessons in the advantages of legal plurality.

Conclusion

In this article, we have attempted to better understand the variegated process of formalizing rain catchment across the United States. Analysis indicates three major trends: (1) the 'codification' of rain, in which water is adjudicated through plumbing codes (administrative law) in contrast to property rights (public law); (2) the institutionalization of rain through market-based instruments like rebates and tax credits; and (3) the rise of policies at different spatial scales, resulting in greater institutional complexity, new bureaucratic actors, and potential points of friction.

Legal tools can facilitate adaptive responses to drought, water scarcity, and climate change (Bruch & Troell, 2011). But as our analysis of Colorado suggests, state water laws may also serve as *disincentives* to rainwater harvesting, particularly in situations where prior appropriation restricts any 'downspout diversions' from upstream water rights. For states that have formalized rain catchment, harvesters must obtain permits or register with the government, bringing rain catchment in line with other water rights. Looking forward, future research should examine the impacts of formalization in practice. It remains unclear how many people harvest rainwater (with or without permits), their specific reasons and motivations (other than environmental stewardship, which policy makers often invoke), their perceptions of rainwater (including issues of quality and ownership), and the extent to which they understand or care about legal regulations and restrictions. Indeed, as water scholars have suggested elsewhere, people collect and manage water regardless of its legal designations (Boelens, 2009a, 2011; Boelens & Doornbos, 2001; Meehan, 2013); and the Utah controversy that opened this article indicates that questions about ownership and property may be officially reconciled, yet ignored, dismissed, or contested in practice. Colorado residents, for example, have called the rain barrel the 'bong' of the backyard garden: a technology prohibited under law but widely utilized nonetheless (Meehan, 2012).

Formalization is highly variable across the United States, illustrative of a fragmented or 'shotgun' approach to policy regarding small-scale water technologies. This work further suggests that socio-legal complexity may actually foster favourable outcomes: states with diverse legal traditions of water, such as Texas, permit more flexible and successful regulatory environments for downspout alternatives. Future research might build on this initial policy inventory, comparing RWH in countries with different legal and socio-economic conditions. For example, in Australia – a federated commonwealth which governs water rights through administrative law at the state level – rain catchment is increasingly popular and even mandated in some cities, such as in new home construction in Brisbane (Haisman, 2005; Moglia, Tjandraatmadja, & Sharma, 2013). As in the United States, Australian governments appear to favour regulation that incentivizes through market-based tools. In 2010, more than 600,000 Australian households received a government rebate or incentive for technologies like rain barrels and cisterns (Australian Bureau of Statistics, 2010). Unlike the United States, Australian water law incorporates principles of integrated water resources management (Haisman, 2005), which allows greater flexibility and formalization of supply alternatives like rainwater harvesting. As water supply alternatives are increasingly mainstreamed, the rain barrel has become a little less renegade.

Acknowledgements

This research was supported by Doctoral Dissertation Research Improvement Grant #0727296 (Meehan) from the National Science Foundation. We thank James Castañeda for his initial work on the database and Nick Perdue for map production. We are grateful to Adell Amos and Manuel Prieto for helpful comments on earlier drafts.

Notes

1. Following the Miller debacle, in 2010 the Utah legislature approved Senate Bill 32, which allows residents to harvest limited quantities of precipitation without obtaining a formal right, but only after registering with the state government. For more detail, see the Utah Division of Water Rights (http://waterrights.utah.gov).
2. For the sake of variety, we use "rainwater harvesting" (RWH) and "rain catchment" interchangeably.
3. For the ARCSA database, see http://www.arcsa.org/?page=273. The NCSL database is at http://www.ncsl.org/issues-research/env-res/rainwater-harvesting.aspx.
4. For potable consumption in Ohio, systems must be small-scale (less than 15 connections or use by 25 individuals daily).
5. Hispanic water law is also present in portions of the US Southwest, such as New Mexico and the Rio Grande Valley (Dobkins, 1959; Perramond, 2013).
6. Arizona Revised Statute §43-1090.01, which provided corporate and individual income tax credits for water harvesting, expired in 2012.

References

Adler, I., Campos, L. C., & Hudson-Edwards, K. A. (2011). Converting rain into drinking water: Quality issues and technological advances. *Water Science and Technology, 11*(6), 659–667.
American Rainwater Catchment Systems Association. (No date). Rainwater Harvesting Laws, Rules & Codes. Retrieved from http://www.arcsa.org/?page=273
Australian Bureau of Statistics. (2010). *Environmental issues: Water use and conservation* (Report No. 4602.0.55.003). Canberra, Australia.
Bakker, K. (2010). *Privatizing water: Governance failure and the world's urban water crisis*. Ithaca, NY: Cornell University Press.
Bath, R. (1999). A commentary on Texas water law and policy. *Natural Resources Journal, 39*, 121–128.
Blomley, N. (2004). *Law, space, and the geographies of power*. New York, NY: Guilford.
Blomley, N. (2008). Simplification is complicated: Property, nature, and the rivers of law. *Environment and Planning A, 40*(8), 1825–1842. doi:10.1068/a40157
Boelens, R. (2009a). Aguas diversas. Derechos de agua y pluralidad legal en las comunidades andinas. *Anuario de Estudios Americanos, 66*(2), 23–55. doi:10.3989/aeamer.2009.v66.i2.316
Boelens, R. (2009b). The politics of disciplining water rights. *Development and Change, 40*(2), 307–331. doi:10.1111/j.1467-7660.2009.01516.x

Boelens, R. (2011). Luchas y defensas escondidas: Pluralism legal y cultural como una practica de resistencia creative en la gestion local en los Andes. *Anuario de Estudios Americanos, 68*(2), 673–703.

Boelens, R., & Doornbos, B. (2001). The battlefield of water rights: Rule making amidst conflicting normative frameworks in the Ecuadorian highlands. *Human Organization, 60*(4), 343–355.

Bruch, C., & Troell, J. (2011). Legalizing adaptation: Water law in a changing climate. *Water International, 36*(7), 828–845. doi:10.1080/02508060.2011.630525

Bruns, B. R., & Meinzen-Dick, R. (2005). Frameworks for water rights: An overview of institutional options. In B. R. Bruns, C. Ringler, & R. Meinzen-Dick (Eds.), *Water rights reform: Lessons for institutional design* (pp. 3–25). Washington, DC: International Food Policy Research Institute.

City of San Diego Public Utilities. (No date). Rainwater Harvesting Pilot Program Rebate Guidelines. Retrieved from http://www.sandiego.gov/water/pdf/conservation/rainbarrelguidelines.pdf

Conca, K. (2006). *Governing water: Contentious transnational politics and global institution building.* Cambridge, MA: The MIT Press.

Cummings, K. (2012). Adapting to water scarcity: A comparative analysis of water harvesting regulation in the four corner states. *Journal of Environmental Law and Litigation, 27*, 539–570.

Dobkins, B. E. (1959). *The Spanish element in Texas water law.* Austin, TX: University of Texas Press.

Doesken, N. J., Pielke, R. A., & Bliss, O. A. P. (2003). Climate of Colorado: Climatography of the United States No. 60. Colorado Climate Center. Retrieved from http://ccc.atmos.colostate.edu/climateofcolorado.php

Ghisi, E. (2006). Potential for potable water savings by using rainwater in the residential sector of Brazil. *Building and Environment, 41*(11), 1544–1550. doi:10.1016/j.buildenv.2005.03.018

Gittins, J. (2009, May 30). Is it illegal to harvest rainwater in Utah? *Utah water law and water rights* [blog]. Retrieved from http://utahwaterrights.blogspot.com/2009/05/is-it-illegal-to-harvest-rainwater-in.html

Haisman, B. (2005). Water rights reform in Australia. In B. R. Bruns, C. Ringler, & R. Meinzen-Dick (Eds.), *Water rights reform: Lessons for institutional design* (pp. 113–152). Washington, DC: International Food Policy Research Institute.

Harris, L. M., & Roa-García, M. C. (2013). Recent waves of water governance: Constitutional reform and resistance to neoliberalization in Latin America (1990-2012). *Geoforum, 50*(1), 20–30.

Haynes, M. (2010, May 27). Press Release from the Office of the State Engineer: Water Harvesting. Colorado Division of Water Resources. Retrieved from http://water.state.co.us/DWRDocs/News/NewsArticles/Pages/RainwaterHarvesting.aspx

Hollenhorst, J. (2008, August 12). Catching rain water is against the law. *KSL.com.* Retrieved from http://www.ksl.com/?nid=148&sid=4001252

Jepson, W. (2012). Claiming space, claiming water: Contested legal geographies of water in South Texas. *Annals of the Association of American Geographers, 102*(3), 614–631. doi:10.1080/00045608.2011.641897

Johnson, K. (2009, June 28). It's now legal to catch a raindrop in Colorado. *The New York Times.* Retrieved from http://www.nytimes.com/2009/06/29/us/29rain.html?_r=0

LA Stormwater. (No date). Low Impact Development. Retrieved from http://www.lastormwater.org/green-la/low-impact-development/

Lancaster, B. (2006). *Rainwater harvesting for drylands, volume I.* Tucson, AZ: Rainsource Press.

Meehan, K. (2012). Water rights and wrongs: Illegality and informal use in Mexico and the US. In F. Sultana & A. Loftus (Eds.), *The right to water: Politics, governance and social struggles* (pp. 159–173). London: Earthscan.

Meehan, K. (2013). Disciplining de facto development: Water theft and hydrosocial order in Tijuana. *Environment and Planning D: Society and Space, 31*(2), 319–336. doi:10.1068/d20610

Melosi, M. V. (2000). *The sanitary city: Environmental services in urban America from colonial times to the present.* Pittsburgh, PA: University of Pittsburgh Press.

Moglia, M., Tjandraatmadja, G., & Sharma, A. K. (2013). Exploring the need for rainwater tank maintenance: Survey, review and simulations. *Water Science & Technology, 13*(2), 191–201.

National Conference of State Legislatures. (2012). State Rainwater Harvesting Statutes, Programs and Legislation. Retrieved from http://www.ncsl.org/issues-research/env-res/rainwater-harvesting.aspx

Pandey, D. N., Gupta, A. K., & Anderson, D. M. (2003). Rainwater harvesting as an adaptation to climate change. *Current Science, 85*(1), 46–59.

Perramond, E. P. (2013). Water governance in New Mexico: Adjudication, law, and geography. *Geoforum, 45*(1), 83–93. doi:10.1016/j.geoforum.2012.10.004

Perreault, T. (2008). Custom and contradiction: Rural water governance and the politics of usos y costumbres in Bolivia's irrigators movement. *Annals of the Association of American Geographers, 98*(4), 834–854. doi:10.1080/00045600802013502

State of Texas (2011). *House Bill 3391.* Retrieved from http://www.capitol.state.tx.us/tlodocs/82R/billtext/html/HB03391F.htm

Steffen, J., Herriman, M. J., Pomeroy, C. A., & Burian, S. J. (2013). Water supply and stormwater management benefits of residential rainwater harvesting in U.S. cities. *Journal of the American Water Resources Association, 49*(4), 810–824.

Texas Water Development Board. (2005). *The Texas Manual on Rainwater Harvesting.* (3rd ed.). Austin, TX: TWDB. Retrieved from http://www.twdb.state.tx.us/publications/reports/rainwater-harvestingmanual_3rdedition.pdf

Wilkinson, C. F. (1985). Western water law in transition. *University of Colorado Law Review, 56*(3), 317–345.

Winters, K. E. (2013). A historical perspective on precipitation, drought severity, and streamflow in Texas during 1951–56 and 2011. U.S. Geological Survey Scientific Investigations Report 2013–5113. Retrieved from http://pubs.usgs.gov/sir/2013/5113/

Disputes over territorial boundaries and diverging valuation languages: the Santurban hydrosocial highlands territory in Colombia

Bibiana Duarte-Abadía[a] and Rutgerd Boelens[a,b]

[a]CEDLA Centre for Latin American Research and Documentation, University of Amsterdam, Netherlands; [b]Water Resources Management Group, Department of Environmental Sciences, Wageningen University, Netherlands

ABSTRACT

We examine the divergent modes of conceptualizing, valuing and representing the *páramo* highlands of Santurban, Colombia, as a struggle over hydrosocial territory. Páramo residents, multinational companies, government and scientists deploy territorial representations and valuation languages that interact and conflict with each other. Government politicians and neo-institutional scientists wish to reconcile diverging interests using a universalistic territorial representation, through game theory. This generates a hydrosocial imaginary that renders invisible actors' power differentials that lie at the core of the territorial resource use conflict. We conclude that this 'governmentality' endeavour enables subtle, silent water rights re-allocation.

Introduction

In Colombia, the *páramos* (Andean highland wetlands) are strategic hydrosocial territories invigorating agricultural production systems, biodiversity conservation practices, water supply for urban centres, and multisectorial activities. These days, demographic changes in combination with neoliberal policies favouring foreign investment in large-scale extractive industries have resulted in the páramos becoming objects of struggle, arenas of conflicting governance interests and disputes about how to manage and value territory and its water. In this battlefield, the rights of local people in hydro-territorial management are increasingly restricted, while the extraction-based production model continues expanding with disregard for socio-environmental impacts (cf. Baud, De Castro, & Hogenboom, 2011; Bebbington, 2009; De Castro, Van Dijck, & Hogenboom, 2014; Hogenboom, 2012).

This article analyzes the illustrative case of the páramo in Santurban, which is located in the departments of Santander and Northern Santander. In response to development and environmental conservation challenges, and engagement in climate change adaptation and mitigation programmes, the Colombian government now proposes to delimit strategic water ecosystems – páramos and wetlands – to exclude them from mining, agricultural and other activities that might affect water provision and regulation (Bermúdez, 2013; Hurtado, 2010).

This delimitation has begun generating conflicts due to the unequal distribution of socio-ecological benefits and damages that this process will entail for societal groups. They have competitive interests and divergent powers, which are expressed in discordant languages of valuation about the territory (Hoogesteger, Boelens, & Baud, 2016; Martínez-Alier, 2004; Saldías, Boelens, Wegerich, & Speelman, 2012). The case of the Santurban páramo shows these confrontations, where the rights of local inhabitants, who work in agricultural production and small-scale mining, confront the powers of large-scale (multi)national mining companies. Simultaneously, these two sectors confront the uncertainty of political-administrative effects of páramo delimitations, which imply excluding certain actors and activities while allowing others. This process is headed by the government, which as the article will show appears to be an ambivalent player in this game. On the one hand, the government needs to respond to claims for environmental conservation, and on the other, it actively pursues an aggressive neoliberal agenda that is at odds with livelihood protection and threatens the páramo's ecology functions. As the article will show, because of fierce popular protests against the 'hard face' of neoliberalism in the recent past, the government has now turned its eye to 'soft face' strategies – of new institutionalism and game theory – in order to convince the population of that same neoliberal program. It plays its subtle games on shaky grounds: recent ministers of environment, since late 2013, had refrained from revealing the new boundaries – the páramo delimitations – because of the high social and political sensitivity, since these (now for the first time with detailed maps at a scale of 1:25,000) would indicate precisely who would be affected. Meanwhile, another important actor, the environmental movement in Bucaramanga, is pressuring for decisions that will curb mining activities in the páramo and guarantee downstream drinkable water supply.

The present article examines how interest groups sustain different values and representations of what constitutes 'the páramo' to legitimize ways of managing and appropriating the Santurban páramo. At the same time, it analyzes how interest groups' socio-economic, political and ecological values and meanings are contested and wielded, according to their position and relationship with the hydrosocial territory. This shows that conceptualizing this hydrosocial territory does not lend itself to 'objectifying' a single truth; it is an area where divergent socio-environmental imaginaries are generated and contested (cf. Boelens, 2014; Crow et al., 2014; Lu, Ocampo-Raeder, Crow, & Romano, 2014; Perreault, 2014; Saldías et al., 2012).

The article is based on field and desk research done from 2011 to 2015. Its basis was laid by the 'Páramos and Life Systems' project of the Alexander von Humboldt Biological Resources Research Institute, which sought to understand páramo community livelihoods in times of severe ecosystem transformation processes. Additional research was carried out under the banner of the Justicia Hídrica alliance to study the páramo's political-ecological relationships. Fieldwork involved participatory action research, production systems analysis, landscape ecology characterization and hydrosocial network analysis as the main methodological approaches. Literature review and fieldwork were used to identify and characterize actors according to their positions, interests, levels of agency, and dependence on the páramo.

The next section discusses concepts relating 'hydrosocial territory' to the (mis)match among diverse valuations of the páramo, used by different stakeholders to negotiate its use and management. The third section of the article compares the representation regimes of people living in the páramo with those of extraction-based companies, environmentalist groups and governmental actors. In all, we examine the representations of how hydrosocial flows are articulated through discourses, materialized by socio-legal and technological structures, and institutionalized through behavioural norms and political and economic establishments. These representations, according to the positions they defend, promote particular ways of distributing resources and decision-making power (cf. Boelens, 2014; Duarte-Abadía, Boelens, & Roa-Avendaño, 2015; Perreault, 2014; Swyngedouw, 2014).

The fourth section illustrates the role of government politicians and their leaning toward objectifying, de-politicizing scientific approaches. It examines the positivistic neo-institutionalism prevailing in environmental economics, presented as the tool to make hydrosocial territories provide water for 'the majority'. As we argue, this theoretical approach seeks to produce and apply universally valid sets of norms and principles to design specific institutional transformations. In particular, we analyze how the governors–scientists link has applied game theory to stress the importance of a 'collective rationality' in managing natural resources for common use. The fifth section presents and reflects on the outcomes of the governmental hydrosocial territorialization project.

We conclude that the government's neoliberal project subtly deploys contradictory discourses to conceal its opposing policy objectives. It closely aligns with neo-institutionalist strategies, which deny the contrasting modes of how actors value territory and pretend that things are commensurable that are not. In the Santurban case this is expressed in governmental decisions to permit large-scale mining operations in ecological protection areas, and to install universalistic 'payment for environmental services' that conceive of nature and territory as a zone for sustainable extraction of water and their inhabitants as individuals who maximize the benefits of collective action. At the same time, theoretical games and official plans seem unable to curb the impact of large-scale mining in the territory – or the voices of protest.

Hydrosocial territories and languages of valuation

Territories are politically organized space constituted by the interaction between their biophysical and social properties and qualities (Baletti, 2012; Bridge & Perreault, 2009). Relations and agreements among stakeholders define the limits and opportunities for actions, uses and control of territory, reflecting diverse actors' power to symbolically appropriate and politically/economically control territorial space. Divergent actor groups seek to install their own 'regimes of representation' to imagine and materialize 'territory': they involve the rules, relationships and social actions that aim to establish how territorial reality should be known, characterized, appropriated and controlled. Since these regimes of representation commonly suit actors' own particular modes and interests in territorial production and reproduction, in a given space there are multiple representations of 'territory', whereby stakeholders have unequal powers to materialize their imaginaries (Boelens, Hoogesteger, Swyngedouw, Vos, & Wester, 2016;

Hoogesteger et al, 2016; Fernández, 2005). Asymmetries regarding access to territorial benefits, in combination with a lack of political participation and cultural and institutional recognition of marginalized groups who aim to foster their ways of seeing and living 'territory', often characterize water and environmental justice conflicts (Bridge, 2014; Perreault, 2014; Schlosberg, 2004; Zwarteveen & Boelens, 2014). Such conflicts are expressed in different valuation systems and languages. The latter tell us how social groups understand, express and relate to the world, place or ecosystem that surrounds them. Farber, Costanza, and Wilson (2002) define valuation systems as sets of norms and moral frames that orient people's action and judgment in order to support their decisions and actions. Languages of valuation, therefore, concretize actors' regimes of representation; they represent actors' worldviews and knowledge systems (epistemology and ontology), socio-economic interests and cultural and political relations, expressed through concepts, discourses and normative frames (see also Escobar, 2008; Martínez-Alier, 2004)

These diverse regimes of representation clash, and transformation of hydrosocial territories[1] reflects the relative power of the different stakeholders and produces new forms of local-national-global management and interrelations (Rodriguez-de-Francisco & Boelens, 2016; Swyngedouw, 2009). In this respect, in current neoliberal policy practice it is common to see the dominant stakeholders impose market-based territorial representations and monetary language on the others, generally disregarding customary knowledge systems, values and meanings that link to context-bound ecological and socio-cultural legacies (see also Crow et al., 2014; Goff & Crow, 2014; Vos & Boelens, 2014).

Imposing such an outright neoliberal policy and market-based environmental governance rationality, however, is a tricky endeavour for Latin American governments in the twenty-first century, first because their countries still bear the deep scars of the aggressive neoliberal privatization and free-marketization policies advocated by the Washington Consensus and the Friedman/Hayek 'Chicago Boys' in the 1980s and 1990s. These policies met with huge peasant, indigenous and popular resistance throughout Latin America, which was usually repressed with horrifying governmental and military violence. To overcome such resistance and foster acceptance while pursuing similar (but now 'greened') neoliberal agendas, many Latin American governments have embraced 'new institutionalism' (or rational choice theory), with game theory as a crucial tool. Ostrom's new institutionalist groundwork is highly influential here (see e.g. Ostrom, 1990; 2009). Ostrom provides a framework to regulate and direct unpredictable human behaviour by means of collective action, based on rational choice theory. Beyond market or government rules, self-organizing institutions are able to define working rules and norms that structure social, political and economic interaction (Forsyth & Johnson, 2014). According to Ostrom, individuals can conserve the common goods – as in our case water and páramos – and engage in collective action when they have credible and reliable information about the cost and future benefits of their actions, and when they are enabled to rationally define the rules of the game. Given the assumed commonalities among people's working rationalities (such as fostering individuals' benefits while lowering their transaction costs), the approach for conservation and water policies is generally presented as exemplary for reconciling conflicting values and interests that converge in the same hydro-territory.

Though this new institutionalist conceptual framework is often (and rightly) presented as a critique of neoliberal economic thinking, paying important attention to informal working rules and 'people's collective arrangements' around common property resources, many studies have shown its deep affinities with the universalist-economicist family and its fallacies, including its similarities to neoliberal presuppositions (see e.g. Boelens & Zwarteveen, 2005; Büscher & Fletcher, 2015; Espeland, 1998; Rodríguez-de-Francisco & Boelens, 2016; Mollinga, 2001; Moore, 1990). Most importantly, the new institutionalist paradigm avoids studying power relationships and understates complexities and the diversity of (water and territorial) cultures and epistemologies in order to be able to devise universally valid principles for (e.g., water) designs and policy solutions. Its efforts, first to 'equate' and 'uniformize' items that are incommensurable[2] and, next, to present them as universally valid definitions and categories, may carry great risks. For example, Espeland (1998, p. 223–224), who studied the application of rational choice theory and game theoretic tools to silence opposition to large dam building in indigenous territories in Arizona, argues that the framework "requires that we value in a resolutely relative way. The commensuration it demands may violate, even obliterate, other social boundaries that help order our lives and define us…. The logic of this form can erase or diminish that which is hard to reconcile with instrumentality: thick, messy context, historical legacies, uncertainty, ambivalence, passion, morality, singularity, the constitutive and expressive salience of symbols." Regarding its universalism and commensuration, Forsyth and Johnson (2014) add that Ostrom's framework predefines the problems that local institutions were seeking to resolve, and thereby puts too much faith in only the economic-rationalist type of political bargaining process to achieve outcomes. Besides, it overlooks for whom and for what purposes the resource is exploited or demanded and the consequences of its socio-ecological distribution. As a result, as Mollinga (2001, p. 733) comments, new institutionalism's appeal for policy makers "lies in its suitability for designing standardised policy prescriptions, and its exclusion, or rephrasing, of the issues of power and politics". As this article examines, these ingredients, largely shared by new institutional and neoliberal frames of policy thought, make game theory into a welcome 'soft face' addition to the 'hard face' neoliberal policies that the Colombian government wants to install in order to exploit and transform local hydrosocial territories.

Proliferation of divergent valuation languages about the Santurban páramo

From pre-Hispanic times, the páramos have been inhabited by indigenous communities, with models of occupation based on "vertical economies" – exchange systems involving control over agricultural production in, and trade among, zones with different altitudinal and climatic properties (Murra, 1972). Indigenous mythologies and cosmogonies conceive of the páramos as sacred places where different gods came from. They controlled water, the origin of life and its continuation (Boelens, 2014; Osborne, 1990). In rural concepts, these referents survive and are expressed in protecting the páramo's lagoons. Nowadays, these water sources supply much of the local drinking water systems, and are places of identity formation, often integrating human, natural and supranatural aspects. Cultural meanings and values regarding water tend to foster self-organization around the objectives of protecting lagoons and ensuring local water supply, as is happening in several municipalities in Santurban. Their valuation languages have site-specific historical, ethical, economic and cultural features, constructed through goals shared by a collective (Penna & Cristeche, 2008). So, the páramos have acquired a deep-rooted social nature, built on place-based knowledge (Echavarren, 2010; Escobar, 2008; Gómez-Baggethun, 2009). The foundations of the páramos' rural economies influence construction of cultural identity, which in turn determines political capacity – changes in the one cause changes in the other (Van der Ploeg, 2010).

The páramos have been occupied by internal migration (driven by civil wars), by dispossession of indigenous peoples driven off their land, and by government colonization policies. In the 1960s and 1970s, government policies facilitated development of the potato and livestock industries. This enabled rural communities to appropriate these territories and build their livelihoods, confronting highly adverse conditions. Páramos are also strategic places to control rural production, roadways, commerce and urban centres, therefore these constitute important arenas confronting armed stakeholders. In south-western Santurban páramo, Berlín region, electrical transmission lines have been installed, along with optical fibre and gas pipelines, as strategic points for commercial relations with Venezuela.

In the 1980s, the Revolutionary Armed Forces (FARC-EP) had control over the Santurban páramo, until the so-called democratic security policies scheme, Operation Berlín, brought the military against FARC–EP and expelled them. In 2003 the military battalion set up there, enabling multinational company Greystart, currently called Eco Oro Ltda., to expand. On various occasions, this company had to suspend its activities because of FARC-EP interventions in the zone. Thus, the páramos represent positions for geopolitical control of the territory and multiple economic interests, introducing and reinforcing the corresponding valuation languages (cf. Bebbington, Humphreys Bebbington, & Bury, 2010).

In this context, Law 685 of 2001 further fostered applications for mining concessions throughout Colombian territory. Earlier legislation included property titles granting the right to use the land through three phases: licences for exploration, extraction, and mining contributions (Law 2655, 1988). This law was amended by Law 685, eliminating environmental requirements and converting no-mining zones into restricted zones, while also cancelling the economic benefits that local communities used to receive from mining (Duarte, 2012). When the Uribe government ended (2009), 9000 mining concessions had been granted in areas of páramos, wetlands and national parks; 416 of these concessions correspond to páramo areas (2014; Bermúdez, 2013). In Santurban, by 2011, there were 65 concessions, 15 of them with environmental licences to begin extraction (Ungar, Osejo, Roldán, & Buitrago, 2014).

Another sector present in the hydrosocial territory is agriculture, which wants the páramo for fertile croplands or range livestock. The sub-region of Berlín, in Santurban, is the country's second-largest producer of scallions, harvesting from 250 to 380 tonnes a day (Franco, 2013). This represents approximately USD 280,000 a month and livelihoods for 5000 families (Quintero, 2014). In the Berlín sub-region, onion growers and sellers are an economic power sector; also, several small farmer groups in the páramo grow onions. Increasingly, their interests are in conflict with environmental discourses and norms, which have grown in importance since the 1990s and, particularly since the early twenty-first century, now constitute a powerful voice.[3] The latter attempt to stop the rapid agricultural encroachment transforming the páramos.

Environmental discourses applied in the region have their historical roots. According to Molano (2012), botanical and scientific expeditions at the end of the colonial period and during the nineteenth century constructed natural science–based productivist knowledge about the páramo and other Colombian ecosystems. Molano says that this knowledge facilitated economic exploitation of the land, so that the Spanish viceroyalty could cover local food supply requirements and export produce. Currently, urban and scientific societies – communities far from these places – value the páramo above all for of its ecological functions, entailing water catchment, holding and regulation. These notions commonly focus on páramos as "natural spaces" (Escobar, 2010), ignoring or misrepresenting their significance for social life, cultural and historical identification, and livelihood production.

Modernistic values for the páramo exist in a context of great competition and demand for water. The north-eastern sub-region of the Santurban páramo supplies 17 municipal drinking water systems, including Cucuta and Pamplona, an irrigation district in the Zulia Basin, and the Tasajero thermoelectric power plant. In the south-western sub-region, water demands for human consumption are concentrated in the metropolitan area of Bucaramanga, Floridablanca and Girón. These compete with the mining interests of foreign companies in the municipalities of California and Vetas. They compete for the waters of the Surata, Tona and Río Frío rivers (Figure 1).

The Colombian government has considered the páramos as zones of great importance to the country's development. This is the approach of conservation legislation and decision making and reflected in the National Natural Resource Code (Law 2811 of 1974). Law 99 of 1993 provides protection of the country's biodiversity and especially páramo zones (No. 2 and 4, Article 1), acquisition of areas of value because of water for municipal and environmental entities (Article 111 and Law 373 of 1997), and resolutions organizing environmental zoning of the páramos (Resolution 0839 of 2003). These legal frameworks currently reinforce and interweave with new discourses about mitigating climate change and the policy measures this will require.

In summary, the historical development of the different rural populations' knowledge systems, valuation languages and regimes of representation has diverse cultural and economic-productive sources. They coexist and are constructed between a cultural legacy rooted in a system of rural traditions and notions of modernity, immersed also in a neoliberal economic model (see also Van der Ploeg, 2010). Therefore, while páramos represent places to 'coexist' in small-farm subsistence livelihoods, they have also been valued and conceived as a source of wealth that must be protected and, in turn, as a place that can be "owned, moved, purchased and sold according to the whims of individual interests and economic power" (Blatter, quoted by Ulloa, 2002, p. 193). Therefore, different territorial projects are not isolated from global market dynamics. Divergent local territorial imaginaries and materialities and their respective valuation languages face off and also interact with dominant imaginaries and languages, as a subset of the contradictions and conflicts produced in the confluence of different societies (Table 1).

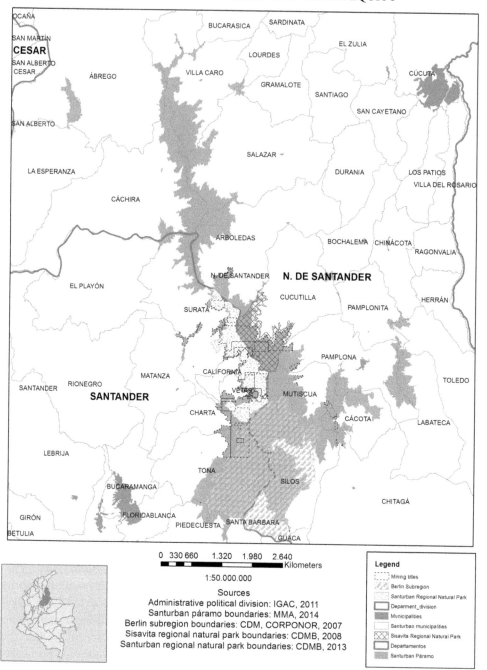

Figure 1. Santurban páramo, protected areas and mining titles. Author: Bibiana Duarte-Abadía

Table 1. Expressions of valuation languages.

Valuation language	Meaning	Actors involved	Illustrative expression
Cultural/metaphysical valuation language	Páramo as a sacred place integrating human–natural–divine relationships	Indigenous and peasants	"The people who live close to the lagoons have looked after them. The lagoons provide the water for local aqueducts, but when these are not protected the water sources may get angry. Therefore, indigenous communities always have tended to worship these sources and make ritual offerings." [Doña Aura, inhabitant of Vetas, páramo de Santurban, 2012, quoted in Buitagro, 2012).
Natural/scientific valuation language	Páramo as highlands wetland ecosystem with unique biodiversity	Academics; environmentalist citizens (Bucaramanga, Cucuta); national government (Ministry of Environment)	"Páramo is an ecological unit with high importance for water regulation.… The páramos are also ecosystems with high capacity to capture carbon; these two functions can contribute to mitigate the effect of climate change. Hence páramos need to be protected from economic activities" (MMA, 2001).
Socio-cultural valuation language	Páramo as rooted dwelling place with cultural identity	Indigenous, peasants and páramo dwellers	"We like to live in the páramo because is a healthy environment for animals, for our families. Here we want to stay, making our life and living" (Villamizar, 2013, cited in Franco, 2013).
Agro-economic valuation language	Páramo as a space for food production	Peasant organizations, agriculture organizations, onion growers, immigrants, food traders	"The páramo [is home to] farmers who by sowing potatoes provide the food to the people who live in the towns and the cities" (Villamizar, 2013, cited in Franco, 2013, p.127).
Socio-economic valuation language	Páramo as means of producing local economy and livelihoods	Mayors; small-scale miner organizations; agricultural organizations; peasant communities	"Unemployment, the indifference of the government, and interventions by foreign mining enterprises generate uncertain future for the families of California [many of whom] can only survive by extracting ore and whose hunger cannot wait" (Mayor of California, 2014, cited in Osejo, 2014, p. 58).
Economic-extractivist and developmentalist valuation language	Páramo as mineral resource to enhance national development	National government (Ministry of Mines and Energy); mining multinationals; military forces; Revolutionary Armed Forces; and small-scale local miners.	"Development research reports that in the last decade the Gross Domestic Product increased from 8% to 15%, which is why the government lists mining activities as a key driving force of development" (Corredor, 2013). "We face the challenge to promote the development of a mining economy in a sustainable way" (environment minister, cited in Hurtado, 2010).

Divergent regimes of representation and their complex interaction: páramo dwellers, government, multinational companies and environmental movements

Since colonial times, in the sixteenth century, mining has been conducted in the municipalities of California and Vetas, known as mining districts, located in páramo zones. For inhabitants of Vetas, gold and water used to be complementary. Historically, small-scale mining has been part of their livelihoods. Gold has driven the growth of towns and, along with water, has configured the territory by organizing actions to use and manage them. This has defined areas for agriculture, mining, livestock, towns and conservation. For the inhabitants of Vetas, gold represents wealth, but also history, legend, symbolizing tradition and knowledge, as well as one of the mainstays of their socio-economic livelihood (Buitrago, 2012; cf. Cremers, De Theije, & Kolen, 2012).

However, the entry of foreign capital has generated a crisis in local mining. In the last 15 years, most of the artisanal mines have been sold to multinational companies. These small-scale or artisanal mining companies were family associations, hiring an average of 20 workers; they used to be called "underground" mining, with low technical sophistication (Buitrago, 2012). Many residents of California, in Santurban, transferred their land ownership rights and mining extraction concessions to large companies, and subsequently went to work for these companies. In Vetas, for instance, Eco Oro has acquired 10% of the territory, totalling 1518 ha (Zapata, 2012, cited in Ungar et al., 2014).

Expansion of foreign capital through multinationals was enabled by various factors. The first was Law 20 of 1969, which declared mining activities to be of public utility and social interest. The second factor has to do with market liberalization policies and the strengthening of neoliberalism since the late 1980s, and later under the government of César Gaviria (1990–1994). The third factor is associated with the 'democratic security policy scheme' mentioned above, which consolidated territories for foreign capital and development of mining and agro-industrial zones. Finally, during the Santos government (2010–214), the mining-energy sector was emphasized in the country's development model.

In response to this situation, the Ministry of Environment enacted Law 1450 of 2011, which prohibits most economic activities. It forbids mining, agriculture, animal husbandry and hydrocarbons exploitation in páramo ecosystems. For this purpose, it called for delimitation of the páramos at a detailed scale of 1:25,000 to protect them more accurately. At the same time, these restrictive frameworks and public utility discourse stimulated land sales, either to the government or to multinational companies. In fact, as in the municipality of Vetas, many residents preferred to sell their land at higher prices to the multinationals rather than to the governor's office for conservation.

The multinationals reconfigured the páramo to suit their interests in massive extraction of resources, changing the rules of play in land and water management (Buitrago, 2012). In Santurban, the recent reterritorialization by foreign capital has blurred borders and reorganized scales, mixing the local and external (Garay, 2013). Swyngedouw (2009) calls this "glocalisation", building strategic multi-scale compositions in response to the commercial flows and geopolitical interests of multinationals. Bauman (2000) refers to the "liquid" modern world, an allusion to the fluidity with which globalization generates a world of generalized circulation in which flows are freed of territorial constrictions.

In Colombia, formally, underground resources may be declared national property to supply public goods (Article 332 of the 1991 Constitution); however, instead of applying this article to protect public assets from deterritorialization by capital flows from the global market, the state uses it to reserve the right to authorize *private parties* to extract underground mineral ores, by granting mining concessions. In a Kafkaesque manner, governmental plans for protecting the territory are undermined by the government's own policies, bending over backwards for its multinational allies.

The government has started a campaign to declare *small-scale* national mining operations illegal. First, the 1988 Mining Code was amended (Law 685 of 2001) to eliminate small-scale mining as a legal category, placing it in the same category as informal mining, which tends to have the connotation of illegal mining (Duarte, 2012; cf. Cremers et al., 2012). Second, environmental authorities in recent years have been quite restrictive of mining activities. According to mines' technical standards, the authorities decide which ones can extract ore; this favours multinational companies and places traditional small-scale miners at a disadvantage. Moreover, water rights are allocated for extraction according to the categorization of the mining activity in terms of its legality, which directly affects small-scale mining. So, technological developments to extract gold have profound legal implications for mining rights, and access to sources of water.

Mining issues in Santurban are an exemplary illustration of how neoliberalism, rather than disempowering the national government according to a 'laissez-faire' discourse, reinforces the government's role, putting it to work for global market strategies. The government intervenes aggressively as the regulatory entity – in social, economic and cultural life, and in territorialization and deterritorialization. The recent entry of multinationals into Santurban's territory has limited free access to major lake complexes in the páramos as the companies buy up property. For the inhabitants of páramo mining districts, multinational companies' exploring for gold, buying concessions and getting environmental licences dispossesses them of not only their livelihoods but also their water.

At the same time, the demographic and economic growth of Bucaramanga and other semi-urban towns such as Cucuta and Pamplona have increased environmental demands on the páramo, particularly for water supply. Taking into account that the mining districts are concentrated at the headwaters of the Tona, Surata and Frío Rivers supplying Bucaramanga, different sectors of the city (academic, political, entrepreneurial, environmental, labour union and others) have joined to defend the Santurban páramo from mining activities by multinationals (Duarte-Abadía & Roa-Avendaño, 2014). These sectors, particularly the urban, gathered under the Committee to Defend the Santurban Páramo, and in February 2011 they reached a consensus to deny a social licence for the Eco Oro company and prevent open-pit gold mining.[4]

In this context, the language these sectors have used in dealing with Eco Oro works to raise consciousness about caring for and respecting páramo as a 'natural space', guaranteeing water quality and supply. Their actions reinforced initiatives to expand the Sisavita regional natural park, north-east of the Santurban páramo, and declare regional natural park areas in Santurban. However, at the same time, such declarations entail strict protection, excluding smallholder activities – restrictions that were supported by the downstream coalitions among dominant water users. The latter comprise the energy sector, the agribusiness sector (represented by the irrigation district's powerful private beverage company) and the public services enterprise.

Páramo inhabitants, small-scale miners and farmers saw these environmental protection frameworks as a threat to their livelihoods, which polarized them, with páramo defenders from urban zones versus inhabitants of these high-altitude zones. The conflict was worsened by the different multinationals' presence, which in addition to co-opting small traditional mining enterprises adopted the discourses of 'defending the rights of the territory and its inhabitants'. Even relationships between the different societal sectors of Bucaramanga and the páramo inhabitants have become conflict-ridden. Under environmental arguments, residents of the high-altitude areas are marginalized, considered water polluters and stakeholders jeopardizing the health of citizenry and ecosystem.

In recent years, because of the investment insecurity that is created by these environmental legislative frameworks, the Eco Oro laid off over 1500 workers after ending the exploration phase and has threatened to sue the Colombian government for USD 200 million dollars if prevented from continuing with mining projects.[5] The economic and social effect of Resolution 0839 of 2003 (which created new categories of protection to organize and zone the páramos environmentally) restricts the water rights (concessions) of farmers for agricultural activities in páramo zones. This situation creates conflicts between Colombia's environmental authorities (Autonomous Regional Corporations) and rural people, especially when their actions and decisions favour some sectors over others (cf. Bebbington et al., 2010; Boelens & Gelles, 2005). Conservation policies regarding páramo use and management tend to increasingly restrict páramo dwellers' livelihoods while allowing multinational mining activities.

Making divergent values and interests commensurate in Santurban: the game of water and life

> We have an ironclad commitment to delimit all the country's páramos, which we expect to finish by next year, and we will restrict activities that can be done around them, to ensure that these natural water factories can provide water catchment, regulation and supply services. (President Santos, World Water Day, March 2013)[6]

> 'The law [to delimit páramos] does not provide for any transition or compromise with local stakeholders, so it ultimately simplifies a reality that requires complex solutions', an expert in páramos told La Silla (preferring, like the other sources queried, to remain nameless because he works on a daily basis with the environmental entities). (Bermúdez, 2014).

Despite the great diversity of representations, values and interests related to páramos as hydrosocial territories, environmental authorities are clear that they are water factories that must be 'known' with objectifying scientific language and that their opposing interests must be matched with the universalistic, equalizing rationality of neo-institutionalism. Apparently, as the second quotation illustrates, the more critical scientists, who are aware of the need to examine local complexity in greater depth, are afraid to raise their voices.

Different research institutes, consultants and environmental authorities have engaged in delimiting the páramos. One of the first was the páramo in Santurban. However, the Ministry of Environment delayed decisions for over a year, because of the socio-economic impacts of delimitation. First there is the huge investment made by the company, entitling them to sue the Colombian state if forced to leave the zone. Then there are the positions of the attorney general of the nation, governors of the region, and municipal mayors, responsible for enforcing acquired rights but also the well-being of the local people.

To find a way out of these confrontations, in early 2014 the environment minister of that time organized a discussion group with the diverse stakeholders' representatives disputing development in Santurban's hydrosocial territory. Small farmers, miners, representatives of multinationals, environmental authorities, citizens of Bucaramanga, environmentalists, researchers and the academic sector met with the expectation to learn about and reach consensus-based decisions on managing the páramo under the new delimitation. The dialogue was mediated to overcome these dilemmas using game theory and experimental economics, stressing the importance of collective rationality and reason to manage shared-use resources (cf. for example, Bromley, 1992; Ostrom, 1990). It assumes an understanding of how human beings universally reason and behave when resolving conflicts. The facilitator of the dialogue in Santurban explained:

> Each chip you have in your hands costs 3000 pesos. Each of you can do whatever they want, keeping or investing your chips. Those that appear in the piggy bank get doubled and distributed in equal parts. If I invest and no one else does, what I invest gets scattered all over and I get nothing, so we need everyone to invest. We got a large proportion – nearly 80% of available resources – invested in the piggy bank, to be redistributed among everyone. (Cárdenas, 2014)

Under this scenario, the different opposing values and interests (represented by the chips) became commensurate, using the universal value of money, to facilitate a consensus among everyone. The assumption is that when incentives are 'correct', the motives of individuals to maximize their profits will ensure that opposing groups automatically try to find the most efficient way to organize the distribution of water, funds and other relevant resources. The game attempts to harmonize everyone's interests:

> You have 40 chips, right? You can invest them in a large fund to produce water. And what we have are five large groups in a sequence who are going to receive the benefits of the water produced by the whole community. The problem is that, to deliver the water, we do it first for one group, we see how much water they take, and how much they leave for the rest.... If shared interests produce an agreement, this should distribute the water not only more efficiently but also more fairly and equitably. (Cárdenas, 2014)

As the game's facilitator explains, this approach aims to forge agreements among the different stakeholders to distribute water more fairly, to maximize societal well-being. For this, it compares individual and group behaviour in coping with problems of equity and efficiency in collective water management. Explaining the thinking underlying this, Cárdenas (2011) refers to equity as grounded in the trust and reciprocity among stakeholders to face their responsibilities for environmental externalities. This makes efficiency the result of efforts by each individual to contribute to maximizing societal well-being.

Faced with the problem of divergent interests, the neo-institutionalist game attempts to solve it through arrangements for cooperation and agreements for trust (Cárdenas, 2009). For example, the stakeholders furthest from water sources must increase their contributions to those higher up, in order to receive more water. The theory of working for collaboration in united exchange and marketing 'among peers' prefers not to speak of the major power inequalities between, for example, the multinational companies and small farmers, or between powerful cities downstream and peasant communities upstream. As if it were natural and automatic, applying game theory in the arena of water battles for the Santurban páramo has the implicit objective to make the participants understand that 'consensus', mediated by commensuration using money's universal value, will lead to more rational, collective, optimal, efficient, just solutions.

The mediation in Santurban takes the neo-institutionalist perspective that conflicts originate in the 'lack of mutual cooperation' in allocating and distributing water. However, it ignores the fundamental *causes* for not generating cooperation on the basis of the presumed 'shared interests'.

These include social groups' opposing interests, and profoundly unequal economic power, in a discriminatory, exclusionary political structure. Another directly related cause is the existence (and juxtaposition) of different worlds with different cultures and worldviews, and incompatible valuation languages. Even though historically their representation regimes flow together or strategically interact in the hydrosocial territory (in, for example, political and economic co-opting of local residents; environmental discourse adopted by the multinationals; and neo-institutionalist conceptualization in the environmental movement), this does not mean that they can all be represented using a single universal valuation language (cf. Goff & Crow, 2014; Martínez-Alier, 2004). Consequently, the key is to ask whether implementing game theory and the theory of collective action, with its neo-institutional approach, actually achieves social justice and consensus as claimed – and if not, who wins and who loses in this game with water and life.

As a result of this 'dialogue', following instructions and foreseen outcomes of the game, stakeholders concluded that the solution to their clash of interests lay in building agreements for cooperation (Cárdenas, 2014). However, none clarified the type of cooperation to be agreed, and they have not even been able to agree about the cooperation mechanisms. Some days later, like a *deus ex machina*, the environment minister announced implementation of a public model of payment for environmental services (PES), but with no concrete scheme, to make conservation profitable and open up the dialogue in Santurban.

PES is established by monetary transactions between users downstream and residents of higher watersheds, for the latter to protect the environment to conserve and enlarge water flows to sustain economic and productive activities in the areas below. PES assumes that commoditizing water plays a harmonizing and homogenizing role with divergent interests, and therefore this universalistic reasoning and language have spread worldwide in the past decade, with strong economic and political backing by international environmental policy agencies (Boelens, Hoogesteger, & Rodriguez-de-Francisco,

2014; Büscher & Fletcher, 2015; Büscher, Sullivan, Neves, Igoe, & Brockington, 2012; Rodriguez-de-Francisco & Boelens, 2014; Rodriguez-de-Francisco et al., 2013). As a result of the presumed 'open dialogue', the PES model was *totally pre-planned* by the government in alliance with neo-institutionalist scholars and was *the only outcome that could 'rationally' emerge* from the game – a game and theory based on the inevitable superiority of collaboration and mercantile exchange among partners. As Foucault reasoned (1980), these ideas are not powerful because they express truth, but rather are true because they are backed by power (see also Robbins, 2004). In response to a scheme of neoliberal, neo-institutionalist governmentality, the language used in the 'game' presents options that are profoundly political (regarding fundamental issues of distribution and exclusion) as if they were neutral or technical. It applies the discourse of scientific objectivity, denying that power relationships permeate the knowledge produced and decisions made about delimitation and exclusions in the páramo. Along this line of thinking, their PES proposal treats human beings as rational indivi-duals seeking aims focusing on their own interests. Accordingly, delimitating the páramo and hydro-territorial configuration, while commoditizing and redistributing water flows, appears and can be portrayed as 'natural', 'inevitable' and scientifically 'rational'. Consequently, at the 'dialogue' in Santurban, facilitators (with all their political, institutional or economic interests) may seem to be mediators without inter-ests or antecedents, who benevolently represent the local well-being and work on behalf of the nation's best interests and universal truth.

For local residents, valuing their territories from the sole viewpoint of mercantile exchange of water has become a factor limiting and delegitimizing their multidimen-sional productive and reproductive territorial relationships and activities. Clearly, power validates certain types of knowledge and disqualifies others, promoting certain narratives and silencing others. So, under the formal discourse of national progress, efficient governance of resources to protect strategic ecosystems, climate change adap-tation and mitigation, and ensuring water service for all citizens, subsistence economies in the páramos have a hard time, while proliferation of large-scale mining, with its 'advanced, clean technologies', seems to be the rulers' hidden agenda for organizing and aligning territories, resources and residents.

Who wins and who loses: delimitation decisions

In December 2014, the new minister of environment, Gabriel Vallejo Lopez, announced the results of the delimitation process (Resolution 2090, Ministry of Environment and Sustainable Development, 2014), strategically blending different valuation languages:

> We need to make balanced decisions in accordance with the socio-economic context; this government focuses on green development and the protection of strategic ecosystems. (Vallejo, 2014)

The delimitation process implements a zoning regime that establishes areas for restoration (25,227 ha), sustainable agriculture (5502 ha), and preservation (98,993 ha). Within "restoration areas", mining activities can continue if the mining titles were acquired before 2010. The municipalities of California, Vetas and Surata are in these "privileged zones". The Berlín sub-region is included in areas of "sustainable agriculture". In the "preservation areas", agricultural and livestock activities cannot be expanded, while mining activities are forbidden. These three areas have to be managed according to ecological criteria in order to guarantee ecosystem services regulation and water supply. In this respect, and in accordance with our analysis in the fourth section, PES and other market-environmentalist instruments are legally installed to promote conservation as a economic activity.

Nevertheless, fundamental questions remain unresolved, such as how to reconcile ecological restoration activities with large-scale mining extraction in these areas. The new measures represent a major step backwards for social and environmental concerns since mining enterprises like Eco Oro have leeway to reactivate their extractive practices. The development of the Angostura Project through underground exploitation is but one example.

> We extend our gratitude to all who participated in this process. We intend that Angostura will become an exemplary mining and investment project in the area of Santurban. We are committed to developing the Angostura Project in a socially and environmentally sustainable manner, abiding by all international mining standards and best practices that will be beneficial for all stakeholders, including our investors and the communities in which we operate. (Eco Oro, 2014)[7]

Deploying a double discursive strategy, the government/transnational company nexus, sustained by market-environmentalist scientists, plays the card of entwining valuation languages to manage differences and keep centralized control of institutional practices; this, to facilitate the perpetuation of capital and existing power relationships.

Rather than 'technical' or 'biological/ecological' criteria for establishing the limits of action in and appropriation of Santurban's páramo, and far beyond presumably open-ended game-theoretical outcomes or predictions regarding societal win-win options, it is the power structures among the stakeholders that define the conditions of access to and control over the hydrosocial páramo territory of Santurban.

Conclusions

This article shows that hydrosocial territories, in addition to resulting from some complex biophysical and political-institutional interactions, also result from the ways they are perceived and interpreted by societies. Simultaneously, different valuations represent different relationships and concepts about the páramo, which are subject to a historical context, social changes, modernization, and expansion of market economies. Therefore, territorial imaginaries lead to technological, political and cultural projects to define their order, and conversely, water control structures and relationships generate and reinforce territorial discourse to legitimize and justify forms of governance.

At this time in the Santurban páramo there is complex interaction, because several processes are becoming more intensive at once: large-scale mining extraction and local-global transformation of the territory; territorial cultivation by multisectorial subsistence economies; the assumed threat of increasing water scarcity in cities due to climate change and extractive industries; and the government and armed forces striving to build their geo-political control over this disputed territory. Neoliberal policies backing the mining and energy sector have drastically influenced the multinationals' territorialization of power. Meanwhile, at lower altitudes, population growth in cities and environmental policies pressure for protection of these ecosystems and restriction of economic activities endangering water security and societal well-being.

So, none of these tensions is separate from the rest. On the contrary, they are the result of globalization phenomena that have broken down the territorial boundaries in Santurban and worked their way into local dynamics. This generates ecological conflicts over distribution, and contradiction in normative frameworks, where water increasingly becomes the bone of contention. This interaction among stakeholders with opposite interests generates epistemological pronouncements and political confrontations among different regimes of representation about 'what the páramo is and should be', each with its own valuation language. Along with the conflicts, this also generates strategic political-discursive coordination among (presumed) allies – páramo inhabitants and mining companies together, defending their access to the páramo using languages of 'territorial-cultural defence'; urban environmental movements and environmental authorities together, representing the páramo as 'water factories' requiring precise delimitation and exclusion of polluters; multinational companies co-opting residents and politicians with the language of money and of national progress and modernization; etc. Each of these discursive coordinations obviously embodies profound contradictions in interests and values.

Amidst these coalitions, divergences and convergences, the government's own ambivalent policy has been forced to juggle a threefold (or more) discursive strategy, with contradictory faces. It tries to ensure environmentally sustainable management by setting limits on extraction, delimiting the páramo and restricting certain territorial stakeholders and activities. At the same time, it seeks to include and involve the different societal sectors and/ or those affected by this management. Third, and fundamentally, it pursues its policy of appealing to foreign capital and mining extraction for 'the nation's well-being'.

To achieve 'consensus' and political stability (without jeopardizing the status quo and the continuity of the extraction-based model), the government has strategically combined with neo-institutionalist science, because of its depoliticizing, universalizing language. Applying game theory and the dilemma of managing the commons is a way to 'convince' and 'include' local residents of the Santurban páramo using norms of rational behaviour and economic truths. These emphasize the commoditization of water resources and the mercantilization of its services to generate 'rational, efficient water use'. This assumes that all inhabitants leave behind their own particular ways of knowing, identifying and valuing, and collaborate with each other to conserve their territory, maximizing every player's individual gains and multiplying their contribution to water conservation. Boelens and Zwarteveen (2005) explain that, in fact, "neo-institutionalist formulae are attractive because of their clarity and the efficiency with which they simplify complex realities and behaviours". As manifested in the case of Santurban, "the beliefs that flows of money and water follow universal scientific laws, and that human beings roughly follow the same rational, utility-maximising aspirations everywhere, are important sources of consolation and relief for policy-makers who are confronted with increasingly complex, seemingly chaotic, and highly dynamic water situations" (p. 736).

The approach actively denies that decisions made about delimiting and reorganizing the territory and redistributing water are profoundly political rather than being just technical or socially optimal; they inherently exclude. The approach also subordinates other modes of valuation and systems of knowledge. It does not consider the factors by which many communities constantly interact dynamically with the hydrosocial territory, including affective relationships, family relations and solidarity, and emotional, moral and cultural values that cannot be expressed in commodities and maximizing profits.

The policy proposal to implement PES seeks to constitute hydrosocial territories whose constituencies' roles and identities have been aligned with the market – water producers and clients – exchanging commodities and cooperating on the basis of universal collective rationality. This assumes that homogeneous groups of producers and consumers exist, under conditions of equal power. However, in Santurban's everyday reality actors have strongly differing power bases and divergent hydro-territorial interests and proposals. The government, therefore, seeks to silence societal conflicts through consensual discourse and through the entwining of multiple valuation languages. But the outcomes of its territorial zoning and economic-productive delimitation decisions evince the firm governmental position that Colombia's neoliberal project should not suffer from lofty socio-environmental ideals and protections.

Meanwhile, growing resistance to large-scale mining in Santander has placed at centre stage the unending contradiction permeating the state, which has enacted two opposite sets of legislation: economic and commercial opening-up to mining, and protection for ecosystems. Páramo inhabitants and communities know that the governmental strategy to present deeply incommensurable issues as if they were understandable through universalist language and solvable through a theory based on games and a depoliticized zoning process, will not solve their real-life problems.

Notes

1. Boelens *et al.* (2016) conceptualize "hydrosocial territory" as "the contested imaginary and socio-environmental materialization of a spatially bound multi-scalar network in which humans, water flows, ecological relations, hydraulic infrastructure, financial means, legal-administrative arrangements and cultural institutions and practices are interactively defined, aligned and mobilized through epistemological belief systems, political hierarchies and naturalizing discourses" (Boelens et al., 2016, p. 2).
2. Incommensurable items have no common measure or standard of comparison, therefore they are impossible to compare in value or quality. Commensurability refers to what can be exactly expressed by some common unit, concept or language (Webster's Dictionary 2016: "having a common measure; capable of being exactly measured by the same number, quantity, or measure").
3. The Berlín sub-region was declared a DMI (integrated management district) in 2007. Decree-Law 2811 of 1974 regulates land use and environmental planning.
4. The Greystart Company, in December 2009, applied for an environmental licence to make an open-pit gold mine, and on 31 May 2011 the Ministry of Environment rejected this application (Duarte-Abadía & Roa-Avendaño, 2014).
5. http://m.vanguardia.com/economia/local/225344-dudas-sobre-unificar-el-limite-del-paramo-con-el-parque-santurban.
6. http://lasillavacia.com/historia/santurban-de-ministro-en-ministro-y-sin-solucion-la-vista-4846.
7. http://www.eco-oro.com/s/NewsReleases.asp?ReportID=688843&_Type=News-Releases&_Title=Eco-Oro-Announces-Boundaries-of-Pramo-of-Santurbn-Declared.

References

Baletti, B. (2012). Ordenamento territorial: Neo-developmentalism and the struggle for territory in the lower Brazilian Amazon. *Journal of Peasant Studies*, 39(2), 573–598. doi:10.1080/03066150.2012.664139

Baud, M., De Castro, F., & Hogenboom, B. (2011). Environmental governance in Latin America: Towards an integrative research agenda. *European Review of Latin American and Caribbean Studies*, 90, 78–88.

Bauman, Z. (2000). *Liquid modernity*. Cambridge: Polity Press.

Bebbington, A. (2009). Latin America: Contesting extraction, producing geographies. *Singapore Journal of Tropical Geography*, 30, 7–12. doi:10.1111/sjtg.2009.30.issue-1

Bebbington, A., Humphreys Bebbington, D., & Bury, J. (2010). Federating and defending: Water territory and extraction in the andes. In R. Boelens, D. H. Getches, & J. A. Guevara Gil (Eds.), *Out of the mainstream: Water rights, politics and identity* (pp. 307–328). London & Washington DC: Earthscan.

Bermúdez, A. (2013). Estos son los efectos de volver al viejo Código Minero. *La Silla Vacia*. 29 April 2013. http://lasillavacia.com/node/43892

Bermúdez, A. (2014). La locomotora minera no despego pero ya tiene rieles. *La Silla Vacia*. 25 June 2014. http://lasillavacia.com/historia/la-locomotora-minera-no-despego-pero-ya-tiene-rieles-47981

Boelens, R. (2014). Cultural politics and the hydrosocial cycle: Water, power and identity in the Andean Highlands. *Geoforum*, 57, 234–247. doi:10.1016/j.geoforum.2013.02.008

Boelens, R., & Gelles, P. H. (2005). Cultural politics, communal resistance and identity in Andean irrigation development. *Bulletin of Latin American Research*, 24(3), 311–327. doi:10.1111/j.0261-3050.2005.00137.x

Boelens, R., Hoogesteger, J., & Rodriguez-de-Francisco, J. C. (2014). Commoditizing water territories: The clash between Andean water rights cultures and payment for environmental services policies. *Capitalism Nature Socialism*, 25(3), 84–102. doi:10.1080/10455752.2013.876867

Boelens, R., Hoogesteger, J., Swyngedouw, E., Vos, J., & Wester, F. (2016). Hydrosocial territories: A political ecology perspective. *Water International*, 41(1), 1–14. doi:10.1080/02508060.2016.1134898

Boelens, R., & Zwarteveen, M. (2005). Prices and politics in Andean water reforms. *Development and Change*, 36(4), 735–758. doi:10.1111/j.0012-155X.2005.00432.x

Bromley, D. W. (Ed.). (1992). *Making the commons work. Theory, practice and policy*, San Francisco: Institute of Contemporary Studies.

Bridge, G., & Perreault, T. (2009). Environmental governance. In: N. Castree et al. (Eds.). *Companion to Environmental Geography* (pp. 475–397). Oxford, UK: Blackwell.

Buitrago, E. (2012). *Entre el Agua y el Oro: Tensiones y Reconfiguraciones Territoriales en el Municipio de Vetas*. Santander, Colombia: Universidad Nacional de Colombia.

Büscher, B., & Fletcher, R. (2015). Accumulation by conservation. *New Political Economy*, 20(2), 273–298.

Büscher, B., Sullivan, S., Neves, K., Igoe, J., & Brockington, D. (2012). Towards a synthesized critique of neoliberal biodiversity conservation. *Capitalism Nature Socialism*, 23(2), 4–30. doi:10.1080/10455752.2012.674149

Cárdenas, J. C. (2009). *Dilemas de lo Colectivo. Instituciones, pobreza y cooperación en el manejo local de los recursos de uso común*. Bogotá: Uniandes.

Cárdenas, J. C. (2011). *Water and economy national meeting: Water, a heritage which circulates from hand to hand*. Conference, Banco de la República, October 6–7. Retrieved from www.banrepcultural.org\\agua\\encuentro-videos.html

Cárdenas, J. C. (2014). Discussion about delimiting the Santurban páramo. Conference, Andes University, Feb. 2. Retrieved from www.uniandes.edu.co\\noticias\\informacion-general\\santurban.

Corredor, G. (2013). Locomotora minera vs. medio ambiente [Mining engine vs environment]. *Dinero*, August 8. Retrieved from http://www.dinero.com/pais/articulo/locomotora-minera-vs-medio-ambiente/181896

Cremers, L., De Theije, M., & Kolen, J. (2012). *Small scale gold mining in the Amazon Basin. Panorama from Bolivia, Brazil, Colombia, Peru, and Suriname. Cuaderno Series.* Amsterdam: CEDLA.

Crow, B., Lu, F., Ocampo-Raeder, C., Boelens, R., Dill, B., & Zwarteveen, M. (2014). Santa cruz declaration on the global water crisis. *Water International, 39*(2), 246–261. doi:10.1080/02508060.2014.886936

De Castro, F., Van Dijck, P., & Hogenboom, B. (2014). *The extraction and conservation of natural resources in South America. Recent trends and challenges.*, Cuadernos del CEDLA, No. 27. Amsterdam: CEDLA.

Duarte, C. (2012). Gobernabilidad Minera: cronologías legislativas del subsuelo en Colombia. Bogotá: Centro de Pensamiento RAIZAL. Retrieved from: https://gobernabilidadminera.wordpress.com/

Duarte-Abadía, B., Boelens, R., & Roa-Avendaño, T. (2015). Hydropower, encroachment and the re-patterning of hydrosocial territory: The case of Hidrosogamoso in Colombia. *Human Organization, 74*(3), 243–254. doi:10.17730/0018-7259-74.3.243

Duarte-Abadía, B., & Roa-Avendaño, T. (2014). El dilema del páramo: Diferentes concepciones en un contexto de justicia hídrica. El caso del páramo de Santurban. *Revista Javeriana, 150,* 71–76.

Echavarren, J. M. (2010). Conceptos para una sociología del paisaje. Universidad Autónoma de Barcelona. *Papers. Revista De Sociología, 95*(4), 1107–1128.

Eco Oro (2014). "Eco Oro announces boundaries of Páramo of Santurbán declared". Retrieved from http://www.ecooro.com (December 22).

Escobar. (2008). *Territories of difference place, movements, life, redes.* Durham and London: Duke University Press.

Escobar, A. (2010). *Una minga para el postdesarrollo: Lugar, medio ambiente y movimientos sociales en las transformaciones globales.* Lima: Desde Abajo.

Espeland, W. (1998). *The struggle for water. Politics, rationality, and identity in the American Southwest.* Chicago: University of Chicago Press.

Farber, S. C., Costanza, R., & Wilson, M. A. (2002). Economic and ecological concepts for valuing ecosystem services. *Ecological Economics, 41,* 375–392. doi:10.1016/S0921-8009(02)00088-5

Fernández, B. M. (2005). *Movimientos socioterritoriales y movimientos socioespaciales. Contribución teórica para una lectura geográfica de los movimientos sociales.* Buenos Aires: Clacso.

Forsyth, T., & Johnson, C. (2014). Elinor Ostrom's legacy: Governing the commons and the rational choice controversy. *Development and Change, 45*(5), 1093–1110. doi:10.1111/dech.12110

Foucault, M. (1980). Power/knowledge. Selected interviews and other writings 1972–1978. In C. Gordon (Ed.), *Power/Knowledge: Selected interviews and other writings 1972 - 1978.* New York: Pantheon Books.

Franco, B. M. (2013). Characterization and analysis of production systems in Guerrero, Rabanal and Santurban páramos. 'Páramos and life system project'. Alexander von Humboldt Institute, Bogotá. Unpublished manuscript.

Garay, L. (2013). Globalización/ Glocalización soberanía y gobernanza. A propósito del cambio climático y extractivismo minero. In V. Saldarriaga., O. Alarcon., & R. Medina (Eds.), *Minería en Colombia: Fundamentos para superar el modelo extractivista* (pp. 9–19). Imprenta Nacional: Contraloría General de la Republica.

Goff, M., & Crow, B. (2014). What is water equity? The unfortunate consequences of a global focus on 'drinking water'. *Water International, 39*(2), 159–171. doi:10.1080/02508060.2014.886355

Gómez-Baggethun, E. (2009). Perspectivas del conocimiento ecológico local ante el proceso de globalización. *Papeles De Relaciones Ecosociales Y Cambio Global, 7*, 57–67.

Hogenboom, B. (2012). Depoliticized and Repoliticized minerals in Latin America. *Journal of Developing Societies, 28*(2), 133–158. doi:10.1177/0169796X12448755

Hoogesteger, J., Boelens, R., & Baud, M. (2016). Territorial pluralism: Water users' multi-scalar struggles against state ordering in Ecuador's highlands. *Water International, 41*(1), 91–106. doi:10.1080/02508060.2016.1130910

Hurtado, R. (2010, December 20). Páramo de Santurban: El agua o el oro. In *Revista Semana, Bogotá*. Retrieved 23 February 2013 from http://www.semana.com/nacion/articulo/el-agua-oro-saturban/334294-3

Lu, F., Ocampo-Raeder, C., Crow, B., & Romano, S. (2014). Equitable water governance: Future directions in the understanding and analysis of water inequities in the global South. *Water International, 39*(2), 129–142. doi:10.1080/02508060.2014.896540

Martínez-Alier, J. (2004). Los conflictos ecólogicos distributivos y los indicadores de sustentabilidad. *Revista Iberoamericana de Economía Ecológica, 1*, 21–30.

Ministry of Environment (MMA). (2001). *Program for the sustainable management and restoration of Colombian high land ecosystems: Páramos*. Bogotá: General Direction of Ecosystems.

Ministry of Environment and Sustainable Development (2014). Resolution 2090 "Through which it delimits the Paramo's jurisdictions - Santurbán-Berlin". www.minambiente.gov.co/index.php/normativa/resoluciones

Molano, J. (2012). *Las Altas Montañas Ecuatoriales de Colombia: Reflexiones y apuestas para su defensa y continuidad*. Unpublished paper presented at the meeting of the Geography Society of Colombia. Bogotá, 20 September 2012.

Mollinga, P. (2001). Water and politics: Levels, rational choice and south indian canal irrigation. *Futures, 33*, 733–752. doi:10.1016/S0016-3287(01)00016-7

Moore, M. (1990). The rational choice paradigm and the allocation of agricultural development resources. *Development and Change, 21*, 225–246. doi:10.1111/j.1467-7660.1990.tb00376.x

Murra, J. (1972). El "control vertical" de un máximo de pisos ecológicos en la economía de las sociedades Andinas. In I. Ortiz De Zúñiga (Ed.), *Visita de la provincia de Léon de Huánuco en 1562* (pp. 429–476). *Peru*: Universidad Nacional Hermilio Valdizán.

Osborne, A. (1990). Comer y ser comido. Los animales en la tradición oral U'WA (Tunebo). *Boletín Museo Del Oro, 26*. www.banrepcultural.org\\node\\26017

Osejo, A. (2014)Characterization of actors' relations and their positioning in regards to use, management and conservation of the Santurban Páramo. 'Páramos and life system project'. Alexander von Humboldt Institute, Bogotá.

Ostrom, E. (1990). *Governing the Commons. The evolution of institutions for collective action*. Cambridge: Cambridge University Press.

Ostrom, E. (2009). Beyond Markets and States: Polycentric governance of complex economic systems. Nobel Prize Lecture, 8 December 2009, Stockholm.

Penna, J. L., & Cristeche, E. (2008). *La valoración de servicios ambientales: Diferentes paradigmas*. Argentina: Publicaciones Nacionales INTA.

Perreault, T. (2014). What kind of governance for what kind of equity? Towards a theorization of justice in water governance. *Water International, 39*(2), 233–245. doi:10.1080/02508060.2014.886843

Quintero, F. (2014). Cebolleros de Santurban se resisten a dejar de cultivar. En *El Tiempo*. Mayo, 2 de 2014. www.eltiempo.com/archivo/documento/CMS-13918957

Robbins, P. (2004). *Political Ecology: A Critical Introduction*. Oxford: Blackwell.

Rodriguez-de-Francisco, J. C., & Boelens, R. (2014). Payment for environmental services and power in the Chamachán Watershed, Ecuador. *Human Organization, 73*(4), 351–362. doi:10.17730/humo.73.4.b680w75u27527061

Rodriguez-de-Francisco, J. C., & Boelens, R. (2016). PES hydrosocial territories: De-territorialization and re-patterning of water control arenas in the Andean highlands. *Water International, 41*(1), 140–156. doi:10.1080/02508060.2016.1129686

Rodríguez-de-Francisco, J. C., Budds, J., & Boelens, R. (2013). Payment for environmental services and unequal resource control in Pimampiro, Ecuador. *Society & Natural Resources, 26*, 1217–1233. doi:10.1080/08941920.2013.825037

Saldías, C., Boelens, R., Wegerich, K., & Speelman, S. (2012). Losing the watershed focus: A look at complex community-managed irrigation systems in Bolivia. *Water International, 37*(7), 744–759. doi:10.1080/02508060.2012.733675

Schlosberg, D. (2004). Reconceiving environmental justice: Global movements and political theories. *Environmental Politics, 13*(3), 517–540. doi:10.1080/0964401042000229025

Swyngedouw, E. (2009). The political economy and political ecology of the hydro-social cycle. *Journal of Contemporary Water Research & Education, 142*, 56–60. doi:10.1111/jcwr.2009.142.issue-1

Swyngedouw, E. (2014). 'Not A Drop of Water…': State, Modernity and the production of nature in Spain, 1898–2010. *Environment and History, 20*, 67–92. doi:10.3197/096734014X13851121443445

Ulloa, A. (2002). La discusión antropológica en torno a la naturaleza, la ecología y el medio ambiente. In G. Palacio & A. Ulloa (Eds.), *Repensando la Naturaleza*. Bogotá: Universidad Nacional de Colombia.

Ungar, P., Osejo, A., Roldán., L., & Buitrago, E. (2014). Caracterización del sistema social asociado al territorio. In C. Sarmiento & P. Ungar (Eds.), *Aportes a la delimitación del paramo mediante la identificación de los limites inferiores del ecosistema a escala 1:25.000 y analisis del sistema social asociado al territorio: Complejo de páramos Jurisdicciones-Santurbán-Berlín*. Bogotá, D.C.: Instituto Alexander von Humboldt.

Vallejo, G. ((2014,). *Delimitan 98.954 hectáreas del páramo de Santurbán*. Conference, December 12, Sostenibilidad Semana.com.

Van der Ploeg, J. (2010). *Nuevos Campesinos. Campesinos e imperios alimentarios*. Barcelona: Icaria.

Vos, J., & Boelens, R. (2014). Sustainability standards and the water question. *Development and Change, 45*(2), 205–230. doi:10.1111/dech.12083

Zwarteveen, M., & Boelens, R. (2014). Defining, researching and struggling for water justice: Some conceptual building blocks for research and action. *Water International, 39*(2), 143–158. doi:10.1080/02508060.2014.891168

Diverging realities: how framing, values and water management are interwoven in the Albufera de Valencia wetland in Spain

Mieke Hulshof[a,b] and Jeroen Vos[a]

[a]Water Resources Management Group, Wageningen University, the Netherlands; [b]Acacia Water, Gouda, the Netherlands

ABSTRACT

The Albufera de Valencia is a coastal wetland in south-eastern Spain that has suffered from low water quality since the 1970s. This article explores two divergent framings or imaginaries of the Albufera as a hydrosocial territory. The first, the agro-economic waterscape framing, focuses on the economic and cultural importance of rice production. The second, the idyllic waterscape framing, emphasizes environmental values. The agro-economic waterscape frame is dominant in current water management. Stakeholders deploy highly diverging realities, and the political playing field is not level. Recognition and empowerment are the first steps towards more sustainable water management in the Albufera.

Introduction

Despite consensus on the need to recover the Albufera de Valencia, a coastal wetland in south-eastern Spain, a conflict over improving water management practices remains unsettled. This article explores how divergent framings define different hydrosocial territories (Boelens, Hoogesteger, Swyngedouw, Vos, & Wester, 2016) and, thereby, determine the formulation of problems and solutions in the conflict.

The modernization of irrigation and development of the Júcar River basin management plan, a European Union Water Framework Directive requirement, have recently increased the attention paid to the Albufera considerably. Conflicts over water transfers, and the associated electoral politics, have further intensified the debate. Despite consensus on the need to recover and preserve the Albufera, deadlock persists: exactly *what* should be recovered and preserved, and *how* to do this, is unclear.

The fuzziness of water management in the Albufera is demoralizing for users, managers and experts. After an ecological crisis in the Albufera Lagoon in the 1970s, all stakeholders agreed upon the need for sustainable management, but the goals and the ways to reach them have remained vague. In 2004, the Ministry of Environment funded a large project for sustainable management of the Albufera, and experts agreed that the objective was to recover 'the Albufera of the 1960s'. Yet the meanings attached to that formulated objective, and to the Albufera, are divergent, because in reality that 'Albufera of the 1960s' never existed. It is a social construct; when discussing the subject, each stakeholder and expert at the roundtable imagined a different reality. Research has shown that divergent framings largely determine differences in the formulation of problems and solutions (Bouleau, 2014; Tversky & Kahneman, 1981), but little has been documented on the presence and role of framing in 'agriculture versus environment' water conflicts in the specific context of wetlands. For example, Jury and Vaux Jr (2007), Laurance, Sayer, and Cassman (2014), and Lemly, Kingsford, and Thompson (2000) provide excellent descriptions of the global agriculture–environment conflict, but are limited in their consideration of the various stakeholders. Others, such as Beilfuss and Brown (2010), Duvail, Médard, Hamerlynck, and Nyingi (2012), Gowing, Tuong, and Hoanh (2006), and Namaalwa, Van Dam, Funk, Ajie, and Kaggwa (2013), are more specific about wetlands, but focus on economic uses and pay little attention to stakeholders' perspectives.

This research draws upon the cognitive approach developed by Minsky (1975). We define frames as distinct, coherent sets of meanings (Snow & Benford, 1988; Snow, Rochford, Worden, & Benford, 1986) and focus on their discursive evidence (Dewulf et al., 2009). We describe the two most important framings by first listing their reasoning and framing devices and then elaborating the corresponding storylines. We also look at framing dynamics and dominance, because these processes co-determine the past, present and future of the Albufera.

The article proceeds with an introductory description of the Albufera, summarizing its history, users, institutional framework and the political arena. The third section provides a theoretical background on framing. The fourth is dedicated to the results of this research. The two framings are presented and described; then, the dynamics and dominance issues and their impacts on water management are developed. The final section presents a discussion and conclusions.

Background on the Albufera de Valencia

History

The Albufera de Valencia (Figure 1) was declared a natural park on 23 July 1986, and consists of the Albufera lagoon (2800 ha, averaging 90 cm deep – from here on, 'the lagoon'), 18,000 ha of rice paddies and a one-kilometre-wide dune system, the Devesa del Saler (hereinafter, the *devesa*).

In previous times, the lagoon was a marine bay, with the Júcar and Túria Rivers discharging into the sea just north and south of it. Sedimentation processes throughout the Pleistocene and Holocene led to the formation of sand barriers. Around 6000 BP these barriers cut the bay off from the sea, forming the lagoon (Rosselló-Verger, 1972). Until the Middle Ages, the lagoon was brackish and primarily used for fishing and hunting (Sanchis-Ibor, 1998). Rice production was restricted to the northern and southern boundaries, and irrigated directly with river water. From the fifteenth century onwards, irrigation systems were developed all around the lagoon, strongly augmenting the inflow of freshwater. The lagoon turned fresh in the seventeenth century, following

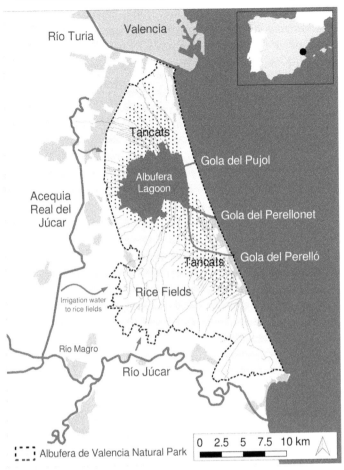

Figure 1. The Albufera natural park and lagoon, the *devesa* and the surrounding rice fields. The rice fields below surface water level are known as *tancats*. (Source: prepared by the authors.)

the construction of a (temporary) fixed closure through the main *gola* (a drainage canal that connects the lagoon to the sea); the lowering of sea level and increased precipitation in the Little Ice Age; completion of the Acequia Real del Júcar Irrigation System; and development of irrigation systems in Cullera and Sueca (Mondría-García, 2004; Sanchis-Ibor, Jégou, & Pech, 2007; Soria, 2006). The new ecosystem resulted in a dramatic loss of revenues from fisheries. To compensate for these losses, the monarchy actively promoted a shift towards rice farming. To expand the farming area, rice farmers started draining parts of the lagoon to create new farming plots, called *tancats*. The transition from fisheries to rice farming was very political (Boelens & Claudín, 2015). In 1761, the king transferred control over the lagoon's water level, crucial for rice production in the *tancats*, from the fishermen to the rice farmers (Sanchis-Ibor, 2001).

The water level for rice production, fisheries, hunting, gathering of herbs and plants and collection of firewood, amongst other practices, were actively managed by those sourcing from the area. In 1911, the lagoon and *devesa* were acquired by the municipality of Valencia. In the 1970s, pollution from exponentially growing urban and industrial compounds around the Albufera, and expansion and intensification of agriculture, led to an ecological crisis in which the lagoon turned hypertrophic (chlorophyll *a* levels exceeded 485 µg/L), all submerged macrophyte vegetation disappeared and many autochthonous fish species died out (Sanchis-Ibor, 2011; Verdú-Vázquez, Sanchis-Ibor, & Marco-Segura, 1999). Since then, construction and expansion of effluent collection and treatment infrastructure have partly resolved these issues.

The Albufera and the challenges it currently faces are a coproduction of nature and man (for comparative purposes, see also Boelens, Hoogesteger, and Rodriguez de Francisco, 2014; Boelens, 2014), the result of socio-ecological processes that have occurred over various spatial and temporal scales (Budds & Hinojosa, 2012): people's activities, uses, decisions and interventions have constituted the Albufera as a material and social object.

Water users

A number of user groups have a stake in the Albufera's water management. These include fishermen, farmers, hunters, tourists and environmentalists.

Fishermen's activities in the Albufera were legally recognized in 1250 when King Jaime I appropriated the wetland for the crown. Common people were allowed to fish, subject to a tax of one-fifth of the catch (Sanchis-Ibor, 1999). Until the mid-twentieth century, the economically most important fish species were eel, mullet and sea bass. The ecological crisis and reduced connection with the sea led fish stocks to collapse. Nowadays, there are 50 to 100 active fishermen in the Albufera. For fishermen, improving water quality, restoring the ecosystem and reopening the *golas* are fundamental (for an in-depth analysis, see Boelens & Claudín, 2015).

Moorish farmers introduced rice to Spain more than 700 years ago. At times, rice production was promoted, as in 1273 when King Jaime allocated some of the wetland's fields to his followers after reconquering Valencia from the Moors. At others, the crown was forced to ban rice production because of the high incidence of malaria (Girona, 1998). Despite these bans, the area under rice production expanded over the centuries. Construction of irrigation canals, siphons, dams and water level regulation infrastructure increased the freshwater available for rice production. Sedimentation and construction of *tancats* made it possible to expand rice fields into the lagoon.

Farmers need water for irrigation at water levels that are adjusted to the rice production cycle. These days, because of high costs and low revenues, rice farmers depend on European Union Common Agriculture Policy (EU-CAP) subsidies to make any profit.

Hunters have been active in the Albufera ever since humans first inhabited its surroundings. For centuries, hunting provided food and extra income for local inhabitants. Up to the nineteenth century, deer, wild boars, wild goats, hares, partridges, ducks and many others were abundant. Nowadays, hunting is allowed only in the rice fields and mainly targets birds. Thus, hunters are very interested in water levels that are favourable to bird populations.

The Albufera has been popular for tourists since the 1970s. City dwellers from Valencia visit the Albufera on weekends to walk and cycle in the *devesa*, lunch in El Palmar, play golf, take a boat trip or lounge on the beach. Other Spaniards come to the Albufera, mainly during holidays. Tourists want a 'beautiful' lagoon and clean beaches.

Environmentalists advocate recovery and preservation of the Albufera because of the variety and importance of its habitats, flora and fauna. The Natura 2000 Assessment Report of the Albufera (CITMA, 2004). lists 16 habitat types and 95 species that should be protected, including many migratory birds. Environmentalists argue that nature should be protected and that, therefore, water management should be adjusted to the natural environment's needs. This would include opening the water control gates in the *golas* more regularly to allow water to flow freely between the Albufera and the sea.

Institutional setting

Several institutions share in managing the Albufera. Formally, three governmental organizations are involved in park management: the Confederación Hidrográfica del Júcar (Júcar River Basin Management Authority), the Dirección del Parque Natural de La Albufera de Valencia (Management of the Albufera Nature Park) and the Servicio Devesa Albufera (Albufera Devesa Service). Alongside this formal state apparatus, fishing and farming associations manage water resources more locally.

In Spain, water is property of the central state, but the Ministry of Environment delegates water management tasks to river basin management authorities. The Júcar River Basin Management Authority (JRBMA) is responsible for managing the Júcar River basin. The JRBMA is responsible for developing a river basin management plan to achieve good ecological status in all water bodies, constructing large-scale hydraulic infrastructure, monitoring water quality, allocating water in the Júcar River basin, and overseeing compliance with EU and national norms and regulations. It is also responsible for allocating and providing water to users in the Albufera.

In Spain, the environment is autonomous communities' responsibility. The Management of the Albufera Nature Park (MANP) is part of the autonomous community (regional government) of Valencia. The MANP manages the park to achieve the Natura 2000 objectives, Ramsar[1] commitments and the nature park's overall sustainability.

As the owner of the lagoon and the *devesa* (Sanchis-Ibor, 2011), the municipality has a technical department, the Albufera Devesa Service (ADS), for management purposes in the park. The ADS is involved in recovering the *devesa* and lagoon, forestry management, preventing wildfires, and all other projects that might have an impact on the lagoon or *devesa*.

Within the Albufera there are three fishing associations, the largest being the Comunidad de Pescadores de El Palmar (Fishing Community of El Palmar). With their roots in the medieval period, fishing associations have their own customary norms and rules. They distribute fishing spots through a yearly lottery and arrange storage and trading.

Rice farmers in the Albufera are organized in associations: the Junta de Desagüe (Drainage Board) and various *comunidades de regantes* (irrigators' associations). The Drainage Board was established in 1862; its main responsibility is to regulate the lagoon's water level according to the rice production cycle in the *tancats*. The water level is regulated by means of sluices and pumping systems installed in the *golas*. Formally, the Drainage Board is chaired by the mayor of Valencia; however, day-to-day management is organized by farmer representatives. The irrigators' associations manage irrigation water in the regular rice fields, negotiate with the JRBMA on allocating water for rice production, control and maintain irrigation infrastructure, mediate product sales and provide farmers with the required fertilizers and plant health products.

Water management and the political arena

In the following sections, 'water management' refers to decision making, investments, operation and maintenance that directly or indirectly influence the Albufera water body, qualitatively or quantitatively. This water management is multi-scale (local, municipal, regional, national and international institutions are involved) and multi-faceted (involving policies, investments, subsidies, regulation and monitoring). The political arena of the Albufera embodies all these scales and facets, including:

Water allocation – applied for by irrigators' associations and environmentalists, planned and granted by the JRBMA, reported to the EU in the Júcar river basin management plan

Inflow of (partly) treated effluents – allocated by the JRBMA, reported to the EU, treated by the autonomous community, used by farmers

Water quality monitoring – done independently by the JRBMA, the MANP and the ADS

Water levels in the lagoon – regulated by the Drainage Board, supervised by the municipality

EU-CAP subsidies – framework and financing provided by the EU, translated to the situation of the Albufera by the autonomous community, applied for by and granted to the farmers.

Water management is high on Spain's political agendas, especially during elections.[2] All formal institutions involved in Albufera's water management are led by politicians. The president of the JRBMA is appointed by the central government in Madrid after national elections; the MANP is appointed by (and thus accountable to) the regional government; and the ADS is liable to the Municipal Board. Up to mid-2015 all three bodies were chaired and governed by the Partido Popular, the conservative Christian-democrat political party. Through this party-based politicization of water management, the electorates of the municipality of Valencia, the autonomous community of Valencia, and Spain are part of the Albufera's political arena. Citizens are informed about the Albufera mainly through education, local press reports, the Spanish historical novel *Cañas y Barro* by Blasco Ibáñez (1902) and the popular TV series based on it, information boards installed throughout the park, demonstrations and protests in the city related to water management, and electoral speeches.

About framing

We draw upon the concept of *framing* to analyze how actors understand the Albufera, and how they formulate problems and solutions according to this understanding. We adopt a constructivist point of view with a Foucauldian take, meaning that we assume that what actors see, know and understand is historically, socially and politically constructed (Bakker, 2012; Boelens, 2014; Feindt & Oels, 2005). Depending on their background and the context, people employ different systems of meanings, symbols, values, precepts and concepts to frame reality to define "the order of things" Foucault

(1970) and, depending on these, highlight and downplay different aspects of a certain complex issue (Boelens, 2014). Each person forms a different image of reality, and thus multiple contested realities develop (Feindt & Oels, 2005; Tversky & Kahneman, 1981). Realities are, thus, products of the human mind.

Frames are locations of tension and struggle; they are not static; they can serve as tools either to create opportunities or to limit freedom. Bakker (2012) describes, for example, how water-related hygiene framing in colonial contexts underscored the superiority of whites, justified segregated water supply systems and helped develop overtly racist population management strategies. At a certain moment in time, one framing is mostly dominant within a certain political arena (Foucault, 1970; see also Gramsci, 2001). In water management, this dominance is reflected in its material, financial and institutional dimensions.

Cognitive approach

The way the political arena understands a reality, in this case the Albufera, has a considerable impact on the formulation of solutions, policies and institutions (Bouleau, 2014; Tversky & Kahneman, 1981). Identification, analysis and understanding of these framings can help identify common ground and sore points, and to prioritize, negotiate and construct positive change (Béland, 2009; Budds & Hinojosa, 2012; Candel, Breeman, Stiller, & Termeer, 2014; Dewulf et al., 2009; Feindt & Oels, 2005).

In this research, we define framing as the construction of a relatively distinctive and coherent set of meanings attributed to a certain concept (Snow & Benford, 1988; Snow et al., 1986). We apply a cognitive approach, i.e., we focus on the discursive evidence that reality is organized and interpreted according to earlier learned schemas (Goffman, 1974, qtd. in Benford and Snow, 2000). The cognitive approach was developed by Minsky (1975), but rooted in ideas on cognitive psychology developed already in 1932 by Sir Frederic Bartlett (Dewulf et al., 2009).

Frame package analysis method

The frame package analysis method was used to identify the two principal frames applied by water users, management institutions and the electorate in the Albufera. To identify and describe each frame, we look at its identity kit of reasoning and framing devices (Table 1) (Van Gorp, 2007). *Reasoning devices* deal with how an actor conceives of a causal chain. In this regard, characterizing Albufera's system, identifying the problem, possible solutions, non-solutions and the underlying moral basis are deduced for each frame (for an earlier application of the methodology, see Candel et al., 2014). *Framing devices* are discursive identifiers of a certain frame. Drawing on the studies of Gamson and Lasch (1983) and Van Gorp and Van der Goot (2012), we look at the following framing devices: key concepts (repeatedly used concepts); verbal devices (concepts with a normative overtone); and metaphors (comparisons used to strengthen an argument). These framing devices can be

Table 1. Empty table used for identification of framing devices.

FRAMING DEVICES			
	Issue	Identity	Process
Key concepts			
Verbal devices			
Metaphors			

spoken, written or depicted (Candel et al., 2014; Van Gorp, 2007) and can relate to (Dewulf et al., 2009):

- Issues – the meanings that are attached to agenda items, events, problems or other substantive issues in a conflict situation
- Identity – the ways parties represent themselves and others
- Process – the interaction process in a conflict situation.

The major difference between reasoning and framing devices is their visibility. Reasoning devices are hidden and can be identified only through careful reading, context analysis and deduction techniques, while framing devices are directly identifiable in texts or figures.

This study used interviews, newspaper articles, websites, blogs, academic papers, project reports, government regulations, minutes of governmental debates, informal talks, and observations to discern framing and reasoning devices. The data were collected between November 2011 and July 2014: four months of continuous field research in 2011–2012 and yearly follow-up field research for three weeks. Semi-structured interviews (recorded and transcribed) were held with user groups, technicians, policy makers and scholars involved in Albufera's water management. The information gathered during these interviews was complemented with informal talks with fishermen, farmers and tourists. The literature studied was variously suggested by interviewees, retrieved from academic journals, accessed via the blog www.albufera.com and provided by technicians involved in Albufera's water management. Observations were collected during field visits and during a conference on rice farming practices in the Albufera in January 2012.

Based on the data collected, Tables 1 and 2 were completed and analyzed based on the above theoretical strategy. Actors are found to deploy one of the frames when they show discursive, documentary or managerial evidence of the corresponding framing and reasoning devices. Historical and current water management practices are described in relation to frame dominance and dynamics.

Results: the two framings of the Albufera hydrosocial territory

This section presents the two most prominent framings present in the political arena of the Albufera and some reflections on framing dynamics and dominance, with the actors found to deploy each frame. The tables describe the framing and reasoning devices used

Table 2. Empty table used for identification of reasoning devices.

REASONING DEVICES				
Characterization system	Identification problem	Possible solutions	Non-solutions	Moral basis

and provide an elaborate description. The section describing framing dynamics and dominance is ordered along the material, financial and institutional dimensions of water management.

Agro-economic waterscape frame

The *agro-economic waterscape* frame focuses on water's agronomic production capacity in the Albufera (Tables 3 and 4). The framing revolves around the economic and cultural value of rice production and water's role therein. The irrigators' associations, the Drainage Board, the JRBMA, the MANP, the municipality and the ADS show evidence of upholding this frame. Transcriptions of interviews with these actors are replete with the concepts, devices and metaphors included in Table 3. In addition, at least one line of reasoning they shared during the interviews and which they used to support their decisions is in concordance with the devices provided in Table 4.

The economic importance which farmers have been attributing for over 700 years to rice production in the Albufera is well illustrated by an historic quote. When in 1753 a prohibition of rice production was decreed, to combat yet another malaria outbreak, the population massively violated this regulation, alleging that they would "rather be ill with a full stomach than be hungry only" (Girona, 1998). Still valid today, the quote shows the dependency of the Albufera's population on rice farming for making a living over the centuries. Nowadays, there are approximately 4000 rice farmers in the Albufera (personal communication, B. Dies, 26 June 2014). Over time, these farmers have arranged the distribution of water and the cycle of water levels amongst them in a way they believe is participatory and democratic. Because their fields are interconnected, farmers depend on each other for field preparation, planting and harvesting. The agro-economic waterscape frame highlights the cultural importance of this

Table 3. Table of framing devices corresponding to the agro-economic waterscape frame.

FRAMING DEVICES			
	Issue	Identity	Process
Key concepts	irrigation, paella, economy, agriculture, tradition, drainage, collective, rice, gastronomy, livelihood, productivity, globalization	self: hard-working, innovative, flexible others: outdated fishermen, naive ecologists, bothersome bureaucrats	trivial, insignificant, boring hassle
Verbal devices	babbling water, tuned over centuries, empirical knowledge, supporting life cycle of birds, cultural heritage, rice straw management, ancient rights	irrigation engineers, irrigators' associations, Drainage Board, social and democratic, collective enterprise, day-to-day managers	We'd rather be ill with a full stomach than be hungry only (protests against prohibition of rice production during the 1753 malaria crisis)
Metaphors	freshwater is farmers' gift, water embraces the soil, natural water treatment plants, natural filters	pro-ducks (about the ecologists/ biologists working in the park), the *Gran Señor* with the money (about the EU)	

Table 4. Table of reasoning devices corresponding to the agro-economic waterscape frame.

REASONING DEVICES

Characterization system	Identification problem	Possible solutions	Non-solutions	Moral basis
Traditional but costly agro-economic production system is essential for farmers to make a living	Cost-efficiency, environmental rules and regulations, globalization of the market, water shortages, salinization	Subsidies and support, water allocation, irrigation with treated effluent	Involvement of outsiders, social agitation	Cultural and economic value of rice paddies should be safeguarded

well-coordinated collective enterprise and emphasizes the extraordinary empirical knowledge of the environment and the production process that farmers have gathered over time.

The agro-economic waterscape frame describes the Albufera's development in terms of agriculture, irrigation and drainage, and stresses how the current freshwater ecosystem is rooted in the rice-growing tradition. The role of irrigation and drainage in turning the lagoon into a freshwater body is cited repetitively: "Freshwater is a farmers' gift" (anonymous farmer, April 2011). With this type of statement, actors emphasize the importance of water surpluses from agriculture for the functioning of the whole ecosystem. Farmers have had legal authority over management of the *golas* since 1761. They conduct water towards the Albufera through their network of irrigation canals and organize day-to-day lagoon water level management. In addition, the water is filtered, purified and oxygenated while flowing through the rice fields. According to proponents of this frame, without these contributions, the wetland would now be an infertile wasteland of swampy pools.

Finally, the cultural aspect of rice production also refers to a gastronomic tradition. It is popular for both Valencianos and (weekend) tourists to visit the villages in the Albufera to enjoy a nice lunch or dinner. Proponents of the frame stress the importance of revenues from these visits for the local economy.

The main challenges to the Albufera agro-economy are the low prices paid for their products and the increasingly high input costs. In past decades, globalization has severely increased competition in the market and lowered the prices. This, in combination with high pumping costs, means that rice production is profitable only thanks to subsidies of €1100/ha per year granted by EU-CAP. However, these subsidies come with environmental restrictions, which are very difficult to comply with (e.g., burning rice straw is prohibited). Other challenges include water shortages and possible salinization problems in the near future, but those are not acute. However, in the long term, this frame argues that continued financial support and allocation of sufficient water are crucial for the agro-economic system. Proponents request review of the environmental interventions underlying the support, and more specifically their translation into specific norms and regulations. They say that special attention should be paid to applicability and feasibility vis-à-vis other legislation and economic profitability. Further, according to the actors engaging with this frame, the JRBMA must allocate sufficient water to Albufera farmers to uphold rice production.

With regard to non-solutions, proponents of this frame especially oppose involvement of outsiders and disproportional social agitation. They say coordination and negotiation are best to come up with a plan to safeguard the Albufera's future economic and cultural value.

Idyllic waterscape frame

The *idyllic waterscape* frame sees the Albufera as a precious natural park that should be recovered and preserved for future generations; it attributes high value to nature. The frame focuses on the Albufera's environmental and cultural value. The framing and reasoning devices corresponding to this frame (Tables 5 and 6) are deployed by environmentalists, the ADS, tourist operators and fishermen.

Regarding the issues, the frame stresses the high biodiversity, the importance of autochthonous species and the existence of exceptional habitats (e.g., the reed islands or *matas*) in the park. Regional, national and international recognition is recurrently highlighted, including the Natural Park Declaration (1986), inclusion in the Ramsar network of wetlands (1989), and appointment of the area as a protected bird sanctuary (*zona especial para protección de las aves*, 1994) and as an important bird area (2000). Sufficient water quality and quantity is repetitively stressed in relation to recovery and preservation. The situation as it was before the 1960s is idealized, and defined as the goal to aim for. Several interviewees, at a certain point, introduced a new subject with the words "When I was a boy..."

The idyllic waterscape frame regards environmental degradation as *the* problem. Proponents of this frame present inadequate water management as the direct cause of this problem, with politics as the underlying driver for this inadequacy. Water quality and water quantity, and especially their interconnection, are seen as problematic; e.g.,

Table 5. Table of framing devices corresponding to the idyllic waterscape frame.

FRAMING DEVICES			
	Issue	Identity	Process
Key concepts	biodiversity, flora, fauna, species, endemic, autochthonous, macrophytes, reed islands (*matas*), springs, preservation, fisheries, tourism, Ramsar, ZEPA, IBA, pollution, eutrophication, sludge, overexploitation, legislation, politics, elections, peace, future	self: tired, frustrated, victims others: ill-advised, traditional, economically important farmers, egocentric and mindless politicians	tedious, frustrating, mediocre, unfair, would like to see respect and compliance
Verbal devices	picturesque scenery, biologic parameters, poor water quality, residual waters, modernization of irrigation, tipping point, environmental crisis, massive fish deaths, touching is lethal, do not eat the fish, decreased water quantity, inflow decreased by 80%, shortage of water resources, management problem, fragmentation of management, coordination is lacking, no data exchange, legal backup is lacking, political left–right dichotomy, ecological base flow	When I was a child...	Estudio de Desarollo Sostenible de L'Albufera de Valencia, river basin management plan, most people are ignorant
Metaphors	bird sanctuary, oasis of peace, smelly dark-green pool	director: natural park is drowning, the first director of the natural park is like the father of the Albufera	we are floating around in a deep black hole, puppet-show

Note. ZEPA = zona especial para protección de las aves (protected bird sanctuary). IBA = important bird area.

Table 6. Table of reasoning devices corresponding to the idyllic waterscape frame.

REASONING DEVICES				
Definition system	Identification problem	Possible solutions	Non-solutions	Moral basis
Degraded natural park that has to be recovered for future generations	Water quantity, politicians, economic crisis, corruption, communication	Research, European Union, protests and demonstrations, collective action	Effluent to solve water quantity problem, restriction of technical interventions	Valencia and Europe should protect their environmental assets and natural heritage

due to accumulation of nitrogen, phosphates and alien material, the system is hyper-trophic, dissolved oxygen cycles are deregulated and pH values are much too high (Soria, 2006). Further, sediments are polluted with heavy metals and cleaning products, and concentrations of toxic un-ionized ammonia are dangerously high Mondría-García (2010). Several interviewees deployed verbal devices, including 'touching is lethal', '80% decrease of inflow', 'massive fish deaths' and 'smelly dark-green pool', to substantiate the problem. The normative tone of these words reflects the urgency that proponents of this frame feel. The frame argues that politics is the culprit; water management is controlled by politicians in water management institutions, i.e., the JRBMA, MANP and ADS. This frame argues that this political involvement and the left–right political dichotomy that is present always and everywhere in Spain are the major challenges to water management in the Albufera.

With regard to possible solutions, proponents believe that there might be three ways out: the EU forces compliance with environmental legislation; collective action reinforces the position of stakeholders holding this frame; or demonstrations and protests by environmental nongovernmental organizations gain momentum and persuade the electorate to vote for politicians who acknowledge the urgency of Albufera's problems. The EU is seen as a hegemon who, if willing, is able to force change. Further, this frame stresses the interdependency of stakeholders and the existence of closely related stakes as a basis for a solution. Collective action could help improve the environment's unacceptably weak, vulnerable position. Demonstrations and protests convincing the electorate to prioritize the Albufera during elections are more difficult to achieve. They expect only the occurrence of an extreme event to be able to create the required momentum for such prioritization to happen.

Non-solutions pointed out by proponents of this frame are allocation of effluents by the JRBMA to the Albufera and construction of new infrastructure. They believe a more integrated, long-term plan is needed.

The idyllic waterscape frame is based upon the moral conviction that Europe and Valencia should protect their environmental assets and natural heritage. This frame uses recurring framing devices such as 'we are floating around in a deep, black hole' and 'political puppet-show' to underline the hopelessness of the situation.

Frame dominance and dynamics

Frames in the Albufera de Valencia are, as elsewhere (Boelens, 2014), neither mutually exclusive nor static. Although the vast majority of the stakeholders clearly tend towards one of the frames, most of them uphold a mixture of the two. Note, for example, that the ADS evidently uses framing and reasoning devices for both the agro-economic and the idyllic waterscape frames. On the one hand, the ADS stresses that it is ridiculous to claim that rice farming is one of the major sources of pollution and that it is unfair that farmers do not know where they stand, because norms and rules are unclear. On the other hand, the ADS recovered La Mata de la Manseguerota to save endemic fauna and flora, coordinated an eel protection project, and is eradicating many invasive species for biodiversity reasons. Since each actor holds a different image of reality, it is not probable that this image will ever fully coincide with one frame; in contrast to the other actors, in the case of the ADS, neither frame was found to be dominant over the other.

Currently the agro-economic waterscape frame is dominant in the political arena of the Albufera. Materially, this is reflected in how water allocation is well organized for rice farming, but remains suboptimal for the lagoon. The Júcar River Basin Management Plan gives the hydrological requirements of the lagoon as 167 hm^3 of water per year, but does not clearly relate this number to the system's environmental requirements. Further, maximum salinity is set according to the threshold value for sustainable rice production; monitoring of water quality is geared towards ensuring that toxic components do not pollute rice fields; and water levels in the lagoon are set according to the rice production cycle in the *tancats*. This also reflects the institutional dominance of the agro-economic waterscape frame. The Drainage Board controls the *golas* and the water level in the lagoon; they decide when, where and how water flows out of the lagoon. The irrigators' associations manage the inflows to the system. The Júcar River Basin Management Plan allocates water to the lagoon, but according to representatives of the irrigators' associations no agreements are in place to convey 'environmental' water through the irrigation canals to the lagoon, which is fed by drainage water from agricultural fields (the so-called *sobrantes*, or leftovers). Finally, from a financial point of view, most actors talk about water in cost-production terms (see also Swyngedouw, 2014), and despite the inclusion of some environmental cross-compliance conditions, pricing mechanisms are fundamentally geared towards supporting agriculture.

Although the data collected cannot fully cover frame dynamics, some interesting examples of frame changes over time were described by interviewees and in documents. Historically, in Spain, technocratic hydraulic frames focusing on agronomic production were dominant (Boelens & Post Uiterweer, 2013; Swyngedouw, 2014). However, changes and shifts in framing are continuous – at times, large and immediate, at other times, small and prolonged. Alerted by the environmental crisis in the 1970s and the adoption of the EU rules and regulations (e.g. EU WFD, Natura 2000), environmental concerns temporarily took a more prominent place in the Albufera's political arena. More recently, with the emergence of the global economic crisis in 2008, the political arena shifted back towards more agro-economic oriented frames, and it continues that way to this day.

'Local' frames are influenced by national, European and worldwide events, and the views that follow from them. One way these dynamics are expressed in the Albufera at present is through increased tension between agriculture and environment. The economic crisis and lower profitability of agriculture favour a frame primarily based on economic valuation, whereas the EU is pushing for more attention to the environment.

Further, we encountered a set of four interesting examples of practices put to work to actively promote the one or the other frame. First, many proponents of the agro-economic waterscape frame trivialize pesticide and fertilizer use in rice paddies. They claim that the paddies function as helophyte filters and absorb excess chemicals, improving water quality before it flows into the lagoon. Second, fishermen employ discursive practices in favour of the idyllic waterscape frame. They claim to be the longest-standing stakeholders in the Albufera, and therefore entitled to a healthy, diverse ecosystem. Third, the municipality of Valencia has submitted an application to UNESCO for the Albufera to become a natural world heritage site. By means of this process, and its eventual approval, the applicants aim to highlight the importance of good water quality, the urgency of recovering cultural heritage, and the need to care for traditional economic activities. The material-symbolic construction of environmental and cultural distinctiveness and UNESCO's status are deployed to persuade others to adhere to the idyllic waterscape frame. And, finally, in the fourth example, the idyllic waterscape frame blames agriculture for being the most voluminous user of freshwater. In a water-scarce area such as Valencia, where water shortages are widely recognized as highly problematic, this is an explicit attempt to pull down the legitimacy of the agro-economic waterscape frame.

The shifts described above affect the order of things – here, the order of water management (Foucault, 1970). The dominant agro-economic waterscape frame prior-itizes socio-economic and cultural components over the environment and helps legit-imize the corresponding water management arrangements. Authority, practices, control, rules and rights contributing to rice farming are supported by this frame which would otherwise not be accepted (Boelens, 2014).

Discussion and conclusions

At the very start of this research, we found that water management in the Albufera de Valencia is highly fragmented and falls short of requirements: the management frame-work is unclear; stakeholders are highly dissatisfied with water management practices and rules; regulations and frameworks are not observed. Our realization that the coordination and tuning necessary for sustainable development are missing, and that stakeholders' realities diverge, defined the focus of this research. This article has explored how actors framed the Albufera hydrosocial territory, which values related to the frames, and how current water management is based upon the most dominant frame. Three general conclusions can be drawn.

First, using a cognitive approach (Minsky, 1975), two major frames were identified, which we have named *agro-economic waterscape* and *idyllic waterscape*. The agro-economic waterscape frame focuses on the economic and cultural value of rice produc-tion, highlighting the historical importance of rice production for Albufera's livelihoods and the role of irrigation and drainage in turning the system into a freshwater wetland. Proponents of this frame make the Albufera's future dependent on continued rice farming, which depends on continued EU-CAP subsidies. The idyllic waterscape frame places environmental value at the forefront for Albufera's system and argues that there is a moral obligation to preserve this value for future generations. This frame identifies an extensive series of acute problems (e.g. pollution, decreasing biodiversity, collapse of fisheries) that must be tackled now if people want to continue enjoying and using the Albufera. A comparison of the reasoning devices underlying the two frames suggests that framing plays an important role in water management (see also the discussion in Groenfeldt, 2006). The problems identified by the two frames are far apart, and so are the proposed solutions. Solutions put forward by one frame are even identified as non-solutions by the other (e.g., allocation of treated effluent for irrigation purposes, and involvement of outsiders). However, explicit recognition of the frames and their role in water management is missing from the discussion. Especially in the more formal documents and large-scale projects, attempts are made to integrate imaginaries and materializations that have not been thoroughly defined, explained or acknowledged first.

The second conclusion concerns the framing dynamics and, more specifically, relative adherence to the one and the other frame. The content and dominance of the agro-economic waterscape frame and the idyllic waterscape frame vary over time. Some of these changes can be traced back to a broader context, to what happened and is happening on other temporal and spatial scales. The adoption of environmentally friendly EU rules and regulations, obliging governments to comply, and the worldwide intensified debate on sustainability, for example, increased adherence to the idyllic waterscape frame in the Albufera. Conversely, the outbreak of the global economic crisis in 2008 contributed to the current dominance of the agro-economic waterscape frame.

Third, we find that the agro-economic waterscape frame is now dominant in the Albufera's political arena. Water management clearly follows the reasoning devices of this frame: water levels are set according to the rice production cycle, and the Drainage Board firmly holds institutional hegemony. This dominance can be explained largely by an historic-contextual thread (rice–food–productivity–economy) and the absence of a level playing field. In line with international trends and EU requirements, multi-stakeholder negotiations are advocated to achieve sustainable development in the Albufera (Edmunds & Wollenberg, 2002; Warner & Simpungwe, 2003). Dialogue and consensus-building to define problems, design solutions, and implement, monitor and evaluate action plans are assumed to be essential to democratic decision making (Hemmati, Dodds, Enayati, & McHarry, 2002). However, multi-stakeholder negotiation approaches assume neutral conditions, while different recognition of their values do exist between Albufera's stakeholders, resulting in different levels of power to legitimize claims and ideas (see also Warner 2006). To level the playing field, Edmunds and Wollenberg (2002) and Warner and Simpungwe (2003) suggest holding meetings at neutral venues, providing transport services, implementing open, accessible ways of communication and information exchange, organizing impartial facilitation and establishing shared rules of order. However, in practice, mutual recognition of values is hard to achieve in this way. Edmunds and Wollenberg (2002), therefore, claim that besides levelling the playing field, it is fundamental to acknowledge the power relations related to legitimacy of frames and actively work to empower the disadvantaged. In the case of the Albufera, currently, this would mean empowering stakeholders holding the idyllic waterscape frame. This empowerment, however, will be difficult and can probably be achieved only by hegemonic, coercive international community intervention.

All together, these conclusions shine a light on the present deadlock. Our findings show that the intended joint, consensual water management is hard to reach because stakeholders deploy highly diverging realities and the playing field is uneven, and cannot be fully levelled. Recognition and empowerment are, therefore, the first steps towards more sustainable water management in the Albufera de Valencia.

Notes

1. The Ramsar Convention is an intergovernmental treaty that provides a framework for conservation and wise use of wetlands and their resources.
2. Contested arenas include the planning of inter-basin water transfers, water allocation plans, construction of desalinization plants, and the need for stricter regulations and compliance in effluent treatment.

Acknowledgements

We are deeply indebted to all our interviewees for sharing their knowledge, vision and feelings with us.

References

Bakker, K. (2012). Water: Political, biopolitical, material. *Social Studies of Science, 42*(4), 616–623. doi:10.1177/0306312712441396

Beilfuss, R., & Brown, C. (2010). Assessing environmental flow requirements and trade-offs for the Lower Zambezi River and Delta, Mozambique. *International Journal of River Basin Management, 8*(2), 127–138. doi:10.1080/15715121003714837

Béland, D. (2009). Ideas, institutions, and policy change. *Journal of European Public Policy, 16*, 701–718. doi:10.1080/13501760902983382

Benford, R. D., & Snow, D. A. (2000). Framing processes and social movements: An overview and assessment. *Annual Review of Sociology, 26*, 611–639. doi:10.1146/annurev.soc.26.1.611

Blasco Ibáñez, V. (1902). *Cañas y Barro*. Valencia: Prometeo.

Boelens, R. (2014). Cultural politics and the hydrosocial cycle: Water, power and identity in the andean highlands. *Geoforum, 57*, 234–247. doi:10.1016/j.geoforum.2013.02.008

Boelens, R., & Claudín, V. (2015). Rooted rights systems in turbulent water: The dynamics of collective fishing rights in La Albufera, Valencia, Spain. *Society and Natural Resources, 28*(10), 1059–1074. doi:10.1080/08941920.2015.1024370

Boelens, R., Hoogesteger, J., & Rodriguez de Francisco, J. C. (2014). Commoditizing water territories: The clash between the Andean water rights cultures and payment for environmental services policies. *Capitalism Nature Socialism, 25*(3), 84–102. doi:10.1080/10455752.2013.876867

Boelens, R., Hoogesteger, J., Swyngedouw, E., Vos, J., & Wester, P. (2016). Hydrosocial territories: A political ecology perspective. *Water International, 41*(1), 1–14. doi:10.1080/02508060.2016.1134898

Boelens, R., & Post Uiterweer, N. C. (2013). Hydraulic heroes: The ironies of utopian hydraulism and its politics of autonomy in the Guadalhorce Valley, Spain. *Journal of Historical Geography, 41*, 44–58. doi:10.1016/j.jhg.2012.12.005

Bouleau, G. (2014). The co-production of science and waterscapes: The case of the Seine and the Rhône Rivers, France. *Geoforum, 57*, 248–257. doi:10.1016/j.geoforum.2013.01.009

Budds, J., & Hinojosa, L. (2012). Restructuring and rescaling water governance in mining contexts: The co-production of waterscapes in Peru. *Water Alternatives, 5*(1), 119–137.

Candel, J. J. L., Breeman, G. E., Stiller, S. J., & Termeer, C. J. A. M. (2014). Disentangling the consensus frame of food security: The case of the EU common agricultural policy reform debate. *Food Policy, 44*, 47–58. doi:10.1016/j.foodpol.2013.10.005

CITMA. (2004). Natura 2000 Assessment Form - L'Albufera. Retrieved from http://www.magrama.gob.es/es/biodiversidad/temas/espacios-protegidos/ES0000023_tcm7-153335.pdf

Dewulf, A., Gray, B., Putnam, L., Lewicki, R., Aarts, N., Bouwen, R., & Van Woerkum, C. (2009). Disentangling approaches to framing in conflict and negotiation research: A meta-paradigmatic perspective. *Human Relations, 62*(2), 155–193. doi:10.1177/0018726708100356

Duvail, S., Médard, C., Hamerlynck, O., & Nyingi, D. W. (2012). Land and water grabbing in an East African coastal wetland: The case of the Tana Delta. *Water Alternatives, 5*, 2.

Edmunds, D., & Wollenberg, E. (2002). *Disadvantaged groups in multistakeholder negotiations.* Retrieved from http://www.wageningenportals.nl/sites/default/files/resource/disadvantaged_groups_in_multistakeholder_negotiation_edmunds_wollenberg_cifor_2002_0.pdf

Feindt, P. H., & Oels, A. (2005). Does discourse matter? Discourse analysis in environmental policy making. *Journal of Environmental Policy & Planning, 7*(3), 161–173. doi:10.1080/15239080500339638

Foucault, M. (1970). *The order of things.* New York, NY: Pantheon Books.

Gamson, W. A., & Lasch, K. E. (1983). The political culture of social welfare policy. In S. E. Spiro & E. Yuchtman-Yaar (Eds.), *Evaluating the welfare state: Social and political perspectives* (pp. 397–415). New York, NY: Academic Press.

Girona, P. (1998, 21-26 septiembre 1998). *Valores agroecológicos de la agricultura tradicional valenciana: el arroz.* Paper presented at the III Congreso de la SEAE: Una alternativa para el mundo Rural del tercer milenio, Valencia.

Goffman, E., (1974). *Frame analysis - An essay on the organization of experience.* Boston, MA: Harvard University Press.

Gowing, J. W., Tuong, T. P., & Hoanh, C. T. (Eds.). (2006). *Land and water management in coastal zones: Dealing with agriculture-aquaculture-fishery conflicts.* Wallingford, UK: CAB International.

Gramsci, A. (2001). *Further selections from the prison notebooks.* London: Electric Book Company.

Groenfeldt, D. (2006). Multifunctionality of agricultural water: Looking beyond food production and ecosystem services. *Irrigation and Drainage, 55*(1), 73–83. doi:10.1002/ird.217

Hemmati, M., Dodds, F., Enayati, J., & McHarry, J. (2002). *Multi-stakeholder processes for governance and sustainability.* Sterling, USA: Earthscan Publications Ltd.

Jury, W. A., & Vaux Jr, H. J. (2007). The merging global water crisis: Managing scarcity and conflict between water users. *Advances in Agronomy, 95*, 1–76. doi:10.1016/S0065-2113(07)95001-4

Laurance, W. F., Sayer, J., & Cassman, K. G. (2014). Agricultural expansion and its impacts on tropical nature. *Trends in Ecology & Evolution, 29*(2), 107–116. doi:10.1016/j.tree.2013.12.001

Lemly, A. D., Kingsford, R. T., & Thompson, J. R. (2000). Irrigated agriculture and wildlife conservation: Conflict on a global scale. *Environmental Management, 25*(5), 485-512`. doi:10.1007/s002679910039

Minsky, M. (1975). A framework for representing knowledge. In P. H. Winston (Ed.), *The psychology of computer vision.* New York, NY: McGraw-Hill.

Mondría-García, M. (2004). *Jornada de Debate sobre el Desarrollo Sostenible de l'Albufera de Valencia. Documento de Conclusiones.* Valencia: Confederación Hidrográfica del Júcar.

Mondría-García, M. (2010). *Infraestructuras y eutrofización en l'Albufera de València. El modelo Cabhal.* (Doctoral Tesis Doctoral), Universidad Politécnica de Valencia, Valencia.

Namaalwa, S., Van Dam, A. A., Funk, A., Ajie, G. S., & Kaggwa, R. C. (2013). A characterization of the drivers, pressures, ecosystem functions and services of Namatala Wetland, Uganda. *Environmental Science & Policy, 34*, 44–57. doi:10.1016/j.envsci.2013.01.002

Rosselló-Verger, V. M. (1972). Los Ríos Júcar y Túria en la génesis de la Albufera de Valencia. *Cuadernos De Geografía, 11*, 7–25.

Sanchis-Ibor, C. (1998). *De la gola a les goles. Canvi ambiental secular a l'Albufera de Valencia.* Valencia: Fundació Bancaixa.

Sanchis-Ibor, C. (1999). *La Albufera de Blasco Ibáñez.* Valencia: Palmart Editorial.

Sanchis-Ibor, C. (2001). *Regadiu i canvi ambiental a l'Albufera de València*. (Doctor), Universitat de València, Valencia.

Sanchis-Ibor, C. (2011). La Albufera de Valencia: Cincuenta años de eutrofia. *Mètode, Universitat De València, 70*, 32–41. http://metode.cat/es/Revistas/Dossiers/La-Albufera-de-Valencia/LAlbufera-de-Valencia. Retrieved from http://metode.cat/es/Revistas/Dossiers/La-Albufera-de-Valencia/LAlbufera-de-Valencia

Sanchis-Ibor, C., Jégou, A., & Pech, P. (2007). L'Albufera de Valencia. *Une Lagune De Médiance En Médiance. Géographie Et Cultures, 63*, 5–22.

Snow, D. A., & Benford, R. D. (1988). Ideology, frame resonance, and participant mobilization. *International Social Mov Researcher, 1*, 197–218.

Snow, D. A., Rochford, E. B., Worden, S. K., & Benford, R. D. (1986). Frame alignment processes, micromobilization, and movement participation. *American Sociological Review, 51*, 464–481. doi:10.2307/2095581

Soria, J. M. (2006). Past, present and future of la Albufera of Valencia natural park. *Limnética, 25* (1–2), 135–142.

Swyngedouw, E. (2014). 'Not A drop of Water…': State, modernity and the production of nature in Spain, 1898-2010. *Environment and History, 20*, 67–92. doi:10.3197/096734014X13851121443445

Tversky, A., & Kahneman, D. (1981). The framing of decisions and the psychology of choice. *Science, 211*(4481), 453–458. doi:10.1126/science.7455683

Van Gorp, B. (2007). The constructionist approach to framing: Bringing culture back in. *Journal of Communication, 57*, 60–78. doi:10.1111/j.0021-9916.2007.00329.x

Van Gorp, B., & Van der Goot, M. J. (2012). Sustainable food and agriculture: Stakeholder's frames. *Communication, Culture & Critique, 5*(2), 127–148. doi:10.1111/j.1753-9137.2012.01135.x

Verdú-Vázquez, A. V., Sanchis-Ibor, C., & Marco-Segura, J. B. (1999). Regadío y Saneamiento Urbano en l'Albufera de València. Análisis Cartográfico. *Cuadernos De Geografía, 65/66*, 61–79.

Warner, J., & Simpungwe, E. (2003). *Stakeholder participation in South Africa: Power to the people?* Paper presented at the 2nd Interntaional Symposium integrated water resources management: Towards sustainable water utilization in the 21st century, ICWRS/IAHS, Stellenbosch, Western Cape, South Africa.

Warner, J. F. (2006). More sustainable participation? Multi-stakeholder platforms for integrated catchment management. *International Journal of Water Resources Development, 22*(1), 15–35. doi:10.1080/07900620500404992

Disputes over land and water rights in gold mining: the case of Cerro de San Pedro, Mexico

Didi Stoltenborg[a] and Rutgerd Boelens[a,b]

[a]Water Resources Management Group, Department of Environmental Sciences, Wageningen University, Wageningen, the Netherlands; [b]CEDLA, Centre for Latin American Research and Documentation, and Department of Geography, Planning and International Development Studies, University of Amsterdam, Amsterdam, the Netherlands

ABSTRACT

This article analyzes different visions and positions in a conflict between the developer of an open-pit mine in Mexico and project opponents using the echelons of rights analysis framework, distinguishing four layers of dispute: contested resources; contents of rules and regulations; decision-making power; and discourses. Complexities in this study manifest how communities' land and water rights are circumvented by governmental bodies and ambivalent regulations favouring the large mining company. This process is importantly reinforced by international trade legislation. Multi-actor, multi-scale alliances may offer opportunities to foster environmental and social justice solutions.

Introduction

In 1996, Minera San Xavier (MSX), Mexican tributary of the Canadian mining company Newgold Inc., announced that it wanted to start a large open-pit gold and silver mine (Figure 1) in the municipality of Cerro de San Pedro, occupying 373 ha of community land. This was subject to great controversy as the scale and type of mining operation would put a heavy burden on the available land and water, not to mention adverse environmental effects. Resistance was fierce, and several opposition groups united themselves in the Frente Amplio Opositor (Broad Opposition Front, or BOF). Despite the opposition, MSX started operating in 2007. As of this writing, its presence is still being disputed.

Though mining is a highly profitable business for some actors, the downsides of mining activity are becoming more and more obvious. Environmental degradation, illegal land acquisition, water contamination, corruption, violence, resistance and conflict are often associated with mining development (Hogenboom, 2012; van der Sandt, 2009; Wilder & Romero-Lankao, 2006). *Campesino* (peasant) and indigenous communities are affected by mining activity in the area, and the livelihood strategies of mine-adjacent communities are often endangered through decreased access to and control over the land (Peace Brigades Internacionales, 2011; van der Sandt, 2009). Frequently, the economic 'benefits' promised by mining companies, for example in the form of

Figure 1. Before and after. On the left, the hill of Cerro de San Pedro before 2007, when operation of the mine started. On the right, the status of the landscape at the same time of year in 2013. The open-pit mine about 200 m from the centre of the village of Cerro de San Pedro has caused a large conflict that continues to date. (Left photo from BOF, 2013; right photo, Jesse Samaniego Leyva, 2013.)

temporary employment, do not outweigh the losses suffered (Perreault, 2014; Sosa & Zwarteveen, 2011; Yacoub, Duarte-Abadía, & Boelens, 2015). These negative effects often give rise to conflicts, and, unfortunately, conflict in a 'miningscape' is generally the rule rather than the exception. In 2013, the Observatorio de Conflictos Mineros en América Latina (OCMAL, 2014) registered 13 large-scale mining conflicts in Mexico, most of which involved foreign companies. One of these conflicts is taking place in Cerro de San Pedro.

This article elaborates how conflict arose over land and water rights between inhabitants of Cerro de San Pedro and MSX. The article examines how this 'natural resources conflict' is not just about rights to access resources, but also about underlying injustice in local, national and international rules and regulations, and about the question of the legitimacy and authority to shape these rules. It also shows how interconnected, powerful actor alliances, discourses, and knowledge claims profoundly influence the struggle over land and water in this municipality.

The article is based on literature and archival investigation throughout 2013 and field research in September–December 2013 in Cerro de San Pedro, with follow-up correspondence and conversations in 2014 and 2015. Semi-structured interviews were conducted with inhabitants of Cerro de San Pedro and surrounding villages, migrants who had left the zone, local municipality and government officials, mine representatives, mine-opposing groups, and journalists and scholars in San Luis Potosí. Next, a series of interviews were conducted (in three meetings each) with particular, representative individuals living in the mining area, whose life histories were compiled. Quotes and comments were taken from several interviews and conversations (all in Spanish and translated by the authors), which were conducted during both formal and informal encounters whilst performing fieldwork in Cerro de San Pedro (names of interviewees have been changed where needed to protect informants). The research for this article forms part of the activities of the international research and action alliance Justicia Hídrica/Water Justice.

The structure of the article is as follows. In the next section Mexico's neoliberal development, which paved the way for MSX to operate in Cerro de San Pedro, is elaborated. In the third section the background of the conflict over natural resources in

Cerro de San Pedro is discussed, after which the conceptual framework, echelons of rights analysis, used to analyze the conflict is explained. In the fifth section the results are presented. The concluding section reflects on how economic interests and associated discursive practices have had severe implications for the inhabitants of Cerro de San Pedro.

The background: Mexico, a protectionist state, takes a neoliberal path

To understand the conflict in Cerro de San Pedro, a brief look at the history of Mexico's laws concerning land, water and mining in the last century is essential. After the revolution of 1910, Mexico created a protectionist state in which land and water rights were a noncommodity. After years of unequal division of land and water under the *hacienda* system, the Mexican government expropriated the large landowners and reallocated the majority of the land and water to former day-labourers. These labourers formed farmer groups that collectively manage the resources to this day: the so-called *ejidos*, or social property sector. Under the *ejido* system, the majority of the allocated land is managed collectively whilst a small part can be cultivated for private purposes (Assies & Duhau, 2009). Under the law of *ejido* tenure, land was a non-negotiable resource. Article 74 of the Mexican Constitution stated that the ownership of common-use land is "imprescriptible, inalienable e inembargable [imprescriptible, inalienable and indefeasible]", i.e., land could not be transferred to third parties, land rights could not expire, and they could not be seized through an injunction (Herman, 2010). Water rights were linked to agricultural property rights under *ejidal* law, which means that they could not be sold, rented out, used on other lands, or used for purposes other than those stated in the grant (Assies, 2008).

However, after 1992, legislation on land and water rights changed. In the 1980s Mexico faced a severe economic crisis, and the World Bank, the International Monetary Fund and the Inter-American Development Bank demanded that Mexico adopt neoliberal policies if the country wanted to obtain international credit, similar to many other Latin American countries in the 1980s and 1990s (Achterhuis, Boelens, & Zwarteveen, 2010; Hogenboom, 1998, 2012; Wilder, 2010). The main focus of the restructuring of the economy was on opening the Mexican market to foreign investment. The social property sector and its regulatory framework of the time did not allow private ownership, as *ejidos* could not legally be privatized. This conflicted with the aim of increased foreign investment in Mexico, as land and water could not be converted into private and transferable commodities. Hence, according to neoliberal policy makers, if Mexico was to increase private (foreign) investment, legislation on land and water rights needed modification (Herman, 2010). Among others, the Agricultural Law, the Mining Law and the Foreign Investment Law were profoundly changed. To open up the mining sector to foreign mining companies, an amendment was made to Article 6 of the Mining Law that enables land to be alienated through "temporary occupancy". This provision enables mining activity to occupy land, and prioritizes mining above any other form of land use. The temporary occupancy permit is granted by Mexico's Ministry of Economics (Bricker, 2009; Herman, 2010). The 1992 market revisions paved the way for the North American Free Trade Agreement (NAFTA), which Mexico joined in 1994 (e.g. Hogenboom, 1998). Through NAFTA, foreign direct investment was stimulated greatly, and it was predicted that Mexico, as a developing country, would economically benefit most from these investments (Krueger, 1999; Ramirez, 2003). Meanwhile, for Canadian and US mining companies it became appealing to invest in Mexico due to the relatively low tax rates. It was shortly after the union with NAFTA that MSX announced its interest in exploiting the minerals in Cerro de San Pedro.

NAFTA has received criticism that environmental standards are easy to circumvent, due to the 'investor-state mechanism' that NAFTA encompasses. NAFTA aims to have investors of different countries treated equally and protected from expropriation by all levels of the (host) government. NAFTA's Chapter 11 gives an investor the right to challenge the government for noncompliance with the agreements made in NAFTA, in an international court, superseding national law. In the design phase, this mechanism was meant as a defensive measure to protect foreign companies against arbitrary and unreasonable government actions. However, it has had several deeply problematic side effects (Hogenboom, 1998, 2012; Solanes & Jouravlev, 2006, 2007). For example, first, it allows foreign companies to operate in the host country, but in case of a dispute they can go directly to the international arbitration process and entirely bypass domestic courts. Second, starting this process is relatively cheap and easy. This makes it an attractive option for foreign companies that wish to protect themselves against restrictions posed by new environmental laws or social security policies which could have a negative impact on their business (Mann & Von Moltke, 1999). The option of appealing to the international court under NAFTA is only available to companies operating under NAFTA, and not, for example, to communities or other non-business stakeholders who fear injustice, unequal competition, or socio-environmental costs (Herman, 2010; Nogales, 2002). On more than one occasion, multinational companies have used the possibility of suing governments for noncompliance with agreements made in NAFTA to prevail against environmental restrictions (Hogenboom, 1998, 2012; Solanes & Jouravlev, 2006, 2007). For example, near Cerro de San Pedro, in 2001, the American corporation Metalclad was awarded $16.5 million dollars by the NAFTA committee after the state of San Luis Potosí refused the installation of a hazardous waste transfer station. This shocked both national actors and international environmentalists (Kass & McCarrol, 2000).

The conflict in Cerro de San Pedro

Cerro de San Pedro has a long mining history. The gold and silver reserves in the area were already exploited by the indigenous inhabitants of Cerro de San Pedro, the Huachichiles, before the Spanish arrived. The Spanish conquistadores started exploiting the first mines in Cerro de San Pedro in the sixteenth century (Reygadas & Reyna Jiménez, 2008; Vargaz-Hernandez, 2006). Over time, several mining companies have come and gone in Cerro de San Pedro. Livelihoods were based on mining and agriculture, the latter practised for subsistence purposes under the *ejido* system.

The last large mining enterprise, before the current mining company, was active until 1948: the American Smelting and Refinery Company. After its shutdown, the majority of the miner families left the town to work in other mines in the north of Mexico; others went to San Luis Potosí in search of jobs. The remaining inhabitants developed new livelihood strategies, for instance based on tourism, making use of the local ecology and cultural heritage opportunities (interviews with local residents, October 2013; Reygadas & Reyna Jiménez, 2008; Vargaz-Hernandez, 2006). As soon as MSX announced that it wanted to exploit the minerals by means of an open-pit mine in 1996, opposition to the project started, involving inhabitants from Cerro de San Pedro and neighbouring villages, as well as relatives and ex-villagers now residing in cities like San Luis Potosí.

The mine covers 373 ha and consists of a large open pit, two waste dumps and a lixiviation area. In the lixiviation area, a water-cyanide solution is applied to the rock debris, dissolving the gold and silver particles. From the bottom of the heap the solution, now enriched with gold and silver particles, is drained. Eventually the water is evaporated, and what remains is a mixture of gold and silver known as doré. For this process, 16 tonnes of cyanide, dissolved in 32 million litres of water, is applied daily to the lixiviation area (Newgold Inc., 2009). MSX has water concessions for 1.3 MCM (million cubic metres) per year, but project opponents claim that the actual water extraction is much higher, and simple calculations show that the actual water needs of the mine are many times this figure (pers. comm., BOF member Eduardo da Silva, November 2014). For comparison, Interapas, which is responsible for the drinking water supply of the four municipalities close to the city of San Luis Potosí (more than a million people), has a total authorized annual extraction of 85 MCM of water (Peña & Herrera, 2008b). This use of cyanide, amongst other issues, has given rise to great opposition, as water is scarce in the area and cyanide is extremely toxic (Lutz, 2010). Years of litigation followed, during which a large number of court cases were filed, rejected, delayed or overruled by other courts. A large number of court cases were filed against MSX, questioning the legitimacy of the environmental/land use change permit, the mine's water use permit, and many other issues. Despite the mine's having lost a number of these court cases, MSX started its mining operation in 2007 (Herman, 2010; Peña & Herrera, 2008a). To date, MSX is still active in Cerro de San Pedro, but in 2015 the company announced it would gradually shut down the mine and accordingly start rehabilitation activities. These rehabilitation activities are described in MSX's project plan. They form part of a larger 'shutdown plan', in which MSX describes the process of closing the mine, and have been approved by the Secretaría de Medio Ambiente y Recursos Naturales (Secretariat of Environment and Natural Resources). Examples of planned rehabilitation activities are sterilization of the lixiviation area and reforestation of the mining pit. However, deep distrust of the mining company and the Mexican government, due to a lack of transparency and suspicion of corruption (as explained later), has caused the BOF to demand that the president of the municipality form a commission of independent experts to supervise the remediation work and its effectiveness (pers. comm., BOF, December 2015). Given the hugely problematic and controversial track record of the company, the government and other mining companies in the country in terms of (in)action regarding socio-environmental rehabilitation, this lack of trust is deep and generalized in the population.

Changing landscapes and waterscapes in Cerro de San Pedro

Unlike earlier decades' (tunnel-based) mining operations, the current open-pit mining practices have had a tremendous impact on the land and waterscape. The hill itself

(Cerro de San Pedro, the Hill of Saint Peter) – the place containing gold and silver particles – has been completely excavated (63 ha); beside it, two new hills of waste material have emerged (145 ha), and a newly constructed hill two kilometres to the south makes up the lixiviation area (120 ha). The new hills have altered the natural drainage pattern, blocking a dam and river in the village. Great amounts of dust cause severe pollution (Gordoa, 2011), and farmers in the area complain of crop failure caused by the pollution (interviews with local farmers, October 2013). The profound changes in the landscape caused by the mine were fiercely opposed. The litigation process in obtaining permits was seemingly never-ending, with courts referring to other courts and rejecting responsibility, creating a vicious cycle of court cases with no solid resolution (Herman, 2010; Peña & Herrera, 2008a). The differing opinions within the village drove a wedge between villagers, and a fully fledged conflict started in Cerro de San Pedro. Project opponents living in Cerro de San Pedro talk about cases of severe intimidation, aggression and violence against them, inflicted on them by both MSX employees and pro-MSX villagers. The economic interests in the realization of the mining project were enormous, and the national government put pressure on local authorities to issue the required permits. Oscár Loredo, the young mayor of Cerro de San Pedro, at first announced that he would not ratify municipal permits, but later changed his mind. He claimed that he was being put under great pressure by MSX, by the state, and even by the president (Vicente Fox), and that he could no longer stand the pressure. He claimed that he felt he had no choice and that his life was at risk. He added that his personal fears had made him change his mind about the matter. When challenged by one of the council members, the mayor responded: "Does my life not matter to you?" (Herman, 2010; Reygadas & Reyna Jiménez, 2008). Shortly afterwards, the municipal permits were ratified. A few years before, the former mayor, Baltazar Loredo (father of Oscár Loredo), was murdered after openly opposing the mining project (Vargaz-Hernandez, 2006).

Water availability in Cerro de San Pedro

Cerro de San Pedro and the city of San Luis Potosí are located in the hydrological watershed of the Valle de San Luis Potosí. This watershed stretches over approximately 1900 km^2 and supplies drinking water for about 90% of the San Luis Potosí population (more than a million people). The Valle de San Luis Potosí aquifer is being over-exploited: yearly, approximately 149.34 MCM is extracted from the deep aquifer, while only an estimated 78.1 MCM recharges it (Santacruz De Leon, 2008). As a way of mitigating aquifer over-exploitation, the Mexican government declared a *zona de veda* in the area.[1] *Vedas* are designed to prevent uncontrolled and unlimited water extraction from the deep aquifer, and thus aim to obtain a sustainable equilibrium, allowing human activities without degrading the environment (Conagua, 2012). Since 1961 the largest part of the Valle de San Luis Potosí aquifer has been subject to the *veda*, including the mine in Cerro de San Pedro (Conagua, 2004).

Another decree, issued 24 September 1993, designates 75% of the municipality of Cerro de San Pedro as a *zona de preservación de la vida silvestre* (zone for the preservation of wildlife). This decree was issued a few years before MSX announced its interest in exploiting the gold and silver reserves of the area (BOF, 2014). The State

Congress assigned Cerro de San Pedro and the surrounding area this protected status due to its ecological function and importance for watersheds. This was formalized in a state decree, which mandated that in 75% of the municipality of Cerro de San Pedro (1) no changes were to be made in the subsoil for a period of 20 years; (2) the area was not suited for industrial activities with high water consumption; and (3) it was acknowledged as having an important function for wildlife preservation (Gordoa, 2011; Vargaz-Hernandez, 2006).

Despite the *veda* and the watershed protection status, MSX has managed to bypass all regulations and court cases by obtaining a temporary occupancy permit and thus to start and continue mining. To illustrate the consequences, we present life histories from two affected families who took diverging paths. These stories show that some families outright opposed the transformation of their livelihoods, while others thought they could take advantage of the economic opportunities the mine would offer. After that, we present brief conceptual notes to examine the natural resource conflicts in Cerro de San Pedro.

Subsistence and opposition: the story of Doña Morena Sanchez Aguilar

"MSX arrived in 1996 and announced it wanted to start an open-pit mine. My husband and I were against the plans from the very beginning; we saw that they would destroy our surroundings, the land with which we are connected. My family descends from the indigenous Huachichiles: I belong here, this is my land. Moreover, we never saw the need to work for the mine. We saw another future for Cerro de San Pedro. Tourism was starting to pick up, and there were plans for the development of Cerro de San Pedro, such as the building of a hotel and restaurants. We didn't need the mine at all! Before MSX arrived, life was very different in Cerro de San Pedro. The town was united: we used to have dinner together, there were masses, sometimes we would dance on the square. This all changed when MSX came. They divided us. In the very beginning, almost everyone was against the plans of MSX. However, when MSX started to pay people for their 'vote', things changed. Our neighbours became our enemies! It got really violent, once they even tried to shoot my husband. The mine tried to silence the people who were against them. Houses were set on fire, our house was boarded up, all to intimidate us. Yet it was never proven that the mine was behind these things. These were really scary times, especially when our mayor was murdered. We stayed in the house and didn't meet up with our neighbours anymore. We were lonely. My husband, who was the fiercest in his opposition against the mine, passed away a year ago. Since then, the relationship with our neighbours has normalized a bit. We greet each other in the streets again. Yet we never interact with those who work for the mine. Sometimes MSX invites us for activities, but we never go. They might be starting to think we're now okay with the presence of the mine, but we are not. Moreover, we will never forget how our neighbours treated and threatened us a few years ago. This village is now a divided place. Life will never be the same here." (Interview with Doña Morena, November 2013)

Joining the opportunities offered by mining? The story of Don Vicente Estrada Diaz

"In 1996 Minera San Xavier came. They started going by our houses and told us that they were planning to start a mine here. They promised us that the mine would bring us a lot of benefits: job positions for everyone living in the village, a medical station, scholarships for our kids, and so on. I wanted a better future for my children; I wanted to give them the opportunity to study, which I never had. So to me this sounded like a great opportunity, and we agreed with the plans. My brothers and I sold about 19 hectares of our land, which we previously used to cultivate, to MSX. I am in favour of the mine out of necessity: I wanted job opportunities for me, my family and my village, and MSX was the only option we had. In other words, I am not in favour of the mine but in favour of a source of work. Of course I don't like the total change of our environment, or the contamination that mining activity brings about. Yet in MSX I saw the only way out of our poverty. Nowadays I doubt whether I should have been so positive about MSX in the beginning. Back then, I was one of the first to be in favour of the mine, and I convinced many of my neighbours. I really thought that MSX would give us a better life, yet did I know that MSX would not live up to all those promises they made? Yes, MSX improved our livelihoods, but not as much as we all expected and hoped. I feel I have let my people down: I was the one most positive about the mine, and look what we have now. We are all very worried about the contamination. The dust pollution and maybe the cyanide have severely affected our harvest. I still cultivate my fields, but the one located close to the lixiviation area hasn't given any yield at all. The plants are often covered in dust: how are they supposed to grow like that? It has rained quite a lot this year, and in other places my *milpa* grew pretty well. It must be the mine affecting my harvest, but what can I prove? I have no money or education to prove all this. We are very worried about the future. MSX is not going to operate here forever. What are we going to do when the mine is gone? The history is going to repeat itself. Everybody will leave our village, since there is no work. But this is where we were born, where we were raised and where we got married. How could we not love our land? Yet most of us stopped cultivating our land a long time ago. The best fields for cultivating were sold to MSX: now their offices are on top of it. Living off the land is a hard life and the young people are not attracted to this lifestyle anymore. On top of that, what can they do with a contaminated area? Only the old people, me, my wife and some other people, will stay here. But we have no option. When MSX goes, the life will go from our village as well." (Interviews with Don Vicente Estrada, October–November 2013)

Conceptual notes: examining the entwined layers of the natural resource management conflict

In mining areas such as Cerro de San Pedro, the use of natural resources such as land and water lies at the root of local inhabitants' livelihoods. Consequently, opposing interests and the struggle over access to and withdrawal of these natural resources has high potential for conflict. When new stakeholders, such as the MSX gold mine, enter the playing field and claim a substantial share of the resources, access rights are commonly reallocated through an interplay of (socio)legal, economic and political power. Redefinition of rights always causes some actors or use sectors to lose, such as less well-accommodated social groups, or the environment, while others reap the benefits and strengthen their political and economic positions – a basic tenet in political ecology (e.g. Forsyth, 2003; Neumann, 2005; Robbins, 2004).

In scrutinizing the conflicts that arise during the reallocation of natural resources, political ecology does not only attempt to focus on which population groups are most affected by these politics. It also aims to clarify the political forces that are at the roots of environmental distribution conflicts (e.g., Boelens, 2009, 2015; Brooks, Thompson, & El Fattal, 2007, 2013; Martínez-Alier, 2003; Neumann, 2005; Peet & Watts, 1996; Robbins, 2004; Turton, 2010; Wilder, 2010). Further, in order to grasp such political forces and their unequal outcomes in terms of resource distribution, there is also the need to focus on the ways in which environmental knowledge itself is produced, how 'knowers' are defined, by whom, and how they conceptualize 'environmental problems' and 'solutions'. All of these often implicitly generate uneven allocation of social costs and benefits. As Hajer (1993, p. 5) commented: "The new environmental conflict should not be conceptualized as a conflict over a predefined unequivocal problem with competing actors pro and con, but is to be seen as a complex and continuous struggle over the definition and the meaning of the environmental problem itself" (see also Forsyth, 2003). Different actors, with different socio-economic, cultural and political backgrounds, will commonly perceive and evaluate environmental transformations differently, and therefore use different frameworks to construct their "environmental imaginaries" (Peet & Watts, 1996, p. 37; see also Feindt & Oels, 2005).

As is common in such water extraction disputes, the conflict in Cerro de San Pedro exhibits many different levels and issues over which actors collide (Adler et al., 2007; Achterhuis et al., 2010; Brooks et al., 2007; Martínez-Alier, 2003; Turton, 2010; Wilder, 2010). To unravel the depths of the conflict, we use the echelons of rights analysis (ERA) framework (Boelens, 2009, 2015; Duarte-Abadía, Boelens, & Roa-Avendaño, 2015; Zwarteveen & Boelens, 2014). ERA can be applied to conceptually and empirically distinguish several mutually linked levels of abstraction within a natural resource management (e.g., water governance) conflict:

- The first echelon is about conflicts over access to and withdrawal of resources. In order to materialize these access and withdrawal rights, technological artefacts, infrastructure, labour and financial resources have to be in place. In this echelon the conflicts regarding access to and distribution of the resource(s) in question are examined.
- The second echelon refers to conflicts over the contents and meaning of the rules and regulations that are connected to resource distribution and management. Conflicts often occur over the contents of rules, norms and laws that determine the allocation and distribution of land, water and other territorial resources. Key elements of analysis in this field are the bundles of rights and obligations, roles and responsibilities of users, criteria for allocation based on the heterogeneous values and meanings given to the resource, and the diverse interpretations of fairness by different stakeholders.
- At the third echelon, conflicts over decision-making power are analyzed. Who is entitled to participate in questions about the division of land and water rights? Whose definitions, interests and priorities prevail? Who is able to exert formal or informal influence, and how?
- The fourth echelon is about the opposing discourses that are used by the different stakeholders to express the problems and solutions concerning land and water

rights. Different regimes of representation claim 'truth' in different ways, and so legitimize their policies, plans, actions and the distribution of the resources. This last echelon seeks to coherently link all echelons together in one convincing framework (see e.g. Boelens, 2009, 2015; Zwarteveen & Boelens, 2014).

Throughout history and across continents and cultures, political and economic elites have often sought to justify their (often highly unequal) use of the environment by making use of a discourse that upholds this use as if it were for 'the greater good'. Opposing groups subsequently challenge these elite groups by forming their own counter-discourse. Hence, as the ERA framework illustrates, environmental conflicts do not concern only material practices but are simultaneously struggles over rules, over authority, and over meaning and ideological structures. In miningscapes, as in other arenas where actors fight over natural resources, rather than a search for absolute truths about environmental problems, we witness a battle about "the rules according to which the true and false are separated and specific effects of power are attached to the true", a struggle over "the status of truth and the economic and political role it plays" (Foucault, qtd. in Rabinow, 1991; see also Boelens, 2014; Boelens & Vos, 2012; Brooks et al., 2013; Forsyth, 2003; Turton & Funke, 2008; Wilder, 2010; Vos, Boelens, & Bustamante, 2006). As we see in this case about how the mining company MSX managed to get access to land and water rights in Cerro de San Pedro, discourses are not innocent tools: they often serve to justify particular policies and practices, and obliterate alternative modes of thinking and acting.

Unravelling the conflict in Cerro de San Pedro: the echelons of rights

Conflict over access to and withdrawal of land and water

In the analysis of the conflict in Cerro de San Pedro we see that on the surface the conflict revolves around access to land and water, the quality of these natural resources, and their use purposes and practices. Conflict over access to land is importantly expressed in the false lease contract that was presented by MSX, and the subsequent temporary occupancy of *ejido* land. Mexican law holds that the surface of the land belongs to the land right holders, in this case the *ejidatarios*, yet the subsoil remains the property of the government. This means that for MSX to obtain access to the land both a mining concession for the subsoil from the Mexican government and a rental agreement with the *ejidatarios* were required (Herman, 2010). Obtaining the mining concession from the government was not a problem. And since most rightful title-holders (*ejidatarios*) had left Cerro de San Pedro after 1948, MSX had the few remaining inhabitants (*avecindados*) sign a lease contract. However, these people did not hold the land title and thus could not legally rent the land to MSX. The false lease agreement was initially accepted, but eventually, in 2004, it was declared void. Meanwhile, between 1996 and 2004, despite the lack of a legal permit to access the land, MSX continued construction activities. Eventually, in 2005, a temporary occupancy permit was granted to MSX (Herman, 2010). The means by which the land was 'temporarily occupied' was also subject to controversy. In practice the occupancy means that local inhabitants are no longer able to use this land for agricultural purposes (for which the land was originally intended under the Agricultural Law), for tourism, or for artisanal mining.

Another part of the conflict is related to water use. MSX requires a large amount of water for operation of the mine, in an area in which water is already a scarce resource. The mining activity also has large impacts on the quality of the environment. Great controversy exists over the negative consequences for the quality of the land and water in the affected area, such as contamination of surface and groundwater; dust pollution; negative impacts on flora and fauna; contamination by heavy metals; and profound change of the surface of the landscape (Gordoa, 2011; Reyna Jiménez, 2009). The question is what quality of land and water will be left to local inhabitants after MSX closes the mine. Already evident is that, as explained above, the mining practices have transformed the territory into a huge open pit, altering its current and potential land uses, changing the village itself, drying out the groundwater wells, and even blocking the small river that once meandered through town but now has been usurped by the mine.

Disputing the content of the rules and regulations

Land rights. In Cerro de San Pedro, and Mexico in general, we see that the conflict is equally about the contents of the rules and regulations linked to mining and land and water use. At the very base of this conflict are two laws, the Agricultural Law and the Mining Law. Mexico's Mining Law considers mining to be of benefit to the entire society. Thus, any kind of exploration, exploitation and beneficiation of minerals should get preference over any other types of land use, including agriculture and housing (GAES Consultancy, 2007; Herman, 2010). However, this is not in accordance with Article 75 of Mexico's Agrarian Law, which states that "in cases where lands have been proven to be of use to the *ejido* population, the common land uses in which the *ejido* or *ejidatarios* participate may be prioritized" (Herman, 2010, p. 84). To ensure that mining activity can eventually take over all other forms of land use, Article 6 of the Mining Law enables land to be alienated through "temporary occupation" (Herman, 2010). Yet the Agricultural Law does not recognize this temporary-occupation instrument. Moreover, the Constitution considers land given to the *ejidos* "imprescribtible, inalienable e inembargable". Nevertheless, the denial of these fundamental rights is precisely what has taken place in Cerro de San Pedro. By denying *ejidatarios* the property of the subsoil as well as the surface, *ejidatarios* are legally being excluded from the game. The Agricultural Law recognizes them as the legal landowners, but the Mining Law considers mining to be in the public interest. So, the threat of having their land expropriated in the name of this public interest is ever-present for local villagers. If landowners do not agree to a lease contract they risk losing everything, with no compensation, through temporary occupancy. This puts them in an unequal negotiation position and forces them to accept unfair lease contracts (Clark, 2003; Ochoa, 2006). The Mining Law's temporary-occupancy provision de facto undermines the land titles of *ejidatarios*. Estrada and Hofbauer (2001) and the BOF state that local inhabitants are even further disadvantaged by the lack of legal follow-up: the Agrarian Attorney is obligated to supervise and assess the process of selling or renting *ejido* land to third parties, yet in practice this is often not done. In Cerro de San Pedro (and the majority of similar cases in which lease or sale contracts were produced between *ejidos* and a mining company), the *ejidatarios* were not informed about their rights and the possible risks of living close to mining activity.

In addition, as we elaborate below, San Pedro's customary rights and national agrarian laws supporting the position of Cerro de San Pedro's landowners are even further hollowed out by NAFTA rules, which stimulate and empower investors' rights and nullify the contents of local socio-legal arrangements protecting the environment and the community.

Water rights. Another subject of fierce conflict is the neoliberal policies that have converted water rights from a noncommodity into a tradable asset, which importantly favoured MSX's opportunities to operate in San Luis Potosí. These changes have allowed purchase and sale of presumably out-of-use water permits and the proliferation of well perforations in the *veda* zone (considered, under the new laws, 'relocation' of the old well), despite the clear objective of reducing over-exploitation of the aquifer. MSX obtained its water permits by making use of the new regulation and thus managed to buy 12 concessions totalling 1.3 MCM annually (Newgold Inc., 2009; Santacruz De Leon, 2008). Project opponents state that, keeping in mind the severe over-exploitation of the aquifer, tradable water rights put extra pressure on the aquifer in San Luis Potosí and endanger future water provision for San Luis Potosí inhabitants. Moreover, opponents claim that the granting of 1.3 MCM of a 'scarce resource' for mining purposes shows that the so-called scarcity is not an environmental condition, but rather the result of priorities that the government assigns to certain uses. They argue that the government decides that for some uses water is 'abundant' whereas for others it is 'scarce' (Peña & Herrera, 2008b). Scarcity in this sense is a social construction and political phenomenon rather than a natural state of the environment.

Besides the *veda*, the 1993 preservation of wildlife decree (preventing industrial water use) was also circumnavigated by MSX, triggering important conflict. In 2005 the *Sala Superior del Tribunal Federal de Justicia Fiscal y Administrativa* (Superior Chamber of the Federal Fiscal and Administrative Federal Tribunal) declared that the 1993 decree speaks of "industrial activity" (which has lower water rights prioritization), and since mining can be considered a "primary activity" (with higher priority), it is not subject to the decree (Herman, 2010). This decision provoked strong controversy, and project opponents objected in another court, starting a vicious cycle of court cases, seemingly without end. Project opponents claim that the location of the lixiviation area, in a zone designated for aquifer recharge according to this decree, besides being illegal, poses an extra threat of contamination of the aquifer (BOF, 2014).

Conflict over decision-making authority

As explained, MSX used a provision within the Mining Law that allows "temporary occupancy" of the land to acquire usufruct rights. This was granted by the Ministry of Economy in 2005, and thus the *ejidatarios'* rights to land and water were circumvented. Temporary occupancy has given birth to a much deeper discussion about the contents of the laws, how they interact and who has legal and/or legitimate power. In this case, the Mining Law was given preference over the Agricultural Law, but a large litigation process started, contesting the decision-making power of Mexico's courts: who is to decide whether the Mining Law supersedes Agricultural Law, or otherwise? This discussion is profoundly connected to power positions, discourse and knowledge, discussed in the fourth echelon.

Similar disputes relate to decision-making authority regarding the three decrees that have been issued in the past (*zona de monumentos, zona de veda* and *zona de preservación de vida silvestre*), which all have been overthrown in favour of mining activity. They were intended to protect the region socio-environmentally and culturally, but recent political power plays have altered and generated reinterpretation of the decrees with the aim of welcoming MSX to the area. However, the authority to overthrow these decrees (varying from the state governor to the national government) is fiercely disputed in the courts. BOF is actively fighting the decisions taken by authorities. BOF member Eduardo da Silva said, "Even when MSX leaves Cerro de San Pedro, our job is not done. There are so many other places in which the same thing is going on. We are not just questioning MSX, but equally the Mexican government: eventually, the government is the one who allows the law to be broken. Our goal is to change this governmental system, full of corruption, and to change the laws and the legal system that make it possible for companies such as MSX to operate in the illegal way they currently are" (pers. comm., October 2013).

The long legal battle and the different courts' rejecting responsibility and consequently referring to other courts have enabled MSX to continue operating while court cases remained pending. Several members of BOF mentioned that they feel that the Mexican government has deliberately adopted a 'from pillar to post' strategy of rejecting responsibility and referring to other courts, to postpone decision making and meanwhile give MSX the chance to operate (pers. comm., BOF member Eduardo da Silva, October 2013). Herman (2010, p. 85), quotes BOF's lawyer Esteban in her research on Cerro de San Pedro: "The legal processes are so poorly managed and the regulations are so vague that there are lots of ambiguities around the Agrarian Registry.... So the ejidatarios are not only against the mine, they're also litigating so that the courts recognize their rights."

International legislation has also put its mark on developments in Cerro de San Pedro, and brings up the question of which type of legislation (national or international) supersedes the other. Through NAFTA's Chapter 11, a foreign company may sue the host government, if it considers that the latter has not complied with agreements made under NAFTA, and thus put the company economically at a disadvantage. As UN's principal water lawyer, Miguel Solanes, writes about NAFTA and its threats to the public interest: "There is a tendency to replace the obligatory jurisdiction of the State with that of international arbitration tribunals.... Two types of economic players are thus created: those having all manner of guarantees, whatever the fluctuations in the economy, and those, usually ordinary citizens, who do not have any" (Solanes & Jouravlev, 2006, p. 63; see also Solanes & Jouravlev, 2007). In Mexican practice, on several occasions, local and national governments were sued by companies over the revocation or cancellation of environmental permits, after which the companies received large compensations for their economic loss from the host government (e.g. Kass & McCarrol, 2000). Solanes explains: "Only investors have legitimacy to request the intervention of investment arbitration courts, and to initiate suits and legal actions. They create the arbitration market, which depends on investors for its existence – the risk of capture and bias is strong. Since they are based on international agreements, investment courts trump national jurisdiction. In addition, other fora such as human rights courts lack the enforcement powers of the decisions of arbitration courts" (pers. comm., 20 December 2014).

MSX has threatened the Mexican government with NAFTA Chapter 11 to obtain the required permits at a time when the process seemed very difficult. Similar to the observation made by Warden and Jeremic (2007), just the threat of use of this provision has already caused a strong chilling effect in the case of Cerro de San Pedro. NAFTA provides an enormously powerful position for MSX vis-à-vis the national and local governmental authorities. Local communities are not allowed to object to resolutions taken within NAFTA, even though they often are the ones facing the greatest impact. Denying local inhabitants and communities the ability to file a complaint under NAFTA repudiates their legal status and stake in the conflict. As in Cerro de San Pedro, this creates enormous power differences between the local inhabitants versus the foreign company (Ochoa, 2006). As Miguel Solanes comments: "International investment agreements and their arbitration courts have made a travesty of local interests and power devolution. An arbitration court, at the international level, beyond local and national judges, ends up adjudicating conflicts between public local interests and global companies and investors. The international investment court does not only perform beyond local reach, but also outside the limits of public interest at the local level. Its mandate is to protect investors' interests, disregarding local problems" (pers. comm., 20 December 2014).

Conflicts among discourses

MSX's discourse is an essential tool in reaching its objectives, and helps understand how the issues explained in the previous echelons were tackled. The Cerro de San Pedro case witnesses how the mine's powerful discursive practices aim to morally, institutionally and politically legitimize their particular interests in using, managing and usurping the local natural resources, thereby arranging the human, the technological and the natural worlds in a 'convenient miningscape', as if these bonds were entirely natural.

Under its Corporate Social Responsibility programme, MSX claims that the company is deeply concerned with the environment, health, safety, and community development in both social and economic terms (Herman, 2010; Newgold Inc., 2012a). MSX states that it will provide jobs, education, healthcare and infrastructure to the local residents. On top of that, MSX claims to work with the newest techniques in order to minimize impact on the environment and reduce chances of contamination. Work and safety standards at work are said to be high; the wages that the mine offers would be high compared to Mexican standards (Newgold Inc., 2012b). The company's discourse explains that MSX is genuinely concerned with the livelihoods of its employees. By strongly advocating their commitment to security, health, environment and sustainability, MSX creates a discursive link between large-scale open-pit mining and positive development of the area. For example, MSX's annual Sustainability Report focuses largely on the job opportunities that MSX created for local inhabitants and on MSX's community development support, e.g. by means of the development of alternative income sources such as cactus nurseries, fish farms and a supply of microcredit for entrepreneurial initiatives to enable the people to sustain themselves after MSX abandons operation (pers. comm., MSX representative, November 2013; Newgold Inc., 2012b). By obtaining internationally recognized certificates confirming their 'sustainable operation strategy', such as the Conflict-Free Gold Certificate, MSX aims to comfort the public and government when it comes to health, society, environment and pollution (see also Vos & Boelens, 2014).

In many of its social activities – such as collective tree-planting days, the museum in Cerro de San Pedro in which the mining operation is explained and the 'benefit' for the local community is emphasized, and workshops on the production of silver jewellery – the company combines its strong power position with the creation of particular mine-convenient knowledge and facts, to make its mining truths become locally accepted truth. MSX has a very powerful position in this sense: it uses its economic position to influence public (e.g. mass-media) and governmental opinion, enhancing its social and political power position. Knowledge is actively created by MSX, as the company itself is in charge of monitoring the quality of water, air and soil. Thus, MSX establishes firm, triangular linkages between the three fundamental elements of Foucauldian discourse – power, knowledge and truth – mutually linked and shaping each other.

While some of the inhabitants have been convinced and have adopted the discourse of the mine's important economic, social and even environmental function for the region, others (for example those united in the BOF) have developed critiques and alternative or counter-discourses. BOF's mission is to stop MSX's activity in Cerro de San Pedro. To reach this, BOF actively spreads information in newspapers, social media and other outlets about the litigation process and the adverse environmental, cultural and economic effects caused by MSX, and organizes a yearly anti-mine music festival that takes place in Cerro de San Pedro.

Clearly, analyzing mining discourses and counter-discourses in Cerro de San Pedro gives insight into how the different groups perceive environmental problems and design solutions. Guthman (1997, p. 45) notes that the "production of environmental interventions is intimately connected to the production of environmental knowledge, both of which are intrinsically bound up with power relations. Therefore, the facts about environmental deterioration have become subordinate to the broader debates on the politics of resource use and sustainable development." Many villagers perceive that the process of knowledge production by the mining company, consultants, and state agencies reflects but also reinforces social and economic inequities in the area.

In everyday life, the discursive struggles and conflicts in the region – together with the diverging interests of villagers in relation to the mine's operations – have fuelled the divide-and-rule strategy that MSX has applied to the villagers since it arrived. When MSX came to Cerro de San Pedro, the village was unanimous in its objection to the mine. However, local inhabitants explain that the ambience in the village slowly changed and opinions on the mine started to diverge. For example, people say that certain families received money in return for their vote and others did not receive anything. The division between Cerro de San Pedro's villagers came to an all-time high when several pro-MSX villagers attacked some anti-MSX villagers, with the anti-mine villagers just able to run for their lives. Effectively objecting to the presence of MSX is more difficult for the anti-mine villagers if opinion is divided, an asset cleverly used by MSX.

Ways forward

Mining conflicts as in Cerro de San Pedro show a common feature, typical of most cases in Latin America as well as in many other regions: mining companies' power positions are reinforced by strong state backing and international investment agreement, producing a profoundly unequal negotiation position for the mining-affected populations. For the latter to obtain fairer and equal access to litigation possibilities, the mining company should be forced back to the negotiation table, and government institutions need to be made accountable for their key role as public service entities. To be successful, various cases provide evidence that this requires forging *multi-actor* alliances that work on *multi-scalar* levels, thus creating civil society networks that are internally complementary while connecting the local, national and global (see e.g. Bebbington, Humphreys-Bebbington, & Bury, 2010, and Boelens, 2008, for cases in Peru, Bolivia and Ecuador; Ochoa, 2006, for México; Urkidi, 2010, for Chile; and Hoogesteger, Boelens, & Baud, 2016, for Ecuador). By linking, for example, local village initiatives, women's groups, and journalists and newspapers with provincial indigenous and peasant federations, national ombudsman and civil rights offices, international research centres, and environmental and human rights NGOs, the negotiation forces (including access to research, information dissemination and possibilities for international arbitrage) can become more balanced and one-sided discourses can be challenged. Getches (2010) describes important opportunities for these multi-actor networks to use international norms and laws that can counterbalance powerful NAFTA-type agreements (see also Solanes & Jouravlev, 2007). Thus, besides more localized first-echelon resource struggles, in particular for second- and third-echelon strategies, the marginalized mining-affected communities may find important support through multi-actor and multi-scalar network action. These can seek to bend discriminatory rules or apply (inter)national protective regulations, and to balance the currently skewed decision-making powers. At the fourth echelon of ERA, such multi-actor network strategies will also strengthen the building of an alternative discursive framework, one that enables challenging the 'official' regimes of representation and can generate broader support for socially and environmentally friendly alternatives.

Currently the leading mining opposition group in Cerro de San Pedro, the BOF, is working on proposals for a new Mining Law, which builds on a more equitable and ecologically sound management of land and water resources. Getting the Mexican government to accept this will require great lobbying skills, a large network of influential partners and a well-balanced discourse. To this end, BOF is forging an alliance of local, national and international environmental organizations and universities, such as Pro San Luis Ecológico, Greenpeace México, and Amnesty International. These alliances provide BOF with access to new strategic-political opportunities, not only in Cerro de San Pedro but also in other mining arenas in Mexico. In the Cerro de San Pedro case, in which extraction activity is almost at its final stages, the main effort is now to try to reduce the damage done to the environment. Demanding ethically and ecologically responsible mining practices and waste cleanup, and also enabling alternative local livelihood opportunities, such as ecological and cultural tourism, might improve future job opportunities for the villagers whilst reducing the environmental impact.

Conclusions

Making use of the ERA framework shows that this mining conflict goes beyond the obvious struggle over accessing or defending land and water resources. In Cerro de San

Pedro, an exemplary struggle is being fought over land and water, yet with underlying struggles over the content of the rules and rights, and disputes regarding the decision-making authority to *make* those rules, which in the end seek to *distribute* the resources in particular ways. The discourses that are developed are not just weapons and counter-weapons in this struggle but also seek, in accordance with each party's interests and worldviews, to convincingly answer the questions and coherently link the issues raised in the first three echelons. In Cerro de San Pedro, they aim to depoliticize and naturalize MSX's miningscape, or, alternatively, show its profound contradictions as well as politically motivated 'mining truth', and arrange for 'alternative truths'.

MSX obtained access to the *ejido* land through the consent of Mexican government institutions which, despite the lack of required permits, allowed MSX to proceed with its operations on communal land. Moreover, the position of *ejidatarios* vis-à-vis the powerful mining company is further weakened by the systemic legal contradictions between the Agricultural Law and the Mining Law. The possibility of a temporary occupancy that can overrule the 'inalienable' land titles of *ejidos* inherently means that in Cerro de San Pedro, and throughout the country, *ejidos* can and will be pushed aside in favour of mining companies. Also, NAFTA has had great influence on the litigation process and on the bargaining position of local inhabitants. NAFTA's Chapter 11 gives foreign companies the opportunity to ignore national legislation concerning environmental and social rights and operate directly under the rules and regulations of NAFTA. NAFTA does not accept complaints from local inhabitants or communities, dismissing local communities' co-decision-making about their own futures.

The decision to overrule the existing decrees (the *veda*-regulations and the zone for the environmental preservation) in favour of MSX shows the Mexican government's eagerness for MSX to exploit the area. Although these decisions are contested by project opponents and remain to be decided in Mexico's courts, the circumvention of these decrees shows the extent to which an international, powerful actor like MSX can influence execution of national environmental legislation. Linked to these decrees is the granting of water concessions to MSX. Governmental allocation of 1.3 MCM annually to a mining industry contrasts with the total lack of water in a few neighbouring villages, with the endangered provision of water quantity and quality to a large city like San Luis Potosí, and with the official argument that in this valley water is generally an extremely scarce resource for which *veda* restrictions need to be obeyed. Water-scarcity declarations in the region clearly refer to political statements and priorities, which in the power context of San Luis Potosi easily bypass the natural state of this resource. The government declares a state of water scarcity when villages claim subsistence water use, yet it can simultaneously declare water abundance when a multinational mining company asks for large volumes of water to produce metals and a toxic environment. The economic interests of the few prevail over ensuring that the mine's neighbouring villages are provided with the most basic human right.

In the end, the changes in land and water rights in Cerro de San Pedro result from a complex interplay between different actors, where the court systems, officials and governments at diverse levels play a double and deeply troublesome role, and where multinational MSX has cleverly used the loopholes in the laws, plus its economic and discursive powers, to reach its objectives. In addition, international agreements as NAFTA have had a profound unethical impact on the litigation process, stimulating encroachment and sidelining social and environmental rights. The real victims of this interplay are the *ejidatarios* and inhabitants of Cerro de San Pedro, who have lost their alternative income-generating activities and access rights to land and water and who, once MSX abandons its operation, will be left without job opportunities, and with a polluted, entirely distorted environment.

Yet, the mine's deeply problematic impact may be reduced and halted in the near future. Through multi-actor networks that creatively engage in multi-scale action, mining-affected population groups, together with a variety of mutually complementary advocacy and policy actors, are striving to balance negotiation power and force the mining company to clean up mining residues and enable alternative local livelihood opportunities.

Note

1. *Zona de veda*: a specific area in which additional water use (on top of those concessions already granted) is not allowed; in short, new water concessions are no longer granted. This provision is designed to preserve the quantity and quality of the water, which can be either of superficial or subsurface nature (Conagua, 2012). However, the previously established amount of water for extraction may be so high as to prevent a healthy equilibrium from being reached. In the Valle de San Luis Potosí the *veda* refers to groundwater exploitation.

References

Achterhuis, H., Boelens, R., & Zwarteveen, M. (2010). Water property relations and modern policy regimes: Neoliberal utopia and the disempowerment of collective action. In R. Boelens, D. Getches, & A. Guevara-Gil (Eds.), *Out of the mainstream. water rights, politics and identity* (pp. 27–55). London: Earthscan.

Adler, R., Claassen, M., Godfrey, L., & Turton, A. R. (2007). Water, mining and waste: An historical and economic perspective on conflict management in South Africa. *The Economics of Peace and Security Journal*, 2(2), 32–41. doi:10.15355/epsj.2.2.33

Assies, W. (2008). Land tenure and tenure regimes in Mexico: An overview. *Journal of Agrarian Change*, 8(1), 33–63. doi:10.1111/j.1471-0366.2007.00162.x

Assies, W., & Duhau, E. (2009). Land tenure and tenure regimes in Mexico: An overview. In J. M. Ubink, A. J. Hoekema, & W. J. Assies (Eds.), *Legalising* land rights. *Local practices, state responses and tenure security in Africa, Asia and Latin America*. Leiden: Leiden University Press.

Bebbington, A., Humphreys-Bebbington, D., & Bury, J. (2010). Federating and defending: Water, territory and extraction in the Andes. In R. Boelens, D. Getches, & A. Guevara-Gil (Eds.), *Out of the mainstream. Water rights, politics and identity* (pp. 307–327). London: Earthscan.

Boelens, R. (2008). Water rights arenas in the Andes: Upscaling Networks to strengthen local water control. *Water Alternatives*, 1(1), 48-65.

Boelens, R. (2009). The politics of disciplining water rights. *Development and Change*, 40(2), 307–331. doi:10.1111/j.1467-7660.2009.01516.x

Boelens, R. (2014). Cultural politics and the hydrosocial cycle: Water, power and identity in the Andean highlands. *Geoforum*, 57, 234–247. doi:10.1016/j.geoforum.2013.02.008

Boelens, R. (2015). *Water, power and identity. The cultural politics of water in the andes*. London and Washington DC: Routledge/Earthscan.

Boelens, R., & Vos, J. (2012). The danger of naturalizing water policy concepts: Water productivity and efficiency discourses from field irrigation to virtual water trade. *Journal of Agricultural Water Management*, 108, 16–26. doi:10.1016/j.agwat.2011.06.013

BOF. (2013). *Website of BOF (Broad Opposition Front)*. Retrieved July 9th, 2013, from http://faoantimsx.blogspot.mx/

BOF. (2014, April 5th). *Carta a Carmen Aristegui*. Retrieved October 23th, 2014, from http://faoantimsx.blogspot.mx/2014/04/carta-carmen-aristegui.html

Bricker, K. (2009). Chiapas Anti-Mining Organizer Murdered. Retrieved April 8th, 2014, from http://narcosphere.narconews.com/notebook/kristin-bricker/2009/12/chiapas-anti-mining-organizer-murdered

Brooks, D. B., Thompson, L., & El Fattal, L. (2007). Water demand management in the Middle East and North Africa: Observations from the IDRC forums and lessons for the future. *Water International, 32*(2), 193–204. doi:10.1080/02508060708692200

Brooks, D. B., Trottier, J., & Doliner, L. (2013). Changing the nature of transboundary water agreements: The Israeli–Palestinian case. *Water International, 38*(6), 671–686. doi:10.1080/02508060.2013.810038

Clark, T. (2003). *Canadian mining companies in Latin America: Community rights and corporate responsibility*. Paper Conference CERLAC/York University and Mining Watch Canada, May 9 - 11, 2002, Toronto.

Conagua. (2004). *Registro publico de derechos de agua*. Retrieved March 25, 2014, from http://www.conagua.gob.mx/Repda.aspx?n1=5&n2=37&n3=115

Conagua. (2012). *Vedas Superficiales*. Retrieved March 25, 2014, from http://www.conagua.gob.mx/ConsultaInformacion.aspx?n1=3&n2=63&n3=210&n0=1

Duarte-Abadía, B., Boelens, R., & Roa-Avendaño, T. (2015). Hydropower, encroachment and the repatterning of hydrosocial territory: The case of Hidrosogamoso in Colombia. *Human Organization, 74*(3), 243–254. doi:10.17730/0018-7259-74.3.243

Estrada, A. C., & Hofbauer, H. (2001). *Impactos de la inversión minera canadiense en México: Una primera aproximación*. México, DF: Fundar, Centro de Análisis e Investigación.

Feindt, P. H., & Oels, A. (2005). Does discourse matter? Discourse analysis in environmental policy making. *Journal of Environmental Policy & Planning, 7*(3), 161–173. doi:10.1080/15239080500339638

Forsyth, T. (2003). *Critical political ecology: The politics of environmental science*. London: Routledge.

GAES Consultancy. (2007). *Mexico - mexican market profile mining*. Ontario Ministry of Economic Development and Trade.

Getches, D. (2010). Using international law to assert indigenous water rights. In R. Boelens, D. Getches, & A. Guevara-Gil (Eds.), *Out of the mainstream. Water rights, politics and identity*. London: Earthscan.

Gordoa, S. E. M. (2011). *Conflictos socio-ambientales ocasionados por la minería de tajo a cielo abierto en Cerro de San Pedro, San Luis Potosí. (Licenciatura en Geografía)*. San Luis Potosí: Universidad Autónoma de San Luis Potosí.

Guthman, J. (1997). Representing crisis: The theory of himalayan environmental degradation and the project of development in Post-Rana Nepal. *Development and Change, 28*(1), 45–69. doi:10.1111/dech.1997.28.issue-1

Hajer, M. (1993). Discourse coalitions and the institutionalization of practice: The case of acid rain in Great Britain. In F. Fischer & J. Forester (Eds.), *The argumentative turn in policy analysis and planning* (pp. 43–76). Durham and London: Duke University Press.

Herman, T. (2010). *Extracting Consent or Engineering Support? An institutional ethnography of mining, "community support" and land acquisition in Cerro de San Pedro*. Mexico: Department of Studies in Policy and Practice, University of Victoria.

Hogenboom, B. (1998). *Mexico and the NAFTA environment debate. The transnational politics of economic integration*. Utrecht: International Books.

Hogenboom, B. (2012). Depoliticized and repoliticized minerals in Latin America. *Journal of Developing Societies, 28*(2), 133–158. doi:10.1177/0169796X12448755

Hoogesteger, J., Boelens, R., & Baud, M. (2016). Territorial pluralism: water users' multi-scalar struggles against state ordering in Ecuador's highlands. *Water International, 41*(1), 91–106. doi:10.1080/02508060.2016.1130910

Kass, S. L., & McCarrol, J. M. (2000). The 'Metalclad' Decision Under NAFTA's Chapter 11. *New York Law Journal. Environmental Law*. Retrieved 27 April 2014, from http://www.clm.com/pubs/pub-990359_1.html

Krueger, A. O. (1999). *Trade creation and trade diversion under NAFTA*. National Bureau of Economic Research.

Lutz, D. (2010). Beware of the smell of bitter almonds. *Newsroom*. Retrieved July 23, 2014, from http://news.wustl.edu/news/Pages/20916.aspx

Mann, H., & Von Moltke, K. (1999). *NAFTA's Chapter 11 and the environment. Addressing the impacts of the investor-state process on the environment*. Winnipeg, Manitoba: International Institute for Sustainable Development.

Martínez-Alier, J. M. (2003). *The environmentalism of the poor: A study of ecological conflicts and valuation*. Cheltenham: Edward Elgar.

Neumann, R. P. (2005). *Making political ecology*. London: Hodder Arnold.

Newgold Inc. (2009). *Manifesto de impacto ambiental. Modalidad regional unidad minera Cerro de San Pedro - Operación y Desarrollo*. Cerro de San Pedro, San Luis Potosí: Minera San Xavier S.A.

Newgold Inc. (2012a). *Mining Project in Cerro de San Pedro*. Retrieved 9 July, 2013, from http://www.newgold.com/properties/operations/cerro-san-pedro/default.aspx

Newgold Inc. (2012b). *Reporte de Sustentabilidad 2012*. Cerro de San Pedro.

Nogales. (2002). The NAFTA environmental framework, Chapter 11 investment provisions, and the environment. *Annual Survey of International & Comparative Law*, 8(1), 6.

Ochoa, E. (2006). Canadian mining operations in Mexico. In L. North, T. D. Clark, & V. Patroni (Eds.), *Community rights and corporate responsibility: Canadian mining and oil companies in Latin America* (pp. 143–160). Toronto: Between the Lines.

OCMAL. (2014). *Conflictos Mineros en México*. Retrieved April 18, 2014, www.ocmal.com

Peace Brigades Internacionales. (2011). Undermining the Land. The defense of community rights and the environment in Mexico. *Mexico Project Newsletter*. London.

Peet, R., & Watts, M. (1996). *Liberation ecologies: Environment, development and social movements*. New York: Routledge Press.

Peña, F., & Herrera, E. (2008a). El litigio de Minera San Xavier: Una cronología. In M. C. Costero-Garbarino (Ed.), *Internacionalización económica, historia y conflicto ambiental en la minería. El caso de Minera San Xavier*. San Luis Potosí: COLSAN.

Peña, F., & Herrera, E. (2008b). Vocaciones y riesgos de un territorio en litigio. Actores, representaciones sociales y argumentos frente a la Minera San Xavier. In M. C. Costero-Garbarino (Ed.), *Internacionalización económica, historia y conflicto ambiental en la minería. El caso de Minera San Xavier*. San Luis Potosí: COLSAN.

Perreault, T. (ed.), (2014). *Minería, Agua y Justicia Social en los Andes. Experiencias Comparativas de Perú y Bolivia*. Justica Hídrica. Cusco: CBC.

Rabinow, P. (1991). *The foucault reader*. London: Penguin.

Ramirez, M. D. (2003). Mexico under NAFTA: A critical assessment. *The Quarterly Review of Economics and Finance*, 43(5), 863–892. doi:10.1016/S1062-9769(03)00052-8

Reygadas, P., & Reyna Jiménez, O. F. (2008). La batalla por San Luis: ¿El agua o el oro? La disputa argumentativa contra la Minera San Xavier. *Estudios Demográficos Y Urbanos*, 23(2 (68)), 299–331.

Reyna Jiménez, O. F. (2009). *Oro por cianuro: Arenas políticas y conflicto socioambiental en el caso Minera San Xavier en Cerro de San Pedro*. San Luis Potosí: COLSAN.

Robbins, P. (2004). *Political ecology: A critical introduction*. Chichester: Wiley.

Sandt, J. van der (2009). *Mining conflicts and indigenous Peoples in Guatemala*. The Hague: Cordaid.

Santacruz De Leon, G. (2008). La minería de oro como problema ambiental: El caso de Minera San Xavier. In M. C. Costero-Garbarino (Ed.), *Internacionalización económica, hierstoia y conficto ambiental en la minería. El caso de Minera San Xavier*. San Luis Potosí: COLSAN.

Solanes, M., & Jouravlev, A. (2006). *Water governance for development and sustainability*. Santiago: UN/ECLAC.

Solanes, M., & Jouravlev, A. (2007). *Revisiting privatization, foreign investment, international arbitration, and water.* Santiago: UN/ECLAC.

Sosa, M., & Zwarteveen, M. (2011). Acumulación a través del despojo: el caso de la gran minería en Cajamarca. In R. Boelens, L. Cremers, & M. Zwarteveen (Eds.), *Justicia Hídrica: Acumulación, Conflicto y Acción Social.* (pp. 381–392). Lima: IEP.

Turton, A. R. (2010). The politics of water and mining in South Africa. In K. Wegerich & J. Warner (Eds.), *The politics of water: A survey.* London: Routledge.

Turton, A. R., & Funke, N. (2008). Hydro-hegemony in the context of the orange river basin. *Water Policy, 10*(S2), 51–70. doi:10.2166/wp.2008.207

Urkidi, L. (2010). A glocal environmental movement against gold mining: Pascua-Lama in Chile. *Ecological Economics, 70,* 219–227. doi:10.1016/j.ecolecon.2010.05.004

Vargaz-Hernandez, J. G. (2006). *Cooperacion y conflicto entre empresas, comunidades, nuevos movimientos sociales y el papel del gobierno.* El caso de Cerro de San Pedro.

Vos, H. de, Boelens, R., & Bustamante, R. (2006). Formal law and local water control in the andean region: A fiercely contested field. *International Journal of Water Resources Development, 22*(1), 37–48. doi:10.1080/07900620500405049

Vos, J., & Boelens, R. (2014). Sustainability standards and the water question. *Development and Change, 45*(2), 205–230. doi:10.1111/dech.12083

Warden, R., & Jeremic, R. (2007). *The Cerro de San Pedro case. A clarion call for binding legislation of Canadian corporate activity abroad.* Toronto: KAIROS.

Wilder, M. (2010). Water governance in Mexico: Political and economic apertures and a shifting state-citizen relationship. *Ecology and Society, 15*(2), 22.

Wilder, M., & Romero-Lankao, P. (2006). Paradoxes of decentralization: Water reform and social implications in Mexico. *World Development, 34*(11), 1977–1995. doi:10.1016/j.worlddev.2005.11.026

Yacoub, C., Duarte-Abadía, B., & Boelens, R. (2015). *Agua y Ecología Política. El extractivismo en la agro-exportación, la minería y las hidroeléctricas en Latino América.* Justicia Hídrica. Quito: Abya-Yala.

Zwarteveen, M. Z., & Boelens, R. A. (2014). Defining, researching and struggling for water justice: Some conceptual building blocks for research and action. *Water International, 39*(2), 143–158. doi:10.1080/02508060.2014.891168

Territorial pluralism: water users' multi-scalar struggles against state ordering in Ecuador's highlands

Jaime Hoogesteger[a], Rutgerd Boelens[a,b] and Michiel Baud[b]

[a]Water Resources Management Group, Department of Environmental Sciences, Wageningen University, the Netherlands; [b]CEDLA (Centre for Latin American Research and Documentation) and Department of Geography, Planning and International Development Studies, University of Amsterdam, the Netherlands

ABSTRACT

Ecuadorian state policies and institutional reforms have territorialized water since the 1960s. Peasant and indigenous communities have challenged this ordering locally since the 1990s by creating multi-scalar federations and networks. These enable marginalized water users to defend their water, autonomy and voice at broader scales. Analysis of these processes shows that water governance takes shape in contexts of territorial pluralism centred on the interplay of divergent interests in defining, constructing and representing hydrosocial territory. Here, state and nonstate hydro-social territories refer to interlinked scales that contest and recreate each other and through which actors advance their water control interests.

Introduction

The 2008 Constitution of Ecuador is seen by many as one of the most progressive constitutions worldwide in terms of water (Harris & Roa-García, 2013; Hoogesteger, 2015a; Roa-García, Urteaga Crovetto, & Bustamante Zenteno, 2013). Its content reflects the remarkable political agency that the peasant and indigenous–based water users movement of Ecuador developed in the mid-2000s in opposition to a state that tries to control water in its territory through various legal and institutional arrangements. In Ecuador, in the early 2000s, following the massive indigenous uprisings of the 1990s, local autonomous water user organizations consolidated several user-based provincial and national federations and networks as a strategy to defend their interests by bridging the divide between state-led interventions, management and control on the one hand, and autonomous, user-based local organizations and governance on the other (Boelens, 2008). The article uses the concept of 'hydrosocial territories' to analyze these grass-roots organizations, their achievements and their relationship to the Ecuadorian state and its policies. Through it we examine the relations between processes of state territorial ordering, scale, and water users' struggles to defend their water, autonomy and voice.

The article is structured as follows. It first explores how hydrosocial territories at different scales are conceived, arranged and (re)created by actors as a means to gain influence over or contest water control. Second, it examines how and why the Ecuadorian state has territorialized water since the 1960s through a series of policy and institutional reforms. Third, it shows how peasant and indigenous water users structure their irrigation systems as specific hydrosocial territories and how these water users have created provincial and national federations and networks through which they demand voice and authority vis-à-vis different state and international actors. A concluding section reflects on these multi-scalar territorial (re)configurations.

This article is based on field research conducted in Ecuador (2008–2014) and ongoing involvement in the field of Ecuadorian water governance and politics over the last two decades. The research consisted of participant observation and action-research. Additionally, extensive and recurrent semi-structured interviews were held with the members and leaders of water user organizations, as well as with state officials, policy makers and NGO staff who were involved in supporting water user organizations. The field data collected were triangulated to obtain consistency and representation.

Hydrosocial territories, scale and 'territories-in-territory'

According to the *Encyclopaedia Britannica* (2012), in ecological terms a territory can be defined as an area defended by an organism or a group of similar organisms for a specific purpose. This definition provides a starting point for analyzing hydrosocial territories, as they too are created by actors who in view of a specific purpose (e.g. political control, conservation, natural resources use, or production) claim and defend authority over a specific space and the related social group(s), resources, technologies and ecological environment in which they exist. Autonomy and self-governance by the specific group that defines/defends a territory (based on ethnicity, language, religion or the use of a natural resource) is quintessential (Simeon, 2015). A territory can therefore be constructed only in relation to other actors (Barth, 1969) and humans' alignment with 'the natural' and 'the built'. In other words, it is based on subjects' strategies and powers to persuade or force others to accept the importance (and consequences) of consolidating a specific territorial arrangement. In this sense, territories determine how space is organized politically by interweaving its biophysical and social qualities (e.g. Baletti, 2012; Bridge, 2014; Meehan, 2013).

Scale, defined as a 'socio-spatial level of analysis' (Perreault, 2003, p. 98), is an important element of territories and how they are understood. This results from the fact that territories are spatially organized at different interrelated and often overlapping levels, such as the national, regional and local. These different nested levels are mutually constitutive even though their boundaries and content are far from neutral or fixed (Howitt, 2007; McCarthy, 2005). Understanding territories and their constitution therefore necessarily also implies understanding their scalar dimensions and interrelations.

Territories, in terms of their discursive and material properties (resources, technologies, boundaries, relationships, objectives), are disputed from 'within' and from 'the outside' as diverse subjects aim to shape territory according to their own interests (Boelens, Hoogesteger, Swyngedouw, Vos, & Wester, 2016). Disputes arise from competing territorial projects, images and interests that aim to consolidate boundaries, socio-natural organization and control for a specific purpose at a specific spatial scale

(Hoogesteger & Verzijl, 2015). The resulting (often conflicting) material and discursive practices and interventions constitute the politics of territorialization and imply processes of empowerment and domination, and of inclusion and exclusion in decision making (Swyngedouw, 2014; Zimmerer, 2000). Therefore, what territory 'is' and what it 'should be' are strongly determined by class, gender, ethnic or institutional interests, and power relations (see Bebbington, Humphreys Bebbington, & Bury, 2010; Boelens, 2014; Castro, 2008). These are closely related to legitimacy and authority, which are often invoked through claims on knowledge (scientific, indigenous/vernacular) (Boelens & Seemann, 2014). and social organization and order (bureaucratic, market, community, associational). It follows that hydrosocial territories are constructed through discourses and practices that shape the interactions, alliances and power struggles in spatially bounded socio-natural networks. How hydrosocial territories are structured can be understood by studying their contents (water control), their size and boundary arrangements (definitions of place, space and socio-natural boundaries as they relate to water), and their relations (ties in and among different actors, resources and natural environments at different scales) (cf. Lynch, 2006; Marston, 2000).

Divergent views, interests and powers over how specific territories are or should be shaped lead to the coexistence of multiple territorial notions and contested construction processes. We use the term 'territorial pluralism' to point out the fact that diverse territories are overlapping, interacting and conflicting in one and the same geographical-political space.[1] These 'territories-in-territory' have partially different, partially similar building blocks, but these are patterned in different and sometimes opposing ways, configurations and meanings (e.g. 'water' as physical-chemical H_2O, as a commoditized resource, or as a sacred being; 'water users' as servants of the nation-state, as clients, or as socio-nature's caretakers). Therefore, different 'territories-in-territory' are structured by different rules and normative frameworks, sources of legitimacy, forms of authority and related discourses. Once a territory (or a particular 'territory-in-territory') is constituted, it becomes an organizing element of human/nonhuman interactions, influencing both nature and society.

When analyzing irrigation systems through the lens of hydrosocial territories, they appear as spatially bound socio-material constructs in which water is controlled by interrelated physical elements (e.g. water sources and flows; hydraulic infrastructure to divert, conduct and distribute water; water provision places), normative elements (rules, rights and obligations regarding access to water and related resources), organizational elements (human organization to operate/sustain the system; capacities and knowledge of the art of irrigation) and financial and agro-productive elements (soil, crops, technology, capital, labour force). These are embedded in cultural and political traditions and powers at different yet interrelated scales (Boelens, 2014; Perramond, 2013; Swyngedouw, 2009). Consequently, in irrigation systems a fundamental interdependence exists among infrastructural, normative and organizational components within the spatial boundaries that are marked by the command area of the system.

State (re)territorialization of water governance

Disengagement from governmental systems is today well-nigh impossible. (Giddens, 1990, p. 91)

State territorialization takes place on different spatial scales and is usually hierarchically organized in a top-down manner in conformity with administrative boundaries (nation, province, municipality, irrigation system) and procedures. These are enforced through hierarchically organized institutions in charge of managing water affairs (for instance, the Ministry of Water Resources, Ministry of Agriculture, or Ministry of Energy). The state 'steers water society' and consolidates its grip depending greatly on its techniques of territorial water governance. Formal mechanisms include water use infrastructure construction and control, water use permit administration and water allocation, legal frameworks, and control over investment programmes in the water sector (Wester, Rap, & Vargas-Velásquez, 2009). This coordinated control over delimited territories is intended to achieve administrative coordination, supported by an apparatus of direct or indirect surveillance (Giddens, 1990). More informal and indirect ones consist of technical assistance and social and political programmes and interventions.

The social and material designs and technical layouts of such hydro-territorial control systems are based on socio-political and normative notions that project specific power relations and related order (Gellert & Lynch, 2003; Saldías, Boelens, Wegerich, & Speelman, 2012; Warner, Wester, & Hoogesteger, 2014). Rationalizing water control in new 'efficient forms of water governance' guided by 'expert knowledge', 'modern irrigation techniques' and 'water policy models' is a fundamental cornerstone of the state's project, which aims to rearrange local hydrosocial territories and their socio-cultural frameworks (cf. Duarte-Abadía & Boelens, 2016). These rearrangements thus take place through coercive forms of power, and the seductive power of expert-modernist norms, knowledge and organization, which naturalize state policy models and control and seek water users' 'self-correction' (Boelens, Hoogesteger, & Baud, 2015; Foucault, 1978/1991; Swyngedouw, 2014). State institutions related to water thus constitute an important source of power and have the legal faculty to mediate amongst competing water claims and interests to 'defend its population' and ensure the greater societal good/benefit. State discourse presents this steering of society and the mediation of conflicts as its source of legitimacy to exercise authority, extract resources and generate income through the (re)arranging of territories and increased state control over local communities and their territories.[2]

Rearranging Ecuadorian water governance through (de)centralization

> Decentralization … may re-distribute decision-making geographically, but the way decisions are made remains hegemonic. (Nina Pacari, Interview, October 2004)

In Ecuador, as in many other Latin American countries, from the 1960s onwards water management was nationalized and centralized. National institutions were created to establish hierarchic control by implementing a water regulatory framework, allocating water concessions at the national level, and constructing and operating hydraulic infrastructure. In Ecuador, this resulted in the consolidation of the National Ecuadorian Water Resources Institute (INERHI) in 1966 and a National Water Law in 1972. INERHI was created at the ministerial level to centralize national water control. This meant control over the allocation of water rights, mediation of water conflicts, enforcement of legal norms, and the expansion and management of the irrigation frontier. It was staffed mostly by civil and agricultural engineers. As a result, most funds were directed towards constructing and expanding the 79 state-managed irrigation systems (Hoogesteger, 2014). INERHI's strong involvement in resources administration, development and management principles profoundly imposed government rationality in local irrigation management.

This imposition of governmental order continued through the decentralization and privatization of the 1980s and 1990s (Boelens et al., 2015; Hoogesteger, 2013a). These were implemented under pressure from the World Bank and other bilateral funding agencies. The reforms aimed to decentralize governance powers to lower layers of government (provincial and municipal) and promote a greater involvement of both water users and the private sector in irrigation management tasks while attempting to introduce a neoliberal market rationality into local water management practices. In 1993, the Law of State Modernization, Privatization, and Provision of Public Services was enacted, opening the doors to far-reaching national water reforms and the privatization of water services. As part of this reform package in 1994, INERHI was replaced by the legally and institutionally weaker National Council of Water Resources (CNRH) and the delegation of irrigation responsibilities to the four existing and five newly created regional development corporations (RDCs) (Cremers, Ooijevaar, & Boelens, 2005). The highly understaffed CNRH was conceived as an inter-ministry council with limited budget and power to carry out policies, projects, or programmes. Its main responsibility was to administer water allocation by granting water use rights, mediating water conflicts and enforcing legal norms concerning water use in all water user systems of the nation through 12 regional water agencies. The RDCs were to coordinate at the regional level with the provincial governments and the other relevant ministries in the design and implementation of development programmes (Cremers et al., 2005). Nevertheless, in practice most RDCs limited their activities to the irrigation sector and especially the operation and management of state irrigation systems. This was done with limited funding from the central state and was largely ineffective, also because many provincial governments established their own irrigation support units. As part of the decentralization process, from 1995 to 2001 the Ecuadorian irrigation management transfer programme, which was financed through a World Bank loan, was implemented by the Executive Unit of the Technical Assistance Project (UEP-PAT).

Not only was the national state restructured and bypassed in this broad effort to 'decentralize' water control and neoliberalize local water territories; the representatives of organized water users, such as the national Confederation of Indigenous Nationalities of Ecuador (CONAIE) and the Inter-institutional Irrigation Forum (FIR), a broad water-user organization and NGO-based civil-society alliance, were sidelined and their proposals and demands ignored. Despite the fact that CONAIE had developed a solid proposal for a new water law and a corresponding institutional framework (CONAIE, 1996) and led the national indigenous movement to large-scale, nation-wide mobilizations and protests that paralyzed the country for weeks (Boelens, 2008), CONAIE was not considered a legitimate stakeholder in water affairs by national and international decision makers. Similarly, when the FIR presented a proposal for user empowerment and democracy in the irrigation sector, a proposal that was created by several of its member institutes (with grounded experience, knowledge and networks), their proposal was likewise ignored. Instead, UEP-PAT hired three international consultancy companies with no contextual knowledge of irrigation to implement the irrigation management transfer programme. This programme consisted of (1) creating water user associations with the capacity to administer, operate, maintain and manage their irrigation systems; and (2) legally turning over the irrigation management responsibilities from the state agencies to the newly created water user associations. The programme boiled down to transferring irrigation systems administration, operation and maintenance tasks and related financial burdens to poorly prepared water user associations in 35 irrigation systems (Cremers et al., 2005; Moscoso, Nieto, Chimurriz, & Díaz, 2008).

Neoliberal 'modernization' dismantled not just the Ecuadorian state bureaucracy but also the institutional capacity to assist user organizations in managing their own water affairs (Cremers et al., 2005). This was clearly recognized by some leaders of social movements. Indigenous leader Nina Pacari (1998, pp. 279–282) observed: "The form of organization to administer irrigation currently proposed by the State destroys traditional forms of organization, generating conflicts and weakening decision-making." In the vacuum left by the understaffed national water institutions, the decentralized governments, the private sector, international donors, and NGOs, all newly empowered by the new legislation, took on greater roles in water project implementation, albeit with their own ideas and interests regarding the territorialization of water at different scales (Hoogesteger, 2013a, 2014, 2015b). This plurality, added to the fact that most of the country's irrigation systems were managed autonomously by landlords or user collectives with hardly any state intervention, importantly contributed to the deepening of territorial pluralism in Ecuadorian water governance in the late 1990s and early 2000s (Boelens, 2015).

In 2007, following President Rafael Correa's promises to the water user organizations which had helped to vote his 'Citizen Revolution' into power, the new left-wing government implemented a set of institutional reforms that aimed at re-establishing the national state, its rationality and control at the centre of water affairs. A National Water Secretariat (SENAGUA) was created as the 'sole governmental water authority', replacing CNRH. It was allocated a broad set of responsibilities in the water domain. A few months after its creation, SENAGUA switched from provincial administrative boundaries to a model based on nine river-basin management units (Warner et al., 2014). From 2007 to 2014, the number of SENAGUA offices increased from 12 to 43, organized hierarchically: 1 national secretariat, 9 river-basin management offices and 33 sub-basin management offices. Through this new bureaucratic structure SENAGUA aimed to 'guide', 'regulate' and 'coordinate' all water-related affairs in the country (Boelens et al., 2015). Next, SENAGUA began the construction of several large multipurpose dams throughout the country (Hidalgo, Boelens, & Isch, 2014), destroying existing local, community-managed hydrosocial territories in the name of 'development'. Legal pluralism and local territorial autonomy are recognized and protected in the 2008 Constitution, but only as long as they do not interfere with 'national progress' and the 'well-being of all Ecuadorians'.

User-based counter-territories and the struggle for autonomy and voice

Existing legal frameworks are not appropriate for us, our own norms and wishes. (Carlos Oleas, irrigator and president of Interjuntas-Chimborazo, Interview June 2005)

271

The consolidation of irrigation systems as (semi)autonomous grass-roots hydrosocial territories lies at the heart of the development of the Ecuadorian water user movement. Therefore, to understand 'the' movement – its origins and demands – it is important to understand how, from the grass-roots perspective, hydrosocial territories are constituted from the bottom up at different interrelated scales. As explored below, local water-use territories form the basic building blocks around which broader grass-roots water territorialization takes place. Interlinking these local hydrosocial territories with broader spatial scales is crucial for water-centred social movements. It enables dispersed and often spatially and politically less powerful water user groups to develop regional and national political agency. Hydro-territorial organization, on scales where infrastructure and water flows do not physically bind with the social, consolidates around shared causes and builds on shared threats, claims and identities, as analyzed below.

To control irrigation systems, water users constantly adapt to arrange the interlinked social, political and natural domains according to their interests and beliefs. In this process agro-ecological properties, water flows and water-control infrastructure strongly determine how water-use systems function. Infrastructure enables water delivery to specific places and actors through canals, dams, division structures, pipes and outlets, which establish spatial and social boundaries. That is why infrastructure reflects and simultaneously embeds social norms (e.g. how to understand, operate and repair technologies, and who benefits, when, where and how much) and requires particular forms of human organization (Hoogesteger, 2013b).

Ecuadorian water user communities and associations tend to follow a different rationality of signifying and patterning these elements than, for example, state agencies or development institutions (see Boelens & Gelles, 2005; Hoogesteger, 2015b; Rodriguez de Francisco, Budds, & Boelens, 2013). The former's rules, rights and obligations shape and are shaped by collective action and social organization around the joint creation and maintenance of irrigation infrastructure and water flows (Hoogesteger, 2013c). These relationships generate strong bonds of shared hydrosocial/territorial identity, linking users to their water sources, infrastructure, territory and user/community organizations.

Autonomous user-managed irrigation systems

In Ecuador, most irrigation systems are managed autonomously and cover areas of up to a few hundred hectares. There are no precise figures about these systems. National statistics show that a total of some 850,000 ha is irrigated. Approximately 200,000 ha is managed by systems officially labelled as 'state-managed' or recently 'transferred'. Some 320,000 ha is under community control, managed by some 2000 common-property systems. The rest, which includes the most productive land, is under private control (Gaybor, 2011).

Community-governed irrigation systems are regulated by rules, obligations and rights that are locally defined. These vary from one irrigation system to another. In these irrigation systems especially strong processes of belonging and place-based identity creation develop around the collective creation and re-creation of infrastructure, local normative structures (rights and responsibilities) and organizational (community or supra-community) arrangements, many of which reach beyond the strict contours of the irrigated area and its users. They sometimes include part of the upper catchment of the source from which water is taken, potential areas for the expansion of the irrigation system and aspirant water users (individuals, groups or communities). Thus, water and its flows are the uniting factor that delineates local hydrosocial territories. They may or may not follow the same social and material boundaries of what is usually defined as an irrigation system. These 'locally rooted' processes lead to water-centred territorialization and determine the organization of user groups. They also form the background of their defence of collective water rights against external actors. Local water user communities often try to use a strategic selection of state rules, resources and procedures in their own favour (Gelles, 2000). They consider formal institutions and structures both a threat to local autonomy and, at the same time, an opportunity and a defence mechanism for their water use and allocation.

For instance, in Gompuene, a customary irrigation system (50 hectares serviced by one main canal) in Chimborazo Province, the defence of water rights and access vis-à-vis state, hacienda and other power groups plays a key role in linking eight communities to each other and to the river and infrastructure and their spatial boundaries. The territory, however, is dynamic and changes over time. Gompuene's irrigation system was hydraulically, organizationally and politically linked to the new, larger Licto system in the late 1990s, incorporating a total of 1700 hectares, 20 communities, and a new (larger and more secure) water source. Jointly, Gompuene and Licto users shape the 'new' irrigation system, its 'territory' and its organization around the material base of water and hydraulic infrastructure. The mutual interdependence arises from the need to cooperate through collective action, which is crucial to construct, operate, maintain and defend the water-use systems. Its rights and property relations are now embedded in Gompuene and Licto user collectives' cultural, agro-ecological and political context. Their collective and individual water rights are literally inscribed in the infrastructure, contours and water flows of the irrigation system, which have been shaped by their specific histories of communitarian cooperation and fierce conflicts with external actors who threaten their access to water and self-governance autonomy (Boelens & Gelles, 2005).

A contrasting example is the neighbouring Chambo irrigation system, which had a long history of state management (Hoogesteger & Verzijl, 2015). As in most state systems, agency personnel designed, planned, constructed and managed the systems top-down, through the imposition of state-based normative frameworks in which users had few decision-making rights (Hoogesteger, 2015b). However, when the state ran out of funding for operational costs and personnel, and retreated, users started to organize to fill management gaps. They networked informally to find external allies to consolidate their organizations, infrastructure and technical skills. Through these processes, the water users discursively, socially and materially took possession of the irrigation system, its management, its water flows and the space that these occupy through a process of grass-roots reterritorialization. At present the water users are organized into 82 local management boards (which includes a management board of aspirant irrigators) and an overarching water user association. Though their organization and water distribution practices are greatly structured by state guidelines, the water users self-govern their irrigation system and advance their claims for water access, self-governance and voice in decision making beyond the irrigation system contours vis-à-vis the state and other external actors by joining forces with Licto and Gompuene, amongst others, on broader spatial scales (Hoogesteger & Verzijl, 2015).

Representing water users' interests at broader scales

In the late 1990s and early 2000s, in response to neoliberal reforms, water users in Chimborazo created the Provincial Water Users Federation, Interjuntas-Chimborazo, joining approximately 300 grass-roots irrigation and drinking-water organizations (Hoogesteger, 2012). This federation enabled them to voice their demands and participate in decision-making processes both provincially and nationally. By creating an extensive network of water user organizations, including not just community but also state-managed systems, Interjuntas-Chimborazo has been able to press for action to meet users' needs through lobbying, formal participation in decision making, and massive popular protest. They have challenged the territorial organization of the state in many ways: in terms of contents, rationality, legitimacy and strategies. As Interjuntas President Carlos Oleas explained:

> We demanded recognition for our ambition to reunite all users in one organization, both for drinking water and irrigation. However, our wish was rejected by the authorities. We are not only cheap labour; we also want to make decisions! (Interview June 2005)

Interjuntas-Chimborazo's ability to mobilize thousands of water users in the province for popular street protests, as evidenced in 1997, 2005 and 2010, significantly legitimized its spatial and social extension. This street-gained legitimacy has also powerfully influenced the federation's position when pressing its claims and negotiating water users' demands with provincial and national state water agencies. At the same time, it has enabled the federation to attract allies to its network, establishing cooperation with nongovernmental and human rights organizations, indigenous federations, and donors, nationally and internationally.

A similar case is the Federation of Peasant Water User Groups of Cotopaxi (FEDURICC), originally formed in 1997 by six user organizations from the largest irrigation systems in Cotopaxi Province. Initially supported by local government, FEDURICC quickly increased its membership and now comprises over 330 grass-roots organizations. Like Interjuntas-Chimborazo, it has developed legitimacy and the capacity to mobilize its members for massive street protests throughout the province. As with Interjuntas, FEDURICC's demands have centred on respect for local autonomy and self-government, the recognition of its right to participate in state decision making concerning irrigation development, and defence of users' water rights. It is widely recognized that the water user federations' vigilance and social audit systems have enhanced the effectiveness of public investments in irrigation development, benefitting marginalized peasant and indigenous water user communities (Hoogesteger, 2014).

Both these federations have gained territorial legitimacy through their capacity to mobilize their constituents, due to their shared cultural identity and a shared cause. The success of both FEDURICC and Interjuntas-Chimborazo is therefore grounded in the effective creation and discursive mobilization of a 'provincial/territorial water identity'. In Chimborazo, the trigger that initially brought the province's water users together was the plan to close the provincial RDC and Water Agency in 1997 (Hoogesteger, 2012). Between 1997 and 2005 problems of corruption and discrimination against peasant/

indigenous water users at the provincial Water Agency increased dramatically. In reaction, user organizations started to work together, and in 2005 they massively mobilized in Riobamba's streets to demand a change of personnel. They occupied the Water Agency offices for 18 consecutive days until they were accepted as legitimate stakeholders and participants in the process of defining how and according to which procedures the new water agency director would be appointed and monitored by the organized water users. In Cotopaxi, continued corruption scandals in allocation of public funds to the irrigation sector (in both the RDC and the provincial government) brought user organizations together and enabled them to audit and oust two different directors of the irrigation agency. Finally they also forced their inclusion in the decision-making process, which led to the naming of a director they backed. In advancing their interests and claims, representatives of these federations strategically use such successes to discursively legitimize their demands and manifest their federative power within the provincial territorial confines.

Today, both FEDURICC and Interjuntas-Chimborazo play important roles in defending the interests of water users at the national level. They have extended their network through the national multi-stakeholder platform, the National Water Resources Forum (WRF), which brings together multi-ethnic groups, grass-roots organizations, individuals, NGOs, state institutions and academics from the whole country. In WRF, proposals in the national water governance arena are analyzed, debated and created. Through its active participation in national water policy debates since the early 2000s, WRF has become an important channel through which water users link to multi-scalar governance structures and national water legislation affecting their water access and rights and, with this, their local water territories and autonomy.

WRF has contributed to some important changes in national water governance. In 2004, it presented a proposal for legal reforms to the National Congress (Hoogesteger, 2015a), starting a broad movement against the privatization of domestic water supply services in the country's urban areas. WRF publicly rejected the free-trade negotiations by the government, as they would significantly challenge national territorial water integrity and the poorest sectors' water security. In subsequent years, WRF presented new, amended proposals to the National Congress and published position papers to influence and shape public policies, including a fair legal framework for water.

There is no doubt that the most important success of WRF and its constituent organizations has been the inclusion of their demands in Ecuador's 2008 Constitution. In this Constitution, which, as we saw, is considered one of the world's most socially progressive, they managed to include fundamental issues such as recognition of water as a human right, water as a strategic national heritage for public use, recognition of water user organizations' autonomy, respect for their water-rights systems, prohibition of private water management, and user participation in water governance decision making. Another crucial issue was the Constitution's establishment of new allocation priorities: (1) human consumption; (2) irrigation for national food sovereignty; (3) ecological flows; and (4) productive uses. These constitutional changes give legal priority to livelihood needs. This breaks with the previous emphasis on production and modernization (for instance, lumping subsistence and export-oriented commercial agriculture in a single category). It also aims at prohibiting unlimited commercialization of water. Unlike the federations, which gained their legitimacy largely through public protests, WRF has won national territorial representation and legitimacy based on its broad base of constituents, its lobbying activities and the quality of its proposals.

These user-based federations and multi-stakeholder platforms link local water user associations at broader scales and in doing so reaffirm the local hydrosocial territories they represent as well as their up-scaled territorial claims vis-à-vis external actors (provincial and national state agencies, NGOs, other user-based federations and networks, and donors). For instance, when relating to outside actors, Interjuntas leaders frame their demands in terms of 'we, the water users of Chimborazo Province' and highlight the large number of member organizations they represent in the province. Based on this broad constituency and its capacity to mobilize 'the water users' of the province, they claim the spatial boundaries of the province of Chimborazo as 'their' territory in terms of the representation of the water users' interests. Similarly, to legitimize and re-create the WRF's claim to represent water users' interests at the national scale, its leaders repeatedly stress that representatives from water user organizations from all provinces of Ecuador actively participate in WRF. Through this discursive framing and claiming of member-ships (cf. Baud & Rutten, 2004), interests and their belonging to a defined spatial entity (a region, a province, a country), these federations and multi-stakeholder platforms define and re-create the spatial contours of the 'territory' they represent.

Finally, there are three fundamental issues that stand out in the demands of these interlinked federations and networks: (1) their concern for the material interests of local user organizations, such as infrastructural investments, transparent resource distribution, and policy implementation for fairer distribution of water resources; (2) recognition of self-governance faculties and identities, with sufficient autonomy for user-based rule-making and enforcement; and (3) their concern to acquire a binding voice in water management decisions at different scales (national, provincial and in basin management units) to ensure that their demands are met. Especially this last point remains a delicate issue within the present government. Although the government has co-opted many grass-roots leaders into its political apparatus and has created formal spaces for water users' participation on different scales, most leaders concur that these spaces represent mere lip service to long-standing grass-roots demands. To them, President Correa's 'Citizen Revolution' is more a government-based 'revolution for the citizens, not with the citizens'. The critical voices of social movements are muted time and again (Boelens et al., 2015). In practice, water-user participation is often limited to mere consultation, with no binding consequences (Hoogesteger, 2014). The government has tried to delegitimize many of these organizations by questioning their constituency and representativeness while repeat-edly calling for 'individualized' rather than 'representational' participation by water users in these spaces. As a result, grass-roots federations and networks are currently struggling to sustain their territorial claims and legitimacy in relation to both their constituency and the state. In doing so, they are trying to hold on to the core of their three basic demands: redistribution for fair access to resources; recognition of local rule-making authority and management autonomy; and representational voice in decision making.

Conclusions

Territorial pluralism, conceived as different notions of how and what hydrosocial territories are and should be in terms of organization and function, leads to processes of struggle and negotiation that mould water governance and its outcomes at different scales. As the case of Ecuador shows, the multi-scaled territorial organization of the state is crucial for its objectives of ordering society but also influences the structuring of counter-movements and their territories. In Ecuador, as in most other Andean countries (Boelens, 2015), organized water users have created local hydrosocial territories by simultaneously evading and strategically using state ordering and its corresponding institutionality. In doing so they consolidate their access to water, technology and voice through forms of self-governance in the territories of self-determination delineated by irrigation systems. To defend their interests at broader scales, water user organizations 'jump scales' and have developed into a social movement with considerable political agency. These organizations have a clear public presence in national and provincial water governance arenas, and have become an important means for irrigators to defend their access to water, to maintain their autonomy and to struggle for a voice in decision making over issues that concern their water rights.

The analytical lens of hydrosocial territories is useful for understanding the relations and confrontations that exist between different notions of how water and its governance should be organized at different spatial scales. Its value lies in the focus on the importance of space and scale on the one hand, and on autonomy and self-determination in and over these specific socio-natural spaces on the other. In local water systems, water flows and infrastructure play a key role in the defence and delimitation of the spatial and social reach of their hydrosocial territories. At a broader scale, the state's organizational structure (e.g. provinces, river-basin management units and the nation) is important, both in terms of how it organizes control within the national territory and how water user collectives try to challenge and influence this system. Therefore, the state's territorial organization has direct bearing on how and on what scales water users create counter-territories to claim their demands.

In spite of the continuous efforts of the state to territorialize water and users according to its rationale, autonomous water users manage to defend their territorial autonomy and practices. They do so by reconstructing local water management practices, but at the same time organizing in provincial and national water user movement organizations. Thus, Ecuador's national water user movement is comprised of linkages and interactions among place-based multi-scalar water user organizations. This movement has been able to resignify the meanings of political representation and participation in Ecuadorian water governance. It has done so through struggles that address different but interlinked fields of water justice, such as transparency in resource allocations and equity in water distribution; recognition of water users' rights and forms of self-governance; and participation and democracy in decision-making processes.

This analysis of the Ecuadorian peasant and indigenous irrigators brings to the fore that movement activities by water justice groups and movements, though often stressing issues of identity and autonomy, seldom concentrate exclusively on identity itself (Baud, 2010; Castro, 2008; Gelles, 2010; Romano, 2012; Terhorst, Olivera, & Dwinell, 2013). The movement organizations presented here highlight the relationship between territory, identity and ethnic discrimination on the one hand, and political exclusion and unequal resource distribution on the other. Rather than demanding to be yet another influential actor in the hegemonic game of water allocation, they demand an alternative governance perspective, one that guarantees grass-roots families a fair share in the distribution of water and water-based resources (socio-economic justice), and recognition of their autonomy and ways of managing water territories (cultural justice). They have made clear that this can only be accomplished if they can take part in decision making over water governance (political or 'representational' justice; see also Perreault, 2014; Schlosberg, 2004; Zwarteveen & Boelens, 2014). Therefore, claims for representational justice combine with demands for redistribution of resources and recognition of water-based territorial identities and rights. These crucially involve sharing in authority on different scales.

Notes

1. This refers both to the existence and interaction, in one particular spatial location, of actors and networks with different objectives and interests (e.g., superimposing mining territory, indigenous territory and State administrative territory), and of actors and networks with similar water-use purposes but with diverging/conflicting conceptions of 'territoriality' (e.g. superimposed formal, State-structured irrigation systems and informal, vernacular community-based irrigation territories).
2. Underlying formal discourses on national progress, efficient resource governance, and overall citizens' well-being, states often reassign water allocations to meet the needs of strong political and economic sectors and actors (e.g. mining, forestry or agri-business corporations, and hydropower development) or to gain control over the surplus produced by community systems – either for the state administrative body or for particular rent-seeking factions within the state (Baud, 2010; Baud, de Castro, & Hogenboom, 2011; Lynch, 2012).

Acknowledgements

We thank the anonymous reviewers for the insightful comments and suggestions. The usual disclaimers apply.

Funding

This research was funded by the Netherlands Organization for Scientific Research (NWO) division of Science for Global Development (WOTRO) under project number W 01.65.308.00: 'Struggling for water security: Social mobilization for the defence of water rights in Peru and Ecuador'; it forms part of the activities of the international Justicia Hídrica/Water Justice alliance (www.justiciahidrica.org).

References

Baletti, B. (2012). Ordenamento territorial: Neo-developmentalism and the struggle for territory in the lower Brazilian Amazon. *The Journal of Peasant Studies, 39*(2), 573–598. doi:10.1080/03066150.2012.664139

Barth, F. (Ed.). (1969). *Ethnic groups and boundaries: The social organization of cultural differences.* London: Allen & Unwin.

Baud, M. (2010). Identity politics and indigenous movements in Andean history. In R. Boelens, D. H. Getches, & J. A. Guevara Gil (Eds.), *Out of the mainstream: Water rights, politics and identity* (pp. 99–118). London: Earthscan.

Baud, M., de Castro, F., & Hogenboom, B. (2011). Environmental governance in Latin America: Towards an integrative research agenda. *European Review of Latin American and Caribbean Studies, 90,* 78–88.

Baud, M., & Rutten, R. (Eds.). (2004). *Popular intellectuals and social movements: Framing protest in Asia, Africa, and Latin America.* Cambridge: Cambridge University Press.

Bebbington, A., Humphreys Bebbington, D., & Bury, J. (2010). Federating and defending: Water territory and extraction in the Andes. In R. Boelens, D. H. Getches, & J. A. Guevara Gil (Eds.), *Out of the mainstream: Water rights, politics and identity* (pp. 307–328). London: Earthscan.

Boelens, R. (2008). Water rights arenas in the Andes: Upscaling networks to strengthen local water control. *Water Alternatives, 1*(1), 48–65.

Boelens, R. (2014). Cultural politics and the hydrosocial cycle: Water, power and identity in the Andean highlands. *Geoforum, 57,* 234–247. doi:10.1016/j.geoforum.2013.02.008

Boelens, R. (2015). *Water, power and identity. The cultural politics of water in the Andes.* London: Earthscan, Routledge.

Boelens, R., & Gelles, P. H. (2005). Cultural politics, communal resistance and identity in Andean irrigation development. *Bulletin of Latin American Research, 24*(3), 311–327. doi:10.1111/j.0261-3050.2005.00137.x

Boelens, R., Hoogesteger, J., & Baud, M. (2015). Water reform governmentality in Ecuador: Neoliberalism, centralization and the restraining of polycentric authority and community rule-making. *Geoforum, 64,* 281–291. doi:10.1016/j.geoforum.2013.07.005

Boelens, R., Hoogesteger, J., Swyngedouw, E., Vos, J., & Wester, P. (2016). Hydrosocial territories: A political ecology perspective. *Water International, 41*(1), 1–14. doi:10.1080/02508060.2016.1134898

Boelens, R., & Seemann, M. (2014). Forced engagements: Water security and local rights formalization in Yanque, Colca Valley, Peru. *Human Organization, 73*(1), 1–12. doi:10.17730/humo.73.1.d44776822845k515

Bridge, G. (2014). Resource geographies II: The resource-state nexus. *Progress in Human Geography, 38*(1), 118–130. doi:10.1177/0309132513493379

Castro, J. E. (2008). Water struggles, citizenship and governance in Latin America. *Development, 51*(1), 72–76. doi:10.1057/palgrave.development.1100440

CONAIE. (1996). *Ley de Aguas.* Quito: CONAIE.

Cremers, L., Ooijevaar, M., & Boelens, R. (2005). Institutional reform in the Andean irrigation sector: Enabling policies for strengthening local rights and water management. *Natural Resources Forum, 29,* 37–50. doi:10.1111/j.1477-8947.2005.00111.x

Duarte-Abadía, B., & Boelens, R. (2016). Disputes over territorial boundaries and diverging valuation languages: The Santurban hydrosocial highlands territory in Colombia. *Water International, 41*(1), 15–36. doi:10.1080/02508060.2016.1117271

Foucault, M. (1978/1991). Governmentality. In G. Burchell, C. Gordon, & P. Miller (Eds.), *The Foucault effect: Studies in governmentality* (pp. 87–104). Chicago, IL: University of Chicago Press.

Gaybor, A. (2011). Acumulación en el campo y despojo del agua en el Ecuador. In R. Boelens, L. Cremers, & M. Zwarteveen (Eds.), *Justicia Hídrica: Acumulación, Conflicto y Acción Social* (pp. 195–208). Lima: IEP.

Gellert, P., & Lynch, B. D. (2003). Megaprojects as displacements. *International Social Sciences Journal, 55*(1), 15–26.

Gelles, P. (2000). *Water and power in highland Peru. The cultural politics of irrigation and development.* New Brunswick, NJ: Rutgers University Press.

Gelles, P. (2010). Cultural identity and indigenous water rights in the Andean highlands. In R. Boelens, D. H. Getches, & J. A. Guevara Gil (Eds.), *Out of the mainstream: Water rights, politics and identity* (pp. 119–144). London: Earthscan.

Giddens, A. (1990). *The consequences of modernity.* Cambridge: Polity Press.

Harris, L. M., & Roa-García, M. C. (2013). Recent waves of water governance: Constitutional reform and resistance to neoliberalization in Latin America (1990–2012). *Geoforum, 50,* 20–30. doi:10.1016/j.geoforum.2013.07.009

Hidalgo, J. P., Boelens, R., & Isch, E. (2014, September 24–25) Sistema Multipropósito de Agua Daule-Peripa: Una reconfiguración tecnocrática del territorio hidrosocial y despojo en la costa ecuatoriana. Paper for the Congress 'Riego, Sociedad y Territorio', Valencia, Spain.

Hoogesteger, J. (2012). Democratizing water governance from the grassroots: The development of Interjuntas-Chimborazo in the Ecuadorian Andes. *Human Organization, 71*(1), 76–86. doi:10.17730/humo.71.1.b8v77j0321u28863

Hoogesteger, J. (2013a). *Movements against the current: Scale and social capital in peasants' struggles for water in the Ecuadorian Highlands* (PhD). Wageningen, The Netherlands: Wageningen University.

Hoogesteger, J. (2013b). Trans-forming social capital around water: Water user organizations, water rights, and nongovernmental organizations in Cangahua, the Ecuadorian Andes. *Society & Natural Resources, 26*(1), 60–74. doi:10.1080/08941920.2012.689933

Hoogesteger, J. (2013c). Social capital in water user organizations of the Ecuadorian Highlands. *Human Organization, 72*(4), 347–357. doi:10.17730/humo.72.4.jv2177g624q35253

Hoogesteger, J. (2014). Building blocks for users' participation in water governance: Irrigators' organizations and state reforms in Ecuador. *International Journal of Water Governance, 2*(1), 1–18. doi:10.7564/13-IJWG2

Hoogesteger, J. (2015a). NGOs and the democratization of Ecuadorian water governance: Insights from the multi-stakeholder platform el Foro de los Recursos Hídricos. *VOLUNTAS: International Journal of Voluntary and Nonprofit Organizations*, 1–21. doi:10.1007/s11266-015-9559-1

Hoogesteger, J. (2015b). Normative structures, collaboration and conflict in irrigation; a case study of the Pillaro North Canal Irrigation System, Ecuadorian Highlands. *International Journal of the Commons, 9*(1), 398–415. doi:10.18352/ijc.521

Hoogesteger, J., & Verzijl, A. (2015). Grassroots scalar politics: Insights from peasant water struggles in the Ecuadorian and Peruvian Andes. *Geoforum, 62*, 13–23. doi:10.1016/j.geoforum.2015.03.013

Howitt, R. (2007). Scale. In J. K. Agnew, K. Mitchell, & G. Toal (Eds.), *A companion to political geography*. Malden, MA: Blackwell.

Lynch, B. D. (2006). *The Chixoy Dam and the Achi Maya: Violence, ignorance, and the politics of blame* (Mario Einaudi Center Working Paper No. 10-06). Ithaca, NY: Cornell University.

Lynch, B. D. (2012). Vulnerabilities, competition and rights in a context of climate change toward equitable water governance in Peru's Rio Santa Valley. *Global Environmental Change, 22*(2), 364–373. doi:10.1016/j.gloenvcha.2012.02.002

Marston, S. (2000). The social construction of scale. *Progress in Human Geography, 24*(2), 219–242. doi:10.1191/030913200674086272

McCarthy, J. (2005). Scale, sovereignty, and strategy in environmental governance. *Antipode, 37* (4), 731–753. doi:10.1111/anti.2005.37.issue-4

Meehan, K. (2013). Disciplining de facto development: Water theft and hydrosocial order in Tijuana. *Environment and Planning D: Society and Space, 31*, 319–336. doi:10.1068/d20610

Moscoso, R., Nieto, L., Chimurriz, R., & Díaz, H. (2008). *Asesoría Jurídica Créditos Multilaterales*. Quito: Comisión de Auditoría Integral del Crédito Pública.

Pacari, N. (1998). Ecuadorian water legislation and policy analyzed from the indigenous-peasant point of view. In R. Boelens & G. Dávila (Eds.), *Searching for equity. Conceptions of justice and equity in peasant irrigation* (pp. 279–287). Assen: Van Gorcum.

Perramond, E. P. (2013). Water governance in New Mexico: Adjudication, law, and geography. *Geoforum, 45*, 83–93. doi:10.1016/j.geoforum.2012.10.004

Perreault, T. (2003). Making space: Community organization, agrarian change, and the politics of scale in the Ecuadorian Amazon. *Latin American Perspectives, 30*(1), 96–121. doi:10.1177/0094582X02239146

Perreault, T. (2014). What kind of governance for what kind of equity? Towards a theorization of justice in water governance. *Water International, 39*(2), 233–245. doi:10.1080/02508060.2014.886843

Roa-García, M. C., Urteaga Crovetto, P., & Bustamante Zenteno, R. (2013). Water laws in the Andes: A promising precedent for challenging neoliberalism. *Geoforum*. doi:10.1016/j. geoforum.2013.12.002

Rodriguez de Francisco, J. C., Budds, J., & Boelens, R. (2013). Payment for environmental services and unequal resource control in Pimampiro, Ecuador. *Society & Natural Resources, 26*, 1217–1233. doi:10.1080/08941920.2013.825037

Romano, S. (2012). From protest to proposal: The contentious politics of the Nicaraguan anti-water privatisation social movement. *Bulletin of Latin American Research, 31*(4), 499–514. doi:10.1111/blar.2012.31.issue-4

Saldías, C., Boelens, R., Wegerich, K., & Speelman, S. (2012). Losing the watershed focus: A look at complex community-managed irrigation systems in Bolivia. *Water International, 37*(7), 744–759. doi:10.1080/02508060.2012.733675

Schlosberg, D. (2004). Reconceiving environmental justice: Global movements and political theories. *Environmental Politics, 13*(3), 517–540. doi:10.1080/0964401042000229025

Simeon, R. (2015). Introduction. In K. Basta, J. McGarry, & R. Simeon (Eds.), *Territorial pluralism: Managing difference in multinational states* (pp. 3–10). Vancouver: UBC Press.

Swyngedouw, E. (2009). The political economy and political ecology of the hydro-social cycle. *Journal of Contemporary Water Research & Education, 142*, 56–60. doi:10.1111/jcwr.2009.142. issue-1

Swyngedouw, E. (2014). 'Not a drop of water...': State, modernity and the production of nature in Spain, 1898–2010. *Environment and History, 20*, 67–92. doi:10.3197/ 096734014X13851121443445

Terhorst, P., Olivera, M., & Dwinell, A. (2013). Social movements, left governments, and the limits of water sector reform in Latin America's left turn. *Latin American Perspectives, 40*(4), 55–69. doi:10.1177/0094582X13484294

Warner, J., Wester, P., & Hoogesteger, J. (2014). Struggling with scales: Revisiting the boundaries of river basin management. *Wiley Interdisciplinary Reviews: Water, 1*(5), 469–481. doi:10.1002/ wat2.2014.1.issue-5

Wester, P., Rap, E., & Vargas-Velásquez, S. (2009). The hydraulic mission and the Mexican hydrocracy: Regulating and reforming the flows of water and power. *Water Alternatives, 2*(3), 395–415.

Zimmerer, K. S. (2000). Rescaling irrigation in Latin America: The cultural images and political ecology of water resources. *Ecumene, 7*(2), 150–175. doi:10.1191/096746000701556680

Zwarteveen, M., & Boelens, R. (2014). Defining, researching and struggling for water justice: Some conceptual building blocks for research and action. *Water International, 39*(2), 143–158. doi:10.1080/02508060.2014.891168

Amazonian hydrosociality: the politics of water projects among the Waorani of Ecuador

Flora Lu[a] and Néstor L. Silva[b]

[a]Department of Environmental Studies, University of California, Santa Cruz
[b]Department of Anthropology, Stanford University, California

ABSTRACT

In Ecuadorian Amazonia, water is a site of contestation between the state, oil companies (both multinational and national), and indigenous communities. The Waorani, a group known for their hostility to outsiders, have been subject to territorial circumscription and practices of governance with the goal of pacification, sedentism, and geographically concentrating once-seminomadic populations. Such disruption of place-based cultural institutions and relations—particularly interactions with water—persist under the "twenty-first-century socialist" vision of the Correa administration. Sedimented histories of contamination and colonization have catalyzed two contrasting potable water projects in the Waorani community of Gareno: one by a state corporation tasked with using a percentage of oil profits to modernize neglected extraction-site communities and the other a NGO effort to install household-level rainwater catchment and filtration systems, an effort funded and operated by North American and European actors. In this book, we consider the forms of territoriality and institutionality—the new forms of hydrosociality—that each project offers to the Waorani.

Introduction

Water is tough to deal with in Gareno. The large *cancha cubierta comunal* at the center of the village—a multipurpose sports arena where people hold community meetings and play soccer, volleyball, and basketball—becomes a concrete island after a rain: a grassy swamp on one side and a field of thick mud on the other. Using two of the three flushing toilets in Gareno, which are located next to the community school—all built by an oil company—requires slogging through that mud. The toilets' bowls are practically dry and flushing requires leaving the stall to get a bucket of water from a plastic barrel where rain is collected from the rooftop. When a toilet clogged, requiring a mortifying admission to the Waorani president of Gareno, we asked where waste goes when the toilet works and were matter-of-factly told, "the river." There is no septic system. Three faucets line a tiled sink outside the bathroom stall. None of them work.

A short walk behind these schoolhouse bathrooms is the Gareno River, only about 15–20 feet wide, a few feet deep, more of a creek than a river. One of Gareno's residents described the tradeoffs between oil extraction and environmental health, "All the fish in the river died, all of them. In 2006, the company [Perenco] dumped *cloro* (chlorine), dirty water, into the Gareno River and we didn't know at the time what happened. When we protested to the company, they gave us four generators" (Lu 2012, p. 88). Children sometimes play in the Gareno River and people occasionally wash their clothes or dishes there, but residents tend to avoid it for bathing and certainly for cooking and drinking. To collect water for these purposes, our host family jury-rigged a pipe transporting water from a forest stream. The pipe juts out into a small gulley behind their home. When it works well, a steady flow from the pipe makes collecting water for bathing or cooking simple, but a hard rain or drought damages the system and water only trickles out. When visiting Gareno, we bring a water filter and large jugs of bottled water purchased from the provincial capital of Tena, a 2-hour drive away. A paucity of clean water seems particularly cruel in the Amazon rainforest.

Located on the southeastern limit of Napo Province, Gareno is the one of three Waorani (also spelled Huaorani) communities (Figure 1) along an oil road, populated with approximately 23 families (extended and nuclear). It was founded in 1996 when a family left the village of Quehueiri-ono on the Shiripuno River, a community deep in the forest and accessible only by river or air. The early migrations to Gareno from Quehueiri-ono and from its splinter village of Huentaro, also along the Shiripuno, coincided with the opening of a road through the oil concession in which Gareno lies.[1] Having known nearly all of the current residents when they lived in the Quehueiri-ono and Huentaro, the first author vividly recalls productive fishing expeditions along the Shiripuno River, where *barbasco* poison was used when the river levels were low and water ran clear, or where hook and line, or *altaraya* (weighted net), when the river ran high and silted after a recent rain. While the Shiripuno is relatively deep and wide, sandy and navigable by dugout canoe, the Huentaro River is rocky, shallower, and runs clearer and cooler. In the latter especially, bathing is a pleasure and river water can be used for cooking and drinking without concern. In Gareno today, the geographic spine of the community is the oil road, along which

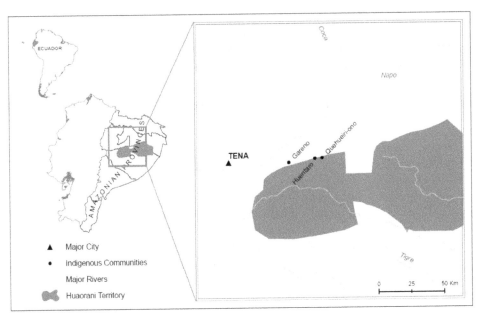

Figure 1.   Map of Waorani study communities.

people travel via motorcycle or flagged-down oil truck when they go on fishing trips where the better part of a day might yield six to ten small *carachama* (armored catfish). Waorani are keenly aware of the degradation of the waterways in their communities.

In these forest communities, rivers are vibrant sites of sociality, of enjoyment, and conviviality. They are sources of increasingly important riverine protein, and respites from the heat of the equatorial sun. In villages only accessible by small plane or river (with the former being prohibitively expensive), rivers are conduits of transport and communication, especially as outboard motors become more prevalent and Waorani trekking less so. As landscape features, rivers are integral to the creation of hydrosocial territories, integral to affective geographies (Navaro-Yashin 2012) through which histories, ancestors, sites of residence, and spaces of contestation are continuously recalled and experienced. Ancestral Waorani territory has been defined by rivers: the Napo on the North and the Curaray on the South, perimeters once enforced by generations of hostility between the Waorani, other indigenous groups, and mestizo settlers.

Between February and June 2014, a research team conducted ethnographic and household socioeconomic data collection in Gareno (as well as two other Waorani villages along different oil roads in the northern Ecuadorian Amazon). The team asked nineteen households to conduct a risk mapping interview where they list and then rank their most common concerns according to the incidence and severity of each one (see Lu et al. 2014). By far, the category with the greatest incidence—mentioned by the most respondents—was "contamination": mostly related to water, and caused by oil company activities and the lack of sewage infrastructure. In other interviews, residents described the kinds of benefits and resources living in Gareno provides. Those benefits with the highest incidence included wage labor and educational opportunities, as well as better access to medical care and transport, understandable since in Gareno—unlike some other Waorani communities—families live primarily sedentary, market-integrated, and cash-dependent lives. Though modernization and development marked the advent of electricity, cell phones, and satellite television in Gareno, potable water, sewage infrastructure, and protection from contamination are yet to arrive.

Two conjunctures, two water projects

In this article we focus on the changing community-level hydrosocial dynamics that reflect larger historical and political processes in Amazonia, characterized, in part, by contestation of hydrosocial territories. In line with call for water studies that are place based, rooted in specific societies, and historically contextualized (Zwarteveen & Boelens 2014), in the following sections, we outline two conjunctures that shape contemporary Ecuadorian political economy and ecology. First is Ecuador's emergence as a petrostate beginning in the 1970s—a transformation undeniably shaped by the operations of Texaco—which ushered in a political economic system characterized by multinational corporate dominance, national indebtedness, neoliberal ideologies, and grassroots resistance; struggles that included profound hydrosocial shifts throughout Amazonia. The second conjuncture on which we focus is the election of Rafael Correa in 2006, which marked an ideological commitment to resource nationalism and profound changes in political ideologies that included new practices of distribution of oil revenues, as well as a rise in social investment and infrastructural development. These ideologies and practices have been described by both Ecuadorian government and scholars as a "revolution" via state-sponsored pursuit of higher quality of life for Ecuadorian citizens.

As other chapters in this book discuss, Ecuador is a country of multifaceted forms of contestation and struggle around issues of water, from the neoliberalization of local water territories (Hoogesteger, Boelens, & Baud 2016), to water commodification through payments for environmental services (Rodríguez-de-Francisco & Boelens 2016). The Correa administration maintains a fundamental contradiction between the slated vision of the Revolución Ciudadana, the professed values of the 2008 Constitution, and resource extraction in fragile ecosystems politically-marginalized people. The juxtaposition of water governance and institutionality in the Andean highlands and that of the Amazonian lowlands, the difference between Kichwa and *campesino* communities compared to Waorani territory; both illustrate the heterogeneity of state institutions, citizen resistance, and water governance in this nation. Their commonalities reflect the strategic importance of water (e.g., for mining, hydroelectricity, and agroindustries) and the Correa administration's view of itself as the only legitimate guardian of the "national interest."

Boelens et al. (2016, p. 1) define hydrosocial territories as configurations of people, institutions, water, technology, and the biophysical environment that revolve around the control of water, and involve struggles over meaning, identity, authority, and discourses. In the Waorani case, the extraction of oil in their traditional homelands led to multiple changes in hydrosocial territoriality including contamination of lands and waterways through oil extraction and territorial circumscription of Waorani settlements by state- and corporate-sponsored means. Here we focus on the hydrosocial shifts resulting from projects for provisioning of potable water run by both governmental and non-governmental entities. Perreault (2014, p. 234) states that "Water is given meaning through cultural beliefs, historical memory, and social practice, and exists as much in discourse and symbolism as it does as a physical, material thing." Waorani hydrosociality is forged not only in the naming and reference of waterways as markers of boundaries and important landscape features, but also in the cooperation of a fishing expedition, of a long canoe trip. Hydrosociality is built on social understandings of fish populations as common pool resources, built on bonds forged during bathing, washing, and playing.

In congruence with its socially and environmentally progressive image, the Correa administration dedicates a portion of oil revenues to the remediation of the effects of decades of neglect of communities in sites of oil extraction. A water project by the state-owned development company Ecuador Estratégico (EE) exemplifies a technocratic approach that "solves" the lack of clean water by employing residents as laborers for an EE subcontractor to construct a community-scale water system. Emerging from protests and legal cases resulting from the social and ecological devastation wrought by Texaco (now Chevron) (see Kimerling 1991a; Sawyer 2004), from frustration with the failure of litigation to provide remedy, the NGO on which we focus implemented a different approach, one based on the amelioration of continued harm through participatory, household-scale technologies promoting family autonomy and ownership. We analyze these two water-provisioning projects in Gareno—one by EE and the other by the non-governmental ClearWater project—focusing on the new forms of hydrosociality, territoriality, and institutionality of water that each project presents to the Waorani of Gareno.

An "Amazonian Chernobyl"

The toxic legacy of petroleum production in the Ecuadorian Amazon, or Oriente, began with Texaco, which was involved in the discovery of large reserves of crude near the town of Lago Agrio in 1967. The company drilled 339 wells in an area of 1,094,124 acres (442,965 hectares; Kimerling 2000). The Ecuadorian government perceived Texaco—the first operator of commercial oil fields in the country—to be a prestigious multinational company with access to "world-class" hydrocarbon extraction technology (Kimerling 2012, p. 240). In its contract with the state, the company agreed to use "modern and efficient" practices of oil extraction (Kimerling 2012, p. 241). Instead, it implemented substandard procedures demonstrating a lack of concern for environmental protection, to which the national government acquiesced. According to Sawyer (2001, pp. 162–163):

> Ecuador imposed few restraints on Texaco in the Oriente and the corporation imposed few on itself. Consequently, 28 years of Texaco's crude exploitation indelibly transformed the northern rain forest, scoring it with thousands of miles of seismic grids, over 300 oil wells, more than 600 open waste pits, numerous pumping stations, an oil refinery, and the bare-bones infrastructure essential for petroleum operations.

It is estimated that Texaco's operations generated up to 4.3 million gallons of hazardous waste daily, over a period of 20 years (Jochnick, Normand, & Zaidi 1994; Kimerling 1991a). Shoddy pipeline construction and a lack of remediation capabilities meant an estimated 16.8 million gallons of crude was spilled into Amazonian waterways from Ecuador's largest oleoduct, while crude spilled by secondary pipelines has not been accounted for (Kimerling 1993). Texaco's responses to spills typically entailed shutting off the flow of crude in the damaged portion of the pipeline, allowing the remaining oil in the line to spill, and then making necessary repairs, without cleanup or compensation to affected communities (Kimerling 2012, p. 242). Overall, between 1972 (when Texaco began oil production in the Sucumbíos and Napo Provinces) and 1993 (when Texaco's concessions expired, it left Ecuador, and the state oil company took over), more than 30 billion gallons of toxic wastes and crude oil was discharged into the land and waterways of the Ecuadorian Amazon (San Sebastián & Hurtig 2004). The legacy of Texaco's operations in Ecuador has been called an "Amazon Chernobyl" (Donzinger 2015).

Oil extraction in the Oriente resulted in widespread contamination of water systems, which profoundly changed the economies, ecologies, and the means of sociocultural reproduction for Amazonian indigenous people, who "have borne the costs of oil development without sharing in its benefits and without participating in a meaningful way in political and environmental decisions that affect them" (Kimerling 2006, p. 416). In addition to contamination, oil-related infrastructure—roads, some public services— sparked internal migration of families from the highlands and coast (Murphy 2001), families who settled along the roads that Texaco built in what are now the Sucumbíos, Orellana, and Napo Provinces. The population of the Oriente increased from around 50,000 in 1960 to 400,000 by 1990 (Murphy 2001). In addition to colonization, oil-related infrastructure facilitated deforestation through mining and logging, further undermining native Amazonian livelihoods and cultural ways of life by devastating important physical resources such as waterways, as well as cultural and symbolic attachment to place (Kimerling 1991b, 1996).

Both Indigenous and mestizo residents of the northern Ecuadorian Amazon were poisoned by petroleum extraction. Drilling and oil exploitation produce toxic chemicals that contain heavy metals, strong acids, and concentrated salts. Crude oils themselves are mixtures of 100 or more hydrocarbons, sulfur compounds, and a range of other chemicals, metals, salts, and even radioactive substances. Polycyclic aromatic hydrocarbons (PAHs) are present in essentially all crude oils, and composed of forms of benzene whose toxicity results in both acute and chronic life-threatening illnesses (Jochnick, Normand, & Zaidi 1994). Mercury, a toxic byproduct of oil extraction, can cause mental disorders and birth defects (San Sebastian & Hurtig 2004). Cadmium, yet another byproduct, has been correlated to lung and prostate cancer (Sorahan & Lancashire 1997; Waalkes & Rehm 1994). Oil toxins, nitrogen oxide, sulfur oxide, and carbon react with sunlight to form ozone, which can be harmful to human lungs. These pollutants interact with other molecules to form even deadlier toxins such as carbon monoxide, which has been proven to cause strokes (San Sebastián & Hurtig 2004). Crude oil may enter the human body via skin absorption, ingestion of food or drink, and inhalation of particulates (Jochnick, Normand, & Zaidi 1994).

Scholars and popular media have addressed linkages between petroleum-related environmental degradation and increased disease incidence among exposed populations (Jochnick, Normand, & Zaidi 1994; Hurtig & San Sebastián 2002; San Sebastián & Hurtig 2004). Communities in the proximity of oil wells are threatened by contaminated water and agricultural land and poisoned livestock (Berlinger 2009). Prolonged exposure to crude oil and its toxic wastes has resulted in heightened rates of miscarriages, birth defects, cancers, learning disabilities, amputation, tumors, and premature deaths (Berlinger 2009). For babies, oil exposure has been linked to low birth weight and childhood cancers (Jochnick, Normand, & Zaidi 1994). San Sebastian and Hurtig (2004) describe a "public health emergency." Women living in communities surrounded by oil fields reported a higher incidence of itchy nose, sore throat, headache, ear pain, diarrhea, and gastritis than those living where no oil fields exist (San Sebastián, Armstrong, & Stephens 2000). Risk of miscarriage was 2.5 times higher in women living in the proximity of oil fields (San Sebastián, Armstrong, & Stephens 2002).

In Waorani territory and Amazonia in general, water is a symbolic and a material site through which contamination is continuously experienced. A Waorani man told us a story of trekking through the forest on a hunt, seeing a capybara step into what both animal and hunter believed to be a clear, shallow pond, then seeing the animal sink into tar-like residue of oil spills underneath a thin layer of water. He asserted that if he had retrieved the animal and butchered it, black spots would have been visible in its flesh which would have smelled of crude. In more than one community, we heard stories of how in the early years of oil exploitation, massive die-offs of fish would follow a wastewater or oil spill and people, unaware of the danger, ate the fish and would begin to feel sick almost immediately. The generation of indigenous people who grew up in the early years of Amazonian oil exploitation die of gastrointestinal cancers they assert were unseen in previous generations, cancers they attribute to drinking contaminated water and eating contaminated fish and game. Those people assert that when their children play in certain rivers, particularly after a hard rain, the children's eyes sting, they develop rashes or stomach problems attributed to contamination. As much as it is a marker of hydrosocial territories, and a source of nutrition, a medium through which social ties are strengthened and reproduced, following Amazonian contamination, water became menacing, a bearer of latent threats and painful memories.

Oil governance under the Correa administration

Despite such ecological and sociocultural costs, Amazonian oil has been and continues to be perceived as the means through which an underdeveloped country attains modernization and progress. The military governments of General Guillermo Rodríguez Lara (1972–1976) and Admiral Alfredo Poveda (1976–1979) used oil revenue to create thousands of jobs in the public sector, keep taxes relatively low, subsidize domestic consumption of petroleum products, and build transport infrastructure (Carriére 2001; Gerlach 2003). In the 1980s, the state incurred significant international debt based on borrowing against the consistently high prices of petroleum to finance development. By 1982, world petroleum prices stagnated and fell, and by 1984, structural adjustment measures were applied to curb the inflation that resulted from reduced petroleum revenues. Deregulation, privatization, and foreign investment—now classic facets of neoliberal doctrine—were offered as the remedy (Lu, Valdivia, & Silva, 2017). Presidential administrations—even those elected on left-leaning platforms—responded to these challenges by adopting neoliberal policies supporting the Washington Consensus, and International Monetary Fund, measures likened to "Andean Thatcherism." Among the neoliberal restructuring practices advanced by the state in the late 1980s and early 1990s was the expansion of a state-run oil industry (Valdivia 2008).

Despite increased oil exploitation in service to a substantial external debt, and dedication to neoliberal reforms as a means of maintaining eligibility for international assistance, Ecuador found itself mired in political and economic instability. During the 1990s and early twenty-first century, Ecuadorian citizens engaged in frequent protests, demanding the resignation of elected presidents perceived to lack legitimacy, engage in corruption, and embody prevailing reasons to distrust government and political parties. In a period of 10 years (1996–2006), Ecuador had seven presidents, none of whom completed a 4-year term. The Correa administration assumed power in 2006 on a leftist tide, a platform including increasing state control over the oil industry, and "protecting" national economies from globalization (Hidalgo Flor 2013). Correa promised an end to *partidocracia*, a term referencing a political system dominated by feckless traditional parties and elites responsible for Ecuador's political and economic crises (Machado Puertas 2007). The alternative was the Correa administration's Citizens' Revolution.

The Revolución Ciudadana is a capital-intensive development project introduced by the Correa administration in 2007, a project based on a vision of improving quality of life for Ecuadorians through investments in housing, education, infrastructure, and health care. It is the ideological crux and media standard of the Correa administration. The *Revolución* finds much of its legislative expression in the 2008 constitution, which Correa and his party Alianza País wrote following a successful national referendum. Considered one of the most progressive in the world, the 2008 Montecristi Constitution recognizes and guarantees the rights of democratic government, such as freedom of speech, equality, and protection from discrimination. It establishes rights of nature, of indigenous peoples, and rights to *buen vivir* ("good living," or *sumak kawsay* in Kichwa). The legislation of indigenous cosmovision in the form of *sumak kawsay* and the definition of Ecuador as a plurinational/multicultural state are widely considered a victory for progressive Ecuadorian politics, representing "a new and historic stage in Ecuador's history" (Becker 2011, p. 158).

Such a utopic vision requires both state projections of power, and also revenue. Ortiz (2015, p. 31) argues that President Correa's political project has created "a complex institutional machinery that seeks disciplinary control over the core components of the Ecuadorian public sphere, particularly civil society, under the justification of having designed a structure for participatory democracy." Similarly, Conaghan (2011, p. 280) notes

> ... it is important to recognize that Correa's project is profoundly state-centric in the scope of its aspirations: the constitution, law, and public policy all drive in the direction of shifting greater economic power to public-sector enterprises, extending the government's regulatory grip, and placing more economic resources in the hands of the central government.

But the Ecuadorian state, like all states, reveals the limits of ability to encompass and govern the nation precisely through institutional and spatialized attempts to assert that power (Ferguson & Gupta 2002). In other works (Lu & Silva 2015), we have analyzed the means by which conflicting practices of border making and enforcement by the Correa administration and state-owned oil companies reveal the limits of the ideological declarations foundational to the 2008 Constitution and the Revolución Ciudadana. We have also explored the contradictions inherent to the developmentalist vision and environmentalist claims which the Correa administration touts as demonstrations of the legitimacy and effectiveness of its *Revolución* (Lu, Valdivia, & Silva, 2017). Sitting at the nexus of both developmentalist and environmentalist concerns, water further reveals the incoherence of "hydraulic citizenship" (Anand 2011) as proposed by the Correa administration.

The Revolución Ciudadana's project of development and elimination of poverty are goals that, according to the president, can only be achieved with strategic resources such as oil (Correa 2012, p. 95). The 2008 Constitution, for all its progressivism, stresses the government's control over the extractive industry, defining the oil sector as a strategic asset and declaring the state's inalienable ownership. In one of his weekly public speeches, Correa (Enlace 365, March 15, 2014) said

> I don't like petroleum, but I like destitution even less. It's not that they [the opposition] like nature and we don't [...] We understand that the most important part of nature is the human being. Our children have the right to a good life: education, health, nutrition. Our families have the right to services. Our people have the right to overcome destitution. And for that we need natural resources.

In his first year in office, Correa doubled poverty assistance payments and credits available for housing loans, subsidized electricity rates for low-income consumers, and rechanneled millions of dollars into other social programs (Conaghan & de la Torre 2008). That increased spending was possible due to various financial mechanisms introduced by the new regime. To the astonishment of multilateral lending institutions, the Correa administration eliminated debt–repayment accounts and redirected the "liberated" oil rents into government spending on social programs. The administration increased taxes on foreign oil companies, raising the oil windfall profits tax from 50 to over 90 percent (Conaghan & de la Torre 2008). In 2007, the Correa administration changed its preferred contractual method to the Service Provider model, which allows the national oil company to remain the primary operator in an oil block while benefitting from the technology and expertise of private or foreign state-owned oil companies contracted to undertake extraction in exchange for a production fee per barrel (Ghandi & Lin 2014). The fact that oil was valued around or above $100 per barrel in 2007 was critical to the Correa administration agenda.

The Revolución Ciudadana entails not only providing public projects and services to densely populated urban regions of Ecuador, but also ameliorating decades of neglect in rural areas, especially oil extraction zones. To execute social projects in areas associated with high poverty and critical natural resources, the state created EE in September 2011, through Executive Decree 870 and according to Article 315 of the Ecuadorian Constitution. EE is a "public firm" with administrative and financial autonomy, and claims commitment to improving the living conditions of peoples historically abandoned by the state. Drawing upon funds guaranteed through the Law of Hydrocarbons and the Law of Mining, EE is tasked with planning, prioritizing, and executing local and infrastructural development projects in 11 provinces (Lu, Valdivia, & Silva 2017). The funds disbursed to EE are equivalent to 12 percent of the profits of oil and mining companies, and are dedicated to "local development and infrastructure" in zones influenced by "strategic sector" projects (Correa 2011, p. 1), to alleviate the burdens suffered by those living in extraction sites.[2] Before discussing the efforts of EE in Gareno, we examine patterns of government intervention in Waorani territory more broadly.

Waorani governance and oil

Called a "dynamic beast with a complex structure and myriad consequences" (Cepek 2012, p. 409), petroleum exploitation has had wide-ranging effects on Ecuadorian indigenous people, especially those living in the Oriente. Until the oil industry and colonization arrived in Ecuador's Amazon, indigenous people were of minimal concern in state policy. Today, although indigenous movements in Ecuador have consistently struggled for and received land rights, which mitigate colonization of indigenous territories, the state retains the right to subsurface materials (Mijeski & Beck 2011, pp. 138–139), such as petroleum and mineral ore. The end result of conflicting superficial and subsurface cadastral systems is a de facto lack of territorial autonomy for indigenous people living in spaces where subsurface resources have been identified or are believed to exist.

A year after Ecuador discovered crude of marketable quality and quantity, the government authorized—under the efforts of the Summer Institute of Linguistics (SIL), a Protestant missionary organization, and with the financial and logistical support of oil companies—the establishment of a Waorani "protectorate," an area approximately 160 square kilometers in the westernmost portion of ancestral Waorani territory.[3] By the beginning of the 1970s, almost 80 percent of the Waorani population (which was around 500 people) was intensely controlled by SIL in the protectorate (Finer et al. 2009, p. 6). What were previously indigenous homelands and hydrosocial territories were transformed into oil concessions. The dramatic nature of such a shift cannot be overstated; the Waorani had previously occupied a territory of 20,000 square kilometers, home to dispersed *nanicaboiri*, residential units of related kin that were economically self-sufficient and autonomous. In the protectorate, the Waorani lived at a population density and with a degree of sedentism that starkly contrasted with their traditional practices. Moreover, kin groups that had been at war with each other were living in close proximity. The undermining of culturally salient borders and the circumscription of indigenous peoples to facilitate oil exploitation was an assertion of state "spatial imperatives" of economic development, enacted through "the spatiality of administrative strategies of disciplining [Waorani] populations" (Moore 2000, p. 675) in Ecuadorian Amazonia. Waterways integral to hydrosociality were among the resources that formed part of these geographies of discipline.

For the Waorani, the new borders drawn by the state in order to facilitate oil exploitation catalyzed the devaluation and insecurity of once critically important and endogenously managed resources—forests, waterways, game, and foraged foods—that had been subject to common property regimes. Defined as "private property for the group" (Lu 2001, p. 427–428), common property regimes serve multiple purposes: they foster connections between group members, create social boundaries, mitigate subsistence risk, and structure interpersonal relationships. Before sustained contact with Western society beginning in 1958, sociocultural cohesion and reproduction for the Waorani depended on a common property regime as a means of managing a self-secured and regulated territory, and distributing vital resources. The loss of the ability to access, control, and regulate those resources—along increasing dependence on exogenously controlled resources—entailed a dramatic shift in the production and maintenance of social and spatial borders (Lu & Silva 2015). Missionary-led relocation and oil development catalyzed adoption of a settlement pattern characterized by more densely populated and permanent nucleated villages with schools and landing strips—often located along roads instead of rivers—as many Waorani oriented themselves toward a search for market goods and services. Limited infrastructural development, colonization by once-urban settlers, deforestation, and an influx of market-produced foods, alcohol, and commodities now permeate Waorani communities (Lu 2007, 2012; Lu et al. 2010). Under Correa's Revolución Ciudadana, although the distribution of oil revenues has perhaps become more equitable, oil exploitation continues unabated and unquestioned, and with it so do the alterations in the spatiality of Amazonia. Water, once a resource whose endemic control and distribution promoted social cohesion among the Waorani, has become one of many sites through which questions of development, modernity, colonization, and resource valuation are being negotiated by indigenous peoples and the Ecuadorian state.

Water waste: a failed effort in Gareno

Since their arrival in the early 2000s, Perenco spent anywhere from US$3,000 to US$11,000 a year on "community and education projects" in Gareno. The Anglo-French company built infrastructure, offered wage labor opportunities, and provided goods and services in Gareno and other indigenous communities in their oil concession: a *cancha cubierta*, medical center, schoolhouse, medications, school supplies, and water infrastructure (e.g., the toilets and faucets mentioned above that no longer work). And every week, on Fridays, the oil company would send a truck to take Gareno residents to the local market, or *feria*, about 12 kilometers away. Until the Correa administration and the resource nationalization it implemented, the presence of the state paled in comparison with that of multinational oil corporations in the lives of residents of Gareno. Locals arrived at contractual "or simply verbal arrangements with these companies that enable[d] them to access capital and other material, technical, and financial resources" (Albán Compaña 2015, p. 22). Companies such as Perenco employed a staff person in charge of community relations to negotiate these local demands, implement development efforts, and facilitate operations by diffusing local resistance.

In the last few years, as part of the Revolución Ciudadana and resource nationalism, the Ecuadorian state has taken over these roles in extraction areas through EE, attempting to show itself as an entity that serves "forgotten" communities such as Gareno. Under Correa, reforms to the Hydrocarbon law required foreign companies to renegotiate their arrangements with the government. Perenco, unlike other international oil companies operating in Ecuador, refused to change its revenue-sharing policies and threatened to shut off its facilities in 2007 when Rafael Correa raised taxes (Mapstone 2009). Correa expelled Perenco for failure to meet its tax obligations. The company abandoned its facilities in Summer 2009, and Petroamazonas, a subsidiary of Ecuador's state-owned oil company Petroecuador, took over crude production. These changing arrangements of governance of and through oil exploitation included certain proscriptions by the Correa administration against multinationals' interactions with extraction-site communities.

This trajectory—initial oil exploitation, community relations, and resource provision managed by a transnational oil company followed by a shift of these services to the state-owned company—mirrors a private-to-public shift happening in other Waorani communities within oil concessions and throughout Ecuadorian Amazonia, as the state expands its control over natural resource extraction. When Perenco left, many Gareno residents were concerned about changes under the new regime. Under the multinational, the Waorani negotiated directly with the Perenco representative tasked with community relations: such direct, person-to-person negotiations with an oil company representative endowed with certain levels of decision-making power and financial resources were amenable to the Waorani, most of whom are not accustomed to navigating bureaucracy. With the departure of Perenco and the advent of Petroamazonas, the once clear pathway to the provision of oil-supported benefits had become muddled.

In early March 2014, EE began an ambitious water development project in Gareno (as well as Meñepare and Koñipare, Waorani villages along the same oil road as Gareno) costing almost $384,000. The plan was to build a piping system to transport water from a clean source on Guayusa Hill, approximately 4 kilometers northeast of Gareno, to all the homes below. The community members would later learn, much to their disappointment, that despite promises to provision all households in Gareno with clean water, the project would only provide water for a fraction of the families, for an unclear reason that had something to do with oversights by community president who negotiated the project planning. This water system, promised in 2012, had been prone to delays attributed to corruption among both municipal government representatives and people in the community. The plan in March 2014 was to have the water system construction completed by July, with construction undertaken using labor from community members.

The work involved was backbreaking: clearing forest, digging a trench for the water pipe, and laying concrete, which entailed carrying 50 kilogram (over 110 pound) sacks of concrete 4 kilometers up a steep hill. EE offered wage labor opportunities for women as well, who dug ditches but did not haul the sacks of concrete. Laborers worked from 7 a.m. until 5 p.m. 6 days a week and 7 a.m. until 1 p.m. on Sundays. While the Waorani were accustomed to clearing the forest and did this work with enthusiasm, ditchdigging and carrying concrete were grueling and led many of the men to abandon the project.

Initially, EE offered each worker $15 per day, plus three meals. The community bargained this up to $20 per day without meals, complaining that the engineers on the project make $4,000 monthly. EE called a meeting in Gareno in May 2014, during which some workers complained that the pay was insufficient. People were upset by rumors that a similar project in another Waorani community was paying $30 per day plus three meals. Yet, one of the women workers stressed the importance of the project for the welfare of the community—this was why she participated, she proclaimed, to applause. In the meeting, no pay raise was negotiated. EE itself does not implement these projects, but exclusively subcontracts them, in this case, to a contractor from the Ecuadorian coast, hundreds of miles away. To receive their pay, workers had to wait, sometimes hours, for the contractor to show up. They were prone to having their pay docked when the contractor claimed that they had shown up late for work or left early. These contractors—connected to the social systems of Amazonia primarily through capitalist social relations—have little incentive to work efficiently, use resources wisely, treat local residents with respect, or demonstrate any accountability when the project fails.

We visited Gareno in July 2014, when the construction of the water project was slated to be completed. We walked up past Guayusa Hill, imagining the hike while carrying over 100 pounds. We saw multitudes of bags of concrete leaned up against trees, stacked in small piles, and dumped haphazardly along the trail. We marveled at the two water tanks, 8 feet in diameter and resting at the bottom of 10-foot deep holes at the top of the hill. We walked past a cement mixer, lying on its side, useless. "Someone stole the motor," a Waorani companion remarked. Clearly much labor had been expended, but the project was nowhere near completion, and likely would never be.

That these Waorani families ended up without a potable water system after a very expensive, much publicized, and incredibly delayed project exemplifies the failure of funds from extractivism to provide tangible, lasting improvements in the quality of life of people who for so long have borne the costs of oil extraction. It exemplifies an unsuccessful attempt to institute new forms of state-sanctioned hydrosociality characterized, in large part, by inadequate management including a lack of accountability by project contractors and flawed community relations, resulting in residents feeling they were exploited through a project doomed to failure from its outset.

Yet the national government does not hesitate to champion its good work for local people in its propaganda. In a section titled "Works to generate *buen vivir* in the Amazon region," the online magazine *Pastaza Habla* states that "the challenge of eradicating the poverty in this region of the country is fulfilled by the political and historical decisions of the Government of the Revolución Ciudadana, focused on prioritizing the areas that have generated so much wealth for the country."[4] EE erected an enormous billboard in the middle of Gareno, which reads: "*¡El Petroleo beneficia tu comunidad!*" (petroleum benefits your community). The families will walk by this billboard and continue to drink water from small gullies long after funding for this water project is finished and EE leaves. They will continue to live in a hydrosocial system that is a symbolic and material negotiation between the ideological claims of the Correa administration, a built environment changing in response to the enactment of these ideologies, and the experience of affective geographies recalling fundamentally different modes of controlling and distributing resources, including water.

ClearWater: a different model of hydrosociality

The toxic legacy of Texaco's operations in the Oriente catalyzed efforts to seek environmental justice and remediation, something exemplified in the long-standing legal battle that began as Aguinda v. Texaco, a $1.5 billion class action lawsuit filed in November 1993, seeking financial alleviation of some of damages caused by over 25 years of contamination (Kimerling 2000, p. 83).[5] Since then, the lawsuit has traveled between Ecuadorian and US courts, evolved into a series of new litigations involving growing cadres of lawyers and judges whose actions, for both the plaintiffs and the defense, have been called ethically questionable (Kimerling 2012). The legal battle to resolve the Aguinda v. Chevron verdict is being waged in dozens of courts in six countries and could drag on for years. While this "legal knife fight" drags on (Barrett 2013), Amazonian residents living in contaminated lands drink polluted water. In both the United States and on a global scale, the Texaco trial has affected the way that people think about environmental politics in general, and hydrosociality in particular.

The trial was the focus of a documentary called "Crude: The Real Price of Oil" (Berlinger 2009), in which Trudie Styler, wife of rock star Sting, travels Ecuador's Amazon on a toxic tour and is visibly moved by the cancer victims and destruction she witnesses. Styler catalyzed an effort between UNICEF Ecuador, the Rainforest Fund, and the Amazon Defense Fund to install rainwater collection and filtration systems and tanks in schools, medical centers, and homes as a stopgap measure to address a public health emergency. Her effort provided a model for the creation of ClearWater,[6] which describes itself as "a movement for clean water, cultural survival and rainforest protection in Ecuador's northern Amazon" (www.giveclearwater.org).

Founded by former Amazon Watch corporate accountability director Mitch Anderson, ClearWater was also inspired by assertions from indigenous people in the Oriente who recognized that they might never see a dime from the Aguinda lawsuit, but desperately need clean water. ClearWater initially partnered with the Union of Peoples Affected by Texaco's Oil Operations (*Unión de Afectados/as por las Operaciones de la petrolera Texaco-Chevron*, UDAPT), and the Front for the Defense of Amazonia (*Frente de Defensa de la Amazonia*). More recently, it helped found an organization called Ceibo Alliance, working with indigenous peoples of the Northern Ecuadorian Amazon to defend cultural and land rights through territorial mapping, media communications, and legal advocacy. Funding has come from other celebrity-related agencies, namely Rea Garvey's Saving an Angel and the Leonardo DiCaprio Foundation. To date, over 500 family-sized rainwater catchment and filtration systems have been installed in Cofán, Siona, Secoya, Kichwa, and Waorani communities, the provision of which has made a material difference in quality of life for the people who received such systems.

ClearWater's approach to bring potable water to indigenous populations affected by contamination from oil extraction, and increasingly, oil palm plantations, reflects a model of governance that contrasts with that of EE. The water systems are implemented at the level of the household, not at the level of the community; also local residents are incorporated in a participatory and collaborative approach that entails appropriate technology transfer and ownership of water infrastructure rather than merely employment as workers. According to the organization's website and interviews with staff members, ClearWater involves indigenous partners as coordinators at the level of nationality and community. In each community, meetings are held in which residents receive an orientation about the project and then elect people to oversee the community effort (i.e., the community water coordinator) as well as workers to undertake the installation, under the tutelage of indigenous water technicians. By design, community residents who help to install the rainwater catchment systems become the local technicians who continue to provide support to their kin and neighbors in maintaining the infrastructure (e.g., rinsing of filters and cleaning of ducts). As the ClearWater website states, "Local technicians are a fundamental element of the ClearWater model; they are the face of the communities working to improve their own situations." Yet those systems, like the EE project in Gareno, introduce forms of hydrosociality that are substantively different from ancestral indigenous forms. And while

ClearWater's rainwater catchment systems obviate the concern over latent hydrological threats via contamination, their systems depend on rainwater and as such depend on an element of climatic precarity. We heard more than one tale of limited access to water during dry spells.

In May 2015, ClearWater undertook installation of rainwater harvesting systems in Gareno,[7] roughly a year after the failed EE water project. Community meetings were held, local coordinator and technicians elected, and visits made to each of the beneficiary households to assess habitation, settlement patterns, roofing quality, and material. The newly elected community coordinator played an active role in the house-to-house visit, explaining in the local language the construction process and asking for the beneficiary's cooperation. Guided by technicians, household members play a role in the installation of the system for their family. This sweat equity serves to foster a sense of ownership and increase the likelihood of upkeep. Once the water system is built (cement pad laid, metal framing constructed, cisterns connected by canals to a roof), it belongs to the beneficiary household. If maintained well, a family-level rainwater system can function for a decade or two, with each unit costing about $1,500.

Water provisioning through ClearWater produces a common-pool resource shared within a household, and from which communities as a whole are excluded, meaning that it does not contribute to the production of inter-household bonds in the same manner that might be done, for example, through localized management of a community-run system, itself an idea fraught with potential problems. ClearWater only installs water infrastructure at individual houses, not for public spaces such as the *casa comunal*, because the latter would be less likely to receive maintenance. ClearWater staff commented that state projects take the other approach in part because of the visibility of the one, central infrastructural effort for the collective good; furthermore, a communal sense of ownership and cohesion is presumed. However, without continued state investment and upkeep, or at least efforts to inculcate a sense of ownership and investment, these efforts fall prey to the tragedy of the commons, and are rendered useless. The notion of one, centralized water system serving a community such as Gareno also runs counter to the system of hydrosociality in contemporary Waorani societies, where nucleated communities do not possess the social and cultural cohesion that *nanicaboiri* once did; where common property resource management regimes have faltered under current challenges (see Lu 2001); and where Waorani cultural values emphasize individuality, autonomy, and self-rule. When asked about the defunct EE water project, Gareno residents were dismissive, using the term "*botado*" ("thrown away"), and speaking as if it never happened. The ClearWater effort was undertaken as a restorative and preventative act in response to widespread contamination and international litigation that has yet to provide any redress. While not reduced to mere manual laborers as in the EE effort, Waorani residents are subject to a new set of normative obligations and expectations imposing new forms of hydrosociality, including interfacing with new types of external water "experts," election of community water coordinator and technicians, workshop participation, knowledge acquisition, and infrastructural maintenance.

Conclusion

As Perreault (2014, p. 235) observes, water "shape[s] social relations, even as those social relations act on and shape water's materiality [...] at once naturally occurring and socially produced, both embodiment of and precondition for social power." Before sustained contact with outsiders, Waorani hydraulic technology was minimal: water was not diverted, channeled, or filtered, nor were weirs or other lasting structures built to catch fish. The mobile, autarkic, and egalitarian kin-groups that constituted Waorani society before sustained contact utilized and understood waterways as sources of subsistence, borders between hostile populations, and as meaningful markers of connection to place. They exemplify a unique and compelling example of hydrosocial territoriality and affective geography. Oil extraction and dispossession have necessitated new hydrosocial regimes, new and fractured visions of the hydraulic citizen. In each of the two cases discussed in this paper the histories of changing hydrosociality in Ecuador are addressed differently by distinct institutions of water provision, each suggesting its own image of how the hydrosociality of indigenous peoples who need potable water might be reshaped according to etic assumptions and interventions by powerful individuals and institutions.

EE exemplified a developmentalist approach to improving local standards of living, and sought to compensate residents for the deleterious effects of oil extraction in an effort to gain or maintain their complicity. Provisioning potable water to households would bring these families into the state project, offering *buen vivir* yet requiring their menial labor and outsourcing the oversight of the project to a contractor not from the region and lacking expertise among the beneficiary population. Shunting a small proportion of oil revenues for projects such as water infrastructure in communities historically marginalized such as Gareno seems additionally inadequate when operations fail to deliver promised results. Moreover, the ad hoc approach to provisioning basic services reinforces the sense among locals of the transitory nature of government investment and the disposable quality of the resulting infrastructure.

The ClearWater project, however, inadvertently reward sedentism and settlement permanence, as only families with existing, actively occupied, houses receive these systems. Those with roofs of metal sheets in good condition enjoy a greater ease of participation; those with homes roofed with palm leaves must construct a separate structure to be granted a rainwater catchment system. Discourses about beneficence and beneficiaries, investments and interventions, elide these forms of social ordering and control, the latest in a protracted process of outside agents and forces creating problems and providing solutions for and among the Waorani. This struggle for and against forms of social ordering through access to potable water—an issue linked to the condition of waterways and histories of contamination—is at the heart of contemporary hydrosocial conditions experienced by the Waorani.

Notes

1. The founding and populating of Gareno begs the question of why the early Waorani residents, who previously inhabited ecologically intact forests, voluntarily decided to relocate along an oil road (see Lu 2012).
2. Article 313 of the 2008 Constitution defines energy, telecommunications, nonrenewable natural resources, hydrocarbon refinement, and biodiversity and genetic patrimony as "strategic sectors."
3. On 13 April 1983, then President Osvaldo Hurtado gave the Waorani title to 665.7 square kilometers bounded by the Nushiño and Manderoyacu Rivers and including the protectorate. On 3 April 1990, Ecuador created the Waorani Ethnic Reserve, encompassing 6,125.6 square kilometers or about one-third of their original territory. To date, the amount of territory conceded to the Waorani is 6,791.3 square kilometers, not including use rights to Yasuní National Park. However, the Waorani have effectively lost control of approximately 350 square kilometers to colonization along both sides of the Vía Auca (Trujillo Montalvo 2011, p. 26).

4. http://pastazahabla.blogspot.com/2015/01/boletin-de-prensa-obras-para-generar-el.html
5. Specifically, damage to the plaintiffs was the result of failure to pump unprocessable crude and toxic residue back into wells, discarding of toxics into oil pits and waterways, burning of crude without temperature or pollution control, spreading of oil on roads, and design and construction of pipelines without adequate safety features (Sawyer 2001, p. 186). These cost-cutting practices saved Texaco $4–6 billion between 1972 and 1991, when the company produced over 1.4 million barrels of crude and earned over $23 billion (Sawyer 2001, p. 163), while causing an estimated 1,400 excess cancer deaths via contamination.
6. ClearWater began as a project under the San Francisco non-governmental organization Groundwork Opportunities.
7. One of the rainwater catchment and filtration systems installed in Gareno was paid for by UC Santa Cruz students as part of a College Nine course, "Global Action." Students held a fundraising event, "Life for the Amazon," on 25 February 2015 that included informational displays about the ClearWater project and water technology, information about Waorani culture, tabling by Amazon Watch and the Terence Freitas Café, and sales of Waorani handicrafts and food. Students raised over $2,000, all of which went to ClearWater.

References

Albán Compaña, D. (2015). Teen pregnancy on the oil road: Social determinants of teen pregnancy in an indigenous community in the Ecuadorian Amazon (Master's Thesis). Department of Anthropology, University of North Carolina at Chapel Hill, Chapel Hill, NC.

Anand, N. (2011). Pressure: The politechnics of water supply in Mumbai. *Cultural Anthropology*, 26(4), 542–564.

Barrrett, P.M. (2013). Patton Boggs's fight against Chevron continues. *Bloomberg Businessweek*, May 14, 2013.

Becker, M. (2011). *Pachakutik: Indigenous Movements and Electoral Politics in Ecuador*. Lanham, MD: Rowman & Littlefield Publishers.

Berlinger, J. (2009). Crude: The Real Price of Oil. Documentary. New York, NY: First Run Features.

Boelens, R., J. Hoogesteger, E. Swyngedouw, J. Vos, & P. Wester (2016). Hydrosocial territories: A political ecology perspective. *Water International*, 41(1), 1–14.

Carriére, J. (2001). Neoliberalism, economic crisis, and popular mobilization in Ecuador. In J. Demmers, J.A.E. Fernández, & B. Hogenboom (Eds.), *Miraculous Metamorphoses: The Neoliberalization of Latin American Populism* (pp. 131–149). London, UK: Zed Books.

Cepek, M. (2012). *A Future for Amazonia: Randy Borman and Cofán Environmental Politics*. Austin, TX: University of Texas Press.

Conaghan, C.M. (2011). Ecuador: Rafael Correa and the Citizens' Revolution. In L. Steven, & K.M. Roberts (Eds.), *The Resurgence of the Latin American Left* (pp. 260–282). Baltimore, MD: Johns Hopkins University Press.

Conaghan, C., & C. de la Torre. (2008). The permanent campaign of Rafael Correa: Making Ecuador's plebiscitary presidency. *The International Journal of Press/Politics* 13(2), 267–284.

Correa, R. (2011). Decreto Ejecutivo No. 870 [Creación de *Ecuador Estratégico*, Empresa Pública] Quito: Ecuador Estratégico EP. Available at: http://www.ecuadorestrategicoep.gob.ec/images/leytransparencia/Base%20legal/A1%20Base%20Legal.pdf.

Correa, R. (2012). Ecuador's path. *New Left Review*, 77, 89–111.

Donzinger, S. (2015). Chevron's "Amazon Chernobyl" in Ecuador: The real irrefutable truths about the company's toxic dumping and fraud. *The Huffington Post* May 27, 2015.

Ferguson, J., & A. Gupta. (2002). Spatializing states: Toward an ethnography of neoliberal governmentality. *American Ethnologist*, 29(4), 981–1002.

Finer, M., V. Vijay, F. Ponce, C.N. Jenkins, & T.R. Kahn. (2009). Ecuador's Yasuní Biosphere Reserve: A brief modern history and conservation challenges. *Environmental Research Letters*, 4(3), 034005.

Gerlach, A. (2003). *Indians, Oil and Politics: A Recent History of Ecuador*. Wilmington, DL: Scholarly Resources, Inc.

Ghandi, A., & C.Y.C. Lin. (2014). Oil and gas service contracts around the world: A review. *Energy Strategy Reviews*, 3, 63–71.

Hidalgo Flor, F. (2013). Posneoliberalismo y Proceso Político en el Ecuador. *Revista Internacional de Filosofía Iberoamericana y Teoría Social*, 18(62), 77–88.

Hoogesteger, J., R. Boelens, & M. Baud. (2016). Territorial pluralism: Water users' multi-scalar struggles against state ordering in Ecuador's highlands. *Water International*, 41(1), 91–106.

Hurtig, A-K., & M. San Sebastián. (2002). Geographical differences in cancer incidence in the Amazon basin in Ecuador in relation to residence near oil fields. *International Journal of Epidemiology*, 31(5), 1021–1027.

Jochnick, C., R. Normand, & S. Zaidi. (1994). *Rights Violations in the Ecuadorian Amazon: The Human Consequences of Oil Development*. New York, NY: Center for Economic and Social Rights.

Kimerling, J. (1991a). *Amazon Crude*. Washington, DC: Natural Resources Defense Council.

Kimerling, J. (1991b). Disregarding environmental law: Petroleum development in protected natural areas and indigenous homelands in the Ecuadorian Amazon. *Hastings International and Comparative Law Review*, 14, 849–903.

Kimerling, J. (1993). *Crudo Amazónico*. Quito, Ecuador: Abya Yala.

Kimerling, J. (1996). Oil, lawlessness and indigenous struggles in Ecuador's Oriente. In H. Collinson (Ed.), *Green Guerillas: Environmental Conflicts and Initiatives in Latin America and the Caribbean* (pp. 61–73). London, UK: Latin America Bureau.

Kimerling, J. (2000). Oil development in Ecuador and Peru: Law, politics and the environment. In A. Hall (Ed.), *Amazonia at the Crossroads: The Challenge of Sustainable Development* (pp. 73–96). Washington, DC: Brookings Institution Press.

Kimerling, J. (2006). Transnational operations, bi-national injustice: ChevronTexaco and Indigenous Huaorani and Kichwa in the Amazon Rainforest in Ecuador. *American Indian Law Review*, 31(2), 445–508.

Kimerling, J. (2012). Huaorani land rights in Ecuador: Oil, contact and conservation. *Environmental Justice*, 5(5), 236–251.

Lu, F. (2001). The common property regime of the Huaorani Indians of Ecuador: Implications and challenges to conservation. *Human Ecology*, 29(4), 425–447.

Lu, F. (2007). Integration into the market among Indigenous Peoples: A cross-cultural perspective from the Ecuadorian Amazon. *Current Anthropology*, 48(4), 593–602.

Lu, F. (2012). Petroleum extraction, indigenous people and environmental injustice in the Ecuadorian Amazon. In F. Gordon & G. Freeland (Eds.), *International Environmental Justice: Competing Claims and Perspectives* (pp. 71–95). Hertfordshire, UK: ILM Publishers.

Lu, F., C. Gray, C. Mena, R. Bilsborrow, J. Bremner, A. Barbieri, C. Erlien, & S. Walsh. (2010). Contrasting colonist and indigenous impacts on Amazonian Forests. *Conservation Biology*, 24(3), 881–885.

Lu, F., & N.L. Silva. (2015). Imagined borders: (Un)bounded spaces of oil extraction and indigenous sociality in "Post-Neoliberal" Ecuador. *Social Sciences*, 4, 434–458.

Lu, F., N. Silva, K. Villeda, & M. Sorensen. (2014). Cross-cultural perceptions of risks and tenables among Native Amazonians in Northeastern Ecuador. *Human Organization*, 73(4), 375–388.

Lu, F., G. Valdivia, & N.L. Silva. (2017). *Oil, Revolution, and Indigenous Citizenship in Ecuadorian Amazonia*. Basingstoke, UK: Palgrave Macmillan.

Machado Puertas, J.C. (2007). Ecuador: El derrumbe de los Partidos Tradicionales. *Revista de Ciencia Política (Santiago)*, 27, 129–147.

Mapstone, N. (2009). France's Perenco oil company leaves Ecuador amid tax dispute. *Americas Quarterly*, July 24, 2009.

Mijeski, K.J., & S.H. Beck. (2011). *Pachakutik and the Rise and Decline of the Ecuadorian Indigenous Movement*. Athens, Greece: Ohio University Press.

Moore, D.S. (2000). The crucible of cultural politics: Reworking "development" in Zimbabwe's eastern highlands. *American Ethnologist*, 26(3), 654–689.

Murphy, L. (2001). Colonist farm income, off-farm work, cattle, and differentiation in Ecuador's Northern Amazon. *Human Organization*, 60(1), 67–79.

Navaro-Yashin, Y. (2012). *The Make-Believe Space: Affective Geography in a Postwar Polity*. Durham, NC: Duke University Press.

Ortiz, A. (2015). Taking control of the public sphere by manipulating civil society: The Citizen Revolution in Ecuador. *European Review of Latin American and Caribbean Studies*, 98, 29–48.

Perreault, T. (2014). What kind of governance for what kind of equity? Towards a theorization of justice in water governance. *Water International*, 39(2), 233–245.

Rodríguez-de-Francisco, J.C., & R. Boelens. (2016). PES hydrosocial territories: De-territorialization and re-patterning of water control arenas in the Andean highlands. *Water International*, 41(1), 140–156.

San Sebastián, M., B. Armstrong, & C. Stephens. (2000). Reproductive health of women living in the proximity of oil fields in the Amazon basin of Ecuador. *Epidemiology*, 11(4), S86.

San Sebastián, M., B. Armstrong, & C. Stephens. (2002). Outcome of pregnancy among women living in the proximity of oil fields in the Amazon basin of Ecuador. *International Journal of Occupational and Environmental Health*, 8, 312–319.

San Sebastián, M., & A-K. Hurtig. (2004). Oil exploitation in the Amazon basin of Ecuador: A public health emergency. *Pan American Journal of Public Health*, 15(3), 205–211.

Sawyer, S. (2001). Fictions of sovereignty: Of prosthetic petro-capitalism, neoliberal states, and phantom-like citizens of Ecuador. *The Journal of Latin American Anthropology*, 6(1), 156–197.

Sawyer, S. (2004). *Crude Chronicles*. Durham, NC: Duke University Press.

Sorahan, T., & R.J. Lancashire. (1997). Lung cancer mortality in a cohort of workers employed at a cadmium recovery plant in the United States: An analysis with detailed job histories. *Occupational and Environmental Medicine*, 54, 194–201.

Trujillo Montalvo, P. (2011). *Boto waorani, bito cowuri: La fascinante historia de los Wao*. Quito, Ecuador: Fundación de Investigaciones Andino Amazónicas.

Valdivia, G. (2008). Governing relations between people and things: Citizenship, territory, and the political economy of petroleum in Ecuador. *Political Geography*, 27, 456–477.

Waalkes, M.P., & S. Rehm. (1994). Cadmium chloride in male DBA/2NCr and NFS/NCr Mice. *Toxicological Sciences*, 23(1), 21–31.

Zwarteveen, M.Z., & R. Boelens (2014). Defining, researching and struggling for water justice: Some conceptual building blocks for research and action. *Water International*, 39(2), 143–158.

Water scarcity and the exclusionary city: the struggle for water justice in Lima, Peru

Antonio A. R. Ioris

School of Geosciences, University of Edinburgh, Edinburgh, UK

ABSTRACT

Water management dilemmas represent a unique entry point into the challenging management of metropolitan areas, as in the case of Lima (Peru). A condition of water scarcity goes beyond the mere physical insufficiency of resources, but vividly contains the inadequacy of social relations responsible for the allocation, use and conservation of water. Lima's experience demonstrates the association between investment priorities, political agendas and corruption scandals leading to selective abundances and persistent scarcities that are perpetuated in a hydrosocial territory. The production of water scarcity has been predicated upon discriminatory practices associated with the reinforcement of uneven development and environmental injustices.

Introduction

Political theory lacks a sense of territory; territory lacks a political theory.

Stuart Elden (2010)

Social and economic development in Latin America has been often undermined by unfulfilled promises and contradictions. That is particularly the case in large metropolitan centres where pockets of urban wealth and ostensible affluence stand amidst vast areas of deprivation, overcrowding, pollution and multiple forms of violence (Jones, 2006). Urban water problems clearly demonstrate the difficulty of many prevailing approaches to the management of metropolitan areas in more inclusive and equitable ways (UNDP, 2006). Since 1990, around 22% of the Latin American population has gained access to better drinking water sources, but there are still significant inequalities between different sections of the cities (UNICEF and WHO, 2012). The sanitation sector has also observed a similar trajectory, but with less impressive results, and there are comparable contrasts between achievements in urban and rural areas. Although large conurbations are the main geographical sites of social activity and capital accumulation in the region, a more dedicated appreciation of their specificities and particular dilemmas remains largely unmet (Hanson, 2003). There is still a demand for analytical methodologies able to reconcile urban processes with wider development pressures that simultaneously shape the territory of large metropolises. In that context, Lima's water services constitute a highly emblematic example of the difficult politico-ecological issues faced by Latin American society.

The Peruvian capital has increasingly come to rely on water coming from vanishing Andean glaciers, due to climate change, and from receding aquifers and degraded catchments, due to uncontrolled mining, over-abstraction and untreated effluents (ANA, 2015). At the same time, the water sector of Lima has been strongly influenced by the reconfiguration of the national state after the heterodox economic experiments of the 1980s and the liberalizing reforms introduced in the 1990s (Ioris, 2012a). Investments in infrastructure and management strategies followed changes in institutional frames, absorbed external liberalizing pressures and necessarily conformed to the balance of political power in the country. The modernization of the water sector – essentially, the adoption of more flexible and normally market-based institutions of water management and for the operation of public utilities – has become a key feature of the expanding business opportunities in Peru. For more than two decades now, water problems have been addressed through an emphasis on rationalization, privatization and the promotion of public–private partnerships, which recast water as an asset with economic value, without regard to its properties as a public good and a substance that is essential for social and household life. However, rather than a straightforward process, the renovation of the water industry of Lima epitomized a range of intricate transformations that, directly or indirectly, incorporate water into the production of new sociospatial configurations. After nearly two centuries of home rule since independence, there are still no universal, reliable public services, and for large sectors of the population, water problems remain a personal, corporeal and collective ordeal predicated upon shortages in other social, economic and political arenas.

Taking into account the contested basis of water scarcity in the capital of Peru, the aim of this article is to examine the territorialization of water scarcity resulting from politicized interactions between government, economic sectors and civil society. A related question is to discuss to what extent the territorial reconfiguration of metropolitan Lima – formed by 43 municipalities in the province of Lima and 6 in the province of Callao – may also have the imprint of creative actions and survival strategies adopted by marginalized groups to cope with the failures of public water services. The analysis is based on qualitative empirical research carried out between 2009 and 2013, involving two field trips and constant follow-up contacts. Policy makers, regulators, politicians, NGO activists and experts were interviewed, together with the analysis of secondary data and the attendance of public meetings and events. In addition, ethnographic research was conducted in three locations – Pachacútec, Huaycán and Villa El Salvador, in the north, east and south sections of the metropolis, respectively – with a focus on personal interests, domestic use of water, and behavioural patterns. The next section introduces the main concepts employed for the study of the water problems of Lima – particularly the socionatural basis of territorialization and the state as the main producer of territorialized water scarcity – which is then followed by an analysis of the territorialized evolution of human-made water scarcity under different political administrations.

Socionature, hydrosocial territories and the genesis of urban water scarcity

Linton (2010) titles his book with a provocative question, "What is water?" The unexpected answer comes already in the first page: "Water is what we make of it." In other words, it is a relational substance that is constituted by myriad relationships between social groups. According to his argument, it is not correct to think about water in the abstract, as its properties and characteristics bear the traces of sociopolitical "relations, conditions and potentials". Linton, among many others, criticizes the conventional thinking that permeates most contemporary hydrology, engineering, economics and planning, which is based on, and constantly reinforces, the artificial and unhelpful separation between society and the rest of nature (Whitehead, 1920). Instead, the world in general, and water systems in particular, are fundamentally sites in which society and nature presuppose and continuously reconstitute each other over space, scale and time. This 'socionatural' characteristic of water systems has important repercussions for the understanding of territorialization as a negotiated process (see also Boelens, Hoogesteger, Swyngedouw, Vos, & Wester, 2016). Because water flows through systems that are essentially social, collaboration and disputes between public and private agents have a direct impact on the biophysical properties of water and, crucially, on the territory produced out of socionatural interactions. Different from the conventional definitions, territorialization should be interpreted as the production of historico-geographical configurations out of the engagements between humans and the rest of socionature. The constant production of territories happens at the macro scale (as in the case of cities and countries) and is accompanied by a myriad of micro-scale territorial strategies (Storey, 2012).

Rather than being fixed in time or in space, the circulation of water follows a range of evolving social demands, practices and discourses that shape the territory (Ioris, 2008). More importantly for the purpose of the current analysis, clashes and collaboration around water directly affect the socionatural course of territorialization and result in the affirmation of certain types of territorialities and the condemnation of others. The multiple strategies to produce and control territories replicate the ability of social groups, business sectors and governments to assert, maintain or resist power (Boelens et al., 2016). As defined by Sack (1986), territoriality is a strategy used by groups and public agencies to exercise power over a portion of space in order to maintain control and systematize activities and services. Territory is ultimately a *political technology* that comprises techniques for measuring land, controlling the terrain and promoting socio-economic agendas (Elden, 2010). Although the predominant rationality of state bureaucracies privileges the administrative boundaries of catchments and river basins, the actual territoriality of water is not given or prearranged in advance. The territories of water are actively produced as hydrosocial networks (Boelens, Hoogesteger, & Rodriguez de Francisco, 2014) between groups and sectors with different understandings of water values, asymmetric power relations and often contrasting cultures and technologies (Ioris, 2011). Water territories, such as catchments and hydrological regions, are spatial networks of socionatural phenomena that receive different interpretations and attract contrasting reactions (Zwarteveen & Boelens, 2014). This dynamic basis of territorialization means that, to a larger or lesser extent, any given territorial formation is transitory and will be eventually superseded. An apparently stable territory is nothing else than a 'territorial fix', and its existence depends on its being accepted internally and externally.

In addition, despite the neoliberal rhetoric of a globalized world, the national state is still the main advocate, interpreter and custodian of determined territorial fixes, which typically favour hegemonic socio-economic and political interests (see also Hoogesteger, Boelens and Baud, 2016). The national state continues to play a decisive role in the allocation and use of water, and in that process it creates situations and spaces of abundance or scarcity inscribed in the phenomenon of territorialization (see Perramond, 2016). The territorialization of water scarcity actually happens at the intersection of the overall problems of development and state regulation and inter-personal relations at the scale of household and locations (Zug & Graefe, 2014). The apparatus of the state, through its initiatives, associations and alliances, effectively territorializes scarcities in the form of smaller or larger territories of water scarcity. For instance, investments in infrastructure or environmental restoration in certain areas may happen at the expense of the lack of equivalent investments in other sites (thus producing or maintaining territorialized water scarcity, which will be opposed by those living in these territories according to the their ability to react). By the same token, regulatory toolkits currently adopted by water agencies around the world – such as water licences, user fees, decision-support systems and payments for ecosystem services – are all rationalized in relation to rising levels of scarcity. Unfortunately, most main-stream public policies normally neglect the social construction of water scarcity by limiting the analysis to the (largely utilitarian) balance between supply and demand (Ioris, 2001).

The territorialization of water scarcity is particularly evident in urban areas, which incorporate territorial disputes happening in the rest of the country as much as the wider pressures of globalization. The city "is a *mediation* among mediations" (Lefebvre, 1996, p. 101) that take place between a near spatial order (relations between individuals and local groups) and a far spatial order (larger and more powerful institutions, such as the state). Water is never scarce in absolute terms, but only under specific allocative and institutional conditions that connect hydrological processes with the preparedness to respond to multiple, and often conflicting, management objectives. What typically exists is a 'geography of multiple scarcities' (Ioris, 2012b), in the sense that numerous agendas and socio-economic pressures converge towards the production and the reinforcement of territories of water scarcity in the constrained cartography of the city. Water scarcities are caused by deliberate attitudes and collective practices that contribute to the dynamics of territorialization. Territorialized urban scarcities do not go unchallenged, but are fully manifested, in material and symbolic terms, in the daily life of the people and in the subtle forms of marginalization and ecological degradation (Ioris, 2014). The coevolution of environmental and social change in the territory of the city offers insights into creative pathways towards more democratic urban environ-mental politics (Heynen, 2014).

The central claim of this article is that the scarcity of water is not a single process caused by the shortage of resources but the outcome of present and past decisions and interventions that produced perverse consequences that affect some groups and loca-tions more than others. Urban water scarcity is more than the mechanical result of unequal urban development, but should be seen as an integral driving force of an exclusionary pattern of development that is manifested both in large-scale injustices and in systematic political control and interpersonal, social discrimination. The territory of the city is a dynamic mosaic of locales where water is stored, processed, conveyed, used, wasted and recollected. This mosaic responds to, but also creates, new scarcities. Because of the politicized nature of its access and use, water scarcity is at the same time a collective problem and the medium of renewed forms of capital accumulation that follow urban development. In the next section, the concrete experience of Lima will demonstrate the politicized and perverse origins of urban water scarcity and its perpetuation.

Territorial inscription and territorialized reaffirmation of water scarcity

Lima was established in 1535 by the Spanish invaders to be the capital of the highly lucrative colonial enterprise based on mineral extraction. Contrary to conventional historiography, the territory of Lima was not merely the product of an autocratic decision of the Spanish crown; the new urban centre that actually emerged was the result of a never-ending struggle for territorialization. The disputes around territorialization centred around the control and use of scarce water reserves as much as the defeat of the resistance offered by the local indigenous groups living along the Peruvian coast. The need to secure a strategic port for the colonial administration put great pressure on an arid region with only three small rivers and staggeringly low rates of rainfall (less than 10 mm annual mean precipitation). Increasing levels of water pollution and environmental degradation as early as the first decades of the colony forced the viceroy to reroute the main pipeline in 1562 to avoid contaminated sources (SEDAPAL, 2003). The shortage of water was mitigated by slow urban growth during colonial and early independence years. In the early years of the twentieth century, Lima was a conurbation with around 100,000 inhabitants (4% of the national population), and the expansion of water pipelines and the opening of irrigated parks were important elements of urban modernization.

However, human-made water scarcity became patently evident with the expansion of the city due to massive internal migration following industrialization beginning in the 1930s and intensifying between the 1950s and 1970s, when the rate of demographic growth was sustained above 5% per year (Matos Mar, 1984). Because of the limited number of domiciles available for the poor migrants and the small pace of investment in affordable housing, an entire illegal ('irregular') city was created within and around the regularized areas (Barreda & Ramirez Corzo, 2004). Irregular communities were initially called *barriadas* (*barrio* = neighbourhood) and more recently have been referred to as *pueblos jóvenes* (new settlements). The *barriadas* were normally established in areas without water infrastructure and, at least at first, relied on expensive and doubtful water sold by private vendors (Zolezzi & Calderón, 1987). In tandem with the fight for political recognition, people in the newly formed *barriadas* developed multiple strategies and formed alliances to secure an often precarious access to land and water (Matos Mar, 2012). Lima grew from 645,000 inhabitants in 1940 (10.4% of the national population) to more than 9.0 million in 2010 – approximately 30% of the Peruvian population, producing almost half of the national GDP (INEI, 2010).

The chaotic urban expansion and reactive public policies amplified and perpetuated water scarcity as an integral feature of the urban territory. Ill-planned, hasty urbanization based on the deposition of people in ever more remote and unauthorized locations was the practical solution to urban problems that spoke the language of scarcity and, essentially, guaranteed the perennial continuation of multiple forms of scarcity. As can be seen in Table 1, differences in income were unmistakeably translated into uneven coverage of water services (see more statistics in Ioris, 2012b). While the established areas of the metropolis had almost universal water provision, low-income municipalities had a much lower rate of service coverage. In the last available assessment (SEDAPAL, 2009), the wealthier areas, attended by the Breña and Surquillo service centres, benefited from a rate of water coverage of 99.59% and 96.56%, respectively, while low-income areas, attended by the Villa El Salvador and Callao service centres, had equivalent figures of 86.42% and 84.96%. The situation has improved in recent years (see below on the last government programmes) and, according to official data, in 2012 the rate of water service coverage was 94.6% and sanitation services were serving 89.9% of the metropolitan population. However, in practice a significant proportion of the population still has no access to public service provision and has to resort to private water vendors and discharge sewage in the streets.

It should be recognized that the particular territorial features of Lima – a vast metropolis with extensive sandy hills, limited water stocks and unplanned human settlements – only increases the cost of water services and magnifies logistical difficulties in periphery zones. In addition, the metropolis continues to expand at around 1.5% per year, and its growing territory encroaches particularly upon remote and more adverse localities (INEI, 2012, quoted in SEDAPAL, 2014). The degradation of the small rivers and the over-exploitation of aquifers also reinforces water scarcity, which is aggravated by changes in the hydrological cycle and the melting of glaciers due to climatic change (Cabrera Carranza, 2010). Even so, it can be theorized that the main reason for the persistent imbalances is that public investments and public-private initiatives have been saturated with political spin and party politics directed towards the locations and moments with the best electoral prospects. Water scarcity is manifestly connected with the scarcity of political influence and socio-economic deprivation, which operate together to form a geography of multiple scarcities. High-income neighbourhoods typically enjoy easy and ample access to enough water at a low price (because of the investments made by the government), whereas poor households and shantytowns in the urban periphery have a much lower availability of water, sometimes lower than 20 litres per capita per day. The significant difference in the impact of domestic water charges on household income across different levels of income can be seen in Table 2 (i.e. from 0.8% to 4.1% of monthly income). As can be observed, there is a positive association between income and water use; higher-income users on average

Table 1. Sociospatial inequalities in selected municipalities of metropolitan Lima.

Income condition	Total number of households	Proportion of households with public water service
High-income municipalities	92,276	99.8%
Medium-income municipalities	184,187	77.9%
Low-income municipalities	712,878	68.1%

Source: INEI and SEDAPAL databases (compiled by the author); year of reference = 2007.

Table 2. Expenditure on water and sanitation in relation to domestic income in selected municipalities.

Municipality	Average household income (PEN)	Water use (m^3 per month)	Water tariff (PEN)	Proportion of household income
Chaclacayo	936.0	20.5	38.0	4.1%
Villa Maria	817.1	16.2	31.4	3.8%
S. Juan Lurigancho	936.0	15.9	30.4	3.3%
Villa El Salvador	881.8	14.2	27.2	3.1%
Ate	1,070.0	16.2	31.1	2.9%
S. Martin de Porres	1,165.3	17.1	32.6	2.8%
Callao	1,173.5	17.3	31.6	2.7%
Cercado de Lima	1,677.3	18.2	34.2	2.0%
Jesus María	2,589.6	19.7	37.8	1.5%
La Molina	6,096.3	31.3	80.3	1.3%
Miraflores	5,386.5	23.4	47.9	0.9%
San Isidro	8,303.4	29.6	68.6	0.8%
Average (metropolitan Lima)	2,062.9	18.8	37.0	2.4%

Source: SEDAPAL (2009).

Table 3. Allocation of water per user sector, metropolitan Lima.

User sector	Proportion of total volume of water supplied by SEDAPAL	Water tariff (lower band), year 2015 (PEN/m^3)
Domestic	75.4%	1.186
Social (public fountains)	2.4%	1.186
Commercial	13.8%	5.164
Industrial	2.7%	5.164
State agencies	5.7%	2.893

Source: SEDAPAL (2014) and SEDAPAL webpage for the water tariffs (consulted in March 2015).

pay a smaller proportion of their income, and those who earn less pay on average a lower absolute charge but a higher percentage of their income.

In addition, Table 3 demonstrates the prominent differences in the volumes of water used by the domestic, industrial and commercial sectors, which means that asymmetries in household supply are translated into large-scale imbalances between groups and locations. The higher charges paid by non-domestic users are consistent with tariffs of other large-scale water utilities in South America (e.g. SABESP in São Paulo). The next sections examine the reasons for the persistent failures of public services and political biases in the actions of the state that lie behind the statistics. It is argued that the scarcity of water has in effect become a central driving force behind interventions and public policies introduced by successive governments primarily to serve selective, non-democratic political and economic interests.

Politics and policy changes that perpetuate territorialized scarcity

In 1990, the neoliberal economic shock promoted by recently elected President Fujimori reinforced the uneven urban development and hierarchical water services experienced in previous decades. In the first moment, the incoming administration had to adopt an emergency plan to cope with the alarming situation of water supply in

Lima. Some ongoing projects involving boreholes and storage tanks were concluded to allow localized supply, but soon the new government started to pay attention to the recommendations laid down by multilateral agencies, which comprised institutional reforms that include the separation between policy making and service provision, operation benchmarking within and between water utilities, and novel forms of management incentives, such as charges and water pricing mechanisms (e.g. Klein & Roger, 1995). The failing performance of water supply and sanitation, which led to an outbreak of cholera in 1991 (aggravated by stopping chlorination), provided the needed justification for the Fujimori government to include the water utility of Lima (SEDAPAL) in the list of public utilities to be privatized. SEDAPAL was described as a company with inadequate system maintenance, a high level of unaccounted-for water, excess staff, low metering rates and poor water quality (Corton, 2003). The acquiescence of the grim condition of the metropolitan water services transformed scarcity from a problem into an opportunity for market-based solutions.

Paradoxically, the scarcity of water services in Lima prompted a sudden abundance of money to prepare SEDAPAL to be privatized. After a turbulent preparation period, three international consortiums prequalified to bid for the concession of water service in November 1994, but the process was unexpectedly postponed many times and was eventually cancelled in 1997 (Ioris, 2012a). The main reason for the indefinite postponement of the bidding was that the political price of privatization was not affordable to the president, particularly when his popularity was declining nationwide and Lima was one of the main political strongholds. In addition, there was declining interest from the private sector itself in the privatization of SEDAPAL after other turbulent experiences in South America (e.g. Tucumán and Buenos Aires in Argentina; Limeira and Niteroi in Brazil). Because of the failure of privatization, after the 1995 re-election of Fujimori SEDAPAL embarked upon a larger programme of operational recovery. In 1998, SEDAPAL was transformed into a 'public limited company' and then incorporated into the sphere of FONAFE (the government corporation in charge of the entrepreneurial activity of the state). Those measures ended up reducing the level of problems and, contradictorily, reducing the appetite for privatization within the national government. The water utility of Lima became the recipient of considerable sums of public funds (estimated at USD 2.44 billion, or 14% of the public investment in the period, or on average 0.5% of GDP), which were spent mainly on pipeline replacement and leakage control. Yet, at that point, water provision was still concentrated in the higher-income areas, with 40% of the population consuming 88%, and the poorer 60% only 12%, of the total water (CENCA, 1998).

With the turbulent end of the Fujimori administration and the timid, gradual return to democracy, SEDAPAL faced a deteriorating financial situation and a disjointed management direction. Tariffs again started to grow significantly and became a main issue for the constant confrontation between SEDAPAL and the regulator SUNASS. For example, in 2005 SEDAPAL formally presented a management plan which included an annual increase of 136.9% in tariffs, which was publicly rejected by SUNASS, followed by an open and fierce debate in the mass media and, eventually, a difficult compromise between the two agencies that resulted in the sacking of the president of the regulatory agency. The tendency of the reforms was maintained by President Toledo without removing the government from the centre of economic activity and social interaction

(Roberts & Portes, 2006). With circumstantial economic problems and a tightening budget, the national level of investments in the water sector had declined from USD 228.9 million per year in the 1990s to USD 166.6 million per year in the period immediately after Fujimori's removal. The National Plan of Sanitation (Law DS-007-2006) observed that SEDAPAL would still require works equivalent to USD 1.211 billion between 2006 and 2015, 41.6% of the national investment budget. Toledo's policies were mainly focused on poverty alleviation in rural areas, particularly in the south of the country, where terrorist attacks were worst. Investments started to rise again at the end of Toledo's administration, especially with the announcement of the Miagua programme, which included projects of around USD 1.3 billion throughout the country (USD 657 million for Lima, funded by foreign loans). But they never recovered to the level of the mid-1990s, when the Fujimori government used public money to advance populist, clientelist schemes.

Graham and Marvin (2001) connect the collapse of the modern urban ideal to the ensuing neoliberalization of public services. This was certainly the case in Lima in recent years. The goal of a coherent and inclusive city was maintained only in the discourse of the politicians, while neoliberal infrastructure 'splintering' resulted in fragmentation, premium services and city enclaves. Despite public investments and the involvement of private operators, the material and symbolic affirmation of scarcity continued to define Lima's water policies. The material and symbolic production of scarcity was predicated upon practices of sociospatial discrimination associated with the production of uneven urban development. The interplay between scarcity and abundance was used as a political device to handle expectations in the impoverished areas of the capital city. A practical example is the constantly moving line of water scarcity, the maximum altitude in the hills where the water service is provided, which has allowed the manipulation of investments and engineering plans as conspicuous electoral devices by generations of politicians. This attests to the close connection between the spatialization of power and the manipulation of water scarcity in the peripheries of the city. The last two governments have tried to mobilize resources and policies to tackle water scarcity, but with only partial results that in the end reinforce patterns of privilege and exclusion, as analyzed below.

Water for all?

Water scarcity was particularly politicized and highly contested as an urban issue by Alan García in the 2006 presidential campaign and then during his government (2006–2011). As a candidate, he promised to bring water to the most distant corners of Lima. In 2007, he launched the programme Agua para Todos (Water for All) with a portfolio of more than 1500 large and small engineering works under a total budget of around USD 2.0 billion in Lima alone. The enthusiastic tone of the statements made by the government and its business partners for the construction programme concealed the origins of water scarcity and the uncertainties about the future of the public water sector of Lima. In a few years it became clear that the temporary containment of water scarcity and the largely localized construction works promoted through Agua para Todos failed to address the long-term water problems of Lima due to the persistent shortage of housing and lack of transparency in the public sector. Water scarcity continued to affect those living in marginal, sandy and hilly areas in the periphery, and 48% of the population of metropolitan Lima (though this figure is disputed by SEDAPAL and some experts) receive piped water of substandard quality or for only a few hours every day (RPP, 2013). The rate of unaccounted-for water remains high, at around 29% of water treated and distributed (SEDAPAL, 2014), varying from 17.3% in Villa El Salvador to 44.9% in Callao (Table 4).

In operational terms, SEDAPAL struggled to oversee and to properly manage the number and complexity of the projects launched simultaneously by the García administration, which were focused on large-scale schemes (e.g. dams and treatment plants) rather than securing a more equitable distribution of water. At that point, while someone living in a low-income area such as Villa María used on average 78 litres per day, a resident of high-income San Isidro was using on average 460 litres (SEDAPAL, 2009). Although SEDAPAL was not officially privatized, a significant proportion of the infrastructure was going to be built and operated through public–private partnerships. That was the case with the construction of the Huascacocha Dam (with the participation of the Brazilian company OAS), the Huachipa water treatment plant (participation of the Brazilian Camargo Corrêa), and the Taboada and La Chira sewage treatment plants (participation of the Spanish ACS España). Large-scale investments, feeble control of public funds, and covert connections between politicians and construction companies created very favourable conditions for corruption on a large scale. The abundant evidence of graft and dishonesty during the García government prompted the formation of an investigative commission, known as the *megacomisión*, once Garcia's party (APRA) lost the power to block its formation after the 2011 election. One of its specific remits was to scrutinize the Agua para Todos programme (El Comercio, 2013). Corruption associated with the wastewater treatment plants at Taboada (investment of USD 342 million) and La Chira (USD 192 million) is discussed in Ioris (2013).

The ongoing work of the *megacomisión* has abundantly demonstrated that the water supply of the capital city has been a politically attractive source of corruption for office holders. One of the concrete consequences so far is that, in June 2014, the National Congress decided to formally accuse García and other members of his administration of corruption and mismanagement, but the battle continues in juridical circuits. The Peruvian judiciary is notoriously, and dangerously, close to Alan García, and probably because of that, at the time of writing (March 2015), the

Table 4. Service coverage by SEDAPAL (per service centre), 2013.

Service centre	Total number of domestic connections (water service)	Total number of domestic connections (sanitation service)	Unaccounted-for water
Comas	375,709	360,176	32.3%
Callao	145,371	128,275	44.9%
Breña	156,522	104,697	26.7%
Ate-Vitarte	185,415	178,297	32.6%
S. Juan de Lurigancho	150,409	144,295	26.6%
Villa El Salvador	213,660	209,037	17.3%
Surquillo	172,061	142,661	25.6%
Large clients	13,158	11,739	–
TOTAL	1,412,305	1,279,177	29.0%

Source: SEDAPAL (2014)

judiciary has not yet started its investigation. García remains a formidable politician, whose political group won the metropolitan election in Lima in 2014 and is probably going to win the next presidential campaign, in 2016. Yet the same process of uneven development and biased water services has continued unabated during the presidency of Ollanta Humala, in office since 2011, who introduced another anti-scarcity plan along similar lines and with promises of universal water supply comparable to those made by Fujimori and García.

The Strategic Plan 2013–2017 of SEDAPAL contained five main goals, including service improvement, financial stability and universalization of public services. Nonetheless, several respondents mentioned in our second round of interviews (conducted in April 2013) that 'service universalization' applies only to those in areas that are already operationally viable. They also repeatedly mentioned that SEDAPAL has always been a highly centralized and bureaucratic public utility, which systematically rejects a more open dialogue with the metropolitan population. A related problem is that the public utility has repeatedly struggled to deliver its formal commitments, as in the case of the promise to invest PEN 836.6 million and PEN 790.1 million in the first two years under the Optimized Master Plan 2010–2015, when in practice it invested only PEN 553.2 million and PEN 433.1 million, respectively (El Comercio, 2014). Such underperformance fuels perceived shortcomings of the public sector as an incompetent administrator that facilitate the regular re-emergence of privatizing tendencies. The market-friendly logic introduced by Fujimori and maintained by García was again reinforced by the Modernization of Sanitation Services Act of 2013 (Law 30045), which established a new technical agency in charge of normalizing construction works and, importantly, promoting public–private associations.

From the above it can be contended that, rather than a physical or incidental phenomenon, the constant reinforcement of scarcity – through a combination of neglect and discrimination – has been the fundamental feature of the public water services. The hydrosocial inequalities that help shape the landscape of Lima certainly go beyond the 'dance of statistics' repeatedly employed by politicians and spokespersons. Under conditions of uneven development and sociospatial inequalities, the scarcity of water is not simply the outcome of ill-conceived projects and interventions; it permeates the decision making around the allocation of resources and identification of priorities. Supposedly pro-poor initiatives designed by SEDAPAL have inadvertently reinforced the discrimination of water services between social groups. That has aggravated the lack of urban planning despite the number of interventions and projects affecting the territory of the metropolis (the last plan was published in 1989, before the Fujimori administration) and the always difficult coordination between Lima and Callao, which inevitably contribute to the widespread sense of imprudence and informality. The experience in Peru is comparable to what has happened in other parts of Latin America, as in La Paz, Bolivia, where the communities have systematically complained about the lack of proper participation and attention from the authorities (Laurie & Crespo, 2007). An important aspect of discrimination, both in Bolivia and Peru, is that the proportion of mestizos and non-whites is significantly higher among those without access to water. As aptly observed by Watts (2000), the politics of redistribution is directly related to the politics of recognition.

Conclusions: The territory of the exclusionary city

The evolution of the public water services of Lima, discussed in the previous pages, offers an emblematic example of the territorialization of water scarcity through the consolidation of an urban landscape fraught with sociospatial inequalities. The reforms and the new interventions of the national state in the last three decades constituted the main force behind multiple mechanisms of water management, including heavy investments in infrastructure and management control, that in the end systematically failed to remove the existing territories of water scarcity. On the contrary, state action has followed political demands and hegemonic interests in a way that preserves the established patterns of territorialization. The long-lasting situations of territorialized water scarcity have been politically appropriated as justification for the introduction of new plans and programmes according to a market-friendly rationality aiming to satisfy, primarily, the requirements of national and international elite groups. For instance, water scarcity has served as a powerful excuse for institutional reforms, foreign loans and cooperation projects, which have produced uneven results and never benefited the totality of the metropolitan population. Those initiatives, rather than overcoming the trend of water management problems, have contributed to preserving lifelong privileges and exclusions.

One main conclusion is that the territory of Lima continues to expand through the territorialization of multiple, interconnected scarcities. The water industry of Lima has been under intense transformation since the adoption of broad macroeconomic and institutional reforms in the early 1990s. Yet, although Lima is one of the best serviced cities in terms of infrastructure in the country, its allocation and distribution problems remain largely unresolved. Although substantial sums of money have been invested in the augmentation of the water services, and regulatory adjustments have attracted more international operators than the company can actually deal with, much less attention has been dedicated to creating specific solutions to the concrete reality of water problems in different parts of Lima or to increasing the long-term resilience of the water system. All these factors lead to the impression that water scarcity is in itself a highly kaleidoscopic concept: in the context of urban development, references to scarcity work like kaleidoscope mirrors, reflecting unstable, intricate connections across locations and social groups. As in the past, recent responses are again centred on the appropriation of scarcity as a productive force for the implementation of elitist political agendas. At the same time, novel peripheries are constantly being formed in more remote and more inhospitable locations because of the problematic development and unfair urban expansion.

Overall, in order to fully understand the geography of multiple scarcities of the large Latin American cities, such as Lima, it is necessary to start neither from the aggregate nor from the micro and fragmented, but to simultaneously tackle the entirety of the lived and experienced territorialized processes that connect the local and personal with the higher scales of interaction. In the present case, the turbulent expansion of water services, in moments both of relative scarcity and of abundance of financial resources, is a compelling indication of the specificity of economic development in Lima, which still connects the colonial legacy with the globalizing pressures on today's Peru. Water problems indicate a synergistic connection between the spiral of scarcity, the intensification of political manipulation, and the selective circulation and accumulation of capital. In consequence, existing hydropolitical arrangements help maintain or even enhance the sociospatial inequalities that produce the exclusionary city. The shortage of water in the Peruvian capital presents itself as a totality of relations: at the same time that the periphery suffers from precarious water services, the insertion into market globalization encourages the purchase of private cars and household appliances in the wealthier neighbourhoods, which inevitably raise water consumption and inequality even further. On the other hand, in their struggle for survival, the majority of Lima's population living in the periphery are daily required to mobilize resilience skills and organizational capacity. Both government initiatives and grass-roots reactions certainly contribute towards territorialization and become inscribed in the territory of the city. Yet, the range of policies so far adopted by various administrations to deal with water scarcity have never really addressed the most basic demands for a fully inclusive city.

References

ANA (National Water Authority). (2015). Programa de recuperación del Río Rimac. Retrieved from http://www.ana.gob.pe/gestion-de-la-calidad-de-los-recursos-hidricos/programa-de-recuperaci%C3%B3n-del-r%C3%ADo-rimac.aspx

Barreda, J., & Ramirez Corzo, D. (2004). Lima: Consolidación y expansión de una ciudad popular. *Perú Hoy*, 6, 199–218.

Boelens, R., Hoogesteger, J., & Rodriguez de Francisco, J. C. (2014). Commoditizing water territories: The clash between Andean water rights cultures and payment for environmental services policies. *Capitalism Nature Socialism*, 25(3), 84–102. doi:10.1080/10455752.2013.876867

Boelens, R., Hoogesteger, J., Swyngedouw, E., Vos, J., & Wester, P. (2016). Hydrosocial territories: A political ecology perspective. *Water International*, 41(1), 1–14. doi:10.1080/02508060.2016.1134898

Cabrera Carranza, C. F. (2010). *Impacto del cambio climático en la margen izquierda del Río Rímac – MIRR*. Lima: IMP/CENCA/IDRC.

CENCA (Urban Development Institute). (1998). *El saneamiento básico en los barrios marginales de Lima Metropolitana*. Lima: PAS/World Bank.

Corton, M. L. (2003). Benchmarking in the Latin American water sector: The case of Peru. *Utilities Policy*, 11(3), 133–142. doi:10.1016/S0957-1787(03)00035-3

El Comercio. (2013, February 20). Megacomisión: Hay indicios de "negligencias graves" en Agua para Todos. Retrieved from http://elcomercio.pe/actualidad/1539962/noticia-megacomision-hay-indicios-negligencias-graves-agua-todos

El Comercio. (2014, February 24). Sedapal: ¿Qué hacer para mejorar y ampliar el servicio? Retrieved from http://elcomercio.pe/economia/peru/sedapal-que-hacer-mejorar-y-ampliar-ser vicio-noticia-1711764

Elden, S. (2010). Land, terrain, territory. *Progress in Human Geography*, 34(6), 799–817. doi:10.1177/0309132510362603

Graham, S., & Marvin, S. (2001). *Splintering urbanism: Networked infrastructures, technological mobilities and the urban condition*. London: Routledge.

Hanson, S. (2003). The weight of tradition, the springboard of tradition: Let's move beyond the 1990s. *Urban Geography, 24*(6), 465–478. doi:10.2747/0272-3638.24.6.465

Heynen, N. (2014). Urban political ecology I: The urban century. *Progress in Human Geography, 38*(4), 598–604. doi:10.1177/0309132513500443

Hoogesteger, J., Boelens, R., & Baud, M. (2016). Territorial pluralism: Water users' multi-scalar struggles against state ordering in Ecuador's highlands. *Water International, 41*(1), 91–106. doi:10.1080/02508060.2016.1130910

INEI (National Institute of Statistics). (2010). *Estadísticas nacionales.* Retrieved from http://www.inei.gob.pe

Ioris, A. A. R. (2001). Water resources development in the São Francisco River Basin (Brazil): Conflicts and management perspectives. *Water International, 26*(1), 24–39. doi:10.1080/02508060108686884

Ioris, A. A. R. (2008). Regional development, nature production and the techno-bureaucratic shortcut: The Douro River catchment in Portugal. *European Environment, 18*(6), 345–358. doi:10.1002/eet.488

Ioris, A. A. R. (2011). Values, meanings, and positionalities: The controversial valuation of water in Rio de Janeiro. *Environment and Planning C, 29*(5), 872–888. doi:10.1068/c10134

Ioris, A. A. R. (2012a). The neoliberalization of water in Lima, Peru. *Political Geography, 31*(5), 266–278. doi:10.1016/j.polgeo.2012.03.001

Ioris, A. A. R. (2012b). The geography of multiple scarcities: Urban development and water problems in Lima, Peru. *Geoforum, 43*(3), 612–622. doi:10.1016/j.geoforum.2011.12.005

Ioris, A. A. R. (2013). The adaptive nature of the neoliberal state and the state-led neoliberalisation of nature: Unpacking the political economy of water in Lima, Peru. *New Political Economy, 18*(6), 912–938. doi:10.1080/13563467.2013.768609

Ioris, A. A. R. (2014). The urban political ecology of post-industrial Scottish towns: Examining Greengairs and Ravenscraig. *Urban Studies, 51*(8), 1576–1592. doi:10.1177/0042098013497408

Jones, G. A. (2006). Culture and politics in the "Latin American" city. *Latin American Research Review, 41*(1), 241–260. doi:10.1353/lar.2006.0009

Klein, M., & Roger, N. (1995). *Back to the future: The potential in infrastructure privatization.* Washington, DC: World Bank.

Laurie, N., & Crespo, C. (2007). Deconstructing the best case scenario: Lessons from water politics in La Paz-El Alto, Bolivia. *Geoforum, 38*(5), 841–854. doi:10.1016/j.geoforum.2006.08.008

Lefebvre, H. (1996). *Writing on cities.* Oxford: Blackwell.

Linton, J. (2010). *What is water? The history of a modern abstraction.* Vancouver: UBC Press.

Matos Mar, J. (1984). *Desborde popular y crisis del Estado: El nuevo rostro del Perú en la década de 1980.* Lima: Instituto de Estudios Peruanos.

Matos Mar, J. (2012). *Perú: Estado desbordado y sociedad nacional emergente.* Lima: Universidad Ricardo Palma.

Perramond, E. (2016). Adjudicating hydrosocial territory in New Mexico. *Water International, 41*(1), 173–188. doi:10.1080/02508060.2016.1108442

Roberts, B., & Portes, A. (2006). Coping with the Free Market city: Collective action in six Latin American cities at the end of the twentieth century. *Latin American Research Review, 41*(2), 57–83. doi:10.1353/lar.2006.0029

RPP. (2013, April 05). Afin: El 48% de limeños no cuenta con agua potable de calidad. Retrieved from http://www.rpp.com.pe/2013-04-05-afin-el-48-de-limenos-no-cuenta-con-agua-potable-de-calidad-noticia_582819.html

Sack, R. D. (1986). *Human territoriality: Its theory and history.* Cambridge: Cambridge University Press.

SEDAPAL (Lima Water and Sanitation Company). (2003). *Historia del abastecimiento de agua potable de Lima 1535/2003.* Lima: SEDAPAL.

SEDAPAL. (2009). *Plan maestro optimizado.* Lima: SEDAPAL.

SEDAPAL. (2014). *Plan maestro de los sistemas de agua y alcantarillado.* Vol. I. Lima: SEDAPAL.

Storey, D. (2012). *Territories: The claiming of space* (2nd ed.). London: Routledge.

UNDP (United Nations Development Programme). (2006). *Beyond scarcity: Power, poverty and the global water crisis.* Human Development Report. New York, NY: UNDP.

UNICEF and WHO (United Nations Children's Fund & World Health Organisation). (2012). *Progress on drinking water and sanitation: 2012 update*. New York, NY: UNICEF/World Health Organization.

Watts, M. J. (2000). The great tablecloth: Bread and butter politics, and the political economy of food and poverty. In G. L. Clark, M. P. Feldman, & M. S. Gertler (Eds.), *The Oxford handbook of economic geography* (pp. 195–212). Oxford: Oxford University Press.

Whitehead, A. N. (1920). *The concept of nature*. Cambridge: Cambridge University Press.

Zolezzi, M., & Calderón, J. (1987). *Vivienda popular: Autoconstrucción y lucha por al agua*. Lima: DESCO.

Zug, S., & Graefe, O. (2014). The gift of water: Social redistribution of water among neighbours in Khartoum. *Water Alternatives, 7*(1), 140–159.

Zwarteveen, M. Z., & Boelens, R. (2014). Defining, researching and struggling for water justice: Some conceptual building blocks for research and action. *Water International, 39*(2), 143–158. doi:10.1080/02508060.2014.891168

Inclusive recognition politics and the struggle over hydrosocial territories in two Bolivian highland communities

Miriam Seemann

GIGA German Institute of Global and Area Studies, Hamburg, Germany

ABSTRACT
This article applies a multi-scalar approach to examine the dominant human–nature interactions that underlie recent formalization policies and the (re)configuration of hydrosocial territories in the Tiraque Valley, Bolivia. From a political ecology perspective, it seeks to examine how hydrosocial territories are (re)configured by Bolivia's representative and inclusive discourses and forms of water 'governmentalities'. It analyses how water territories are locally materialized by technological designs, legal structures and power relations that may promote unequal distribution of resources, water rights and decision-making power in conflict resolution processes. This article challenges 'pro-indigenous' and inclusive discourses that promote formal recognition of customary 'water territories'.

Introduction

The struggle for autonomy, territory and natural resources has been a key demand of indigenous movements during the last decades in Bolivia. It has resulted in a series of social protests in which the discourse of 'indigenous identity' and 'customary rights' has played a prominent role. Whereas state control with respect to water management had been quite low in comparison with other countries in the region (Bustamante, 2002), a wave of neo-liberal policies introduced from the 1980s onwards posed a new threat to the water resources found in territories or under the control of peasant and indigenous communities. As a result, irrigators whose livelihoods depended on irrigation water began nationally to protest in defence of their right to access and manage water. Intending to increase future water security, the irrigators' movement, with the support of non-governmental organizations (NGOs), drafted a proposal that contained the legal recognition of their local water rights according to *usos y costumbres* ('uses and customs'), which was passed as the new Irrigation Law in 2004 (Perreault, 2008). In 2013, however, although the Irrigation Law has been in force for almost 10 years, fewer than 10% of the total 5000 irrigation systems in Bolivia had been issued a water registration (Representative of the National Irrigation Service – SENARI, 14 March 2012).

Numerous studies on the formal recognition of water rights in the Andes by, among others, Boelens and Seemann (2014); de Vos, Boelens, and Bustamante (2006); Zwarteveen, Roth, and Boelens (2005), have shown how state policies to institutionalize water management rights and norms have often led to altering dynamic socio-legal normative systems and to subordinating plural water rights under national laws. Yet, little scholarly attention has been paid to the question of how inclusive forms of water 'governmentalities' (re)configure hydrosocial territories.

'Inclusive recognition policies' are state policies that formally recognize specific socially and culturally embedded rules and norms in society (such as, for example, local water management and control rights, which include, *inter alia*, decisions regarding water distribution, irrigation schedules or organizational functions by constituting participatory and inclusive power strategies (Foucault, 1979). They recognize the 'extra-legal' and integrate it into the formal property system, which sustains the ideology of modern water policies while simultaneously neglecting the right to diversity anchored in local water rights. This paper will discuss the Bolivian inclusive recognition policy with two case studies. Through this analysis it gives insights into the process through which particular 'hydrosocial territories' are produced. These territories organize inequality and are by no means automatically equitable in how they grant access to water to users.

Adopting a political ecology perspective, this article argues that the recognition and formalization of local water rights are always political in character and, therefore, contested and contestable (Robbins, 2004). While at first glance both the political development reached by the irrigators' movement and the multicultural policies and discourse of 'good living' (*vivir bien*) introduced by the government of President Evo Morales are quite significant, they require a careful and critical analysis with respect to power structures and water justice as in many cases they may counteract the diversity of local concepts of 'hydrosocial territories'.

In particular, the paper will discuss two opposing communities, Cochimita and Sank'ayani Alto, and examine their highly divergent functions, values and meanings about what are and should be their 'water territories'. Their two opposing concepts of hydrosocial territories not only lead to conflictive situations among the two communities but also are challenged by Bolivia's legal concepts and policies.

This article has two central objectives. The first is an attempt to understand the ways in which local contested concepts of hydrosocial territories are locally materialized by technological designs, legal structures and power relations that may promote unequal distribution of resources, water rights and decision-making power in conflict-resolution processes. Second, it seeks to challenge presumably neutral 'pro-indigenous' and inclusive discourses and forms of water 'governmentalities'. Thus, by applying a multi-scalar approach, the national, regional and local levels are analysed and viewed as interconnected entities. It is argued that in practise uncritical political and legal concepts and policies on hydrosocial territories may lead to local conflicts among water user communities and increase socio-economic and cultural water injustices among Andean water communities. This research forms part of the activities of the international Justicia Hídrical/Water Justice research alliance (see www.justiciahidrica.org).

The empirical data in this article were collected in 2012 through fieldwork in the Tiraque Valley, Bolivia, in which Cochimita and Sank'ayani Alto are located. Data collection consisted of 58 semi-structured in-depth interviews held with community members and national water experts as well as field observations. The study is based on qualitative data analysis and follows an interpretive approach (Mason, 2004; Smith, 1983).

The article begins by providing the background to Bolivia's inclusive recognition politics and conceptualizes 'hydrosocial territories'. The second section introduces the cases of Cochimita and Sank'ayani Alto and their divergent concepts of hydrosocial territories. The third section introduces Bolivia's legal concepts and discourses on hydrosocial territories. The paper concludes by asserting that to acknowledge hydro-social territories as highly contested and divergent entities among water users, institutional discourses and political agendas would be a significant step to engage critically with inclusive recognition politics and a step towards increased local water justice.

Inclusive recognition politics, water 'governmentalities' and contested 'hydrosocial territories'

The irrigators' movement of Bolivia arose in late 1999 as a response to the privatization of Cochabamba's municipal drinking water company, SEMAPA [Servicio Municipal de Agua Potable y Alcantarillado], gave rise to the irrigators' movement of Bolivia in late 1999, and to a social protest mainly organized by the umbrella organization Coordinator for Defense of Water and Life (Coordinadora de Defensa del Agua y de la Vida). Following the Water War of 2000, many Bolivian irrigators saw their water rights unprotected within state law. Four years later, the irrigators' movement effectively fought to have their traditional water rights recognized and formalized in the new Irrigation Law Nr. 2878, which was promulgated in 2004. This law and the discourse on water rights formalization in Bolivia was strongly influenced by the Inter-American Development Bank (IDB), NGOs, a newly consolidated leadership of irrigators and international cooperators. For the first time Bolivian peasants had the possibility of legally recognizing their customary water rights in forms of 'water registrations' (*registros*) (Ministerio de Agua, 2006).[1] By demanding legal security, the irrigators' movement actively contributed to the advancement of uniform rule of state law in often remote peasant and indigenous communities in which the Bolivian state had had little presence. Many irrigation leaders felt the need and desire to protect their water sources legally and therefore aspired to have their customary rights recognized and 'included' by calling upon the powers of (universalistic) state law. As Boelens, Hoogesteger, and Baud (2015) point out (cf. Dean, 1999; Foucault, 1991):

> Governmentality, the art of government, is deployed as the 'conduct of conduct', based on efforts that – though contested and mediated in practise – envision to arrange and align citizens, institutions, infrastructure and practices in strategic ways; in order to steer and control societal development. (p. 3)

Here, it is important to distinguish between modes of 'sovereign power' that work through top-down and hierarchic state law (such as state sanctions, outright violence, legal force, etc.) and 'inclusive' forms of governmentality, which seek to govern the people by promoting social modalities of 'participation', 'integration' and 'recognition' and by exercising bottom-up power mechanisms. An example is how discourses (or 'regimes of truth') are implemented that produce knowledge about water and environment, and that morally justify related behavioural norms, principles and rights. To a large extent the Bolivian Irrigation Law no longer exerts just a top-down power modality, but exercises inclusive power by following the liberal ideology and by inviting all Bolivian peasants to participate in the ordinary justice system 'under equal terms'. The majority of the irrigator's movement perceived this rather favourably and saw it as a way of liberation and security over their water sources, based on customary rights (Albó, 2008). The universalistic discourse claiming that formalized water rights supposedly provides legal security exerted through 'normalizing power' (Boelens, 2014; Foucault, 1979) and worked towards new forms of 'water governmentality'. As such, in the case at hand the register creates new hydrosocial territories through the legalization of space, which is deeply connected to hydraulic infrastructure and water sources; consolidated through historical appropriation (customary rights) by one group of irrigators.

In fact, the territorial aspect is central to the issue of water governance and governmentality in both urban and rural areas (cf. Bebbington, Humphreys Bebbington, & Bury, 2010; Boelens et al., 2012; Mazurek, 2010). The concept of territory, thus, plays a fundamental role in understanding the struggle over water control, since it is ultimately linked to water-power structures and control over water, socio-nature and people themselves, including property rights, technological designs, legal structures and discourses (Baletti, 2012; Peluso & Lund, 2011; Saldías, Boelens, Wegerich, & Speelman, 2012). Agnew (1994) and later Elden (2010), among others, have challenged normative notions of territory as a state-centred fixed, bounded, spatially coherent entity congruent with administrative or national boundaries. These authors reconstructed the concept as intersecting biophysical, ecological, socio-cultural and political–economic domains (cf. Baletti, 2012). It is precisely the socio-technological and cultural–political strategies that different actors deploy to 'dominate these domains' and shape hydrosocial territories according to their interests, which can be understood as efforts to 'governmentalize' space (Boelens, Hoogesteger, & Rodriguez de Francisco, 2014).

Both, political ecology and science and technology studies (STS), reconceptualize 'water' by overcoming 'the dualism of the socio/nature divide' (Swyngedouw, 1999, p. 445) and by recognizing how water combines society, nature, material and discursive issues in dialectical and contradictory ways (Bouleau, 2013; Budds, 2009; Castree & Braun, 2001; Latour, 1993). As such, local hydrosocial territories, which inclusive recognition politics often tend to recognize as natural geographical environments, are socially constructed, linked to micro- and macro-scales, are mediated by power relations and human intervention. As Budds (2009, p. 420), following (Harvey, 1996; Worster, 1992) puts it: 'social relations – played out through diverse artefacts and institutions such as hydraulic infrastructure, water laws and policy discourses – shape how water flows through the waterscape, yet are also themselves shaped by water'. In accordance with the introductory chapter of this journal, hydrosocial territories are, therefore, not objective entities, but should be understood as contested socio-natural imaginaries and constructions (Boelens, Hoogesteger, Swyngedouw, Vos, & Wester, 2016; Swyngedouw, 2007). This acknowledges that they are socially constructed and thus highly contested among water users, institutional discourses and political agendas. Therefore, socio-natures are perceived differently depending on the context, time and actor with a multitude of serious implications (Budds, 2009; Castree & Braun, 2001; Latour, 1993; Robbins, 2004). In accordance with Boelens et al. (2014), this article views the diversity of the concept 'hydrosocial territory' as a strength rather than seeking a fixed definition because 'it is precisely the (divergent) material and discursive production of the concept that gives insight into the field of water control processes and water-power structures' (Boelens et al., 2014, p. 1).

Yet, mainstream literature on water policies and the recognition of local water rights, which are often rooted in positivism and liberal individualism (Green & Shapiro, 1994), tend to ignore the complexity and diversity of hydrosocial territories and water distribution systems (Seemann, 2014; Zwarteveen & Boelens, 2014). Instead most policy documents conceptually neutralize geopolitical spaces, water distribution systems and hydrological cycles. This, however, can have severe outcomes in terms of water (in) justice, as Castree and Braun (2001, p. 5) argue: 'the all-too-common habit of talking of nature 'in itself', as a domain which by definition is non-social and unchanging, can lead not only to confusion but also to the perpetuation of power and inequality in the wider world'. We can fend against such outcomes by understanding inclusive recognition politics as based on socio-natural politics. So doing provides us with opportunities, on the one hand, to challenge critically 'pro-indigenous' water 'governmentalities' and the power-laden formalization policies. On the other hand, it enables one to examine critically how local contested concepts of hydrosocial territories are locally materialized by technological designs, legal structures and power relations. Both form fundamental elements in the struggle for more equitable and 'just' water governance policies.

Local communities in conflict over hydrosocial territories: the cases of Cochimita and Sank'ayani Alto

Cochimita and Sank'ayani Alto are neighbouring communities located in the Tiraque Valley, which is about 65 km from Cochabamba city and is part of the watershed Pucara. The community of Cochimita is situated at 3300 masl and is about 45 minutes from Tiraque town by car along the mountain slope. The community has a total area of approximately 300 hectares of land, of which 70% is arable. As such, the majority of the population is involved in agricultural production for which water is indispensable in the semi-arid region. The community bases its claims to hydrosocial territory and water rights on hydraulic property rules, which are based on investments made by families (such as material resources, time, capital or ritual contributions) in order to gain water rights, which are thus embedded in complex, hybrid social and historical structures (Barrios, 1997).

The main irrigation water source of Cochimita is the system Kayarayoj T'oqo. At 4000 masl it comprises two water dams, Kayarayoj T'oqo I and II, about 13 km north of Cochimita and located on the territory of the community Sank'ayani Alto. The first dam is of stone and was built in 1984 under a project funded by Oxfam America and the Belgium–Bolivia Association, and managed by a Bolivian NGO, while the second dam is more rustic, built by the users themselves. Their main objective is to store water during the rainy seasons to provide water for irrigation during the dry season. Both dams together have a capacity of approximately 536,000 m^3 to irrigate approximately 240 hectares of the total area of the community Cochimita. The total Lluska Qhocha canal length from the dam to the irrigation area is 7 km. A significant portion of this canal, about 4 km, is shared with other irrigation systems, which makes good dialogue between all communities a requirement (Barrios, 1997; Cossio, Soto, & Skielboe, 2010; Cruz Flores, 2010) (Figure 1).

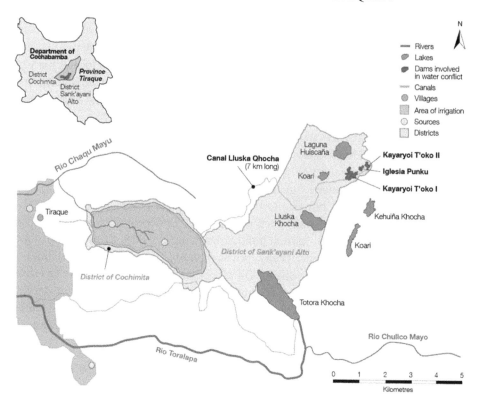

Figure 1. District of Cochimita and Sank'ayani Alto.

As in most cases of Bolivian irrigation, local water rights in Cochimita are framed at the collective and individual level (Bustamante, 2002). As such the community of Cochimita has collective water use rights over its water resources and the dam Kayarayoj T'oqo, which are rooted in the *usos y costumbres* (uses and customs), as one water user explains: 'We have been using the water since a long time ago, since the land reform; our grandparents constructed the dam after the reform, from that moment their water rights have been inherited to the next generation' (water user, 3 February 2012). Simultaneously, each recognized individual member of the water users' organization has the right to use a certain amount of water, independently from the size of land. Besides the individual water rights, however, the decision-making, for example concerning the length of a water shift, is collective and authorized by the irrigation committee. We see that in Cochimita hydrosocial territories and distribution practises become manifest, simultaneously, in water infrastructure constructions, normative arrangements and organizational frameworks to operate and maintain the local water control systems, which with the exception of the closed character (it is impossible to register new water users after the initial registration process) are respected by the entire community.

Even though nowadays their dams are located on the territory of Sank'ayani Alto, Cochimita argues that in the past the land was only pasture and did not belong to any specific community (water user, 2 February 2012). Since it is only in recent years that the land has been distributed and territorialized, the users of Cochimita view themselves as the legitimate users of their hydrosocial territory, which they strive to get officially recognized by the new Irrigation Law (water user, 27 January 2012). This came through on 20 June 2008 when after a long administrative period and field verification the state issued the water registration for the community of Cochimita acknowledging their hydrosocial territory, including their water sources, infrastructure, irrigation area and water management according to their *usos y costumbres* (Registro de Cochimita, 2007). The main initiator of the process was the Directorate of the Irrigation Committee of Cochimita, who for the first time heard about the register in the Association of Irrigation and Services of Tiraque (ARST). The main aim of the community members was to gain an official document to secure their rights to their hydrosocial territory. In a general assembly all community members voted in favour of the registration, unanimously (water user, 3 February 2012).

The Cochabamba Departmental Federation of Irrigators (FEDECOR) together with a national NGO supported Cochimita in the bureaucratic process to produce all documents required by the National Service for Irrigation (SENARI). Today Cochimita belongs to one of the four communities out of the total 141 communities in Tiraque in possession of a water registration. The water register recognizes the hydrosocial territory of the community, including its irrigation infrastructure, but it does not change the internal water distribution, use or control rights. Therefore, community members of Cochimita keep irrigating their fields as they have before. Consequently, the majority of the water users are not aware of or informed about the existing state recognition (*registro*) since it does not affect their daily routine. However, as a direct consequence of now having a registration that assigns legal security regarding their hydrosocial territory, Cochimita is in conflict with its neighbouring community of Sank'ayani Alto on whose territory the water dams of Cochimita are located.

Sank'ayani Alto, the small neighbouring community located in upper Tiraque, does not recognize the water registration of Cochimita. Both communities signed a consensual agreement on 26 January 1990 in which they agreed that the dam Iglesia Punku belongs to the community of Sank'ayani Alto and the dams Kayarayoj T'oqo I and II belong to the community of Cochimita. Nevertheless, community member of Sank'ayani Alto have a contrasting hydrosocial territory concept. They claim their hydrosocial territory rights according to socio-territorial rules and see themselves as legitimate water users, since all three water dams are located on their territory and the irrigation canals run through their community. More specifically, Sank'ayani Alto requested to merge the small dam Kayarayoj T'oqo II with their dam Iglesia Punku to construct a single, larger dam.

The community members of Sank'ayani Alto argue that due to population growth and extended agricultural areas their demand for water for irrigation has increased in recent years. The expansion of agriculture is associated with warming in higher elevations due to climate change.[2] Besides the small dam of Iglesia Punku, Sank'ayani Alto has only two additional water dams that are shared with neighbouring communities with a rather insignificant water portion for Sank'ayani Alto (water user, 7 March 2012). Yet, while Sank'ayani Alto defends their traditional community authority and asserts their decision-making power as fundamental in conflict-resolution processes, community members of Cochimita see their position confirmed and legitimated by its legal recognition and backed up by state authorities, as the following statement underpins:

> I think the registry is good, because with the registration we own the dam, the document helps us before the law. Even the authorities say, if we have a registration, than we own the dam. It tells our neighbour communities not to touch our water, because it is ours and because we have the document. This is how SENARI, the Central Provincial, the mayor and authorities have told us and they support us. As such the water registration is very important for us. (former president of the committee, 3 February 2012)

Since 1990 the two communities have been struggling occasionally over the small dam Kayarayoj T'oqo II and the right to new canal constructions. Since then, state authorities have always mediated in conflict-resolution processes and agreements among the two communities were first signed but also later broken. This conflict intensified after 2008 when Cochimita obtained official registration, and peaked in 2011. After the dam Kayarayoj T'oqo II was repeatedly damaged, Cochimita called the police of Tiraque, leaders of the peasant union of Tiraque (ARST), FEDECOR and other local authorities to solve the problem. However, their involvement did not lead to an agreement between the two communities. Hereafter, in July 2011, Cochimita submitted an official complaint against five community members of Sank'ayani Alto and accused them on four counts: sabotage, public incitement, water theft and bombing public facilities (Official Agreement, 1990; former president of the committee, 3 February 2012).

By suing Sank'ayani Alto within the official jurisdiction system, the underlying struggle between the increased power of state authorities versus local justice mechanisms and local decision-making power came to the surface. The diverging discourses to defend their particular hydrosocial territories and water hierarchies constitute a main level of conflict. The main discourse of Cochimita's state authorities is in line with the policy discourse of the Irrigation Law that recognizes customary hydrosocial territories and is supposed to provide water security over customary hydrosocial territories. On the one hand, Sank'ayani Alto instead opts for water authorities and rules that are enforced locally as tradition goes and constitute their hydrosocial territory according to their socio-geographical territory, on which the water sources are located (water user, 7 March 2012). On the other hand, in order to confront Cochimita, Sank'ayani leaders opted for a strategy that is also supported by formal law: they sought new alliances with other communities and irrigators organization in the highlands and created the so-called 'Indigenous Irrigation Federation of the Highlands' (Federación de Regantes Indígenas de las Alturas – FRIA). As a federation they assume a higher status than the irrigator's associations and consider themselves at the same level as FEDECOR. As such, at least on paper, they obtain greater representation and power to confront the traditional irrigators. By using the 'indigenous discourse', they align with the political agenda of the political party MAS (Moviemiento al Socialismo) and thus strengthen their political relation with central government. They thereby have the possibility to control their hydrosocial territories, regardless of the endless discussions and conflicts with the irrigators of the valley.

This section has seen how the conflict over hydrosocial territories among two Bolivian communities goes beyond the struggle over the resource itself (access to water and the dam Kayarayoj T'oqo II). Instead day-to-day contestation takes concrete shape over norms and rules (territorial jurisdiction, *usos y costumbres*), over decision-making authority and discourses used, for example to legitimize recognition policies and conflict positions (Boelens, 2009; Zwarteveen et al., 2005). In this case the ongoing presupposition of modernist water policy programmes is not valid, which presumes that recognition policies would reduce conflict over resources and that standardized rule-making will lead to mutually beneficial exchange. Instead, we see that institutionalizing pluralistic and divergent hydrosocial territories deepens rather than weakens conflict dynamics, since the power balance between the communities changes when one community obtains a registration and the other does not: 'Especially those who want to keep their *status quo* are not willing to discuss their water rights, they do not care about other communities without access to water' (Director of the Integrated Watershed Management Program – Gobernación de Cochabamba, 9 January 2012).

Bolivia's inclusive water governance discourse and the (re)configuration of hydrosocial territories

With the election of President Evo Morales, the discourse and policy changed from neo-liberal, decentralized water policies towards a sustainable, participatory and integrated water resources management. Morales' election campaign was shaped by a discourse that combined ethnic and class claims. This ensured the support of indigenous and non-indigenous voters in order, among others, to restore the country's sovereignty, to nationalize natural resources, to release the country from neo-liberal policies and to promote social justice, especially for the most marginalized parts of the population, by 'ending the colonial and racist era'. Overall, both 'indigenous rights' and 'sovereignty' became essentially discursive elements of the Movement for Socialism (Movimiento al Socialismo – MAS) (Anria, Cameron, Goenaga, Toranzo, & Zuazo, 2010; Postero, 2010; Stavenhagen, 2009).

Newly defined policies in the agricultural sector are a fundamental pillar in the National Development Plan that strives to change Bolivia from a primarily export-oriented country to a productive country where the state has the power and control over the development and its surplus. In this respect, the government of Morales introduced the new development paradigm of a 'dignified, sovereign, productive and democratic Bolivia' and introduced the indigenous concept of 'good living' (*vivir bien*) as a fundamental principle, which can be understood as the 'encounter and progress through diversity and interculturality, harmony with nature, social and fraternal life, national sovereignty in all fields and internal accumulation with quality of life' (MMAyA, 2007, p. 5; Ley N° 300., 2012). A central part of the *vivir bien* paradigm is the current state policy 'water for all' that views access to water as a 'human right, legitimate, and fundamental of all living beings' (p. 1) and which calls for a fair, sustainable, participatory and integrated water resources management in the country (MMAyA, 2007).

In the international arena, the government's discourses on inclusion, participation and recognition of customary water rights and territories became a prime example for the struggle for more water justice in the region and beyond and inspired worldwide anti-globalization and anti-privatization movements. At the global level, Bolivia also played a leading role in securing the right to access to clean water since the third World Water Forum in Kyoto and the successful United Nations recognition of water and sanitation as a human right in 2010. Additionally, the Bolivian government opposes the dominant economic model that propagates the commercialization and privatization of water and launched an international campaign for a United Nations declaration against water privatization. At the local level, however, the struggle for water justice by means of inclusive recognition of customary hydrosocial territories has proved a significant challenge. In at least three instances it clearly contradicts the inclusive water governance discourse promoted by the Bolivian government.

First, even though the Irrigation Law in Article 21 clearly stipulates that a registration constitutes a right to *use* and *exploit* water resources, in the case of Cochimita and Sank'ayani the water users of both communities interpret and view the registration over their hydrosocial territories as a property title. Several interviewees confirm this by saying that they view the water registration as a 'land title', as a 'water licence', and as proof that they are the 'confirmed owners' (personal communication with several water users, February 2012). Even though there is no legal water privatization in Bolivia, the case of Cochimita shows that issuing formalized water use rights leads to a perception of natural resource and hydrosocial territory *appropriation*. This (mis-)perception of factual resource privatization increases power imbalances between the communities who enjoy a formalized right (e.g., Cochimita) and those who do not (e.g., Sank'ayani Alto).[3] This is not just a unique scenario in the Tiraque Valley. SENARI is aware of the tendency to misinterpret the notion of the registration by perceiving it as title over their hydrosocial territory: 'That is very dangerous since it causes many problems and conflicts among the communities' (representative of SENARI, 14 March 2012). Consequently, SENARI puts great emphasis on explaining to the communities very clearly that water is a public good and that registration is only about recognizing the *use* of water resources according to *usos y costumbres*:

> Nevertheless, when issuing the community with the official document during a public act, many times we have heard in their speeches that community leaders believed that they had been certified as the proprietor of the water source, saying: 'Now we are the owners of the water resources!' (representative of SENARI, 14 March 2012)

Such asymmetrical power relations (between communities with and communities without an official state register) become particularly manifest in cases of local water conflict among peasant and indigenous communities. At this point we see how new rules for water management at the national level have a direct effect at the local level regarding to indigenous autonomy and justice. It shows how local water users strategically choose between different legal systems to legitimize and rationalize their individual or collective action and claims (Benda Beckmann, 1981).

Second, while the national Irrigation Law, based on bottom-up power mechanisms, emphasizes the implementation of registrations throughout the country, whereby all peasants are assured to benefit *equally*, the heterogeneity among Bolivian peasants and agriculturalists has been overlooked. Moreover, the case of Cochimita and Sank'ayani Alto reveals that the attempt by the promoters to naturalize the Irrigation Law interests and recognition rights framework as 'just' denies overall distributional inequalities on the *inter*-communal scale:

> This is a very critical issue of the Irrigation Law, which has no comprehensive vision of equity and of future resource distribution. It has been mainly based on resource

accumulation by the few who already have access to water. How can the communities in the highlands not be allowed to use the water located on their own territory? (Director of the Sustainable Agricultural Development Program, 17 January 2012)

Here it is important to emphasize that the Irrigation Law has been promoted by a small group of irrigators (together with some NGO representatives), which represents a privileged minority of Bolivia's peasants. This is not only the case in the region of Tiraque but also has been confirmed to be a general issue by the director of SENARI:

> Due to climate change the communities in the highlands demand more water. They argue that the valleys have been allowed to take advantage of the water during the last years, now it is the turn of the highland communities. Thus in many instances the water registry has no value anymore. In many cases the highland communities do not accept the registry arguing that since the water is located in their jurisdiction they can claim the right to use it. (personal communication, 5 January 2012)

Third, the water conflict between Cochimita and Sank'ayani Alto reveals that the registry may increase state power in conflict-resolution processes over hydrosocial territories. Since state institutions were not playing a major role in Bolivian water control, historically local traditional water authorities of peasant and indigenous communities assumed the task to organize and control the activities of hydrosocial territories (Bustamante, 2002; Hendriks, 2006; Oré, Castillo, Orsel, & Vos, 2009). Yet, the Irrigation Law vests state institutions for irrigation (SENARI) with the power to take over decision-making in case communities fail to find agreements on their own. We see that Cochimita decided to apply legal state mechanisms, mainly because they see their interests protected by the state and are no longer willing to search for a consensual agreement often based on *usos y costumbres* and orally transmitted cultural values and traditions (Hendriks, 2006). They switch from the level of local/indigenous jurisdiction to the level of the ordinary justice system. To try to solve the conflict between Cochimita (with a water registration) and Sank'ayani Alto (without) in before the court holds a considerable likelihood that state officials will only value the formal water use right, which has been issued by the state. This, however, would contradict the logic of local and indigenous justice that according to the new Constitution of 2009 exercises the same hierarchical power as the ordinary justice system (Article 179 II). The director of SENARI emphasizes:

> While starting to apply ordinary justice, the communal structures of justice are being destroyed, since former ways of conflict resolution are not respected anymore. Communal justice has had to survive the history of the Bolivian Republic and nowadays under the provisions of a Plurinational State it is tremendously weakened. This is quite paradoxical. (personal communication, 5 January 2012)

The weakening of social and communal structures contradicts the indigenous fight for autonomous resource management over their hydrosocial territories expressed during the Water War of 2000 (former member of the technical committee, personal communication, 16 January 2012). Whereas before, Cochimita and Sank'ayani Alto relied on tradition to solve conflicts among each other according to consensual agreements now, the registry leads to more participation of state institutions in local conflict resolution, water management and the capacity to intervene, but also to control regulation and monitoring. In the case at hand, the formal recognition of customary hydrosocial territories increases (for better and worse) coercive power mechanisms in an area – rural irrigation management – where traditionally the state has been almost absent previously.

We see how uncritical recognition of local and customary hydrosocial territories may fail to bring greater social justice. On the one hand, the new paradigm of inclusive water governance has been quite favourable towards collective and communal water rights and in conceptualizing water as a common and public good and a human right. On the other hand, it idealized 'communal harmony and neglected conflict and class divisions, gender, etc., within community systems, between community systems or between a community and external actors' (Achi, 2010, p. 101). The new discourse on natural resources, such as to promote water as a human right and public good and to recognize local water rights according to *usos y costumbres*, neglects divergent and opposing representations of hydrosocial territories, which are present among Bolivian peasants. As exemplified by the cases of Cochimita and Sank'ayani Alto, the state vision of hydrosocial territories clearly intervenes within this controversy and disempowers marginalized peasants who cannot document their hydrosocial territories according to customary rights.

Conclusions

This article challenges mainstream recognition policies. It argues that inclusive policies of water rights recognition and formalization in Bolivia are not in opposition to but rather support disciplinary water governance. To a large extent, the Bolivian Irrigation Law no longer exerts a top-down power modality, but exercises inclusive power by following the liberal ideology and by inviting *all* Bolivian peasants to participate in the ordinary justice system 'under equal terms' (Boelens, 2009). The Cochimita and Sank'ayani case testifies how local communities actively strive to have their local rights legally recognized.

We witness that the formal recognition of 'water territories' and rights in two Bolivian communities, Cochimita and Sank'ayani Alto, unquestionably implies changes in access to and control over water, which entail unevenly distributed burdens and benefits (cf. Leach, Mearns, & Scoones, 1999; Robbins, 2004). Because of their profound impact, recognition politics necessarily involve political contestation, negotiation and struggle. The Cochimita and Sank'ayani Alto case shows that these unevenly distributed outcomes are constantly shaped by social and political choices and power struggles and demonstrates that they go beyond the mere conflict over access to water and the infrastructure; they also concern the content of rules and norms, regulatory control and regimes of representations (Boelens & Zwarteveen, 2005). As demonstrated above, uncritical recognition politics reconfigure hydrosocial territories and control rights. A majority of water users who are not defined as 'irrigators' and have no access to water according to customary water rights are then excluded. This is particularly the case because such policy fails to address issues of redistributive resource access and justice, and because the formal recognition of local hydrosocial territories necessarily implies the non-recognition and illegalization of a variety of non-formalized hydrosocial territories. In the case presented here, while the official regulations take some local water rights repertoires into consideration (such as the contested concept of *usos y costumbres* and water rights claims based on hydraulic property rules) they deny other local norms and rights (such as socio-territorially based water claims). This contradicts the hydrosocial territorial perception defined by some local water users and thus is likely to cause conflict among communities. Moreover, by so doing, the Irrigation Law contradicts the National Constitution of 2009 that recognizes the access to water as a human right (Article 20), and the Law of Mother Earth and Integral Development for Living Well.

Paradoxically, while Bolivia's inclusive water governance discourse is based on inclusion and calls for 'water for all', it disregards the politically contested nature of hydrosocial territories. The 'pro-indigenous' water 'governmentality' introduced by the Morales government ignores the complexity and contested character of hydro-social territories and instead neutralizes geopolitical spaces and rights frameworks, as if they were objective entities. Presumably value-free formalization policies not only overlook existing inequality in access and control over hydrosocial territories but also lead to misrecognition of numerous marginalized people. Thus, they depoliticize water use and actively deepen socio-economic and social injustice among marginalized users. Therefore, this paper argues for more critical recognition policies that address questions of resource redistribution to and decision-making power by marginalized water users. Policy-makers need to acknowledge that hydro-social territories are far from being objective entities but instead highly contested among water users, involving institutional discourses and political agendas. In practise, attention should be paid to developing context-grounded standards and regulations that avoid leading to the exclusion of particular peasant and indigenous communities. A critical recognition of water control space also implies adaptation and upkeep of rights and room for flexibility, since material realities are dynamic and hydrosocial territories contested.

Notes

1. Article 21 of the Irrigation Law defines registration as follows: 'Registration: Administrative act by which the State through SENARI recognizes and assigns the right to use and exploit the sources of water for irrigation to indigenous and aboriginal peoples, indigenous and peasant communities, associations, and organizations and peasant unions, legally guaranteeing in a permanent manner, the water resources according to uses and customs.'
2. Potatoes that used to be grown at a maximum height of 3800–3900 masl can now be grown above 4000 masl. This increases water demand in the highlands. New crops promoted by the market can also be produced at high altitudes. Consequently, highland communities in Cochabamba that in the past lived mainly from cattle rearing today have become engaged in agriculture production (Director of the Integrated Watershed Management Program, 9 January 2012).
3. Some social activists and leaders of the Water War in Cochabamba also observed that many communities understand water registration as a water title, even though water within the communities is managed as a collective right. This development has been one of the main concerns by the opponents of the Irrigation Law.

Acknowledgements

I am grateful for the generous support of colleagues from the *Centro Andino para la Gestión y Uso de Agua* (Centro Agua) in Cochabamba, Bolivia and for the cooperation of the peasants and families of the Tiraque Valley. This article was inspired in part by my interaction with the Justicia Hídrica (Water Justice) network, based in Wageningen, the Netherlands.

Funding

This work was supported by the German Friedrich-Ebert-Stiftung.

References

Achi, A. (2010). El agua como bien comunal: Síntesis de la investigación en Bolivia. In R. Bustamante (Ed.), *Lo colectivo y el agua: Entre los derechos y las prácticas* (pp. 75–108). Lima, Peru: Instituto de Estudios Peruanos.

Agnew, J. (1994). The territorial trap: The geographical assumptions of international relations theory. *Review of International Political Economy*, 1(1), 53–80. doi:10.1080/09692299408434268

Albó, X. (2008). *Movimientos y poder indígena en Bolivia, Ecuador y Perú*. La Paz, Bolivia: CIPCA – Centro de Investigación y Promoción del Campesinado.

Anria, S., Cameron, M., Goenaga, A., Toranzo, C., & Zuazo, M. (2010). Bolivia: Democracia en construcción. In M. Cameron & J. P. Luna (Eds.), *Democracia en la Región Andina* (pp. 243–272). Lima, Peru: IEP – Instituto de Estudios Peruanos.

Baletti, B. (2012). Ordenamento territorial: Neo-developmentalism and the struggle for territory in the lower Brazilian Amazon. *Journal of Peasant Studies*, 39(2), 573–598. doi:10.1080/03066150.2012.664139

Barrios, F. (1997). Curso de postgrado diplomado 'gestión campesina de sistemas de riego'. In Programa Nacional de Riego (PRONAR) (Ed.) *Riego en las alturas de Cochabamba, caso de la comunidad de Cochimita*. Tiraque: Asociación de Riego y Servicio 'Tiraque', PRONAR.

Bebbington, A., Humphreys Bebbington, D., & Bury, J. (2010). Federating and defending: Water territory and extraction in the Andes. In R. Boelens, D. H. Getches, & J. A. Guevara Gil (Eds.), *Out of the mainstream: Water rights, politics and identity* (pp. 307–328). London: Earthscan.

Benda Beckmann, K. von. (1981). Forum shopping and shopping forums: Dispute processing in a Minangkabau village in West Sumatra. *Journal of Legal Pluralism and Unofficial Law*, 13, 117–159. doi:10.1080/07329113.1981.10756260

Boelens, R. (2009). The politics of disciplining water rights. *Development and Change*, 40(2), 307–331. doi:10.1111/j.1467-7660.2009.01516.x

Boelens, R. (2014). Cultural politics and the hydrosocial cycle: Water, power and identity in the Andean highlands. *Geoforum*, 57, 234–247. doi:10.1016/j.geoforum.2013.02.008

Boelens, R., Duarte, B., Manosalvas, R., Mena, P., Roa Avendaño, T., & Vera, J. (2012). Contested territories: Water rights and the struggles over indigenous livelihoods. *The International Indigenous Policy Journal*, 3(3). Retrieved from http://ir.lib.uwo.ca/iipj/vol3/iss3/5

Boelens, R., Hoogesteger, J., & Baud, M. (2015). Water reform governmentality in Ecuador: Neoliberalism, centralization and the restraining of polycentric authority and community rule-making. *Geoforum*, 64, 281–291. doi:10.1016/j.geoforum.2013.07.005

Boelens, R., Hoogesteger, J., & Rodriguez de Francisco, J. C. (2014). Commoditizing water territories: The clash between Andean water rights cultures and payment for environmental services policies. *Capitalism Nature Socialism*, 25(3), 84–102. doi:10.1080/10455752.2013.876867

Boelens, R., Hoogesteger, J., Swyngedouw, E., Vos, J., & Wester, P. (2016). Hydrosocial territories: A political ecology perspective. *Water International, 41*(1), 1–14. doi:10.1080/02508060.2016.1134898

Boelens, R., & Seemann, M. (2014). Forced engagements: Water security and local rights formalization in Yanque, Colca Valley, Peru. *Human Organization, 73*(1), 1–12. doi:10.17730/humo.73.1.d44776822845k515

Boelens, R., & Zwarteveen, M. (2005). Prices and politics in Andean water reforms. *Development and Change, 36*(4), 735–758. doi:10.1111/j.0012-155X.2005.00432.x

Bouleau, G. (2013). The co-production of science and waterscapes: The case of the Seine and the Rhône Rivers, France. *Geoforum.* doi:10.1016/j.geoforum.2013.01.009

Budds, J. (2009). Contested H$_2$O: Science, policy and politics in water resources management in Chile. *Geoforum, 40*(3), 418–430. doi:10.1016/j.geoforum.2008.12.008

Bustamante, R. (2002). Water legislation in Bolivia. In *Indigenous water rights, local water management, and national legislation,* WALIR Studies, vol. 2. Wageningen, The Netherlands: Wageningen University and ECLAC.

Castree, N., & Braun, B. (2001). *Socializing nature: Theory, practise, and politics.* Malden, MA: Blackwell.

Cossio, V., Soto, L., & Skielboe, T. (2010). *Case studies on conflict and cooperation in local water governance: Report no. 1.* The case of the Tiraque highland irrigation conflict Tiraque, Bolivia. Copenhagen, Denmark: Danish Institute for International Studies (DIIS).

Cruz Flores, R. (2010). *Modelo de gestión del agua de la cuenca Pucara.* Cochabamba, Bolivia: Universidad Mayor de San Simon (UMMS), Centro AGUA.

Dean, M. (1999). *Governmentality: Power and rule in modern society.* London: Sage.

Elden, S. (2010). Land, terrain, territory. *Progress in Human Geography, 34*(6), 799–817. doi:10.1177/0309132510362603

Foucault, M. (1979). *Discipline and punish: The birth of the prison* [c.1977]. New York, NY: Vintage.

Foucault, M. (1991). Governmentality. In G. Burchell, C. Gordon, & P. Miller (Eds.), *The Foucault effect: Studies in governmentality* (pp. 87–104). Chicago, IL: University of Chicago Press.

Green, D. P., & Shapiro, I. (1994). *Pathologies of Rational choice theory: A critique of applications in political science.* New Haven, CT: Yale University Press.

Harvey, D. (1996). *Justice, nature and the geography of difference.* Cambridge, MA: Blackwell.

Hendriks, J. (2006). Legislación de aguas y gestión de sistemas hídricos en países de la región Andina. In P. Urteaga Crovetto & R. Boelens (Eds.), *Derechos colectivos y políticas hídricas en la región Andina* (pp. 47–111). Lima, Peru: IEP Instituto de Estudios Peruanos.

Latour, B. (1993). *We have never been modern.* Cambridge, MA: Harvard University Press.

Leach, M., Mearns, R., & Scoones, I. (1999). Environmental entitlements: Dynamics and institutions in community-based natural resource management. *World Development, 27*(2), 225–247. doi:10.1016/S0305-750X(98)00141-7

Ley N° 300. (2012, October 15). *Ley marco de la madre tierra y desarrollo integral para Vivir Bien.* La Paz, Bolivia: Estado Plurinacional de Bolivia.

Mason, J. (2004). Semistructured interview. In M. S. Lewis-Beck, A. Bryman, & T. F. Liao (Eds.), *The sage encyclopedia of social science research methods* (pp. 1020–1021). Thousand Oaks, CA: Sage.

Mazurek, H. (2010). Introducción: Gobernabilidad y gobernanza: el aporte para los territorios y América Latina. *En* H. Mazurek (Ed.), *Gobernabilidad y gobernanza de los territorios en América Latina.* Lima, Peru: IFEA.

Ministerio de Agua. (2006). *Ley de Riego: Nr. 2878.* La Paz, Bolivia.

MMAyA. (2007). *Plan nacional de desarrollo del riego para 'Vivir Bien': 2007–2011.* La Paz, Bolivia: Ministerio del Agua, Viceministerio de Riego, Servicio Nacional de Riego.

Official Agreement. (1990). *Official agreement on the operation of the reservoirs Iglesia Punku, Sakayani Alto and Kayarayoj Tóqo.* Unpublished. Tiraque, Cochabamba, Bolivia: Sindicato Agrario de Cochimita.

Oré, M. T., Castillo, L. d., Orsel, S. V., & Vos, J. (2009). *El Agua, ante nuevos desafíos: Actores e iniciativas en Ecuador, Perú y Bolivia*. Lima, Peru: Oxfam Internacional.

Peluso, N., & Lund, C. (2011). New frontiers of land control: Introduction. *Journal of Peasant Studies, 38*(4), 667–681. doi:10.1080/03066150.2011.607692

Perreault, T. (2008). Customs and contradiction: Rural water governance and the politics of Usos y Costumbres in Bolivia's irrigator's movement. *Annals of the Association of American Geographers, 98*(4), 834–854. doi:10.1080/00045600802013502

Postero, N. (2010). The struggle to create a radical democracy in Bolivia. Special Issue. *Latin American Research Review, 45*, 59–78. Retrieved from http://muse.jhu.edu/login?auth=0&type=summary&url=/journals/latin_american_research_review/v045/45.S.postero.pdf

Registro de Cochimita. (2007). *Carpeta de usos y costumbres comité de riego Kayarayoj Tocko*. Unpublished. Cochabamba, Bolivia: SENARI.

Robbins, P. (2004). *Political ecology: A critical introduction*. Malden, MA: Blackwell.

Saldías, C., Boelens, R., Wegerich, K., & Speelman, S. (2012). Losing the watershed focus: A look at complex community-managed irrigation systems in Bolivia. *Water International, 37*(7), 744–759. doi:10.1080/02508060.2012.733675

Seemann, M. (2014). *Water security and the politics of water rights formalization in Peru and Bolivia The struggle for water justice in Andean water user communities*. Unpublished Promotionsarbeit. Hamburg, Germany: Universität Hamburg.

Smith, J. (1983). Quantitative versus qualitative research: An attempt to clarify the issue. *Educational Researcher, 12*(3), 6–13. doi:10.3102/0013189X012003006

Stavenhagen, R. (2009). Report of the special rapporteur on the situation of human rights and fundamental freedoms of indigenous people. UN Human Rights Council. Retrieved July 8, 2013 from http://www.refworld.org/docid/4a1d07ec2.html%20[accessed%209%20July%202013]

Swyngedouw, E. (1999). Modernity and hybridity: Nature, regeneracionismo, and the production of the Spanish waterscape 1890–1930. *Annals of the Association of American Geographers, 89*(3), 443–465. doi:10.1111/0004-5608.00157

Swyngedouw, E. (2007). Technonatural revolutions: The scalar politics of Franco's hydro-social dream for Spain, 1939–1975. *Royal Geographical Society*, 9–28. Retrieved from http://de.scribd.com/doc/203721908/Swyngedouw-2008-Technonatural-Revolutions

Vos, H., de Boelens, R., & Bustamante, R. (2006). Formal law and local water control in the Andean region: A fiercely contested field. *International Journal of Water Resources Development, 22*(1), 37–48. doi:10.1080/07900620500405049

Worster, D. (1992). *Rivers of empire: Water, aridity, and the growth of the American West*. Oxford: Oxford University.

Zwarteveen, M. Z., & Boelens, R. (2014). Defining, researching and struggling for water justice: Some conceptual building blocks for research and action. *Water International, 39*(2), 143–158. doi:10.1080/02508060.2014.891168

Zwarteveen, M., Roth, D., & Boelens, R. (2005). Water rights and legal pluralism: Beyond analysis and recognition. In D. Roth, R. Boelens, & M. Zwarteveen (Eds.), *Liquid relations: Contested water rights and legal complexity* (pp. 254–268). New Brunswick, NJ: Rutgers University Press.

Virtual water trade and the contestation of hydrosocial territories

Jeroen Vos[a] and Leonith Hinojosa[a,b]

[a]Department of Environmental Sciences, Wageningen University, Wageningen, the Netherlands; [b]Earth and Life Institute, Université Catholique de Louvain, Louvain-la-Neuve, Belgium

ABSTRACT

Growing trade in virtual water – the water used to produce exported products from agriculture and mining sectors – affects local communities and the environment, and transforms hydrosocial territories. National and international water regulations reshape communities' hydrosocial territories by changing water governance structures to favour export commodity sectors, often inducing strong contestation from local communities. Transnational companies formulate and enforce global water governance arrangements oriented toward strengthening export production chains, often through asymmetrical relationships with local groups in water-export regions. These arrangements compromise political representation and water security for both local communities and companies.

Introduction

Water use for export agriculture and industrial production often affects local communities and the environment. The fresh water 'embodied' in a commodity, known as 'virtual water', refers to the volume of water consumed or polluted to produce a commodity and is measured over its full production chain (Allan, 1998). Virtual water is no longer available for alternative uses. The source of water can be surface or groundwater ('blue virtual water') or the water in the soil ('green virtual water') (Hoekstra & Chapagain, 2008). Virtual water can also include the contamination of water during the production process, i.e., the amount of clean water needed to dilute a contaminated water body to reach an environmental standard ('grey virtual water').

Some scholars have proposed virtual water import to resolve national water scarcity (Allan, 1998; Zhao, Liu, & Deng, 2005). Yet, observing that virtual water export has doubled during the last decade (Dalin, Konar, Hanasaki, Rinaldo, & Rodriguez-Iturbe, 2012) and constituted nearly 30% of the world's direct water withdrawal (Chen & Chen, 2013), others have documented the negative effects of water-intensive trade patterns (cf. Dauvergne & Lister, 2012; Khan & Hanjra, 2009). Direct negative effects include water resource depletion and pollution (Dabrowski, Murray, Ashton, & Leaner, 2009; Hoekstra & Chapagain, 2008) and concentration of water use rights, which might jeopardize local livelihoods.

Small farmers integrated into export markets can also be harmed by increased horizontal and vertical integration of export chains dominated by powerful agro-business and retailer companies, receiving relatively little benefit and heavy exposure to risks: crop failure, low market prices, debt burden and possible land dispossession (GRAIN, 2012, 2014; La Via Campesina, 2013).

Sojamo, Keulertz, Warner, and Allan (2012) and Vos and Boelens (2014b) argue that the increased export of water-intensive commodities also changes water governance, shifting control over water use from local and national actors to those who dominate global production chains. Sojamo et al. (2012) argue that the increase in international virtual water trade sharply decreases the political power for local water governance. This often lessens local authorities' control over the water resources in their territory. Instead, multinational companies and retailers set the terms of trade and establish mechanisms for local water resource use and protection. These water governance rules encompass corporate discourses about water efficiency and productivity and the benefits of international trade, aligning with rhetoric about a technology fix to increase water efficiency and productivity through, for instance, drip irrigation (Boelens & Vos, 2012).

International companies more readily gain access to local water resources when national governments promote natural resource extraction through free-trade agreements and privatization of state-owned and community-held resources, subsidizing international transport infrastructure, and allowing tax havens and permissive environmental legislation. This has enabled companies to set the terms of trade and establish private water stewardship standards as part of their corporate social responsibility (CSR) policies. Recent cases of water grabbing have despoiled local water users of control over their water resources and means of livelihoods (GRAIN, 2012, 2014; Mehta, Veldwisch, & Franco, 2012). These new definitions of 'good' water management and 'fair' distribution of water replace local values, norms and imaginaries with new ones presented as 'universal' and 'natural'. Contentiously, local stakeholders' interests are left out of policy, rules and regulation for water resources management.

The adverse effects of increased virtual water export are contested by local communities. The extent to which contestation interacts with these international trade-driven water stewardship initiatives, to define new forms of water governance, is only recently being debated. How this affects water justice by inducing loss of territorial control of and access to water resources is also a topic of recent research. This article uses the concept of hydrosocial territories to discuss the links between virtual water export and local water control, and contribute to such a debate. After this introduction, the next section identifies the drivers of increasing virtual water trade, the effects on local water users and the environment. The third section analyses the relationship between virtual water export and hydrosocial territories' transformation, taking into account the emerging corporate sector's responses to water conflicts through water stewardship and examples of contestation from below. The paper concludes by drawing some reflections for policy-making.

Drivers of increasing virtual water trade

Virtual water import has been suggested to provide food security for relatively dry countries (Allan, 1998, 2003). Accordingly, neo-liberal water policies (i.e., governmental policies deregulating the water sector and promoting international trade through social and environmental deregulation, among other mechanisms) postulate that virtual water export taps the comparative advantage of relatively water-sufficient countries or production zones, meaning more efficient food production globally (Hoekstra & Chapagain, 2008). However, Seekell, D'Odorico, and Pace (2011) and Suweis et al. (2011) show that virtual water flows do not address relative water scarcity, but export stimulation, global financial structures and consumption. Similarly, Wichelns (2010) asserts that net virtual water trade is not related to water-related relative comparative advantages in producing and receiving regions, but to other factors of relative production advantages and consumers' demand.

Virtual water export more than doubled in the last two decades. Dalin et al. (2012) calculate that total international virtual water trade embedded in the top five agricultural and three livestock products doubled from 259 km^3 in 1986 to 567 km^3 in 2007, mainly soy export for dairy and meat production from South America to Asia and Europe. The literature on trade and globalization (e.g., Clapp & Fuchs, 2009; Gibbon & Ponte, 2005; McMichael, 2009) highlights five main drivers of such increase: growing purchasing power and diet changes by numerous consumers in emerging markets (e.g., increased meat and dairy consumption in Asia has spurred soy exports from South America to Asia, as shown by Dalin et al., 2012); expanding free-trade agreements; increasing international financing for export agriculture; national export policies; and horizontal and vertical integration of companies within the agro-export chain.

Since the mid-1990s, the World Trade Organisation (WTO) has facilitated over 60 free-trade agreements among its member states. Additionally, over 2500 bilateral trade and investment treaties have been signed (UNCTAD, 2007). This spread of free-trade agreements, driven by international organizations and northern countries, is also supported by southern countries, eager to attract foreign direct investment (FDI). Since the early 1990s, many national governments have adopted neoliberal policies for natural resource management in order to facilitate trade agreements. Some 20 countries signed free-trade agreements with the United States, which include provisions to protect foreign investors, sometimes affecting the environment and water-user communities, particularly those labelled 'informal', i.e., with no allocation of formal water rights (Solanes, 2010).

At the same time, national neoliberal policies have also included legislation to incorporate communal water into the natural resources market and apply permissive environmental regulation and monitoring to export companies (e.g., in Chile, Peru and Mexico). Moreover, neoliberal government policies and programmes in countries such as Chile, Peru, Ecuador and Colombia granted government funds to large-scale export agriculture by subsidizing dams, canals, electricity for pumping, etc.

International financing of export agriculture has played an important role in increasing global trade (Burch & Lawrence, 2009; Vestergaard, 2012) enabling international food trade (e.g., see the FAO's 2010 report on the drivers of rising food prices) and purchase of farm land (Cotula, 2012). This has facilitated land and water concentration by large transnational companies and local elites operating in the agribusiness sector. Their strategy of horizontal and vertical integration within the agro-export chain has increased their power to control flows of goods and finance around the world (Baines, 2014; Burch & Lawrence, 2009; Carolan, 2012; Fuchs, Kalfagianni, & Arentsen, 2009; McMichael, 2009; Murphy, 2008).

Examples of major companies that control agro-food chains are given by Sojamo and Larson (2012). Debbané (2013) provides the example of the Ceres Valley in South Africa where financialization played an important role in the accumulation of water by export-oriented large fresh fruit producers:

> The financial and institutional arrangements for the Koekedouw Dam exemplify the neoliberal thrust underpinning national water policy reforms. This is reflected in three important ways: the withholding of government subsidies and state-backed loans for irrigation projects; the expanded role given to the private sector in planning and building large-scale infrastructure projects; and the expanded role given to private commercial banks in financing water infrastructure. This marked a significant departure from previous practices, where the government heavily subsidized the costs of irrigation development. (pp. 2560–2561)

Effects of virtual water export on local water users and the environment

Promoters of international trade advocate that agricultural, mining, energy and industry exports generate income, employment and foreign exchange. In addition, under particular conditions, strengthening global value chains could have a positive impact on local producers (cf. EC, 2012; Goger, Hull, Barrientos, Gereffi, & Godfrey, 2014). Among other arguments, it has been suggested that increasing trade can successfully insert local producers into international markets and create new sources of income and employment for the local population. In territories where customary water rights were protected by law, such as in Chile where most water user associations register water rights, these registers contributed to protecting local rights. Nevertheless, in many instances, virtual water export has had negative effects on water resources and rural populations, especially in relatively arid areas. Increased virtual water trade has led to overexploitation and contamination of rivers and aquifers and/or regions where the political elite has grabbed water away from local communities' livelihoods (Roth & Warner, 2008).

In general, the share of virtual water exported from a country does not directly indicate the negative or positive social and environmental effects in the producing/importing region. This is because the abstract notion of virtual water hides the realities of closing basins (rivers that no longer reach the ocean), large agribusinesses' water rights accumulation, the 'race to the bottom' of ever-deeper tube wells, drying wetlands, vast agricultural areas that become unproductive due to salinization, and especially water user communities' loss of control over water resources. To assess these effects better in exporting countries, Lenzen, Bhanduri, et al. (2012) combine the calculation of net virtual water trade with indices of relative water scarcity and water exploitation. However, this fully measures neither the environmental impact nor the distributive effect on income and access to water. That said, while assessing impacts of net virtual water flows globally can be too complex, therefore inaccurate, the effects of virtual water trade can be assessed more rigorously on a local scale.

Many surface and groundwater sources are overexploited, and lack of drainage leads to waterlogging and eventually salinization (Shah, Burke, & Villholth, 2007; Wada et al., 2010). Total worldwide extraction of groundwater has increased from some 100 km^3 in 1950 to some 1000 km^3 in 2000. Thus, while world population doubled, extraction 10-

folded. This growth in agricultural pumping is concentrated in Asia, foremost in India, China and Bangladesh (Shah et al., 2007). However, aquifers are also overexploited for export agriculture in many other parts of the world. In Central and Northern Mexico, agricultural production for export to the United States deprives many smallholders of access to groundwater (Peña, 2011). Similar cases can be found in the Palestine territories (Zeitoun, Messerschmid, & Attili, 2009), India and Pakistan (Chapagain, Hoekstra, Savenije, & Gautam, 2005), and East Africa (Awange et al., 2013).

Large-scale export agriculture, mining, oil extraction and industry can also cause water contamination and reduce biodiversity (Defries, Rudel, Uriarte, & Hansen, 2010; Lenzen, Moran, et al., 2012; Longo & York, 2008). Agricultural, oil and industrial exports entail grey virtual water export. An example is the large-scale soy production in Brazil, Argentina, Bolivia and Paraguay for export to Asia and Europe, where Palau, Cabello, Naeyens, Rulli, and Segovia (2007) found severe health effects associated with groundwater contamination by agro-chemicals used to grow soy. Further, all over the world, regions specializing in mining or agriculture export show labour and living conditions at critically low levels; women in particular are affected negatively (Langan, 2011; Pearson, 2007).

Even in successful cases, such as Chile, protection of water rights has been constrained by the legal system. The government administration's operational weaknesses failed to enforce pro-sustainability measures and reduce externalities. Limitations in procedures for conflict adjudication by courts, which is expensive, have facilitated water dispossession of small water users who cannot afford the cost of litigation (Bauer, 2005). In many other countries, lack of formal registration of water use, non-recognition and/ or little knowledge of customary use, takeover of water governance structures by elites, weak water management operational capabilities, and the conditional requirements (international financing, FDI, trade agreements and international arbitration courts) have contributed to dispossession and pauperization of local water users (Solanes, 2010; Solanes & Jouravlev, 2007).

Virtual water export and transformation of hydrosocial territories

The analysis in this section, of increasing virtual water exports transforming hydrosocial territories, builds on the definition of hydrosocial territories introduced by Boelens, Hoogesteger, Swyngedouw, Vos, and Wester (2016, p. 2), for whom:

> hydrosocial territories are the contested imaginary and socio-environmental materialization of a spatially bound multi-scalar network in which humans, water flows, ecological relations, hydraulic infrastructure, financial means, legal–administrative arrangements and cultural institutions and practices are interactively defined, aligned and mobilized through epistemological belief systems, political hierarchies and naturalizing discourses.

A hydrosocial territory melds meaning, actors, political power, water flows, water technology and biophysical elements (see also Del Moral & Do O, 2014; Delaney, 2008).

Increasing virtual water export creates new forms of hydrosocial territories. Water use for production that is financed and consumed within a region establishes hydrosocial territories in which water governance is mostly local. Virtual water export creates hydrosocial territories that go beyond this level. Multinational institutions such as financing organizations and major retailers and large multinational agribusiness companies introduce rules and regulation in national regulatory frameworks, and they also influence norms and values regarding water governance.

We identify two main mechanisms by which virtual water trade and changes in hydrosocial territories become associated: 'formal' access to and control of local water sources by international companies, and imposition of water use standards on small producers.

The mechanism concentrating water access

Control over local water resources can be lost many ways. International financialization enables large companies to invest in water high-tech and equipment (e.g., for deep drilling, drip irrigation and water decontamination) and expand their operations. As this has some positive effects on local economies in host regions, it increases their political clout (e.g., to gain water concessions, pay lobbyers and the like). However, large land deals forcing indebted smallholders to sell their land, unfair sharecrop arrangements and permissive environmental regulations applied to large agricultural export companies have been documented in the land- and water-grabbing literature (e.g., GRAIN, 2014; Mehta et al., 2012). For example, in Ecuador, companies that produce export crops and sugarcane have accumulated nearly 75% of formal water rights and concentrate much more water informally and illegally, especially to export bananas (Gaybor, 2011) and flowers (Zapatta & Mena, 2013). In Peru, agribusiness companies exporting fresh fruits and vegetables to the United States and Europe have accumulated water rights and overexploit groundwater on the dry coast (Progressio, 2010; Van der Ploeg, 2008).

The less powerful local water-user groups tend to lose out when confronted with powerful new actors, such as agribusiness enterprises and mining companies (Castro, 2008). This concentration of political and economic power by major agribusiness and retailer companies establishes new hydrosocial territories in which natural resources come under control by multinational companies (cf. Van der Ploeg, 2008, 2010). Governments have shown little interest in improving their knowledge and records concerning water availability and use in order to protect local users' rights (Solanes, 2010).

The mechanism of corporate water stewardship initiatives to produce global hydrosocial territories

The fundamental change in hydrosocial territories' configuration, driven by growing exports of virtual water worldwide, is enabled by an emerging institutional setting of international trade rules, changing national policies and corporate sector codes of conduct. Hydrosocial territories are transformed through FDI, export-oriented national policies and imposition of corporate water use standards on small producers who want to export (Carolan, 2012; Vos & Boelens, 2014b).

Large companies' power over local water resource governance creates new forms of water control that redefine hydrosocial territories at a supranational governance level;

in this process, local development options, values, imaginaries and knowledge are glossed over. A mechanism of this process is the CSR standards for water stewardship often set by international retailers and producer companies, which eventually shape the practices, norms, values and imaginaries of local producers, who adopt the new regulatory and values framework in order to be able to export.

Private companies increasingly take pledges of water stewardship. A water steward-ship certification scheme is formulated by, and self-enforced among, transnational retailer companies. Broadly defined, stewardship is conceived as actions taking care of public goods depending on collective management for sustainability. There are three main reasons for this engagement: to guarantee sufficient production; to reduce the risks of reputational damage; and to gain legitimacy and critical consumers' support (Hazelton, 2014; Vos & Boelens, 2014b; Waldman & Kerr, 2014).

Water stewardship standards can include protection of water sources (in quantity and quality), efficient water use, use of certain water technology and registration of water use (Vos & Boelens, 2014b). Adherence to water stewardship schemes is wide-spread. According to Fulponi (2007), all growers that produce for supermarkets in Europe and the United States are obliged to subscribe to one or more certification schemes. Examples of these schemes are GlobalGAP, BRC, GFSI, FOODTRACE, IFS and also some other non-corporate sector schemes such as the Rainforest Alliance. The trade-off between the income derived from virtual water export by local economies and the 'external' influences on the local hydrosocial territories is the external control over water use standards. For example, GlobalGAP defines what is 'efficient' and 'sustain-able', yet with little participation by or consultation with small local producers and water users (Vos & Boelens, 2014b). Compliance by local suppliers of supermarkets and food companies with the international standards is certified by third-party auditors. Growers are audited annually by auditors who must be accredited by a certifying body under a certain scheme.

In the mining sector, the International Council on Mining & Metals (ICMM) (2014) has adopted a number of standards for its member mining companies. In one of these standards, known as 'the 10 principles of sustainable mining', principle 6 reads: '[mining companies are supposed to] seek continual improvement of our environmen-tal performance. Assess the positive and negative, the direct and indirect, and the cumulative environmental impacts of new projects – from exploration through closure'. Mining companies have to report publically on their performance regarding these principles and provide evidence through third-party verification.

A point of concern in all these initiatives is the low transparency and democracy in formulating standards and monitoring procedures, which are set by dominant market players (Amekawa, 2009; Campbell, 2005). Although roundtables involve different stakeholders in specific sectors (sugarcane, biofuels, cotton, soybeans), large transna-tional companies and supranational policy networks dominate the negotiation table; organizations representing local populations, particularly small farmers, have far greater difficulties to participate and stake claims.

Furthermore, from a political point of view, standards reinforce (economic, political and discursive) power in the Global North (Fuchs et al., 2009). Certifying and audit companies are also almost exclusively from the north. Farmers have to pay for audits and this can be exclusionary, hindering small farmers (cf. Blackmore et al., 2012). In this process of institutional change, the power to regulate water use and water quality implies shifts from local communal and national public authority to international organizations (cf. Van der Ploeg, 2008). This reveals weaknesses in the discourse on global devolution and decentralization. As in any other global space, universal rules and regulation also apply in global hydrosocial territories, leaving little space or no space at all for the intrinsic characteristics of local hydrosocial territories. Actors who dominate the decision-making processes within global and regional organizations, the corporate sector and equity funds are better positioned to influence national governance structures, discourses and norms that determine and influence access to local water sources, their use and local capacities to control virtual water export.

Contestations from below

The emerging global hydrosocial territories are contested. The politics and plans to transform hydrosocial territories and the problems brought by water use concentration and contamination of water sources have led to resistance, protests and conflicts at different scales. Resistance and protests have been related to water grabbing and contamination by agribusiness (Hall et al., 2015; Mehta et al., 2012; Smaller & Mann, 2009; Sojamo et al., 2012), construction of hydropower dams (e.g., Scudder, 2005), and water issues of minerals and oil extraction (Helwege, 2015). Protest has multiple motives and often involves diverse groups representing different local interests, attaching importance to different issues at stake. Issues and intensity of protests also vary over time. Furthermore, the form and intensity of protests are influenced by issues such as political freedoms, local leaderships, articulation of discourses and local economic dynamics.

Notwithstanding the reported strong increase in land grabbing in Africa, Asia and Latin America (GRAIN, 2012; Mehta et al., 2012) and the large volumes of water used in the agribusiness sector, relatively few cases of grassroots protests against 'water grabbing' by export agribusiness are documented. Given that public protest can be dangerous for the protesters due to adverse political and institutional contexts, protest have taken the form of anonymous, non-organized and hidden acts of sabotage, something labelled by Scott (1985) as 'the Weapons of the Weak'. For instance, Moreda (2015) describes resistance of Gumuz communities against land and water grabbing in western Ethiopia. Acts of sabotage and violence allegedly included setting fire to agricultural equipment and 700 hectares of maize ready to harvest and intimidating migrant agricultural labourers. An agribusiness company left the region because the project manager was killed by a Gumuz arrow. This form of resistance is often dismissed by local and national media, and therefore by scholars as well.

In other regions, protest is carefully organized and becomes public. Although more visible, this type of protest is seldom systematically documented, either. Furthermore, despite accusation from industry and national governments about external influences, particularly non-governmental organizations (NGOs), it is relatively rare for local, grassroots' protests to connect with national NGOs to form national alliances or even connect to international networks of activists' movements. We briefly review some illustrations of local protest against water injustices associated with expansion by agribusiness and extractive industries. These examples were taken from academic papers and news media websites, and illustrate the point on local reaction to the effects of virtual water export:

- La Ligua (Northern Chile): in 2011, regional organization MODATIMA (movement for the defence of water, land and the environment) started its protest against overexploitation of the La Ligua and Petorca rivers and aquifer by a few large agricultural export companies producing avocadoes for export to Europe and the United States (Budds, 2009; also see http://modatimapetorca.wix.com/wwwwixcommodatimapetorca).
- Tabacundo (north-east Ecuador): flower export agribusinesses in the Tabacundo valley have grown enormously during the last decade. Nowadays some 3000 hectares of roses are cultivated in the highlands for the US, European and Russian markets. Significant amounts of water for the greenhouses come from the Acequia Tabacundo irrigation system, which also serves small farmers for subsistence agriculture and cattle raising. Given that the 'water left-over' was too little for the smallholders, in 2006 the canal was taken over by a protest march of 3000 small farmers (Zapatta & Mena, 2013).
- Northern Mexico: in this dry region, agribusiness companies extract groundwater to export fresh produce to the United States, which has mobilized many local groups to protest against unequal access to sanitation and irrigation water. For example, in March 2015, the Yaqui People from the state of Sonora defended their water and territory in a march to the capital city to protest against the construction of the Independence Aqueduct (Conn, 2015). Local communities also protest against the drilling of new wells by agro-export companies (Quintana, 2013).
- Senegal: a collective of pastoralists organized protests against a company that planted 20,000 hectares of sugarcane. The Senhuile–Senéthanol project is financed partly by Italian and partly by Senegalese companies, and the national government granted access to the land. Local pastoralists are denied access to their pastureland, firewood, migration routes for cattle and water wells. Local communities organized into the Collective for the Defence of the Ndiaël Reserve. They organized protests and demanded the company's withdrawal from the zone (Word, 2014).
- Huancavelica (Peru): local protest emerged against the allocation of water to the large-scale export agriculture developed in the neighbouring desert coastal region, Ica. Irrigation water for the agribusiness sector is taken from the Ica River, which is fed by the Andean watersheds of Huancavelica. Given the exhaustion of Ica's aquifers, largely due to the expansion of export agriculture, recent plans to divert more water into the Ica River through the Incahuasi canal would negatively impact access to water for the highland community of Carhuancho (Hoogesteger & Verzijl, 2015). Carhuancho's population protested regularly in Ica, but because their complaints were dismissed by Peruvian water institutions, they presented their case at the Public Hearing of the International Water Tribunal held in Guadalajara, Mexico in 2007 (see http://tragua.com).
- Cordova (Argentina): the soya boom in Argentina increased agrochemical water contamination. The area cultivated with soya tripled in 15 years to some 18 million

hectares. The expansion excluded peasants from land and wells they have historically used for goat herding. Protests of farmers located in the soya region and Cordoba's citizens have been going on for a decade, including marches, massive rallies, roadblocks and court cases to stop agrochemical use (Cáceres, 2015).

- Cajamarca (Peru): the gold-mining region of Cajamarca has been the scene of a vast social protest movement against water use and pollution by open-pit gold mines. Peasant communities have been protesting for a decade. Mining expansion into the Quilish mountains and opening of the new Conga mine are contested by inhabitants of the hydrosocial territory, as those mines will affect negatively their water quality and quantity (Yacoub, Blazquez, Pérez-Foguet, & Miralles, 2013). In 2010 the army was sent to Cajamarca to repress massive protest rallies violently. These protests have received support from national and international NGOs (Sosa & Zwarteveen, 2014).

These illustrations of resistance and protests show a great diversity in topics, and in their degree of connectivity with larger networks. Mining projects seem to trigger more protests than does water grabbing by agribusiness, probably due to the relatively large number of labourers who are employed by agribusiness operations (see also Vos & Boelens, 2014a).

As many protests and acts of resistance remain invisible, it is difficult to gauge the magnitude of local resistance and protests. For an idea of water-related protests and conflicts, the online EJOLT Atlas of Environmental Conflicts presents over 1000 environmental conflicts around the world, most related to negative effects on water (see https://ejatlas.org/). Water-related conflicts are also described by the Latin American Water Tribunal (Tribunal Latinoamericano del Agua; see http://tragua. com/). This tribunal has organized a total of seven public hearings between 2000 and 2012. Approximately one-third of the 60-plus water conflict cases presented at the hearings involve socio-environmental impacts of large-scale extractive industries. Other case studies on water-related conflicts can be found in the books and website of the Justicia Hídrica Alliance (see http://justiciahidrica.org/?lang=en).

The above examples show grassroots' action against water extraction and contamination by increasing virtual water export. While many local protests focus on the injustices of water concentration and dispossession, some also involve concern with the transformation and reconfiguration of local and regional (sub-national) hydrosocial territories. Changes in material control over water bodies (the spatial biophysical component of hydrosocial territories) and power relationships underlying such control (the non-material component) are directly or indirectly denounced by these social protests. These cases do not represent any statistical evidence as they use particular definitions of conflict and present cases mainly related to exposure that NGOs or media give conflicts. In reality, there are many more hidden, 'everyday' forms of conflicts, many of which never get any NGO or media attention.

Two examples illustrate the contestation from below against transformations of hydrosocial territories due to changing national forms of water governance:

- In Spain, a vast protest movement emerged against transferring water from northern Spain to the south for export horticulture. The National Hydrological Plan approved in 2001 by Spain's central government would transfer 860 million m^3 per year from the Ebro River in the relatively wet north-east of Spain to the relatively dry south of the country. The south grows fruits and vegetables for export to northern Europe. Ebro–Segura environmental organizations from northern Spain organized massive street protests against the transfer. In 2001 and 2002 in Barcelona, Zaragoza, Valencia and Brussels, over 200,000 people marched against the water transfer. In the new 2005 National Hydrological Plan, the transfer plan was abandoned (Swyngedouw, 2013). This 'Nueva Cultura de Agua' (New Water Culture) movement was against water allocation for export agriculture, and backed increasing 'regionalism', informed by political party politics and further reinforced by the central idea of the 2001 European Water Framework Directive that takes river basins as the primary unit of water governance (Lopez-Gunn, 2009).
- The 2011 Arab Spring movement in Tunisia illustrates how hydrosocial territory issues can be among the multiple issues of very heterogeneous national protest movements. In Tunisia, water governance favouring private companies allegedly was part of the mix of multiple issues fuelling the Arab Spring protest movement. Gana (2012) argues that:

> processes of agricultural restructuring during the past 20 years contributed importantly to fuel the revolutionary dynamics, thus giving a political dimension to food issues. As demonstrated by the rising farmers' protest movement (land occupations, contestation of farmers unions, refusal to pay for irrigation water) [...]. (p. 201)

Gana identified a relationship between the uprising and the past Tunisian policy bias favouring private companies:

> What these multiple forms of protests clearly reveal is the rise of social struggles in the countryside and a profound contestation of former State policies, but also a differentiation of farmer demands, according to the different social groups. Actually, there is a consensus among farmers that agricultural development should be given a renewed and increased attention in State policies, policies that farmers consider to have been biased in favour of the industrial and the touristic sectors. (p. 209)

Further, national and international public and NGOs (like the Third World Network, the Transnational Institute, Via Campesino, Food First, the Businesses and Human Rights Resource Centre, the Justicia Hídrica Alliance, EJOLT and GRAIN) protest against the emerging undemocratic global water governance. They work together with regional and local organizations and engage in lobbying and advocacy work.

Conclusions

This paper explores the relationship between virtual water export and hydrosocial territories. The central argument is that increasing virtual water export over the past 20 years has been accompanied by local contestation against changes in local hydrosocial territories aiming to turn them over to global water governance structures.

Increased virtual water trade has negative effects in regions that export agricultural and mining commodities, where virtual water trade has led to over-exploitation and contamination of rivers and aquifers. Resource capture by the political elite (water grabbing) has taken away local water user communities' livelihoods. However, producers' practices, norms, values and imaginaries also change when they accept the rules and values imposed in order to be able to export.

This paper has defined hydrosocial territories as a fusion of meaning, actors, political power, water flows, water technology and biophysical elements. Hydrosocial territories are co-constituted by the material elements (land, water, ecosystems) within that space and the social power relationships between, and interests of, the people related to that space.

Increasing virtual water export creates new forms of hydrosocial territories. Water use for production that is consumed within a region establishes hydrosocial territories in which water governance is local. Virtual water export creates hydrosocial territories that are supra-regional. Multinational institutions such as financing organizations, major retailers, large multinational agribusiness companies, and multinational NGOs impose rules and regulations, but also influence norms and values related to water governance. The power these organizations exercise creates supra-regional hydrosocial territories that ignore or sideline local values, imaginaries and knowledge. A clear example is the multiple CSR standards for water stewardship set by international retailers and companies. To study the newly emerging hydrosocial territories, the strategies of international food-chain companies should get more attention. Paraphrasing James Scott's (1998) 'Seeing like a State' approach, we suggest that research on virtual water and territory also requires 'seeing like a multinational food-chain company'.

This implies looking at how emerging global hydrosocial territories are contested. Local water users protest against loss of control over local water resources, which is echoed by national and international public and NGOs.

Policy implications for governments, civil-society organizations and companies are that governments, at different levels, can actively protect local water sources, promote just water distribution and activities to counter the negative effects of virtual water export from vulnerable territories. Local communities and their organizations, as well as local, national and international NGOs, can engage in alliances to protect local resources. Companies can protect local communities and the environment, not with standardized water stewardship schemes but by developing and implementing protective measures in cooperation with communities, local production associations, water users' associations, local and national labour unions, environmental NGOs, water basin organizations and other stakeholders.

Funding

The research for this article was carried out under the umbrella of the international Justicia Hidrica/Water Justice Alliance (www.justiciahidrica.org) and the Transnationalization of Local Water Battles research programme, financed by the Netherlands Organization for Scientific Research (NWO).

References

Allan, J. A. (1998). Virtual water: A strategic resource, global solutions to regional deficits. *Groundwater, 36*(4), 545–546. doi:10.1111/j.1745-6584.1998.tb02825.x

Allan, J. A. (2003). Virtual water – The water, food, and trade nexus. Useful concept or misleading metaphor? *Water International, 28*(1), 106–113. doi:10.1080/02508060.2003.9724812

Amekawa, Y. (2009). Reflections on the growing influence of good agricultural practices in the Global South. *Journal of Agricultural and Environmental Ethics, 22,* 531–557. doi:10.1007/s10806-009-9171-8

Awange, J. L., Forootan, E., Kusche, J., Kiema, J. B. K., Omondi, P. A., Heck, B., . . . Gonçalves, R. M. (2013). Understanding the decline of water storage across the Ramser–Lake Naivasha using satellite-based methods. *Advances in Water Resources, 60,* 7–23. doi:10.1016/j.advwatres.2013.07.002

Baines, J. (2014). Food price inflation as redistribution: Towards a new analysis of corporate power in the world food system. *New Political Economy, 19*(1), 79–112. doi:10.1080/13563467.2013.768611

Bauer, C. J. (2005). In the image of the market: The Chilean model of water resources management. *International Journal of Water, 3*(2), 146–165. doi:10.1504/IJW.2005.007283

Blackmore, E., Keeley, J., Pyburn, R., Mangus, E., Chen, L., & Yuhui, Q. (2012). Pro-poor certification: Assessing the benefits of sustainability certification for small-scale farmers in Asia, Natural Resource Issues No. 25. London: IIED. Retrieved from http://www.european-fair-trade-association.org/efta/Doc/Propoorcert-IIED.pdf

Boelens, R., Hoogesteger, J., Swyngedouw, E., Vos, J., & Wester, P. (2016). Hydrosocial territories: A political ecology perspective. *Water International, 41*(1), 1–14. doi:10.1080/02508060.2016.1134898

Boelens, R., & Vos, J. (2012). The danger of naturalizing water policy concepts: Water productivity and efficiency discourses from field irrigation to virtual water trade. *Agricultural Water Management, 108,* 16–26. doi:10.1016/j.agwat.2011.06.013

Budds, J. (2009). Contested H2O: Science, policy and politics in water resources management in Chile. *Geoforum, 40,* 418–430. doi:10.1016/j.geoforum.2008.12.008

Burch, D., & Lawrence, G. (2009). Towards a third food regime: Behind the transformation. *Agriculture and Human Values, 26,* 267–279. doi:10.1007/s10460-009-9219-4

Cáceres, D. M. (2015). Accumulation by dispossession and socio-environmental conflicts caused by the expansion of agribusiness in Argentina. *Journal of Agrarian Change, 15*(1), 116–147. doi:10.1111/joac.12057

Campbell, H. (2005). The rise and rise of EurepGAP: The European (re)invention of colonial food relations? *International Journal of Sociology of Food and Agriculture, 13*(2), 1–19. Retrieved from http://www.ijsaf.org/archive/13/2/campbell.pdf

Carolan, M. (2012). *The sociology of food and agriculture.* Abington: Routledge.

Castro, J. (2008). Water struggles, citizenship and governance in Latin America. *Development, 51*(1), 72–76. doi:10.1057/palgrave.development.1100440

Chapagain, A., Hoekstra, A., Savenije, H., & Gautam, R. (2005). The Water Footprint of Cotton Consumption. Value of Water Research Report Series No. 18. Delft: UNESCO-IHE. Retrieved from http://waterfootprint.org/media/downloads/Report18.pdf

Chen, Z.-M., & Chen, G. Q. (2013). Virtual water accounting for the globalized world economy: National water footprint and international virtual water trade. *Ecological Indicators, 28,* 142–149. doi:10.1016/j.ecolind.2012.07.024

Clapp, J., & Fuchs, D. (Eds.). (2009). *Corporate power in global agrifood governance.* Cambridge: MIT Press.

Conn, C. (2015, May 12). Mexico's Yaqui people launch defense of water and territory. *Tele Sur.* Retrieved from http://www.telesurtv.net/english/news/Mexicos-Yaqui-People-Launch-Defense-of-Water-and-Territory-20150512-0033.html

Cotula, L. (2012). The international political economy of the global land rush: A critical appraisal of trends, scale, geography and drivers. *Journal of Peasant Studies, 39*(3–4), 649–680. doi:10.1080/03066150.2012.674940

Dabrowski, J., Murray, K., Ashton, P., & Leaner, J. (2009). Agricultural impacts on water quality and implications for virtual water trading decisions. *Ecological Economics, 68*(4), 1074–1082. doi:10.1016/j.ecolecon.2008.07.016

Dalin, C., Konar, M., Hanasaki, N., Rinaldo, A., & Rodriguez-Iturbe, I. (2012). Evolution of the global virtual water trade network. *Proceedings of the National Academy of Sciences, 109*(16), 5989–5994. doi:10.1073/pnas.1203176109

Dauvergne, P., & Lister, J. (2012). Big brand sustainability: Governance prospects and environmental limits. *Global Environmental Change, 22*(1), 36–45. doi:10.1016/j.gloenvcha.2011.10.007

Debbané, A.-M. (2013). Dis/articulations and the hydrosocial cycle: Postapartheid geographies of agrarian change in the Ceres Valley, South Africa. *Environment and Planning A, 45*(11), 2553–2571. doi:10.1068/a45693

Defries, R. S., Rudel, T., Uriarte, M., & Hansen, M. (2010). Deforestation driven by urban population growth and agricultural trade in the twenty-first century. *Nature Geoscience, 3*(3), 178–181. doi:10.1038/ngeo756

Del Moral, L., & Do O, A. (2014). Water governance and scalar politics across multiple-boundary river basins: States, catchments and regional powers in the Iberian Peninsula. *Water International, 39*(3), 333–347. doi:10.1080/02508060.2013.878816

Delaney, D. (2008). *Territory: A short introduction.* Malden, MA: Blackwell.

EC – European Commission. (2012). 10 Benefits of trade for developing countries. Retrieved from http://trade.ec.europa.eu/doclib/docs/2012/january/tradoc_148991.pdf

FAO. (2010). Agricultural investment funds for developing countries. Rome: FAO, Hardman. Retrieved from http://www.fao.org/fileadmin/user_upload/ags/publications/investment_funds.pdf

Fuchs, D., Kalfagianni, A., & Arentsen, M. (2009). Retail power, private standards, and sustainability in the global food system. In J. Clapp & D. Fuchs (Eds.), *Corporate power in global agrifood governance.* Cambridge: MIT Press.

Fulponi, L. (2007). The globalization of private standards and the agri-food system. In J. Swinnen (Ed.), *Global supply chains, standards and the poor.* Wallingford, UK: CABI.

Gana, A. (2012). The rural and agricultural roots of the Tunisian Revolution: When food security matters. *International Journal of Sociology of Agriculture and Food, 19*(2), 201–213. Retrieved from http://www.ijsaf.org/archive/19/2/gana.pdf

Gaybor, A. (2011). Acumulación en el campo y despojo del agua en el Ecuador [Capital accumulation and water dispossession in rural Ecuador]. In R. Boelens, M. Zwarteveen, & L. Cremers (Eds.), *Justicia Hídrica. Acumulación, Conflictos y Acción Social* [Water justice; accumulation, conflicts and social action] (pp. 195–208). Lima: IEP & Fondo Editorial PUCP.

Gibbon, P., & Ponte, S. (2005). *Trading down: Africa, value chains, and the global economy.* Philadelphia, PA: Temple University Press.

Goger, A., Hull, A., Barrientos, S., Gereffi, G., & Godfrey, S. (2014). *Capturing the gains in Africa: Making the most of global value chain participation.* Center on Globalization, Governance & Competitiveness, Durham, NC: Duke University.

GRAIN. (2012). Squeezing Africa dry: Behind every land grab is a water grab. GRAIN Report June 2012. Retrieved from www.grain.org/e/4516

GRAIN. (2014). The many faces of land grabbing: Cases from Africa and Latin America. EJOLT Report No. 10. Retrieved from http://www.grain.org/fr/article/entries/4908-ejolt-report-10-the-many-faces-of.pdf

Hall, R., Edelman, M., Borras, S. M., Scoones, I., White, B., & Wolford, W. (2015). Resistance, acquiescence or incorporation? An introduction to land grabbing and political reactions 'from below'. *The Journal of Peasant Studies, 42*(3–4), 467–488. doi:10.1080/03066150.2015.1036746

Hazelton, J. (2014). Corporate water accountability – The role of water labels given non-fungible extractions. *Pacific Accounting Review, 26*(1/2), 8–27. doi:10.1108/PAR-07-2013-0074

Helwege, A. (2015). Challenges with resolving mining conflicts in Latin America. *The Extractive Industries and Society, 2*, 73–84. doi:10.1016/j.exis.2014.10.003

Hoekstra, A., & Chapagain, A. (2008). *Globalization of water: Sharing the planet's freshwater resources*. Malden, MA: Blackwell.

Hoogesteger, J., & Verzijl, A. (2015). Grassroots scalar politics: Insights from peasant water struggles in the Ecuadorian and Peruvian Andes. *Geoforum, 62*, 13–23. doi:10.1016/j.geoforum.2015.03.013

ICMM – International Council on Mining & Metals. (2014). Sustainable development framework. Retrieved December 9, 2014, from www.icmm.com/our-work/sustainable-development-framework/10-principles

Khan, S., & Hanjra, M. (2009). Footprints of water and energy inputs in food production – Global perspectives. *Food Policy, 34*, 130–140. doi:10.1016/j.foodpol.2008.09.001

La Via Campesina. (2013). No to WTO and free trade agreements: Deal a decisive blow to Neoliberalism. Retrieved from http://viacampesina.org/en/index.php/actions-and-events-main menu-26/10-years-of-wto-is-enough-mainmenu-35/1526-deal-a-decisive-blow-to-neoliberalism

Langan, M. (2011). Uganda's flower farms and private sector development. *Development and Change, 42*(5), 1207–1240. doi:10.1111/j.1467-7660.2011.01732.x

Lenzen, M., Bhanduri, A., Moran, D., Kanemoto, K., Bekchanov, M., Geschke, A., & Foran, B. (2012). The role of scarcity in global VW flows. ZEF-Discussion Papers on Development Policy No. 169.

Lenzen, M., Moran, D., Kanemoto, K., Foran, B., Lobefaro, L., & Geschke, A. (2012). International trade drives biodiversity threats in developing nations. *Nature, 486*, 109–112. doi:10.1038/nature11145

Longo, S., & York, R. (2008). Agricultural exports and the environment: A cross-national study of fertilizer and pesticide consumption. *Rural Sociology, 73*(1), 82–104. doi:10.1526/003601108783575853

Lopez-Gunn, E. (2009). Agua para todos: A new regionalist hydraulic paradigm in Spain. *Water Alternatives, 2*(3), 370–394.

McMichael, P. (2009). A food regime genealogy. *Journal of Peasant Studies, 36*(1), 139–169. doi:10.1080/03066150902820354

Mehta, L., Veldwisch, G. J., & Franco, J. (2012). Introduction to the special issue: Water grabbing? Focus on the (re)appropriation of finite water resources. *Water Alternatives, 5*(2), 193–207. Retrieved from http://www.water-alternatives.org/index.php/alldoc/articles/vol5/v5issue2/165-a5-2-1/file

Moreda, T. (2015). Listening to their silence? The political reaction of affected communities to large-scale land acquisitions: Insights from Ethiopia. *The Journal of Peasant Studies, 42*(3–4), 517–539. doi:10.1080/03066150.2014.993621

Murphy, S. (2008). Globalization and corporate concentration in the food and agriculture sector. *Development, 51*(4), 527–533. doi:10.1057/dev.2008.57

Palau, T., Cabello, D., Naeyens, A., Rulli, J., & Segovia, D. (2007). *Los Refugiados del modelo agroexportador. Impactos del monocultivo de soja en las comunidades campesinas paraguayas* [The refugees of the agroexport model: Impacts of the soya monoculture on peasant communities in Paraguay]. Asunción, Paraguay: BASE-IS.

Pearson, R. (2007). Beyond women workers: Gendering CSR. *Third World Quarterly, 28*, 731–749. doi:10.1080/01436590701336622

Peña, F. (2011). Acumulación de derechos de agua y justicia hídrica en México: El poder de las élites [Accumulation of water rights and water justice in Mexico: The power of the elites]. In R. Boelens, L. Cremers, & M. Zwarteveen (Eds.), *Justicia Hídrica: Acumulación, Conflictos y Acción Social* [Water justice: Accumulation, conflicts and social action] (pp. 209–224). Lima: IEP & Fondo Editorial PUCP.

Progressio. (2010). *Drop by drop. Understanding the impacts of the UK's water footprint through a case study of Peruvian Asparagus*. London: Progressio, in assoc. with CEPES and WWI.

Quintana, V. (2013). The new global agri-food order and water disputes in Northern Mexico. *Apuntes, 40*(73), 131–158.

Roth, D., & Warner, J. (2008). Virtual water: Virtuous impact? The unsteady state of virtual water. *Agriculture and Human Values, 25*, 257–270. doi:10.1007/s10460-007-9096-7

Scott, J. A. (1985). *Weapons of the weak: Everyday forms of peasant resistance.* New Haven, CT: Yale University Press.

Scott, J. A. (1998). *Seeing like a State: How certain schemes to improve the human condition have failed.* New Haven, CT: Yale University Press.

Scudder, T. (2005). *The future of large dams: Dealing with social, environmental, institutional and political costs.* London: Earthscan.

Seekell, D., D'Odorico, P., & Pace, M. (2011). Virtual water transfers unlikely to redress inequality in global water use. *Environmental Research Letters, 6*(2). doi:10.1088/17489326/6/2/024017

Shah, T., Burke, J., & Villholth, K. (2007). Groundwater: A global assessment of scale and significance. In D. Molden (Ed.), *Water for food–Water for life. Comprehensive assessment of water management in agriculture* (pp. 395–423). London: EarthScan.

Smaller, C., & Mann, H. (2009). *A thirst for distant lands: Foreign investment in agricultural land and water.* Winnipeg: International Institute for Sustainable Development (IISD).

Sojamo, S., Keulertz, M., Warner, J., & Allan, J. A. (2012). Virtual water hegemony: The role of agribusiness in global water governance. *Water International, 37*(2), 169–182. doi:10.1080/02508060.2012.662734

Sojamo, S., & Larson, E. A. (2012). Investigating food and agribusiness corporations as global water security, management and governance agents: The case of Nestlé, Bunge and Cargill. *Water Alternatives, 5*(3), 619–635.

Solanes, M. (2010). Water, water services and international investment agreements. In C. Ringler, et al. (Eds.), *Global change: Impacts on water and food security.* Berlin: Springer.

Solanes, M., & Jouravlev, A. (2007). Revisiting privatization, foreign investment, international arbitration, and water, ECLAC/CEPAL, Serie Recursos naturales e infrastructura, No. 129.

Sosa, M., & Zwarteveen, M. (2014). The institutional regulation of the sustainability of water resources within mining contexts: Accountability and plurality. *Current Opinion in Environmental Sustainability, 11*, 19–25. doi:10.1016/j.cosust.2014.09.013

Suweis, S., Konar, M., Dalin, C., Hanasaki, N., Rinaldo, A., & Rodriguez-Iturbe, I. (2011). Structure and controls of the global virtual water trade network. *Geophysical Research Letters, 38*(10). doi:10.1029/2011GL046837

Swyngedouw, E. (2013). Into the sea: Desalination as hydro-social fix in Spain. *Annals of the Association of American Geographers, 103*(2), 261–270. doi:10.1080/00045608.2013.754688

UNCTAD. (2007). *Development implications of international investment agreements, United Nations conference of trade and development.* New York, NY: UNCTAD/WEB/ITE/IIA/2007/2.

Van der Ploeg, J. D. (2008). *The new peasantries. Struggles for autonomy and sustainability in an era of empire and globalization.* London: Earthscan.

Van der Ploeg, J. D. (2010). The food crisis, industrialized farming and the imperial regime. *Journal of Agrarian Change, 10*(1), 98–106. doi:10.1111/joac.2010.10.issue-1

Vestergaard, J. (2012). Disciplining the international political economy through finance. In S. Guzzini & I. Neumann (Eds.), *The diffusion of power in global governance, international political economy meets Foucault* (pp. 172–202). Houndmills, UK: Palgrave Macmillan.

Vos, J., & Boelens, R. (2014a). Ríos de oro. La exportación del agua virtual y la responsabilidad social empresarial de las empresas mineras y agro-exportadoras [Rivers of gold: Mining and agro-export companies – Exportation of virtual water versus corporate social responsibility]. In T. Perreault (Ed.), *Mineria, agua y justicia social en los Andes: Experiencias comparativas de Perú y Bolivia* (pp. 203–230). Cusco, Peru: CBC.

Vos, J., & Boelens, R. (2014b). Sustainability standards and the water question. *Development and Change, 45*(2), 205–230. doi:10.1111/dech.12083

Wada, Y., Van Beek, L., Van Kempen, C., Reckman, J., Vasak, S., & Bierkens, M. (2010). Global depletion of groundwater resources. *Geophysical Research Letters, 37*(20), 1–5. doi:10.1029/2010GL044571

Waldman, K. B., & Kerr, J. M. (2014). Limitations of certification and supply chain standards for environmental protection in commodity crop production. *Annual Review of Resource Economics, 6*, 429–449. doi:10.1146/annurev-resource-100913-012432

Wichelns, D. (2010). Virtual water: A helpful perspective, but not a sufficient policy criterion. *Water Resources Management, 24*, 2203–2219. doi:10.1007/s11269-009-9547-6

Word, J. (2014). *Surrendering our future: Senhuile-Senéthanol plantation destroys local communities and jeopardizes environment.* Oakland: Oakland Institute.

Yacoub, C., Blazquez, N., Pérez-Foguet, A., & Miralles, N. (2013). Spatial and temporal trace metal distribution of a Peruvian basin: Recognizing trace metal sources and assessing the potential risk. *Environmental Monitoring and Assessment, 185*(10), 7961–7978. doi:10.1007/s10661-013-3147-x

Zapatta, A., & Mena, P. (2013). Acumulación de agua y floricultura en un mosaico de territorios de riego: el caso Pisque, Ecuador [Accumulation of water and flower production in a mozaic of irrigation territories]. In A. Arroyo & R. Boelens (Eds.), *Aguas robadas: despojo hídrico y movilización social* (pp. 167–183). Serie Agua y Sociedad, 19, Quito: Justicia Hídrica, IEP & Abya Yala.

Zeitoun, M., Messerschmid, C., & Attili, S. (2009). Asymmetric abstraction and allocation: The Israeli-Palestinian water pumping record. *Groundwater, 47*, 146–160. doi:10.1111/j.1745-6584.2008.00487.x

Zhao, J. Z., Liu, W. H., & Deng, H. (2005). The potential role of virtual water in solving water scarcity and food security problems in China. *International Journal of Sustainable Development and World Ecology, 12*(4), 419–428. doi:10.1080/13504500509469651

From Spain's hydro-deadlock to the desalination fix

Erik Swyngedouw and Joe Williams

School of Environment, Education and Development, University of Manchester, Manchester, UK

ABSTRACT

The inception of Spain's 'new water politics' in 2004 elevated seawater desalination from supplementary water supply to an alleged panacea for the country's recurrent water crises. Desalination became the subject of an extraordinary and delicate consensus that strategically aligned disparate (and sometimes unlikely) actors. This movement, the paper argues, represents a techno-managerial attempt to remove political dissent from the sphere of water governance, and to build regional and national consensus around a re-imagined productionist logic for Spain's hydraulic development. The paper outlines six contradictions of desalination, however, that together form a potential terrain for a repoliticization of the Spanish waterscape.

Depoliticization by desalination

The oceans that cover two-thirds of the Earth's surface have always been important political spaces of cultural significance, resource and food collection, transportation, and territorial and military dispute (Steinberg, 2001). Until recently, however, the salty waters of the oceans could not be considered an extractable resource. The prospect of one day turning seawater, one of the planet's most abundant properties, into freshwater, one of its most contested and (relatively) scarce, has inspired engineers and political leaders for many years. Over 50 years ago, US President John F. Kennedy said that if ever a cost-effective method were devised to purify seawater, the achievement would 'really dwarf any other scientific accomplishment' (Kennedy, 1961). Through the development of, at first, distillation techniques and, more recently, reverse osmosis (RO) membrane technologies, which remove salt and impurities from salt and brackish water – or more precisely, extract freshwater from salt water – Kennedy's dream is being realized. In a few short years seawater desalination has grown into a multi-billion dollar global industry (Global Water Intelligence, 2010). Desalination, proponents are quick to point out, offers a drought-resistant, rainfall-independent source of freshwater (Fritzmann, Löwenberg, Wintgens, & Melin, 2007; Shannon et al., 2008). Ocean water is seen by many as unlimited, uncontested and free. Ostensibly, desalination provides the 'silver bullet' solution to the combined problems of growing demand for water and an unreliable and dwindling supply.

In Spain, desalination rose to prominence when, in March 2004, the socialist PSOE, fronted by José Luis Rodriguez Zapatero, unexpectedly won the national elections. The new government immediately suspended the more controversial components of the Second National Hydraulic Plan, which had been instituted by the previous conservative government led by the Partido Popular in 2001. The plan had centred around the contentious Ebro Transfer, through which 1050 hm^3 of water would have flowed from basins with 'surplus' water in the north to places of 'deficit' in the semi-arid southeastern regions of the Spanish Mediterranean coast and to Barcelona (Swyngedouw, 2015). The Ebro Transfer had enflamed inter-regional disputes, expostulated by those who espoused less centralized modes of governance and condemned by environmental activists as ecologically destructive. A month after work had begun, the transfer was suspended and the National Hydraulic Plan declared 'finished' (Downward & Taylor, 2007). In its place the government proposed a new policy, Programa AGUA (AGUA Programme), in which the shortfall from the abandoned Ebro scheme was balanced through the planned development of Spain's coastal desalination capacity. The old transfer-based hydraulic paradigm was replaced, and the future of desalination was cemented as the 'fix' for the contested politics and fears of scarcity that beleaguered traditional modes of water governance in Spain (Swyngedouw, 2013). The marine solution was staged as a panacea for the country's terrestrial water woes. Nonetheless, the desalination proposals continued through new techno-managerial practices to focus on the supply side by expanding freshwater supply.

Yet, the presentation of desalination as a panacea, we argue, does not point to this new techno-hydraulic configuration as being somehow uncontested and unproblematic. Instead, it betrays an attempt to build consensus around a particular hydro-modernist vision by removing discontentment and disagreement from established spheres of public engagement with politics. In other words, desalination has become a techno-managerial tool in a broader trend of depoliticization that a growing group of political philosophers and theorists are calling the post-political or post-democracy (e.g., Mouffe, 2005; Rancière, 2006; Swyngedouw, 2011a; Žižek, 1999). Key hallmarks of the post-political era, these scholars argue, include the perceived inevitability of market capitalism as the only possible economic and social organizational structure, the use of technocratic forms of management organized around problem-focused governance, and a focus on consensus-building within established political parameters and the corresponding relegation of dissenting voices to the margins (Wilson & Swyngedouw, 2014). Through the depoliticization of spheres of environmental governance, ecological crises – climate change, water scarcity, pollution, destruction of habitat and so on – are increasingly managed through consensual technocratic 'solutions' that are both consistent with and advance a broadly neo-liberal agenda. Although dispute between conservative and socialist parties about the right mix of techno-managerial dispositives still animates policy arguments, dissensus about the trajectory of future socio-ecological relations and the position of water therein is foreclosed, thereby silencing alternative or different socio-ecological futures.

Through an analysis of the development of Spain's waterscape in the post-dictatorship era, we argue that the assimilation of desalination into the political mainstream represents a form of depoliticization. Rhetorically, desalination is presented as a radical departure from traditional water management solutions, which relied on large terrestrial infrastructures to transport water substantial distances, while in essence merely reproducing an expansionist hydro-modernist vision for development (March, Saurí, & Rico-Amorós, 2014; Swyngedouw, 2015). The confluence of numerous factors, including escalating regional conflict, European-wide trends away from large state-led developments, growing fears of scarcity and concern over the ecological destructiveness of large terrestrial hydro-engineering, contributed to discredit inter-basin transfer as a viable hydrological future for Spain (del Moral, van der Werff, Bakker, & Handmer, 2003; Saurí & del Moral, 2001). The reconfiguration of the country's hydraulic paradigm within the same broad modernist developmental framework, then, represents an attempt to build political consensus through a market-based techno-managerial fix that normalizes growth at all costs. Fundamentally, the desalination 'fix' deploys extraordinary new techno-social configurations, whilst preserving the same underlying logics of developmental, growth-oriented water governance. The emerging emphasis on consensual water governance and the diffusion of political tension through techno-managerial projects, then, constitutes a form of political disavowal. Or, in other words, the desalination edifice is presented by ecological modernizing forces as a radical, if not revolutionary, socio-technical alternative that nonetheless assures that the dominant political–economic water-guzzling development model can be sustained for a while longer. It is precisely the absence of a contentious political debate over possible hydrosocial futures and their territories (Boelens, Hoogesteger, Swyngedouw, Vos, & Wester, 2016) and the framing of the alternative in purely techno-managerial terms that produces a decidedly depoliticizing trajectory, one that solidifies a post-democratic mode of socio-ecological governance (Swyngedouw, 2011a).

The effort to advance desalination as a consensual, politically uncontested form of governance and supported by an ecologically modernizing water industry (see below) is, however, constantly undermined by a diverse group of disillusioned activists. The recent rise of new radical political movements (like PODEMOS) in the aftermath of the M-15 'indignado' movements that animated a deeply political wave of political contestation over the past few years in Spain is precisely a response to such techno-managerial consensual policies. These movements advocate a more profound, democratizing and egalitarian socio-ecological transformation (Swyngedouw, 2014). Indeed, beneath the veil of consensus lies the basis for an emerging dissensus, and within the apparent panacea of the desalination 'solution' emerge the seeds of new contradictions. The final section of this paper outlines six central contradictions of large-scale desalination. These revolve around: first, growing concerns over the ecological credibility of, on the one hand, the energy intensity and associated CO_2 emissions of the purification process, and the destruction of marine life, on the other; second, the high cost of water production and its implications for environmental justice; third, criticism of the commodification and corporatization of water supply; and fourth, a critique of a growth-centred model of water governance. These concerns, although exemplified in the case of Spain, are consistent with a growing international body of critical scholarship that highlights the deeply contradictory characteristics of the rollout of large-scale desalination (e.g., Feitelson & Rosenthal, 2012; March, 2015; March et al., 2014; McEvoy, 2014; Swyngedouw, 2013) within a depoliticizing context (Swyngedouw, 2011b). Such contradictions, we argue, form the basis for a potential repoliticization of the Spanish waterscape.

Promises of the desalination 'fix': building a consensual hydro-vision

Dreams of making pure the salty waters of the Mediterranean, of mobilizing the sea to quench the thirst of the land, have long lurked in the conscience of Spanish technocrats. For example, in 1973 a prominent Spanish engineering journal insisted that through the development of the necessary technologies, 'The sea will be our greatest reserve resource, and we have it in abundance' (Valdés, 1973, p. 409). The adoption of Programa AGUA in 2004 sealed the future of desalination as the new panacea for Spain's endemic and recurrent water crisis. The programme outlined plans for the construction of 21 high-volume RO desalting facilities along the Spanish Mediterranean coast, with a combined production capacity of 1063 Hm^3/year (Figures 1 and 2, and Table 1). This is almost the same amount of water that would have been delivered by the hotly contested Ebro transfer project, which was suspended in this same year. An update and review in 2007 by the Department of the Environment included the

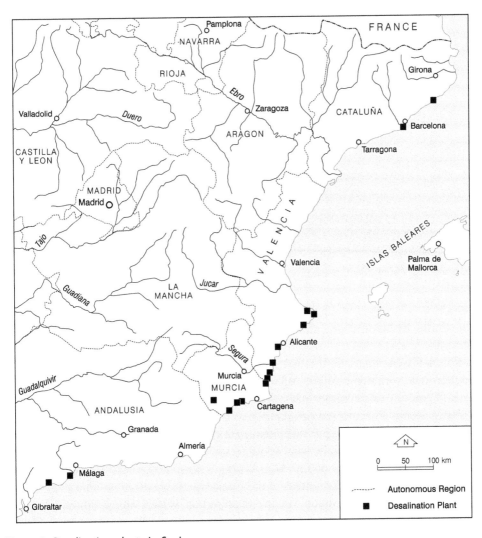

Figure 1. Desalination plants in Spain.

Figure 2. Desalination plant in Alicante.

addition of a number of further desalination plants within the remit of the AGUA programme. Its full implementation will give Spain the largest desalting capacity of any Western county. Table 2 shows the current status of the various desalination plant developments included in the AGUA programme, a little over a decade after its adoption. The combined capacity of completed projects and those still under construction still amount to significantly less than half the initial planned development. While the PSOE resolutely endorsed desalination as a hydrosocial fix to replace the contested inter-basin water transfers, but still pursuing a consensual policy that revolved around increasing the total available volume of freshwater, the Conservative Party quietly relinquished the objectives of the AGUA programme. References to the programme disappeared from official documents of the Conservative government after the elections of 2011, and the implementation of the desalination project either slowed or implemented projects operated at less than full capacity (Table 2). There is still clear dispute over the optimal techno-managerial mix to deal with Spain's water supply problems,

Table 1. Programa Agua: additional new water capacity, estimated cost and planned desalination facilities.

Province	Total additional capacity (hm³/year)	Total estimated cost (€ millions, 2005)[a]	New capacity mainly desalination: new plus upgraded (hm³/year)	Desalination plants	Cost of new capacity (€ millions, 2005)	Demand management (capacity gain) (hm³/year)
Girona	10	47	10	1	25	0
Barcelona	135	848	60	1	176	75
Tarragona	0	215	0	0	0	0
Castellón	78	173	33	2([b])	94[b]	32
Valencia	110	428	0	0	0	107
Alicante	212	618	141	7	292	71
Murcia	204	876	140	5	402	64
Almeria	189	352	165	5	226	24
Málaga	125	227	50	2	70	75
Albacete	0	14	0	0	0	8
Total	1063	3798	599	23	1285	456

Note: [a]Total estimated costs combines three measures: additional capacity production, demand management and environmental improvement/flood protection.
[b]Added in 2006 and 2007 to the AGUA Program.
Source: MMA 2007 (see http://www.mma.es/secciones/agua/entrada.htm; accessed on 25 July 2011).

Table 2. Current status of the major desalination projects included in the AGUA programme.

Location	Capacity (hm³/year)	Investment (€ millions)	Current state
Telde II (Canarias)	5.6	16	Finished in 2011. Remains unused
El Prat de Llobregat	60	230	Finished and operational since 2009
Bahía de Alcudia (Mallorca)	4.9	24	Finished in 2010. Operates at 15% of capacity
Andratx (Mallorca)	4.9	51.9	Finished in 2010. Remains unused
Torrevieja (Alicante)	80	300	Pre-operational (testing phase)
Bajo Almanzora (Almería)	15	88	Operational until 2012
Valdelentisco (Murcia)	48	224	Operational
Campo de Dalías (Almería)	30	130.3	Under construction
Oropesa del Mar (Castellón)	18 (potential expansion to 43)	55.4	Under construction
Moncofa (Castellón)	11 (potential expansion to 21)	49.1	Under construction
Sagunto (Valencia)	8	37.3	Pre-operational (testing phase)
Mojón (Murcia) (expansion)	6	30	Operational
Águilas (Murcia) (expansion)	70	238.3	Pre-operational
Denia (Alicante)	5.8	27.1	Denia city council cancelled the project in 2011
Mutxamel (Alicante)	18	90	Pre-operational
Costa del Sol (Mijas, Málaga)	21	62	Mijas city council blocked construction in 2011
Total (operational and under construction)	406.2	1653.4	

Source: Adapted from March et al. (2014).

but the focus remains nonetheless squarely on expanding supply rather than opening alternative possible futures trajectories.

The promises of desalination, the solution that came from the sea, are many and ostensibly propitious. A promise, no less, of abundant, unclaimed and consensual water supply, free from the contested property rights that beleaguer traditional terrestrial sources, unlimited in potential (March Corbella & Saurí, 2008). Broadly, the panacea of desalination is articulated through two themes. First, the unclaimed and (currently) uncontested sea offers a solution to the problems of contested traditional terrestrial supply. Second, the lure of 'new' water provides a technological solution to the problem of inadequate, allegedly insufficient or unreliable supply. These two themes, which can loosely be understood as the 'scalar fix' and the 'scarcity fix', coalesce into a meta-narrative, geared towards consensus-building around a techno-managerial hydro-modernist vision. Despite being the subject of often fierce opposition, and notwithstanding ongoing debates around water transfers, large-scale seawater desalination has become an emblematic discursive and material vehicle, complementing and partly replacing the older hydraulic paradigm through which tension and conflict in the Spanish waterscape are mediated and managed. Yet, we argue that desalination, and the large-scale terrestrial infrastructure that it replaces, although distinct in some respects, both represent similar techno-natural visions focused on water supply (rather than possible other forms of socio-hydraulic management, particularly demand management), and consistent with a particular modernizing political–ecological development approach with a broadly neo-liberalizing logic.

The scalar fix: from transfers to desalination

The mobilization of the seas through the application of high-technology RO purification techniques radically reconfigures traditional notions of water governance. Freshwater now flows directly from the sea, rather than merely to it (Feitelson & Rosenthal, 2012). The 'new politics of water', which purport to consider both its economic and its ecological value 'thereby optimizing its use and restoring the associated ecosystems' promised by Zapatero, the then Prime Minister of Spain (Zapatero, 2006, p. 264), rely on desalination to provide a 'scalar fix' (Smith, 1984) to the intense socio-environmental conflict, political tension and proliferating discontent that characterized the country's terrestrial structural hydraulic politics in preceding decades. Embodied in the AGUA programme is a fundamental shift away from inter-basin water transfers, which are associated with an authoritarian, top-down and bureaucratic political imaginary, towards desalination, which is seen as 'local', democratic, decentralized, market efficient and ecologically sustainable. The so-called 'new politics', then, describes the disassembling of one particular hydro-developmental vision and the reassembling of another through new scalar configurations (Swyngedouw, 2015). Two imperatives have driven forward the rescaling of water governance in Spain.

The first concerns growing discontentment with the established hegemony of inter-basin transfers on ecological grounds. Since the 1970s, dominant discourses and practices, which considered aridity and water scarcity as a deficit, an obstacle to national progress and sought to utilize every drop of terrestrial water available, have been challenged. Recent re-imaginations of water have begun to value aridity as part of the sublime ecology and aesthetic of the Spanish landscape (González Bernáldez, 1981, 1989). There has been, consequently, a shift from descriptions of aridity as being 'defective', 'infertile', 'worthless' and so on, to a discursive enchainment that uses signifiers like 'beautiful', 'valuable' and 'sublime'. This has been concomitant with growing concerns over the fate of Spain's wetlands, which have come to be seen as indispensable in the movement to preserve specific cultural and ecological landscapes (García Novo, Toja Santillana, & Granado-Lorencio, 2010). Increasing emphasis is now placed on the importance of restoring heavily modified hydrological landscapes, and on preserving 'basic ecological water flows' (Manteiga & Olmeda, 1992; Palau, 2003). Explicit in this paradigm shift is a rejection of large-scale terrestrial water engineering. Dams and transfers are seen as archaic and ecologically destructive. Desalination, by contrast, is presented as a win–win high-tech, modern and environmentally friendly solution that 'frees up' water for environmental preservation and satisfies the escalating demands of developers and irrigators.

The second imperative of the 'scalar fix' involves the resolution of regional conflict. Since democratization, a new 'scalar gestalt' has emerged through a process of regionalization and greater regional autonomy, which has been accompanied by the creation of regional water agencies with partially or fully devolved hydraulic responsibilities (Lopez-Gunn, 2009). Additionally, the establishment in 2010 of state-organized, but market-based, 'Sociedades Estatales' (state societies), which replace the earlier river basin companies, add a further layer to the complex reconfiguration of water governance. The increasingly complex interplay between, on the one hand, the central state and devolving regions and, on the other hand, growing conflict between those regions has been articulated and played out through the mobilization of water as a contested terrain of scalar reconfiguration. This is expressed clearly in the Guerra del Agua (water war) between the northern regions (and the waters of the Tajo and Ebro rivers) and those of the Levant – put simply, the source regions and the receiving regions. For example, immediately after the approval of the Ebro Transfer project in 2001 the 'Platform for the Defence of the Ebro River' was established by the source regions of Catalonia and Aragón. With the emblem of a knotted water pipe in protest of inter-basin transfers, this campaign proved highly effective and ultimately contributed to the suspension of the transfer plan. Such contestations intersect with other forms of regionalism, such as conflict between Catalonia and Aragón for Ebro water entitlements (Arrojo & Visa, 2009), and the advent of 'hydraulic nationalism' in Murcia (Foro Ciudadano, 2005). Again, desalination, which 'guarantees the elimination of all the uncertainties of availability that necessitated the transfers' (Castro Valdiva, 2007, p. 8), is presented as a win–win solution. Ultimately, desal represents here a techno-managerial strategy to diffuse political tensions, and to build regional and national consensus around a re-imagined productionist logic for Spain's hydraulic development.

The scarcity fix: on climate change and Malthus

An extraordinary confluence of socio-ecological concerns over water scarcity is emerging, which converge around, on the one hand, fears of reduced supply as a result of global climate change and, on the other, of increased demand through population growth and economic development. These combined discourses provide a powerful imperative for water managers to seek out 'new' sources. As such, desalination, with its capacity to 'guarantee [...] water, rain or no rain, independent of the climate' (ICEX, 2010, p. 3), promises the ultimate climate adaptation strategy *and* growth potential.

Over the past decade, the global environmental conundrum has crystallized around climate change as the material and symbolic condition around which our socio-ecological predicament circulates (Swyngedouw, 2010, 2011b). At the same time there has been much consternation from business leaders, politicians and other economic elites over how to sustain capital accumulation and economic growth under conditions of rapidly changing socio-ecological relations, which are so often communicated in catastrophic terms. That said, climate projections for Spain's future hydrological cycle paint a somewhat bleak picture. The Fourth Assessment Report of the Intergovernmental Panel on Climate Change (IPCC) projects a decrease of precipitation in the already highly water-stressed southern regions of Spain of up to 40% by mid-century compared with average 1961–90 levels, and a small increase in the northern regions (Pachauri & Reisinger, 2007). This will be exacerbated by reduced river flows as the result of increased evaporation consistent with rising average temperature – an effect that is already putting pressure on freshwater resources (Martín Barajas, 2010). The Centro de Estudios y Experimentación de Obras Públicas (2011) predicts indeed a generalized reduction of precipitation of –5% for the period 2011–41, rising to –9% between 2041 and 2070, and a whopping –17% between 2071 and 2100. The greatest variability will take place along the Mediterranean coast and in the south-east. Overall, the IPCC's expected inevitable temperature increase of 1.5°C will contribute to an estimated 5–7% reduction in reservoir levels, whilst the European Environment Agency (2007) predicts that rising temperatures will lead to increased demand for agricultural water as well as declining availability and quality. The predicted temperature rise will, furthermore, increase evaporation and evapotranspiration, and decrease groundwater recharge and runoff (Vargas-Amelin & Pindado, 2013).

The voices of the climate protagonists are joined by a vocal contingent of irrigators and developers who mobilize growth-centred neo-Malthusian language to argue the case for increasing water supply. Economic development, particularly in south-eastern regions, at least until the beginning of the recession in 2008, was predicated on agricultural exports and coastal property (real estate) development to support the burgeoning tourism industry. Both factions are represented by strong lobby groups, and both demand the expansion of water supply as prerequisite to a growing tourist industry and expanding economy. Such calls are bolstered by international concerns over the effect of a growing population on water resource allocation, and the prevailing assumption that such processes necessitate the augmentation of water supply. For example, in the journal *Science*, Elimelech and Phillip (2011, p. 712) argue that the combined pressures of 'population growth, industrialization, contamination of available freshwater sources and climate' constitute the principal challenges for water provision in the 21st century. Implicit (and often explicit) in such logic is an assumption that there is no other option than to expand supply in order to sustain economic growth.

Here, then, is the central conundrum: how might supply be expanded under conditions of dwindling availability? Once again, desalination becomes the subject of an extraordinary consensus that strategically aligns disparate (and sometimes unlikely) actors, including economic elites, irrigators, ecological modernizers, policy-makers, local businesses, scientists and international figures. The international water industry, too, has rallied around desalination as a panacea. The Spanish company Abeima (2011) has, for instance, argued that 'desalination is an alternative and inexhaustible source of water at a moment when the supply of this resource has become a real problem in many places of the world, aggravated by the condition of climate change'. The dual pressures of increasing demand and diminishing supply contribute towards a hegemonic discourse that favours techno-managerial supply-side solutions. Desalination is presented as an inevitable socio-technical 'fix', and normalizes a growth-at-all-costs model. In short, desalination, and the hydrosocial vision it embodies is the same, in essence, as earlier modes of hydro-modernity, and different primarily in its technical and scalar configuration.

Accumulation by desalination

Thus far, we have demonstrated how disparate and variegated interests, which include growing environmental concern over large terrestrial hydraulic infrastructure, the articulation of regional tensions in the post-democratization era and the politics of scarcity, have fused with a new managerial and technocratic rationale. The disassembling of previous hydro-modernist visions are, through decentralized desalination projects, being replaced with new modes, consistent with neo-liberal market-oriented logic. Questions of the ecological and economic value of water, issues of who pays and who benefits from water, and discursive themes around efficiency and democracy came to prominence in Spain's 'New Water Culture' (NWC) (Aguilera Klink, 2008; Arrojo, 2001, 2005; Martínez Gil, 1997). Environmental concern, demand-side management and calls for locally driven 'sustainable development' coalesced with market imperatives for full cost recovery, 'real' pricing, decentralization and the mobilization of market principles to form a hegemonic hydrosocial meta-discourse. Large-scale terrestrial engineering was seen as authoritarian, state-subsidized (and, therefore, inefficient), and lacking in ecological sensitivity (Lopez-Gunn, 2009). The NWC provided the underlying principles of the AGUA programme and the vision for hydro-environmental modernization it embodied. Desalination was billed as the 'fix' that would simultaneously save the environment and open up spaces for capital accumulation, the frontier of an emerging sustainability–industrial paradigm.

Spain's desalination programme, in current form, represents the commodification of water, but not yet its full privatization (Bakker, 2010). In desalting 'factories' the means of production, comprising seawater and high-technology purification systems, are combined to produce H_2O in commodity form, which contains surplus value (i.e., the output, H_2O, is of greater value than the input – seawater, high-technology purification systems, labour force and energy). Projects are delivered through public–private partnerships in that they are state led, but built, operated and often financed by private corporations for profit. Thus, desalinated water takes the form of a commodity, but one that can currently only be sold through the apparatus of the state under its statutory legal, institutional and concessionary rules. Water supply, then, remains under public control, albeit with substantial privatized elements. This form is described in a process the Spaniards call 'mercantilización' (Bakker, 2002). Due to the technological complexity of RO facilities, and the huge capital costs associated with such developments, only a handful of global corporations have the technical and financial capacity to deliver large projects, either individually or (more commonly) through public–private partnerships. In this way, decentralized, local water supply is connected to multi-scalar financial flows, which link the European Union to national and regional government, and multinational private companies.

The desalting industry is eagerly stepping up to the challenge of water security in water-stressed countries across the globe. Global Water Intelligence (2010) estimated in 2010 that the desalination sector is projected to be worth US$16 billion by 2020. In 2014, Global Industry Analysts (2014) estimated the projected 2020 investment to be almost US$50 billion. As the technology continues to improve and concerns over water security intensify, desalination becomes an ever-more-popular option. Indeed, the industry has expanded exponentially over the past few decades, growing from a total installed capacity of 5 million m³/day in 1965 to more than 14,000 facilities producing 65 million m³/day in 2010 (Figure 3). Spain is at the forefront of this expansion in terms of both domestic installed capacity and industrial capability. After Saudi Arabia, the United States and United Arab Emirates (UAE), Spain is placed fourth in the world in terms of current installed desalination capacity, and even ranks ahead of the United States for newly installed capacity since 2005 (Global Water Intelligence, 2010). Of the top 20 leading desalination contractors in the world by water production volume, eight are Spanish or mainly Spanish companies. Moreover, these companies also lead the way in RO technology, which has emerged as the overwhelmingly favoured option for new facilities because of its relative efficiency and lower cost than alternatives, with seven of the world's leading RO membrane producers based in Spain.

Such is the potential of the oceans' salty waters as a new frontier of capital accumulation and industrialization that the Spanish Institute for Foreign Trade (ICEX) identified desalination as one of eight activities that showcase Spanish innovation and technological excellence internationally. The Spanish desalination industry has even assembled its own lobbying organization, the Spanish Desalination and Water

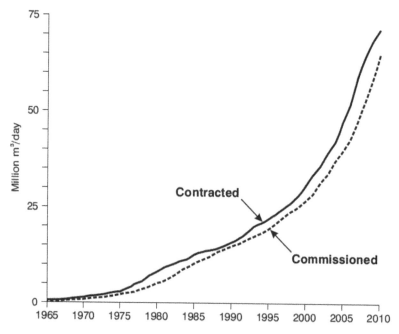

Figure 3. Global cumulative contracted commissioned desalination capacity, 1965–2010. Source: Global Water Intelligence, *IDA Desalination Yearbook 2010–2011* (see http://www.desalyearbook.com/)

Reuse Association (AEDYR), aimed at promoting their economic interests. Off the back of its success with the AGUA programme, the industry is looking to mobilize the Spanish state to promote its credentials as an environmentally sensitive and climate-friendly solution to water scarcity on the international market. Indeed, ICEX. (2010, p. 3) notes that '[t]he announcement of the plans to develop these new desalination plants has been a boon for desalination companies'. The rise of desalination as *the* supply-side water solution heralds the decline of traditionally state-led engineering elite, and its replacement with a new private (and international) technocratic and economic elite. The national focus on high-technology desalination technologies becomes a clear strategic means to enhance the global competitiveness of Spain's hydro-industrial complex.

Rescaling water governance

The decentralization of the Spanish state-space following the fall of national dictator-ship, of which the localization of water supply through the desalination 'scalar fix' is an inextricable part, was in fact part of a double movement in the rescaling of water governance in Spain. The greater regional autonomy administered in the shift from national pipelines to decentralized desalination is accompanied by an 'up-scaling' of significant policy domains to the European level in the democratic era. The rescaling of authority 'down' to local and regional actors, on the one hand, and 'up' to international institutions, on the other, is characteristic of the 'glocalization' (Swyngedouw, 1997) of the Spanish waterscape since joining the European Economic Community (EEC) in 1986. In 'becoming European', the adoption of 'modernizing' principles was intended to solidify democracy and foster desires of progress and development, whereby 'national sentiments smoothly intertwined with the process of European Integration' (Jáuregui & Ruiz-Jimenéz, 2005, p. 73). Despite the social and environmental regulatory frameworks laid down in the Maastricht Treaty of 1992 and consolidated in the 2007 Lisbon Treaty, tacit in the process of Europeanization in Spain was an adoption of neo-liberal economic doctrine. Put another way, the particular form of socio-ecological modernization that accompanied democratization entrenched a multi-scalar market-led and technocratic form of environmental governance that combined the commodification of nature with new forms of market-rationality (Birch, Levidow, & Papaioannou, 2010; Furlong, 2010; Heynen, McCarthy, Prudham, & Robbins, 2007).

In the water sector, these principles were expressed and sanctioned under the European Water Framework Directive (WFD) of 2000, which provides a generic European Union-wide framework for national water policy. A foundational imperative of the framework is to promote a full-cost recovery model for infrastructure develop-ment and water supply. The aim here is twofold: first, to eliminate structural subsidies, which are argued to encourage inefficient use and allocation; and second, to encourage responsible water use with a view to maintaining and/or restoring 'good' environmental flows. Whilst full-cost recovery of water provision does not necessarily imply commo-dification, it points at the need to manage water in market-based manners. Moreover, market-accurate pricing is a necessary precondition for the extraction of profit from water services. Furthermore, under the framework, water governance should be coordinated predominantly at the river basin level by organizations whose remit should ensure inclusive participation of all relevant stakeholders (CEC, 2000; Kaika, 2003).

The impact of the WFD in shaping Spain's hydraulic development was all the more pivotal because its translation into Spanish law in 2003 coincided precisely with the time when debate on the country's water future was at its most fervent (Grindlay, Zamorano, Rodríguez, Molero, & Urrea, 2011; Saurí & del Moral, 2001). The implementation of the framework, then, was carried out under conditions of existing conflict and fierce political debate (Thiel, Sampedro, & Schröder, 2011). The compulsion for accurately priced, regionally governed water, as outlined in the WFD, bolstered those who argued for an end to the traditional hydraulic paradigm. Effectively, this added European Union-level authority to existing critiques of state-centred, hierarchical, autocratic water transfers, and reinforced calls for environmentally sensitive and market-led decentralized 'participatory' forms of water governance. The principles of the WFD were successfully mobilized to undermine further the capacity of transfers and the National Hydraulic Plan to provide economically, politically and environmentally sustainable socio-technical fixes to the country's water woes. In contrast, the framework provided an imperative for new technological solutions to manage Spain's water challenges. Despite not mentioning desalination directly, the marine 'solution' was further legitimized by the WFD as an option that could manage existing conflicts whilst maintaining a reasonably coherent, albeit re-imagined eco-modernizing development trajectory.

Contradictions of desalination: the spectral return of the political?

Large-scale desalination has been hugely successful in asserting the marine solution as a panacea for Spain's water challenges and in reimagining a hydraulic future in which supply is guaranteed in abundance, whilst the conflicts that beleaguered pipeline transfers are placated. This has essentially allowed Spain to continue along its hydro-modernist development trajectory, reconfigured in accordance with new socio-ecological conditions. Fundamentally, the desalination 'fix' represents an attempt to build consensus around a techno-managerial solution that promises unlimited supplies of freshwater, albeit at a cost, whilst simultaneously pacifying the intense political debates that have traditionally saturated water governance. In a word, the rollout of desalination is intimately connected with broader processes of depoliticization. Yet, desalination has increasingly become a terrain of contestation. The complex and heterogeneous political and material relationships that have emerged around the desalination edifice are indeed precarious. The precarious consensus constructed around Programa AGUA, which presented desalination as uncontested, was of course made up of many and divergent agendas. For instance, the alliance between environmentalists wishing to preserve Spain's delicate ecology and developers pursuing economic growth, whilst instrumental in promoting desalination over transfers, is systematically undermined in other respects (e.g., Burballa Noria, 2014; Hulshof & Vos, 2014). Thus, the marine 'fix' contains within itself the kernel of a new set of contradictions, and out of the delicately constructed consensus emerges dissensus. Here we identify six central contradictions that lie at the heart of an effort to repoliticize the desalination debate.

The energy and climate contradiction

The energy requirements for RO seawater desalination are around 4 kWh/m^3, although this varies somewhat depending on local factors like sea temperature, salinity, pumping costs, specific pretreatment conditions and intake/outfall technologies (Cooley & Heberger, 2013). This is significantly more than any other available water source option, including potable reuse and brackish water desalination, both of which utilize the same RO techniques. Energy costs account for between 25% and 50% of the total price of desalted seawater, depending on plant efficiency and other contextual factors. Proponents of desalination proclaim the benefits of having a 'reliable', 'rainfall-independent' water source, but this can only be achieved through the application of vast amounts of energy. In overcoming the uncertainties of rainfall, water supply becomes vulnerable to different pressures, principally energy price increases and fluctuations. The barriers of traditional water sources are effectively overcome, but at the cost of new contradictions and pressures. Moreover, despite being touted as a climate change adaptation strategy, concerns have been raised over the energy intensity of desalination, and its associated greenhouse gas emissions, notwithstanding some currently largely tokenistic attempts by the industry to utilize 'green' energy. Indeed, by increasing reliance on fossil fuel energy, desalination may exacerbate the problems it is intended to solve, and, as such, represent a form of climate 'maladaptation' (Barnett & O'Neill, 2010).

The environmental contradiction

In addition to the problem of CO_2 emissions associated with energy-intense desalination and therefore its viability as a form of climate adaptation, environmental groups have also been vocal on issues of localized impacts to marine environments (Meerganz von Medeazza, 2005). Concerns have been raised over both the seawater intake systems, which suck in small marine life and larvae, and the point-source pollution of highly saline brine discharge, which can collect at the seabed if not dispersed. Although the industry has responded by developing technologies such as subsurface intake wells and discharge dispersers, which can mitigate the destructive aspects of intake/discharge, the marine environment issue has become a rallying point for the contestation and repoliticization of desalination.

The governance contradiction

The desalination 'fix' is deployed through the mobilization of multi-scalar financial flows, which connect the European Union to national and regional government, and multinational private companies. Concerns over the use of public funds to nurture particular interests through the development of desalination have raised questions around democratic water governance. Such concerns coalesce with rising general discontent against the depoliticization of environmental governance and the political marginalization of alterative voices, the speculative boom of the past decade, and the undemocratic character of neo-liberal governance (Heynen et al., 2007).

The growth contradiction

Implicit in the propagation of seawater desalination is an assumption that economic growth is unquestionably desirable and necessary and, moreover, that water resources must be expanded as prerequisite. In many ways, large-scale desalination embodies a growth-at-all-costs paradigm of water governance. Conversely, a small (but growing) group of academics and activists have mobilized around a critique of the growth agenda, arguing that a zero-growth or controlled de-growth model would foster a more ecologically and socially just economy (D'Alisa, Demaria, & Kallis, 2014; Jackson, 2011; Kallis, 2011). Although this discourse has not focused explicitly on water, its arguments fundamentally challenge many of the assumptions on which the desalting industry has flourished, particularly the compulsion for expanding water resources to support economic growth. The de-growth argument insists on the necessity to slow down economic growth while focusing on redistribution of the available resources as the necessary trajectory towards a more benign socio-ecological relationship. 'Sustainable' resource mobilization can, according to the protagonists of de-growth, only be achieved by radically altering the compound-growth trajectory of resource extraction and use (D'Alisa et al., 2014).

The cost contradiction

Purifying seawater to potable standard is, of course, a very costly process. Indeed, because the oceans are in all practical respects inexhaustible, it would be inconceivable for a coastal society to utilize a more expensive or energy-intense water source option. To produce 1 m^3 of freshwater from seawater costs around €0.50 compared with €0.20/m^3 for water transfers and €0.30/m^3 for groundwater pumping (Rico Amorós, 2010). Irrigators, distributers and ratepayers have been unwilling to pay for desalinated water, and indeed in almost all agricultural contexts desalination is simply not a cost-effective option. As a result, irrigators have often illegally expanded cheaper groundwater pumping. Moreover, rising water costs resulting from the development of desalting capacity have in some cases encouraged water conservation amongst users (as well as illegal aquifer pumping), thereby stabilizing or reducing demand for desalted water. This is reflected in Alicante and Murcia, where lower demand has resulted in new desalination facilities being operated consistently below capacity.

The ownership contradiction

It was argued above that one of the major virtues of desalination is its promise of unclaimed and uncontested water, free from the disputes that have marred the mobilization of terrestrial sources. The free pumping of seawater by private companies, legitimized by state permits and concessions, however, is unlikely to continue unchecked indefinitely. Indeed, debates are already underway over who has the legal right to and jurisdiction over the seas, and whether seawater might have a 'cost' like other resources. As the 'privatization of the seas' becomes a more pertinent political issue, and the process of accumulation by dispossession extends beyond the shore, these discussions are likely to intensify.

Conclusion

The mobilization of large-scale seawater desalination as the technological cornerstone of Spain's 21st-century hydraulic vision emerged out of an extraordinary confluence of political, economic, cultural, technical and environmental factors. The 'glocalization' of Spain's political landscape associated with processes of democratization and Europeanization, and growing regional conflict amid calls for greater regional auton-omy and decentralization of water governance, fused (apparently) seamlessly with an environmental movement that sought to maintain and preserve Spain's delicate ecology. Combined, these factors contributed to a widespread and vocal rejection, particularly in exporting regions, of the traditional hydro-engineering model of pipeline transfer, which was seen as ecologically destructive, nationalistic and authoritarian, and generally outmoded. Desalination, by contrast, was presented as a technologically advanced, 'environmentally friendly' and uncontested local water source, and as such was touted as being a win–win 'scalar fix' to Spain's water challenges. The explicit aim was to diffuse political tension through techno-managerial solutions.

In addition to these imperatives for an alternative to inter-basin transfers, fears of imminent scarcity brought together climate change protagonists, irrigators, industrial leaders and developers. Drawing on neo-Malthusian notions of resources scarcity and limits to economic growth, these actors argue that Spain's water supply must be expanded to facilitate growth, on the one hand, and compensate for dwindling tradi-tional supply resulting from climate change, on the other. Desalination, which is prized as the only true rainfall-independent source of 'new' water, is again presented as a panacea for the combined pressures of growing demand and diminishing supply. The confluence of these numerous factors drew together disparate and (often) unlikely allies, forming a hegemonic discourse around a reassembled vision for hydro-moder-nity, with desalination at the centre. Indeed, the pro-desalination agenda has been highly effective in capitalizing on this extraordinarily powerful meta-narrative. Central to this process, we have argued, has been an effort to build consensus around a pro-growth supply-driven governance model that opens up new spaces for capital accumulation, and where dissenting voices are relegated to the political margins.

The presentation of desalination as a panacea, unproblematic and uncontested, very much reflects and exemplifies the ongoing depoliticization of spheres of environmental governance, consistent with a post-political, post-democratic era. The rapid adoption of large-scale desalination in Spain indeed bears all the hallmarks of a post-political techno-managerial project, as outlined above. First, the AGUA programme normalizes a supply-driven growth model, deployed through capitalist socio-ecological organiza-tional structures and consistent with a broadly neo-liberalizing logic. Second, the shift to desalination represents, fundamentally, a technology-driven project geared towards problem solving within established political parameters. And finally, the strategic alignment of disparate actors and agendas under one techno-managerial solution constructed a delicate but hegemonic consensus, the function of which was to expel disagreement from established political structures and processes. Together, these pro-cesses represent a form of depoliticization by desalination. Yet, the six contradictions outlined above, it has been argued, not only undermine the hegemonic consensus constructed around the desalination fix and its viability as a techno-managerial solution but also form the basis for a repoliticization of water governance more broadly. Indeed, growing discontentment around the desalination debate is, at the same time, a reflection of discontentment with market-driven technocratic forms of environmental governance generally. Desalination, which at first served as a form of political disavowal, may become a rallying point for renewed political dissent.

Indeed, the disputed techno-managerial trajectories that have characterized the Spanish hydro-territorial configuration presume a socio-political condition that ignores or disavows radical contestation, mutually exclusive perspectives and imaginations and often profoundly varying social and political power positions of the interlocutors in the process. It implies a form of 'democratic' governance that nonetheless leaves fundamental issues and questions about how to frame socio-natural relations beyond dispute. It invariably points towards a consensus-based model that can be assessed neutrally on the basis of its efficiency, productivity and inclusiveness. It is precisely such a mode of consensual techno-managerial management within an assumedly undisputed frame of market-led efficiency that a growing number of interlocutors identify as 'post-democratic' or 'post-political' (Swyngedouw, 2011a). Rapidly rising discontent with such neo-liberal techno-managerial frames, manifested in Spain through the very vocal 'indignado' movement that erupted with rarely seen intensity on the streets and square of Spain's big cities and later politicized further through the remarkable success of PODEMOS as a political movement, has begun to challenge radically the techno-managerial consensus and questions framing urgent and difficult problems in managerial or technical terms. In contrast, these movements nurture a more foundational political dissensus, one that points towards a more fundamental transformation of the socio-ecological frame that has sutured the political and environmental landscape.

Funding

Research for this paper benefited from the People Programme (Maria Currie Actions) of the European Union's Seventh Framework Programme; under REAS agreement No. 289374 – 'ENTITLE'

References

Abeima. (2011). Desalacion construccion. Retrieved August 23, 2011 from http://www.abeima.es/corp/web/es/construccion/agua/desalacion/index.html

Aguilera Klink, F. (2008). *La Nueva Cultura del Agua*. Madrid: Los Libros de la Catarata.

Arrojo, P. (2001). Hacia una Nueva Cultura del Agua Coherente con el Desarrollo Sostenible. In J. Araújo (Ed.), *Ecología: Perspectivas y Políticas de Futuro* (pp. 117–163). Sevilla: Junta de Andalucía-Fundación Alternativas.

Arrojo, P. (2005). *El Reto Ético de la Nueva Cultura del Agua: Funciones, Valores y Derechos en Juego*. Barcelona: Ed. Paidós.

Arrojo, F., & Visa, L., 2009. Nueva 'Batalla' por el Agua del Ebro – La Generalitat propone abastecer del río a cuatro localidades que se hallan fuera de su cuenca, lo que provoca reacciones diversas en Aragón y Valencia. *El País*, 1 December. Retrieved July 21, 2011 from http://www.elpais.com/articulo/cataluna/Nueva/batalla/agua/Ebro/elpepiespcat/20091201elpcat_2/Tes

Bakker, K. (2002). From state to market? Water *Mercantilización* in Spain. *Environment and Planning A, 34*, 767–790. doi:10.1068/a3425

Bakker, K. (2010). *Privatizing water – governance failure and the world's urban water crisis.* Ithaca: Cornell University Press.

Barnett, J., & O'Neill, S. (2010). Maladaptation. *Global Environmental Change, 20,* 211–213. doi:10.1016/j.gloenvcha.2009.11.004

Birch, K., Levidow, L., & Papaioannou, T. (2010). Sustainable capital? The neoliberalization of nature and knowledge in the European 'knowledge-based bio-economy'. *Sustainability, 2*(9), 2898–2918. doi:10.3390/su2092898

Burballa Noria, A. (2014). The transformation of rural areas in the Spanish state via mega water projects: The case of the Segarra-Garrigues canal. International Conference, Irrigation Society Landscape. Valencia, 25–27 September 2014.

Boelens, R., Hoogesteger, J., Swyngedouw, E., Vos, J., & Wester, P. (2016). Hydrosocial territories: A political ecology perspective. *Water International, 41*(1), 1–14. doi:10.1080/02508060.2016.1134898

Castro Valdiva, J. P. (2007). Comparecía de Don Juan Patricio Castro Valdiva, Vicerrector de Economía e Infraestructuras de la Universidad Politécnica de Cartagena. Murcia: Asamblea Regional de Murcia, Comision Especial de Estudio sobre El Pacto del Agua.

CEC. (2000). *Directive 2000/60/EC of the European parliament and of the council of 23 October 2000 – establishing a framework for community action in the field of water policy.* Brussels: Offical Journal of the Commission of the European Communities.

Centro de Estudios y Experimentación de Obras Públicas. (2011). *Evaluación del Impacto del Cambio Climático en los Recursos Hídricos en Régimen Natural. Resumen Ejecutivo.* Madrid: Centro de Estudios y Experimentación de Obras Públicas (CEDEX), Dirección General del Agua & Oficina Española de Cambio Climático (OECC).

Cooley, H., & Heberger, M. (2013). *Key issues for seawater desalination in California: Energy and greenhouse gas emissions.* Oakland: Pacific Institute.

D'Alisa, G., Demaria, F., & Kallis, G. (2014). *Degrowth: A vocabulary for a new era.* London: Routledge.

del Moral, L., van der Werff, P., Bakker, K., & Handmer, J. (2003). Global trends and water policy in Spain. *Water International, 28,* 358–366. doi:10.1080/02508060308691710

Downward, S. R., & Taylor, R. (2007). An assessment of Spain's Programa AGUA and its implications for sustainable water management in the province of Almería, Southeast Spain. *Journal of Environmental Management, 82,* 277–289. doi:10.1016/j.jenvman.2005.12.015

Elimelech, M., & Phillip, W. A. (2011). The future of seawater desalination: Energy, technology, and the environment. *Science, 333,* 712–717. doi:10.1126/science.1200488

European Environment Agency. (2007). Climate change and water adaptation issues. In *EEA technical report 2/2007.* Copenhagen: European Environment Agency.

Feitelson, E., & Rosenthal, G. (2012). Desalination, space and power: The ramifications of Israel's changing water geography. *Geoforum, 43,* 272–284. doi:10.1016/j.geoforum.2011.08.011

Foro Ciudadano. 2005. El Nacionalismo Hidráulico. *El Verdad.* Retrieved September 29, 2011 from http://www.comunidadescristianasdebase-murcia.com/documentos/el_nacionalismo_hidraulico.pdf

Fritzmann, C., Löwenberg, J., Wintgens, T., & Melin, T. (2007). State-of-the-art of reverse osmosis desalination. *Desalination, 216,* 1–76. doi:10.1016/j.desal.2006.12.009

Furlong, K. (2010). Neoliberal water management: Trends, limitations, reformulations. *Environment and Society: Advances in Research, 1,* 46–75.

García Novo, F., Toja Santillana, J., & Granado-Lorencio, C. (2010). The state of water ecosystems. In A. Garrido & R. M. Llamas (Eds.), *Water policy in Spain* (pp. 21–28). London: CRC Press.

Global Industry Analysts. (2014). *Desalination Technologies – A Global Strategic Business Report.* Retrieved March 24, 2015 from http://www.strategyr.com/Desalination_Technologies_Market_Report.asp

Global Water Intelligence. (2010). *IDA desalination yearbook 20102011.* Oxford: Global Water Intelligence.

González Bernáldez, F. (1981). *Ecología y Paisaje.* Madrid: H. Blume Ediciones.

González Bernáldez, F. (1989). Ecosistema Áridos y Endorreicos Españoles. In Real Académica de Ciencias Exactas, Físicas y Naturales (Ed.), *Zonas Áridas en España* (pp. 223–238). Madrid.

Grindlay, A. L., Zamorano, M., Rodríguez, M. I., Molero, E., & Urrea, M. A. (2011). Implementation of the European water framework directive: Integration of hydrological and regional planning at the Segura River Basin, Southeast Spain. *Land Use Policy*, 28, 242–256. doi:10.1016/j.landusepol.2010.06.005

Heynen, N., McCarthy, J., Prudham, S., & Robbins, P. (Eds.). (2007). *Neoliberal environments: False promises and unnatural consequences*. London: Routledge.

Hulshof, M., & Vos, J. (2014). Diverging realities: How frames, values and water management are interwoven in the Albufera de Valencia wetland, Spain. International Conference, Irrigation Society Landscape. Valencia, 25-27 September 2014.

ICEX. (2010). New technologies in Spain – desalination. *Technology Review* Advertising Supplement to MIT's Technology Revuew Magazine.

Jackson, T. (2011). *Prosperity without growth: Economics for a finite planet*. London: Routledge.

Jáuregui, P., & Ruiz-Jimenéz, A. M. (2005). A European Spain: The recovery of Spanish self-esteem and international prestige. In A. Ichigo & W. Spohn (Eds.), *Entangled identities – nations and Europe* (pp. 72–87). Aldershot: Ashgate.

Kaika, M. (2003). The WFD: A new directive for a changing social, political and economic European framework. *European Planning Studies*, 11(3), 299–316. doi:10.1080/09654310303640

Kallis, G. (2011). In Defence of Degrowth. *Ecological Economics*, 70, 873–880. doi:10.1016/j.ecolecon.2010.12.007

Kennedy, J. F. (1961). 119- the President's news conference 12th April 1961. Washington DC. Retrieved April 4, 2014 from http://www.presidency.ucsb.edu/ws/?pid=8055

Lopez-Gunn, E. (2009). *Agua para Todos*: A new regionalist hydraulic paradigm in Spain. *Walter Alternatives*, 2(3), 370–394.

Manteiga, L., & Olmeda, C. (1992). La Regulación del Caudal Ecológico. *Quercus*, 78, 44–46.

March Corbella, H., & Saurí, D. (2008). Crisis-ridden water governance: The drought of 2008 in metropolitan Barcelona. Paper read at RGS-IBG Annual Conference, 27-29 August, at Manchester, 27-29 August.

March, H. (2015). The politics, geography, and economics of desalination: A critical review. *WIREs Water*, 2, 231–243. doi:10.1002/wat2.1073

March, H., Saurí, D., & Rico-Amorós, A. M. (2014). The end of scarcity? Water desalination as the new cornucopia for Mediterranean Spain? *Journal of Hydrology*, 519, 2642–2651. doi:10.1016/j.jhydrol.2014.04.023

Martín Barajas, S. (2010). Reducción de Recursos Hídricos en España. *Ecologista*, 65, 60–62.

Martínez Gil, F. J. (1997). *La Nueva Cultura del Agua en España*. Bilbao: Bakeaz.

McEvoy, J. (2014). Desalination and water security: The promise and perils of a technological fix to the water crisis in Baja California Sur, Mexico. *Water Alternatives*, 7(3), 518–541.

Meerganz von Medeazza, G. (2005). 'Direct' and socially-induced environmental impacts of desalination. *Desalination*, 185, 57–70. doi:10.1016/j.desal.2005.03.071

Mouffe, C. (2005). *On the political*. London: Routledge.

Pachauri, R. K., & Reisinger, A. (Eds.). (2007). *Contribution of working groups I, II and III to the fourth assessment report of the intergovernmental panel on climate change*. Geneva: Intergovernmental Panel on Climate Change.

Palau, A. (2003). *Régimen Ambiental de Caudales: Estado del Arte*. Madrid: Universidad Politécnica de Madrid.

Rancière, J. (2006). *Hatred of democracy*. London: Verso.

Rico Amorós, A. M. (2010). Plan Hidrológico Nacional y Programa A.G.U.A.: Repercusión en las Regiones de Murcia y Valencia. *Investigaciones Geográficas*, 51, 235–267. doi:10.14198/INGEO2010.51.10

Saurí, D., & del Moral, L. (2001). Recent developments in Spanish water policy. Alternatives and conflicts at the end of the hydraulic age. *Geoforum*, 32, 351–362. doi:10.1016/S0016-7185(00)00048-8

Shannon, M. A., Bohn, P. W., Elimelech, M., Georgiadis, J. G., Mariñas, B. J., & Mayes, A. M. (2008). Science and technology for water purification in the coming decades. *Nature*, 452, 301–310. doi:10.1038/nature06599

Smith, N. (1984). *Uneven development: Nature, capital, and the production of space.* Oxford: Basil Blackwell.

Steinberg, P. E. (2001). *The social construction of the ocean.* Cambridge: Cambridge University Press.

Swyngedouw, E. (1997). Neither Global Nor Local: 'Glocalization' and the politics of scale. In K. Cox (Ed.), *Spaces of globalization: Reasserting the power of the local* (pp. 137–166). New York, NY: Guilford.

Swyngedouw, E. (2010). Apocalypse forever? Post-political populism and the spectre of climate change. *Theory, Culture & Society, 27*(2–3), 213–232. doi:10.1177/0263276409358728

Swyngedouw, E. (2011a). Interrogating post-democracy: Reclaiming egalitarian political spaces. *Political Geography, 30,* 370–380.

Swyngedouw, E. (2011b). Depoliticized environments: The end of nature, climate change and the post-political condition. *Royal Institute of Philosophy Supplement, 69,* 253–274. doi:10.1017/S1358246111000300

Swyngedouw, E. (2013). Into the sea: Desalination as hydro-social fix in Spain. *Annals of the Association of American Geographers, 103,* 261–270. doi:10.1080/00045608.2013.754688

Swyngedouw, E. (2014). Insurgent architects, radical cities and the promise of the political. In J. Wilson & E. Swyngedouw (Eds.), *The post-political and its discontents: Spaces of depoliticization, specters of radical politics.* Edinburgh: Edinburgh University Press.

Swyngedouw, E. (2015). *Liquid power: Contested hydro-modernities in twentieth century Spain.* Cambridge: MIT Press.

Thiel, A., Sampedro, D., & Schröder, C. (2011). Explaining re-scaling and differentiation of water management on the Iberian peninsula. In Fundación Nueva Cultura del Agua (Ed.) *VII Congreso Ibérico sobre Gestión y Planificación del Agua 'Ríos Ibéricos +10. Mirando al futuro tras 10 años de DMA'.* Talavera de la Reina.

Valdés, J. M. (1973). El Futuro de las Presas en España. *Revista De Obras Públicas,* June, 403–410.

Vargas-Amelin, E., & Pindado, P. (2013). The challenge of climate change in Spain: Water resources, agriculture and land. *Journal of Hydrology.* In press. doi:10.1016/j.jhydrol.2013.11.035

Wilson, J., & Swyngedouw, E. (Eds.). (2014). *The post-political and its discontents: Spaces of depoliticisation, spectres of radical politics.* Edinburgh: University of Edinburgh Press.

Zapatero, J. L. R. (2006). Discurso de Investidura de José Luis R. Zapatero, 15 de abril 2004. In F. De Haro Izquierdo (Ed.), *Zapatero, en Nombre de Nada – Crónicas y Conversaciones sobre una Deconstrucción* (pp. 247–270). Madrid: Ediciones Encuentro.

Žižek, S. (1999). *The Ticklish subject – the absent centre of political ontology.* London: Verso.

VIEWPOINTS

Santa Cruz Declaration on the Global Water Crisis

At least one billion people around the world struggle with insufficient access to water. However, the global water crisis is not, as some suggest, primarily driven by water scarcity. Although limited water supply and inadequate institutions are indeed part of the problem, we assert that the global water crisis is fundamentally one of injustice and inequality. This declaration expresses our understanding of water injustice and how it can be addressed.

Crisis manifested

The global water crisis has multiple causes, dimensions and manifestations. One can observe the crisis in rural and urban areas across the global South. We have, for example, observed the following in our fieldwork:

- Peasants impelled to draw water from a spring, when large nearby pipes carry water to a mine in Peru
- People in Lesotho lacking access to clean drinking water as the government exports water to South Africa
- Community water managers excluded from the Nicaraguan water law
- Young girls in rural Nepal carrying water barrels up long mountain trails at night because climate change and hydroelectric projects have made village taps intermittent
- People bathing in a toxic river in Cambodia
- Residents of Dar es Salaam lacking access to water because the pipes fail to reach the informal settlements where most residents live
- Multinational agribusiness companies growing asparagus for export to the industrialized world, in the desert of Peru, with water taken from indigenous communities in the Andes

Environmental injustices are not limited to the global South. They are also manifested in the global North where marginalized communities live in similar conditions. For instance, in California's Central Valley, running from Sacramento to Bakersfield, residents in low-income communities pay high prices for contaminated water for domestic and garden uses, and then have to buy bottled water to drink. Clean water from the Sacramento Delta travels in canals, bypassing these communities, for the benefit particularly of large-scale agriculture in Southern California.

These are a few of the ironies and inequities that make up the global water crisis. They are inequities of access, illustrations of exclusion and misuse, not the consequences of water shortage. They arise from the tendency of water to flow to the powerful and privileged, and often result from larger processes, including those highlighted below.

Urbanization and inequality. In many of the burgeoning cities of the global South, inequities of water provision have been inherited from the spatial segregation between rulers and ruled established by colonial-era city planning. Injustices are frequently embodied as physical infrastructure, because allocational decisions become fixed in infrastructural investments and designs, producing exclusion and poverty for some, and provision and accumulation for others. The fully provisioned cities built for colonial rulers still house those with the most wealth and power. Informal settlements, the product of historically unprecedented migration from rural areas to cities since independence, surround the provisioned core. Lack of access to water and sanitation generates unproductive work and degrading conditions that limit the enterprise and creativity of much of the population in these informal settlements, particularly women and children. People living in informal settlements may be marked as undeserving and trapped in interacting spirals of poverty and marginality. Equitable access could bolster people's capacities, liberate their creative energies, enhance social status and validate citizenship.

Irrigation and injustice. The coexistence of formal laws and pre-existing laws and practices for irrigation water in many parts of the world signals the centrality of water for survival, economic accumulation and political influence. Legal plurality also signals that water access is often highly contested, with varying ideas about how to best or most fairly distribute it. Indigenous practices are often ignored or suppressed by governments which, in the name of production and efficiency and camouflaged by technocratic language, undermine the rights and lifeways of vulnerable and marginalized communities. Large irrigation schemes in many parts of the world become mechanisms for accumulation by large landholders, while small farmers are dispossessed with only residual access to water that ignores domestic needs. An equitable allocation of water should raise living standards, revive rural communities, and recognize productive uses of water for livelihoods and the environment, even when these benefits do not have clear monetary valuation.

Mining. Mining requires substantial quantities of water. Mining companies can bring development (and justice) as they develop the infrastructure required for mineral extraction. At the same time, the companies degrade water sources, territories and cultures. While there are cases where mining companies have engaged communities at the negotiating table, the overwhelming power of the companies may prejudice the outcome of negotiations. Large-scale mining activities often destroy both the physical-hydrological as well as the institutional 'waterscape', altering the courses of rivers and polluting water and soils, as well as transforming existing systems of water rights and responsibilities. Mining companies often become the de facto managers of water, but without systems to hold them accountable for their actions. Water justice requires the voices of the powerless to be heard above corporate and state actors.

Land and water grabbing. Access to water is a critical component of land deals in Sub-Saharan Africa and elsewhere. Investors prefer to acquire land with reliable access to water and the potential for irrigation. Only a minority of investment is for rainfed agriculture. While long-term land leases in Sub-Saharan Africa often guarantee the investors the right to access water, water rights are frequently transferred without concern for downstream users or the environment. Small-scale cultivators who have managed to retain their land find that they no longer have access to the water necessary to put it to productive use. In this context, water justice and water equity are contingent upon transparent leasing procedures and protecting the interests of all stakeholders.

Tensions over international rivers. Borders established by colonial rule, infrastructures of storage, distribution and flood mitigation, and the uneven distribution of benefits from water use routinely lead to injustice on many of the world's international rivers. We suggest that the growth of multi-track diplomacy, involving not just governments but citizen organizations and business enterprises, may increase pressures for justice in the allocation of the benefits of international rivers, while possibly also reducing the risk of violent conflict.

The nature and culture of water

Contemporary water allocations have come about through long, winding, co-evolutionary processes consisting of interactions between different actors, technologies and institutions. Decision making around water occurs in relation to other overlapping land and wildlife management practices (for example, agriculture/aquaculture, gathering of aquatic resources, irrigation and mining) that can be occurring at the same time and place but carried out by different stakeholders, under different governance regimes, and assigned to different institutions (for example, mining activities may be regulated by one ministry and wildlife by another). An assessment of what is equitable or fair often depends on the observer's (political and situational) perspective and identification.

Water is not only a natural resource but is also an element imbued with spiritual, social, cultural and symbolic meaning. Indeed, water and society are mutually constituted. Efforts to promote equity in water governance thus cannot be achieved if these complex contexts, facets and interconnections are diluted or overlooked.

Justice and equity

Equality in some dimension is sought by almost all political philosophies. Philosophies of the Right seek equal freedoms and liberties, sometimes emphasizing freedom from government and taxation. Philosophies of the Left seek equal opportunities and outcomes, particularly with respect to material capabilities, income, assets and education. Water justice can address both sets of concerns, those of freedoms and those of capabilities.

Water justice encompasses questions of distribution and cultural recognition, as well as political participation. Justice may extend from demands for equal access, to demands for recognition of difference and autonomy in how water is used, to demands for full participatory democratic rights and citizenship. These demands are connected in complex and sometimes contradictory ways, linking resource access to questions of identity, belonging and territory.

The growth of environmental concern, and the recognition of large-scale ecological imperatives, have raised awareness that access to natural resources, and social interaction with them, can be a nexus of inequality. In the industrialized world, the banner of 'environmental justice' has proved valuable in struggles against the siting of toxic waste close to low-income and minority communities. The idea of water justice and equity is beginning to be raised in the non-industrialized world. It provides a space, and suggests a set of metrics, for leverage and struggle centred on a relationship to nature constituting human society. Nonetheless, demands for justice cannot be based on some outside, transcendent view of what justice is or should be; they need to be connected to and informed by demands for equity and justice articulated by movements or struggles for water justice.

Water justice can be conceived as equitable or comparable access for particular water uses and deliberated fairness between uses. Within uses, or sectors (domestic, agricultural, industrial, mining), equitable access or allocation can be determined straightforwardly. In cities, for example, households with full provision of domestic water can be distinguished from those excluded, and equitable levels of service established. Equity between uses may be more difficult to establish. The injustice of water allocation and access *between uses*, concretized in laws, agencies and infrastructure, would be subject to the criterion of fairness. Deliberation in the light of development goals, and the questioning of entrenched rights and inequities, are required to achieve a fair or just distribution between mining and communities, between cities and farming, and between domestic and irrigation water.

In many cases, promoting justice also requires broadening political participation, extending citizenship, guaranteeing democratic rights and recognizing cultural differences. Injustices do not just become manifest in how water is distributed. Rather, they inhere within the structures through which rights to water are defined, and by whom, as well as in who has the ability to make and benefit from water investments.

Implications for action and research

Existing water discussions and much research assume that water questions can be resolved either with straightforward, globally applicable technological interventions or with generic changes in government and policy. If, as we suggest, water questions are about inequities arising from diverse intertwined processes involving a range of actors, ecologies and technologies and influenced by questions of territory, identity and belonging, then solutions may not be simple, global, and primarily technical or governmental. Inequalities are embedded in particular histories, reflecting the character of that place, and its boundaries and conflicts. Actions and research for more justice therefore need to be explicitly connected to and grounded in people's experiences of injustice and their strategies and struggles to contest and remedy it. Diverse and plural conceptions of equity and justice emphasize the need for critical pluralism and critical engagement, rather than unthinking application of global ideas.

If inequalities arise from diverse co-evolutionary processes, neither innovative technologies in filtration, pumping, distribution and storage, nor generalized governance remedies like privatization, will be sufficient to resolve established injustices. They may, in fact, skew access even further and create new injustices. Equity requires the development of new and improved insights into how actual distributions of water – and of water-related powers, rights and authorities – come about. Creative institutions and practices may best be generated through engagement with the excluded, impoverished and dispossessed. Water justice requires that all stakeholders can find ways to act collectively in their own best interests. New forms of engagement are required with those who directly experience, and struggle against, injustices.

Remedies for injustice

We, the undersigned scholars, community members, activists, officials and citizens, declare that the principal form of the water crisis is not a shortage of water, nor failures of government, but the many injustices in access to, the allocation of, and the quality of water. The global water crisis is not likely to be resolved by the provision of more water. Redressing injustice is a more promising approach. That requires a critical rethinking and transformation in how water, water rights and authority are distributed. We recognize and build upon work that has gone before, including notably the work of the Justicia Hídrica/ Water Justice Alliance, and the work to implement the human right to water and to include water in corporate social responsibility certification initiatives.

An understanding of the multidimensional causes of injustice, including historical decisions about infrastructure, unnoticed aspects of technologies, the diversity of ecological constraints, and the use of water to accumulate wealth and power, may each suggest possible openings for the redemption of inequities.

We suggest that this work can be furthered through some of the following portfolio of measures to mitigate inequities and to seek a wider water justice.

Policy dialogue could be instigated with diverse stakeholders to examine persistent water inequities. There could be harmonizing mechanisms to redress imbalances of power. This mode of action has been pioneered on a range of questions by community-based organizations in several countries. The object of such dialogues on water justice would be to open up long-ignored injustices for collective action by government, judicial process and social protest. Active and conscious efforts to include those who most directly experience injustices are important here.

Local actions, multi-scalar mobilizations and democratic assessment. Mobilizations by marginalized household members, water user families, environmental justice organizations, and grass-roots communities and federations often raise significant questions of water equity. *Resistance* to large hydroelectric and irrigation structures, for example, has sometimes led to multi-stakeholder and democratic discussion. On a global scale, the World Commission on Dams is perhaps the most substantive example of such discussion. Comparable initiatives are required to evaluate the influence of new combinations of physical infrastructure and the social and environmental choices they embody.

Academic and reportorial investigations. Both scholarly and journalistic investigations, in a wide range of academic disciplines and by those in the media specializing in questions of poverty or the environment, could examine the implications of established as well as new infrastructure and institutional boundaries. Water access could be understood under this framework but expanded to include multi-scalar processes and situations where boundaries are complicated by the politics of space.

Santa Cruz, California, 15 February 2014

Signatories

The undersigned endorse the principles of the declaration as it appears above. It was prepared by a small group of participants (marked * below), building on the work of many others, in the NSF-sponsored workshop on Equitable Water Governance.

Rutgerd Boelens*, Jessica Budds, Jeffrey Bury, Christopher Butler, Ben Crow*, Brian Dill*, Adam French, Leila M. Harris, Colin Hoag, Seema Kulkarni, Ruth Langridge, Flora Lu*, Timothy B. Norris, Constanza Ocampo-Raeder*, Tom Perreault, Sarah Romano, Susan Spronk, Veena Srinivasan, Catherine M. Tucker, Margreet Zwarteveen*

***

Comments on the Santa Cruz Declaration

Salman M.A. Salman

Fellow, International Water Resources Association and a former Lead Counsel and Water Law Adviser with the Legal Vice Presidency of the World Bank

The Santa Cruz Declaration: a new, bold perspective on the global water crisis

When the world woke up in the mid-nineties of the last century to the growing challenges facing its water resources, the immediate and almost unanimously agreed-upon 'felon' was scarcity. Water, every expert kept repeating, is a scarce and finite resource, with no alternative, and upon which there is a total dependence for survival. We have been constantly reminded that of the 1400 million cubic kilometres of global water resources, only 2.5% is freshwater; and of that amount, 99% is permanent ice or in deep aquifers. Thus, there is not enough water on the planet, particularly with the escalation in population growth (from 1.6 billion to 6.1 billion during the last century alone). Other causes to blame include urbanization, environmental degradation and climate change. Spatial and temporal variations – too much water in the wrong place at the wrong time – are other major causes to which the global water resources problems have been attributed.

With the beginning of this century, attention shifted to management as another main cause of the problems, and examples of poor management practices worldwide were cited. The irrigation sector, which consumes about three-quarters of the world's water, wastes far more than it actually uses for food production because of the absence of incentives for rational utilization and conservation. Similar is the fixed-rate structure for urban water users, regardless of how much they consume. Participation of users in management and water governance has been widely flagged and discussed as the ultimate solution for the challenges facing water resources. However, that approach did not answer the questions with regard to water allocation to the poor and to vulnerable groups. Subsequently, the concept of the 'human right to water' started emerging, and some experts even argued that it is part and parcel of water governance.

Indeed, the emergence of the concept of the human right to water through General Comment No. 15 in 2002, and its further strengthening by the resolutions of the Human Rights Council and the United Nations General Assembly in 2010, highlighted the issue of injustice in allocation. However, this approach has not addressed the root causes of the injustice. The concept tries to address the consequences rather than the causes. And even its attempt to lay the foundations for just allocation is met with various challenges, ranging from lack of legislation on the human right to water in most countries, to poor implementation due to the other competing priorities. After all, the International Covenant on Economic, Social and Cultural Rights only requires states "to take steps … to the maximum of their available resources with a view of achieving progressively the full realization of the rights" under the covenant.

And here comes the relevance and contribution of the Santa Cruz Declaration, because it attempts to look into the root causes of the global water crisis that the concept of the human right to water parried. The declaration boldly asserts that the global water crisis is fundamentally one of injustice and inequality. It lists a number of examples to prove that the crisis is one of inequity of access, and of exclusion and misuse, as a result of "the tendency of water to flow to the powerful and privileged". The examples that the declaration lists to prove its assertion of inequality and injustice are quite vivid and varied. They include urbanization, where allocation decisions become fixed in infrastructural investments and designs, producing exclusion and perpetuating poverty and injustice. This injustice extends to the irrigation sector, where in the name of increasing production and efficiency the water rights of smallholders and indigenous communities are constantly encroached upon and weakened, to the benefit of the large irrigation schemes of the strong and privileged. The same approach is followed in mining, as well as in land grabs (mainly irrigated land), where land lease or sale means water reallocation to the new multinationals or foreign governments at the expense of the local, tribal and indigenous communities.

Building on these concrete examples of injustice and inequality, the declaration rightly pronounces that the water crisis is not likely to be resolved by the provision of more water (whether through diversion, storage or more pumping) but rather through redressing injustice. Indeed, the assertion would apply even when the concept of the human right to water is fully adopted and implemented.

Nevertheless, this bold pronouncement should be the beginning. More research and analysis of the root causes of injustice and inequity in water allocation (favouring the strong and privileged) are still needed. Such research can and should address questions such as: Should water be a commercial product and a commodity for profit (to the detriment of the poor), or is it "a heritage which must be protected, defended and treated as such" (European Union Water Framework Directive)? Has the private sector played any role in enhancing and improving water service delivery (compared with the services of publicly owned utilities), or has it instead widened the gap of injustice and inequity, and strengthened the 'haves'? What lessons should we draw from the Cochabamba experience and outcome?

These and other questions will no doubt need to be addressed as the bold and innovative contents of the Santa Cruz Declaration are developed, strengthened and advanced.

Malin Falkenmark

Professor, Stockholm International Water Institute and Stockholm Resilience Centre

The global water crisis referred to in this declaration relates primarily to the wide-spread societal water allocation crisis ('water supply' crisis) and problems involved of justice and equity within uses and between uses.

That is a very different global water crisis from the one my colleagues and I have been focusing on since the 1970s and most recently analyzed in the forthcoming book, *Water Resilience for Human Prosperity* (Rockström et al., Cambridge University Press, due March 2014). This is a water-scarcity-related resource crisis looked at from the perspective of the next few decades and paying particular attention to the prospects of feeding the world's population by 2050, when world population is expected to stabilize.

When looking at water in this perspective, *a global water crisis is clearly in view, primarily driven by water scarcity*. It is related to, *inter alia*, increasing levels of water crowding and water stress in dry-climate regions with continuing rapid population growth (doubling or even tripling by 2050) and particularly vulnerable to decreasing water availability and increasing frequency and severity of droughts. The fact that these regions depend on protective irrigation and therefore consumptive water use for increased food security implies that avoiding a global water crisis will be essential to address in the forthcoming 2015 Sustainable Development Goals programme.

I think that it is essential to be very clear about the distinction between the globally widespread societal water allocation crisis referred to in the Santa Cruz Declaration and the growing water resource and security crisis. It is in other words essential to get conceptually clear by distinguishing between societal water allocation issues (basically 'water supply' services) within urban, industrial and agrobusiness/irrigation areas, on the one hand, and basin-scale water resource management issues, on the other hand, where such problems as water pollution, consumptive water use, ongoing river depletion and closing river basins combine into global-scale issues of fundamental importance for the future of humanity.

What we can see is in other words two parallel global water crises emerging:

> a *water supply service* crisis, not always driven by water scarcity, which can be alleviated by improved water governance and management; and
>
> a *water resource* crisis, driven by increasing water scarcity, which has to be adapted to by mental shifts, resilience-based approaches and adaptive water policies.

Summarizing, I feel that the title of the Santa Cruz Declaration on the Global Water Crisis is misleading. A more adequate title would refer to a Global-Scale *Water Supply* (or *Water Allocation*) Crisis. Otherwise, its intent might get internationally confounded by a rapidly increasing global awareness of the increasing challenges related to the sharpening global-scale water scarcity and the shrinking leeway within the global fresh-water constraint for increased biomass production, primarily biofuels and food, for an increasing humanity.

Yoram Eckstein

Fulbright Professor of Hydrogeology, Tomsk Polytechnic University, Russian Federation; Associate Editor, Water International

I read this 'declaration' in the morning, and since I got hot under my collar I decided to put it aside. Then I read it again in the evening, and got 'dismayed' again. So, here is the 'beef'.

The 'declaration' opens with the following stunning statement summarizing the authors' "understanding of water injustice and how it can be addressed":

> The global water crisis is not, as some suggest, primarily driven by water scarcity. Although limited water supply and inadequate institutions are indeed part of the problem, we assert that the global water crisis is fundamentally one of injustice and inequality. We, the undersigned scholars, community members, activists, officials and citizens, declare that the principal form of the water crisis is not a shortage of water, nor failures of government, but the many injustices in access to, the allocation of, and the quality of water. The global water crisis is not likely to be resolved by the provision of more water.

It never fails to amuse me to see when lawyers and 'social justice activists' talk (or write) about injustice in allocation of a resource without any understanding of the physical nature of the resource. I do not know if the group authoring this 'declaration' is driven by self-promotion in stating that "the global water crisis is fundamentally one of injustice and inequality". As a physical scientist, I am stunned, as I am sure are most if not all of my peers, by the opening statement that "the global water crisis is not ... driven by water scarcity". Tell this to a Jordanian, Israeli, or Palestinian, just to mention a few. How can they consider injustice in the agricultural sector (e.g. irrigation) when disregarding unequal distribution of water on the globe? How they can bunch the injustices and inequalities in the agricultural sector of the Middle East or Sub-Saharan Africa with the injustices and inequalities in the same sector of Nicaragua or Cambodia?

In my humble opinion, any discussion of 'water justice' on a global scale and not in the context of at least climatic regions is a waste of time. This particular 'declaration' will join myriads of meaningless documents with no practical implications.

Jerry van den Berge

Policy officer for Water, Waste and European Works Councils, European Federation of Public Service Unions

It is a relief to read the Santa Cruz Declaration and to remember that the global water crisis is not a crisis of water shortage or scarcity but an outcome of inequality and injustice. Good to be reminded, I say, because this was already noted in the United Nations Development Programme's *Human Development Report 2006*. The report argued that the roots of the crisis in water can be traced to poverty, inequality and unequal power relationships, as well as flawed water management policies. It also made clear that there is a real crisis:

> Access to water for life is a basic human need and a fundamental human right. Yet more than 1 billion people are denied the right to clean water and 2.6 billion people lack access to adequate sanitation. These headline numbers capture only one dimension of the problem. Every year some 1.8 million children die as a result of diarrhea and other diseases caused by unclean water and poor sanitation.

Then the "Declaration" ends with the following assertion:

> We, the undersigned scholars, community members, activists, officials and citizens, declare that the principal form of the water crisis is not a shortage of water, nor failures of government, but the many injustices in access to, the allocation of, and the quality of water. The global water crisis is not likely to be resolved by the provision of more water.

These figures have only slightly improved over the past seven years. Besides these dimensions, there are other aspects of the global water crisis related to the management of fresh-water resources. More and more, the rights and ownership of these resources are accumulating in the hands of fewer but bigger corporations, leaving poor communities without access to water or with polluted water. Inequalities increase, while corporations seek higher profits and knock out 'competitors' in what they see as a market.

So, where do we stand now as the crisis continues? Are inequality and injustice persistent? I would not accept this as a conclusion. It basically shows that injustice, poverty and inequalities are hard to fight. This always reminds me of the famous words of many people who stood up to fight for their rights in Latin America: *la lucha continua*. I think these words address the crisis better than any so-called Millennium Development Goal or the newly invented Sustainable Development Goal. Let me be clear on this: each well-meant effort to reduce poverty must be undertaken. But they have to be seen in the light of a fight for justice.

That is where the Santa Cruz Declaration helps and is so useful: it identifies injustices as roots of the crisis. People generally tend to turn away from injustice, because it confronts them with a dilemma: they should act against injustice, but most of the time people don't know what to do or they are afraid it would require of them a big change in behaviour. Take for example Nestlé, exploiting water resources worldwide at the expense of local communities and the environment. Should people stop drinking Nespresso, or boycott their chocolate? People don't like to fight, but on the other hand people like to do good. So the fight for justice must be linked to a 'doing-good feeling'. The remedies that are suggested in the declaration should be turned into actions: actions that give people a feeling they are doing something 'good'.

But this is limited to individual action. Added up, they make a change, but a bigger, societal change is needed to achieve water justice for all. The paradigm of competition, economic growth and market logics has to change. This paradigm is the basis of inequality, and in the unequal global competition the (powerful) winners remain winners and the (powerless) losers remain losers.

This means that large-scale political action is needed that can only follow from the critical engagement of people. Awareness of power relations in water and mobilizing people to stand up for their rights is the first step towards overturning existing inequalities and injustices. The power balance must be brought to a fair equilibrium; and to know how to reach this, we must identify the forces that maintain inequalities and the status quo in power balance. The Santa Cruz Declaration is a tool to raise awareness and should be brought to the attention of a wide audience. Its message must be repeated the way corporations repeat their annoying advertisements.

In Europe, people led by a coalition of trade unions and NGOs have used another tool, the European Citizens' Initiative, to raise awareness on the right to water and to spark the fight for water justice. A huge mobilization made a change in European policies with regard to water, and it still continues to change the political discourse. We need more of these kinds of declarations and mobilizations of masses to generate actions that slowly but surely shift the balance of power in water from injustice and inequality to water justice around the world.

Maude Barlow

Council of Canadians, Ottawa, Ontario, Canada; Food and Water Watch, Washington, DC, USA and formerly Senior Advisor on Water to the 63rd President of the United Nations General Assembly

I am very pleased to offer a comment and strong support for this declaration. I have been deeply involved in the struggle to protect water as a public trust, a public service and a human right for many years and I cannot strongly enough stress the need for dialogue on the principles upon which we must move ahead to build a water-secure future for all.

The declaration challenges one of the fallacies of modern water thought, and that is that technology will fix everything and we therefore have no need to curb our consumption of water or the way we move water around the world for our convenience. Modern societies view water as an endless resource for our personal pleasure and profit and not as the most essential element of a living ecosystem. We have disassociated ourselves from the cultural, historic, ecological and spiritual aspects of water and see it as something to be tamed for the service of a modern economy based upon unlimited growth.

So we pollute, mismanage and displace water, redirecting local water supplies to the wealthy and to industry and out of the reach of millions of small farmers, indigenous people and those living in urban slums in the global South, and increasingly in the global North as well. Privatizing water services and trading water on the open market are manifestations of a mindset that sees water as property that rightly privileges the powerful.

The declaration clearly places inequality and injustice as the core issues of the water crisis, and I agree. I would add that even the ecological crisis – and we are truly a planet running out of accessible clean water – is a legacy of the cavalier way in which we have allowed water to be used for development. Massive amounts of land-based water are dumped into the rising oceans every year, after they have 'served' the needs of big cities and industry, instead of being returned to local communities. Virtual water imbedded in the global trade in food, clothing and electronics keeps the global market profitable for an elite but takes a terrible toll on water supplies removed from watersheds never to be replenished.

In my new book, *Blue Future: Protecting Water for People and the Planet Forever*, I argue that we must create a new water ethic that puts water and its preservation and restoration at the heart of all we do if we and the planet are to survive. This new ethic must be based on four principles: that water is a human right; that water is a common heritage and public trust; that water has rights, too; and that water can teach us how to live together.

As the authors of the Santa Cruz Declaration understand, we cannot tackle the environmental crisis in a vacuum. To do it right, we must confront the issues of environmental injustice and bring the voices of the marginalized into the centre of decision making. As I said in *Blue Future*:

> The quest to protect water forever is also inextricably bound with human rights. If we want to transform conflict into peace and create new ways to govern that honour watersheds and ecosystems, it is essential to recognize that lack of access to clean water is a form of violence and that there can be no peace or good governance without justice. That means putting the human right to water and sanitation as well as the right to engage in the process at the centre of a new, more collaborative form of watershed governance. Conflict transformation goes beyond the concept of conflict resolution in that it requires confronting unjust social structures that underlie the conflict.

K.J. Joy

Senior Fellow, Society for Promoting Participative Ecosystem Management and Facilitator of the Forum for Policy Dialogue on Water Conflicts in India

The Santa Cruz Declaration on the Global Water Crisis: a good beginning to repoliticize the water discourse

The Santa Cruz Declaration, outcome of a workshop organized by the University of California, Santa Cruz, that brought together a range of academics and others to seek ways of making water governance more equitable, is much needed and timely. In the present-day mainstream discourse on water, centred around scarcity, supply augmentation, efficiency, institutional reforms and different modes of privatization, by characterizing the present water crisis as primarily one of "many injustices in access to, the allocation of, and the water quality of water", the declaration has the potential to repoliticize the water discourse. Since the 1990s, the water discourse was becoming increasingly depoliticized and sanitized, and it is high time that we take a break from such a discourse. I do think that by centre-staging water justice and equity as the core issues, the declaration provides a departure point from the mainstream discourse and a rallying point for all those who have been critical of the mainstream water discourse led by supranational agencies like the World Bank, the Asian Development Bank and the Global Water Partnership, and the national governments – all wedded to a neoliberal agenda.

Having said the above, I do think that there are a few problematic formulations and propositions, and in the rest of this brief note I would like to discuss a few of them. These are not to be seen as criticisms of the declaration or an effort to belittle the declaration in any way. I am making these comments and observations more as a co-traveller of those who drafted this declaration, with the hope that these comments and suggestions would help in further refining and sharpening it.

First, I really did not understand the compulsion or desperation to bring the Right and the Left onto the same page. For example, see the statement:

> Equality in some dimension is sought by almost all political philosophies. Philosophies of the Right seek equal freedoms and liberties.... Philosophies of the Left seek equal opportunities and outcomes.... Water justice can address both sets of concerns, those of freedoms and those of capabilities.

According to me, this is problematic on two counts. One, the equal freedoms and liberties the Right talks about are primarily within the market and commodity exchange framework and not the same as of the Left's transformative framework. Two, along with historically embedded inequalities, the capitalist and neoliberal state is responsible for the wide-spread injustices and inequities in the water sector. It is like trying to reconcile the irreconcilable, and in the process the political edge of the declaration has been significantly blunted. I think it is important to recognize that the social justice and equity the declaration talks about cannot be fully realized within the capitalist framework or state and are possible only within a radical social transformative framework. And this needs to be stated clearly in the declaration if it is to inspire new forms of political action. It also has implications for agency that would take the social justice and equity agenda contained in the declaration forward. I was rather surprised that the word 'capitalism' is not mentioned even once in the declaration.

Second, though one would agree broadly with the statement that "large irrigation schemes in many parts of the world become mechanisms for accumulation by large landholders, while small farmers are dispossessed with only residual access to water that ignores domestic needs", inequity is not a hallmark of large projects alone. The smaller ones too are very often iniquitous. In other words, inequity and injustice cut across the size of the projects. Experience in India shows that very often it is easier to fight against injustice in relation to larger, centralized projects because the enemy is outside and far removed from the day-to-day life of the toiling masses. In the case of smaller projects, very often the fight has to be against the local elites who control and monopolize water, and this is all the more difficult because the deprived depend on them for their day-to-day life. Thus, my suggestion would be to take the water justice and equity issue beyond the sterile debate of 'large versus small' and similar binaries. Of course this is not to belittle the struggles against the destructive content of large hydroelectric projects.

Third, as the declaration rightly points out, "Water justice can be imagined as equitable or comparable access for particular water uses and deliberated fairness between uses." In fact, equity within a particular use shows how egalitarian and fair the distribution is. However, when it comes to allocation across different uses, then the fairness of allocation very much depends on the developmental trajectory adopted by the state. For example, India, and also many other countries with similar developmental trajectories, with the preoccupation with high growth rate through the industrialization and urbanization route, has been re-allocating water from agricultural use to industrial use and from rural to urban areas in a nontransparent, undemocratic manner. Allocations across uses, their re-allocations and resultant trade-offs are embedded in the political economy of the country, and I do not find much of an explicit discussion of the political-economy angle of water allocations and injustices in the declaration.

Fourth, of course the demand for justice "needs to be connected to and informed by demands for equity and justice articulated by movements or struggles for water justice". However, there seems to be an underlying disdain in the declaration for any larger normative principles around the question of justice and equity, and this could be problematic. I would argue that what happens on the outside has a bearing in shaping the world-view of the local, and we also need to accept that, for historical reasons, very often the world-views around justice and fairness may not be all that egalitarian, even within local communities. In such situations, commitment to certain larger normative principles – for example the human right to water – can help in working with the people in evolving a socially just water agenda. So the issue is not of choosing perceived world-views over larger principles of justice, but one of problematizing the relationship between the two, and in this, efforts need to be made to see that lived experience and articulations are not compromised. In fact, this is important in the context of how we harmonize the "local actions" and "multi-scalar mobilizations" that the declaration talks about.

Finally, the declaration talks about "policy dialogue", "local actions, multi-scalar mobilizations and democratic assessment", and "academic and reportorial investigations" as a possible portfolio of measures to mitigate inequities and to seek a wider water justice. It adds that, in the case of dialogue amongst multiple stakeholders with differing and conflicting interests, there should be certain "harmonizing mechanisms to redress imbalances of power". This is all the more true in the case of India and other similar countries where the power relations amongst different stakeholders are very skewed. Here I would argue that creating a level playing field for all is an important precondition if such dialogues are to result in socially just outcomes. In my many years of work in the water sector in India in general, and water conflicts in particular, I have found that lack of access to reliable data and information asymmetry amongst the different stakeholders is a major impediment. Thus, democratization of data should be an important demand. I would very much like it if a demand for making all water-related data easily accessible and in the public domain were also reflected in the declaration.

Notwithstanding the above critical comments and suggestions, I do endorse the basic proposition of the declaration that water justice and equity are at the core of the water crisis, and not water scarcity alone. The declaration is a good beginning in the much-needed repoliticization of the water discourse, and we need to further build on it.

Erik Swyngedouw

Professor of Geography, School of Environment and Development, Manchester University

From water justice to eco-political equality: politicizing water

As the Santa Cruz Declaration attests, water is an extraordinarily contested and problematic thing. First, while water accessibility has improved in many parts of the world, the reality remains that progress is painfully slow, with more than a billion people worldwide still living with inadequate access to water and/or sanitation services. Unsatisfactory access to water remains the number-one cause of premature mortality in the world, and of serious health problems, poverty, stalled development, and conflict. Despite detailed information and sophisticated insights into key drivers and bottlenecks, the simple fact remains that too many people die unnecessarily because of water-related conditions that are easy to remedy and have invariably to do with uneven power relations and the perverse geographies of uneven development. The situation on the wastewater side of the cycle is even more disastrous: 80% of wastewater is not collected and treated, with negative consequences for health, ecological service provision and socio-economic development. Indeed, water keeps flowing uphill, to money and power.

The declaration highlights the variegated processes of water injustice and inequalities, water struggles and strategies of water dispossession that choreograph many of the world's diverse hydro-social constellations. In a context of proliferating accumulation by dispossession, of unchecked concentration of resources in the hands of the few – often nurtured by managerial objectives that consider the techno-managerial organization of optimal market forces as the only horizon of the possible – and of a rapid deepening of unequal social, political and economic power relations, all manner of socio-ecological struggles that revolve around the signifier of 'justice' are actively resisting the often violent theft, not only of water but of a wide range of common-pool resources. These struggles, despite their radical heterogeneity, nonetheless share a concern with a more equitable and solidarity-based organization of access to, appropriation and transformation of the commons.

While 'justice' remains an ethically grounded and deeply humanitarian concern that galvanizes much of the social struggles against the often forced dispossession of hydraulic resources, the notions of 'equality' and 'solidarity' are of course more openly political terms. I would like to suggest the opening up of the debate from an ethical concern with 'justice' to a political vision aimed at ega-libertarian collective management. The idea of water-as-commons organized through democratic being-in-common may permit the shifting of the terrain from ethics and justice and a focus on struggles of resistance (against the dispossessing intruder) to more directly and openly political visions and imaginaries that might nurture and galvanize political struggles aimed at the ega-libertarian transformation and collective management of the commons instead of the dominant exclusive and private regimes of managing the commons.

The declaration's aspiration is to advocate a scholarly and practical move from considering water as a predominantly techno-managerial concern to one that focuses decidedly on socio-biological life and well-being, and consequently a concern with internal social tensions and conflicts over uneven access to and control over water resources. This of course presumes a perspective that does not ignore or disavow radical contestation, that explores mutually exclusive perspectives and imaginations, and that acknowledges the often profoundly varying social and political power positions of the interlocutors in the process. Achieving equitable access implies a form of 'democratic' governance that includes considering different political constellations of organizing the hydro-social cycle. It points inevitably towards a perspective that destabilizes consensus-based models that can presumably be assessed neutrally on the basis of efficiency, productivity and inclusiveness. It is precisely such a mode of consensual techno-managerial management within an assumedly undisputed frame of market-led efficiency that reproduces the existing water inequalities.

In sum, this declaration opens up a vital agenda for anyone concerned with water and political equality. The water issue is indeed an emblematic issue, one that expresses in its variegated meanderings the functioning of democracy, not just as a system of governing but as a set of principles articulated around equality, freedom and solidarity.

Miguel Solanes

Senior institutional expert, Madrid Water Institute, Spain; former water law advisor, United Nations

The inequities affecting present water allocation and sustainable utilization are compounded by international investment protection agreements. Both these agreements and the international arbitration courts that enforce them are intended to protect the rights of foreign investors. Public-interest considerations are secondary matters, and in a number of cases countries have been forced to pay compensation when regulations affected the returns of foreign investors. Consequently, many nations are wary of protecting the environment and water, since protections might result in penalties or condemnations.

The customary rights of local populations are also secondary to the interest of foreign investors. Investment arbitration relies on written law and formal records, which in most developing countries are not available. As a result, waters used through the millennia are considered vacant and granted in block, without research or inventory, to foreign investors. Local populations lose their means of survival, turn landless, and enter the long lines of illegal migrants. Their impact and fate are illustrated by the deaths in Lampedusa and Gibraltar and in the transit from Central America to the United States.

Further research on the reformulation of investment treaties to afford greater protection to local rights and interests is needed. Simply adjusting them to the national legal practices of nations would be an important, albeit not sufficient, step ahead.

Laureano del Castillo

Executive Director, Peruvian Centre for Social Studies, Peru

It is not easy to comment on a document like the Santa Cruz Declaration because it is a very comprehensive and well-structured text. Indeed, the declaration reviews some of the processes that are the basis of the current water crisis, defines what justice in access and use of water means, and suggests some lines of action to address the problems discussed.

For example, the cases of Peru related to mining and production of asparagus for export cannot be considered in isolation, because they reflect how development is understood in many countries, including Peru: that is, resources are leveraged without looking at the consequences for the environment or for the poorest people. Certainly this is short-sighted. However, it has become the dominant point of view in governing circles: that natural resources should be extracted, exploited and put on the global market before other countries do so. This is all carried out in the name of common interest, although the benefits actually accrue only to a few companies.

The declaration mentions some processes leading to the current problems of inequity. These include urbanization, irrigation, land and water concentration and the use of international rivers (although in this last case lakes and other sources of transboundary freshwater should also be mentioned). What makes the situation more delicate is that these different processes are usually going on simultaneously in many countries.

At a time when our concern for the preservation of the environment has grown, the declaration also clearly defines the underlying nature of this crisis. It is a crisis of water management, i.e. how use rights are allocated to water, how water is used and how the various systems of water management are viewed, beyond the aspects of technical or economical efficiency. Keeping in mind that the effects of climate change may exacerbate problems around water, the declaration emphasizes the main cause of these problems: it is related to the way water rights are allocated and how they are used. This creates more inequality in an increasingly unequal world.

In this regard, the Santa Cruz Declaration is in contradiction with the illusion that emerged during the industrial revolution, that science and technology are able to dominate nature and overcome any challenge. This conception is still present, and interestingly can be seen in the advertisements for a new private university in Peru, which, after noting that some water is running out, says in response, "They are not engineers." Without denying the importance of efforts to better use water, with more efficiency and avoiding contamination, the problems posed by the declaration are broader in nature. We have known for a long time that water is a resource that serves many different uses, and that for that reason it must be managed socially. The technical and economic aspects must be addressed together, without forgetting that water belongs to everyone and therefore should not be a source of injustice for anyone.

Having said that, one cannot ignore the fact that modern agriculture consumes about 70% of available freshwater. In many parts of the world, family farming, which also includes peasant and indigenous communities, is the most important supplier of food from agricultural sources. This should oblige governments to support efforts towards improved efficiency in water use in modern agriculture, which would be to the benefit of all.

Over 20 years ago, the Dublin Declaration, the result of the International Conference on Water and Environment, urged that steps be taken to address the challenges of water management. A quick look at the global picture, as the Santa Cruz Declaration points out, shows that we have regressed instead of moving forward. It is time to start changing the way we manage water. The declaration is a good start in new discussions about this issue.

Territories and imaginaries, contestation and co-production

Ben Crow[a], Flora Lu[b], Rutgerd Boelens[c], Jaime Hoogesteger[d], Erik Swyngedouw[e] and Jeroen Vos[f]

[a]Sociology Department, University of California, Santa Cruz, CA, USA; [b]Department of Environmental Studies and Provost of College Nine and College Ten, University of California, Santa Cruz, CA, USA; [c]CEDLA, University of Amsterdam and Department of Environmental Sciences, Wageningen University, the Netherlands; [d]Department of Environmental Sciences, Wageningen University, the Netherlands; [e]School of Environment, Education and Development, University of Manchester, UK; [f]Department of Environmental Sciences, Wageningen University, the Netherlands

At first glance, the chapters of this book paint a bleak picture. Social forces, nurtured by the vicissitudes of capitalism and power, embed patterns of exclusion and deprivation into the actions of companies and state agencies, diverting and polluting water, appropriating water rights and constructing projects and territories with profound consequences for livelihoods and identities of marginalized groups. That bleak picture can be correct. Nevertheless, this book considers analytics and practices that can challenge such injustices: theoretical insights that offer much broader understanding of the many worlds co-constituted by water and society, and a hopeful set of examples describing innovative social movements arising from the contested terrains anchored in the social and natural arrangements of water. Furthermore, Chapter 24 is a declaration and a set of responses and commentaries that suggest ways to act more effectively on the several dimensions of the equitable governance of water.

Analytical insight

The idea of a *hydrosocial territory* is utilized to illuminate the contested visions of society and nature underlying proposals for public action and establishing authority over territory, water and people. Hydrosocial territories often reflect the priorities and ideas of specific social groups and carry notions of who should own, use and control water, and who is equipped with the rights and legitimacy to speak for water and its governance. Many contributing authors describe different imagined and lived worlds constructed and carried by marginalized and underrepresented communities and classes, understandings of water that do not align with those prevalent in the discourses of government institutions or capitalist social relations. These imagined alternative orders illuminate resistance and challenge prevailing ideas about what should be done to water and who should do it.

In the Andean highlands of Ecuador, the implementation of policies centered around a plan to establish payment for environmental services; such a depoliticized, hydrosocial territory replaces the existing order of society, water and community (Rodríguez-de-Francisco and Boelens, Chapter 6). In the valleys of the Bolivian Andes, complex local irrigation systems are constructed by peasant farmers, though their historical claims and organizational capacities are overlooked (Saldías, Boelens, Wegerich and Speelman, Chapter 7). In the Páramo highlands of Santurban, Colombia, differences of power, representation and valuation are revealed in the ostensibly apolitical hydrosocial territories promoted by technocrats, companies and government (Duarte-Abadía and Boelens, Chapter 15). In the Tiraque Valley of Bolivia, socio-natural interactions and hydrosocial territories are reconfigured through technological design, legal structures and unequal power relations (Seemann, Chapter 21).

After the book's introduction, Chapter 2 introduces the idea that water justice is built from local, grassroots ideas of justice. An analytical tool kit of key concepts in this chapter – including situated knowledges, socionatures and a broad focus on contestation over allocation, rules, authority and discourse – enable understanding of the diversity of water injustice and identify the many challenges facing activists, community organizations, government officials, journalists and academics seeking more equitable governance of water.

Zwarteveen and Boelens (Chapter 2) outline a broad conception of water justice that recognizes issues beyond scarcity and allocation. Recognizing that 'struggles over natural resources become struggles over meaning, norms, knowledge, identity, authority and discourse', the chapter suggests four elements to be considered in water justice: distribution, recognition, participation and justice for the nonhuman world. Many chapters illustrate the breadth of issues of justice (Hoogesteger, Rutgerd Boelens and Michiel Baud in Chapter 18, Seemann in Chapter 21, Ioris in Chapter 20 and many others).

In contrast to these ideas of local, situated and four-dimensional justice, top-down approaches are described in many chapters. Most governments and international agencies deploy technocratic, apolitical ideas of water governance where inequities are seen as pragmatic questions to be addressed with technical fixes and economic innovation, including new water storage, treatment and distribution systems, and novel regulatory principles and organizations.

These technologies, reforms and institutions reflect specialized understandings of water coming from economists, engineers, officials and public health professionals. Policies emerging from these ideas can be constrained by the needs of office holders, wealthy and influential interests and well-represented communities. Whether dealing with surface water, groundwater, rainwater, 'virtual water' or even water from inland or open seas, the contributions by Srinivasan and Kulkarni (Chapter 8), Dill and Crow (Chapter 9), Perramond (Chapter 13), Meehan and Moore (Chapter 14), Hulshof and Vos (Chapter 16) and Swyngedouw and Williams (Chapter 23) make clear that these modernist understandings of water and technocratic proposals tend to ignore the historical roots and embedded social contexts of water injustices, while privileging dominant water technologies, rationalities and politico–institutional frameworks.

At the global level, international agencies use a comparable technocratic discourse that spreads top-down, purportedly apolitical ideas. At this scale, 'safe drinking water' has been a common focus of ideas of equity in relation to household water in the global south. This notion considers only health aspects of domestic water to the exclusion of water access that could facilitate escape from poverty (Goff and Crow, Chapter 5). While increased trade in water-dense cereal crops has defused some international conflicts over water, trade in 'virtual water' has also enabled export-oriented food production in dry areas, with often devastating impacts on local water and food securities (Vos and Hinojosa, Chapter 22).

Inklings of a more equitable order: networks and committees, alliances and federations

Chapters in the book describe the building of networks, committees, federations and alliances that pioneer alternatives to centralized, technocratic water governance reforms. While such alternatives and the communities they organize are diverse, each facilitates nego-tiation across divides of power, wealth and social position, building cultural and strategic processes that may give voice to communities, their needs and organizations. The chapters do not argue that 'local' or 'vernacular' water norms, organizational forms and modes of knowledge are necessarily 'better' or more equitable; they criticize the fact that the latter do not even get the chance to prove, manifest or maintain themselves. Modernist policies, state-biased laws and regulations, capitalist public–private partnerships and the prevailing rational-technocratic disciplines of scientific water knowledge production simply exclude, omit or annihilate them by relegating these context-based water organizations, authorities, relationships and forms of rule-making to the domains of 'non-knowledge', 'superstition' and 'backwardness'.

Water committees and innovation

In Venezuela, technical water committees co-produce public service delivery and use situated knowledge to build new relations and accountability between citizen and state (McMillan, Spronk and Caswell, Chapter 10). In Honduras, committees, networks and alliances painstakingly constructed across obstacles of power and understanding have established a reserve that is equitable and sustainable (Tucker, Chapter 11). In Nicaragua, water committees not recognized by the government have begun to change ownership, autonomy and state regulation of water (Romano, Chapter 12).

In New Mexico in the United States, local action has encouraged approaches to community-based administration of rainwater. Meehan and Moore (Chapter 14) ask, who owns the rain? And they find that perverse and unenforceable water regulations and laws make rainwater harvesting illegal. Gradually and unevenly, nonetheless, states and municipalities have been seeking, against sometimes-risible legislative difficulties, to catch up with the impulse of residents to use rainwater more effectively in water-scarce states. Diversity in legal traditions in this case allows greater innovation and more effective rainwater harvesting.

Networks and alliances

Community-managed irrigation systems in the inter-Andean valleys of Bolivia include hydrosocial networks, organization and infrastructure across watershed boundaries (Saldías, Boelens, Wegerich and Speelman, Chapter 7). Government water policies, however, tend to misrecognize these complex territories and their context-based ideas of space, place and water system development, preferring to draw the boundaries of water systems according to geological or bureaucratic, rather than social, formations.

Hundreds of community-managed irrigation and drinking water associations in Ecuador have been able to defend their hydrosocial territories and gain voice through two large water user federations organizing mass protests, lobbying and participation in state agencies. Water users' alliances have used a National Water Resources Forum – bringing together multiethnic groups, grassroots organizations, academics, non-governmental organizations (NGOs) and state institutions – to represent and legitimize their territorial claims and gain inclusion of their demands in the 2008 Ecuadorian constitution (Hoogesteger, Boelens and Baud, Chapter 18).

This case of water federations in Ecuador illustrates how hydrosocial territories, conceptualized as plural, partially overlapping and mutually conflicting, assist in the studying of water governance and equity. Social movements, built through the aggregation of water user associations, strengthened their legitimacy, collective identities and shared strategies, and so were able to gain recognition as stakeholders and influence laws and the constitution. And while Ecuador's government has made significant discursive advances in the participatory governance of water, and nature in general, we see that in the context of Amazonian oil extraction, indigenous hydrosocial territories are sites in which the material outcomes of government activity challenge ideologies that purport to be widely beneficial to local communities (Lu and Silva, Chapter 19).

In sum, the chapters of this book make two advances in our understanding of water and society. First, the concept of plural, overlapping and contested hydrosocial territories illuminates the tensions and misunderstandings that can arise from a cultural–political and power-laden mismatch between community, government and private capitalist enterprises' understandings of and plans for water. Here, incommensurate visions, power disparities and forms of management can perpetuate persistent inequalities and injustices. The chapters highlight, conceptually and empirically, how relationships of class, gender, ethnicity and caste drive and become embedded in the materiality and institutionalization of socionatural, technopolitical, territorial configurations. Second, the authors convey a broader conception of water justice, contestation and change. In this view, justice for marginalized communities seeks more than questions of distribution to include cultural recognition, political participation and concern for the nonhuman world. And, social movements, contestations and struggles over water are engaged in arenas of discourse, meaning, knowledge and authority. These two advances illuminate the diverse movements, alliances, water user associations, federations, committees and networks that seek change toward equitable governance of water and toward many forms of water justice demanded by marginalized, disenfranchised and deprived communities.

IWRA Executive Board Election Results

A new Executive Board of IWRA for 2016-2018 has been elected and will assume office on 1 January 2016 for a three year period to terminate on 31 December 2018.
The officers elected are from 15 different countries, representing five continents and different disciplines.

Executive Board of IWRA 2016-2018

President
Patrick Lavarde (France)

Past President
Dogan Altinbilek (Turkey)

Vice-Presidents
Naim Haie (Portugal)
Yuanyuan Li (China)
David Molden (Nepal)

Secretary General
Guy Fradin (France)

Treasurer
Renee Martin-Nagle (USA)

Committee Chairs
Awards Committee, James E. Nickum (Japan)
Membership Committee, Carl Bruch (USA)
Publication Committee, Henning Bjornlund (Australia)

Directors
Aziza Akhmouch (Morocco)
Mukand S. Babel (Thailand)
Bruce Currie-Alder (Canada)
Guillermo Donoso (Chile)
Jun-Haeng Heo (Rep. of Korea)
Gary Jones (Australia)
Y.S. Frederick Lee (China)
Ahmet Mete Saatçi (Turkey)
Raya Marina Stephan (Palestine)
Philippus (Flip) Wester (Netherlands)

Index